CW00762234

114 50

Anthropometry of the Head and Face

Second Edition

Anthropometry of the Head and Face

Second Edition

Editor

Leslie G. Farkas, M.D., C.Sc., D.Sc., F.R.C.S.(C)
Associate Professor, Department of Surgery
University of Toronto
Director, Craniofacial Measurements Laboratory
Craniofacial Program, Division of Plastic Surgery
Research Consultant, Plastic Surgery Research Laboratory
The Hospital for Sick Children
Toronto, Ontario
Canada

Raven Press New York

Raven Press, Ltd., 1185 Avenue of the Americas, New York, New York 10036

© 1994 Raven Press, Ltd. All rights reserved. This book is protected by copyright. No part of it may be reproduced, stored in a retrieval system, or transmitted, in any form or by any means, electronic, mechanical, photocopying, or recording, or otherwise, without the written permission of the publisher.

Made in the United States of America

Library of Congress Cataloging-in-Publication Data

Anthropometry of the head and face/edited by Leslie G. Farkas.
p. cm.
Includes bibliographical references and index.
ISBN 0-7817-0159-7
1. Face–Measurement. 2. Cephalometry. 3. Anatomy, Surgical and topographical. 4. Anthropometry. 5. Surgery, Plastic. I. Farkas, Leslie G.
[DNLM: 1. Cephalometry–methods. 2. Face. 3. Racial Stocks. WE 705 A6285]
QM535.A65 1994
573'.692—dc20
DNLM/DLC
for Library of Congress 94-991

The material contained in this volume was submitted as previously unpublished material, except in the instances in which credit has been given to the source from which some of the illustrative material was derived.

Great care has been taken to maintain the accuracy of the information contained in the volume. However, neither Raven Press nor the editor can be held responsible for errors or for any consequences arising from the use of the information contained herein.

9 8 7 6 5 4 3 2 1

Cover photograph provided by Superstock, Inc.
BIRTH OF VENUS, Sandro Boticelli

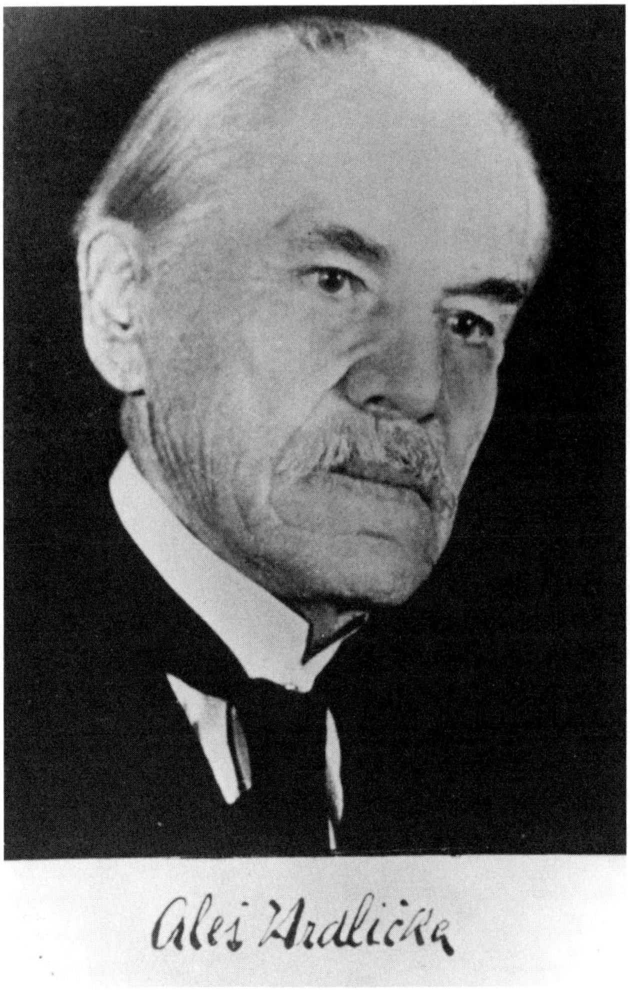

To the memory of

Aleš Hrdlička, M.D. (*1869–1943*),

physician and physical anthropologist,

pioneer in the application of anthropometry in medicine,

Smithsonian Institution, Washington, D.C.

Contents

Contributors

David E. Altobelli, D.M.D., M.D. *Facial Engineering and Morphology Laboratory, Harvard School of Dental Medicine, 188 Longwood Avenue, Boston, Massachusetts, 02115*

Daniel B. Bača, M.D. *Facial Engineering and Morphology Laboratory, Harvard School of Dental Medicine, 188 Longwood Avenue, Boston, Massachusetts 02115*

Janusz Bardach, M.D. *Professor Emeritus of Plastic Surgery, University of Iowa Hospitals and Clinics, Iowa City, Iowa 52246*

Bette Clark *Police Artist, Computer Enhancement Technician, Forensic Identification Services, Metropolitan Toronto Police, 40 College Street, Toronto, Ontario, Canada M5G 2J3*

Ralph B. D'Agostino, Jr., M.A. *Staff Statistician, Department of Behavioral Sciences/Medical Genetics, Eunice Kennedy Shriver Center, Harvard Medical School, 200 Trapelo Road, Waltham, Massachusetts 02254*

Rollin K. Daniel, M.D., F.R.C.S.(C)., F.A.C.S. *Division of Plastic Surgery, Hoag Presbyterian Hospital, 1441 Avocado, Suite 308, Newport Beach, California 92660; and Research Associate, Department of Surgery, McGill University, Montreal, Quebec, Canada*

Curtis K. Deutsch, Ph.D. *Investigator, Department of Behavioral Sciences/Medical Genetics, Eunice Kennedy Shriver Center, Harvard Medical School, 200 Trapelo Road, Waltham, Massachusetts, 02254*

Leslie G. Farkas, M.D., C.Sc., D.Sc., F.R.C.S.(C) *Associate Professor, Department of Surgery, University of Toronto, Toronto, Ontario, Canada; Director, Craniofacial Measurements Laboratory, Craniofacial Program, Division of Plastic Surgery, The Hospital for Sick Children, 555 University Avenue, Toronto, Ontario, Canada M5G 1X8; and Research Consultant, Plastic Surgery Research Laboratory, The Hospital for Sick Children, 555 University Avenue, Toronto, Ontario, Canada M5G 1X8*

Ananda V. Gubbi, Ph.D. *Biostatistician, Western Psychiatric Center, University of Pittsburgh Medical Center, 3501 Forbest Street, Pittsburgh, Pennsylvania, 15213*

Karel Hajniš, Rn.Dr., Ph.D. *Professor and Head, Department of Anthropology, Charles University, Vinična 7, 12844 Prague 2, Czech Republic*

Tania A. Hreczko, Ph.D. *Adjunct Assistant Professor, Department of Pediatrics, Faculty of Medicine, University of Calgary, Calgary, Alberta, Canada T2T 5C7; and Dysmorphologist, Behavioural Research Unit, Alberta Children's Hospital, 1820 Richmond Road SW, Calgary, Alberta, Canada T2T 5C7*

Marko J. Katic, B.A. *Biostatistician, Department of Research Design and Biostatistics, Sunnybrook Health Science Center, 2075 Bayview Avenue, North York, Toronto, Ontario, Canada M4C 4E6*

S. T. Lee, M.D., F.A.M.S., F.R.C.S.(Edin) *Consultant, Department of Plastic Surgery, Singapore General Hospital, Outram Road, Singapore 0316*

Rexon C. K. Ngim, M.D., F.A.M.S., F.R.C.S.(Edin) *Consultant, Department of Plastic Surgery, Singapore General Hospital, Outram Road, Singapore 0316*

Jeffrey C. Posnick, D.M.D., M.D., F.R.C.S.(C)., F.A.C.S. *Associate Professor, Departments of Surgery and Pediatrics, and Director, Georgetown Craniofacial Center, Georgetown University Medical Center, 3800 Reservoir Road NW, Washington, D.C. 20007*

Peter S. Reid, B.A. *Senior Medical Photographer, Graphics Center, Division of Visual Education, The Hospital for Sick Children, 555 University Avenue, Toronto, Ontario, Canada M5G 1X8*

J. David Sills, P.C. *Police Artist, Computer Enhancement Technician, Forensic Identification Services, Metropolitan Toronto Police, 400 College Street, Toronto, Ontario, Canada M5G 2J3*

Govindasarma Venkatadri, M.D., Ph.D. *Craniofacial Center of Western New York, 219 Bryant Street, Buffalo, New York 14222*

Richard E. Ward, Ph.D. *Associate Professor and Head, Department of Anthropology, Indiana University, 425 University Boulevard, Indianapolis, Indiana 46202*

Acknowledgments

This book was made possible by the collective effort and untiring help received from various institutes, municipalities, agencies, colleagues, and close members of my family. Collection of population norms for subjects below 6 years of age was made possible by financial support from the Easter Seal Society Foundation in Toronto and by the Alberta Children's Hospital Foundation of Calgary. Funds for the Facial Attractiveness Study were provided by the Society of Aesthetic Surgeons, USA. Manuscript preparation expenses were covered by funding from The Hospital for Sick Children Foundation in Toronto. Raven Press provided the funds for the artwork in the monograph. Directors and managers of Daycare Centers in Toronto and Calgary offered the valuable technical help during the examination of the children. I am grateful, too, to Hans Lichtenberg, Director of Zoom Professional Photography, for much help given in developing contacts with fashion agencies in Toronto, for assistance in obtaining measurements from the fashion models, and for providing photo documentation of the examinees.

Statistical analysis of the newly collected population data from Toronto was carried out by Tünde Szatmáry, statistician. The Calgary material was analyzed by Marko J. Katic, biostatistician in the Department of Research Planning and Biostatistics of Sunnybrook Hospital in Toronto. His professional help was invaluable in the preparation of tables and in the interpretation of their findings.

The responsible task of manuscript editing was carried out by Neal Thompson of Toronto. Professional help was also provided by editors Sharon Nancekivell and Frank Quinlan, at The Hospital for Sick Children in Toronto. Drawings showing the various anthropometric methods of measurement were made by F. B. Fodor of Montreal, Quebec. I am also greatly indebted to Linda Power, Maria de Vera, Gigi Concepcion, Donna Stead, Nancy Taylor, and Judy Edwards, at The Hospital for Sick Children in Toronto, all of whom readily offered their help when needed. My very special thanks go to Oedill Daniel for her untiring assistance while the material for the monograph was being gathered.

L.G.F.

Foreword

Blaise Pascal made an intuitive anthropometric estimate when he said "The nose of Cleopatra—if it had been shorter, the face of the earth would have changed." Intuitive estimates of the anthropometrics of the human face are part of daily clinical practice in which the plastic surgeon decides "normal or abnormal," "beautiful or aesthetically disadvantaged," "improved or not improved," and so on, and intuition must still be the final standard. Provide a quantitative foundation for this process, however, and a whole new world opens up, in which comparisons over time, comparisons with "ideal" subjects, and facial constructions/reconstructions are possible—with pencil, drafting tools, and anthropometric tables or with digitizer, computer, and software. With such a "Lord Kelvin-approved" foundation, we can apply science and engineering at a basic level to the deformed or unattractive face, and we can free our intuition and imagination to work on advanced frontiers at an accelerated pace. Fortunately for plastic surgeons (and for other surgical specialists, clinical geneticists, forensic investigators, and artists), Leslie Farkas, plastic surgery's *anthropologist laureate,* has made the development of such a quantitative foundation his life work. *Anthropometry of the Head and Face* brings us the core of this work. It shows us how to apply this core in clinical practice, and then it stimulates our imagination with examples of new applications and new fields of investigation. This anthropometric core is based on meticulous direct measurements of an extensive group of Caucasians of different ages and ethnic origin, now augmented by a large group of Asians of various ages and a group of young African-Americans.

Dr. Farkas tells us how to look at a human face, how to measure a human face, how to compare measurements, and how to make clinical judgments.

Anthropometry of the Head and Face tells the reader how to use direct anthropometric measurements in clinical practice and how to utilize indirect measurements from photos. It tells how to set up our photo studio for accurate standardization and maximum photogrammetric benefit. It provides us with a background of scholarly commentary to aid in research, distinguishing strengths and sources of error. And it introduces us to those other fields that use facial anthropometrics, such as the Metropolitan Toronto Police's Computer Assisted Recovery System (CARES), in which family snapshots are metamorphosed into life-sized standard facial photos, and in which the face is "aged" to account for the time elapsed since the last snapshot. Closely related is reconstruction of a life-sized skull and facial images from skull remnants; this process is known as "computer photographic skull reconstruction."

Maureen Mullarkey, the artist, speaks of "the splendid living design" of her artists' model . . . "the drama of the flesh . . . an architectonic system of skeleton and muscle, a musical arrangement of ellipsoids and undulating arcs." Dr. Farkas bravely explores the quantification of the elusive, ephemeral, yet timeless elements of facial beauty in his chapter describing anthropometry of the attractive North American Caucasian face. This work brings out the overriding role of facial proportions as the key to facial harmony or disharmony.

Paul Tessier has in the past commended Leslie Farkas's work as providing us with "a compass map and route marker to assess whether or not the position is correct and the direction is true." *Anthropometry of the Head and Face* suggests that Dr. Farkas will not be content until he has developed an inertial navigational system to guide us.

David W. Furnas, M.D.

Preface

Slightly more than a decade after the publication of *Anthropometry of the Head and Face in Medicine,* this new edition reflects a substantial expansion both in new anthropometric measuring techniques and in the amount and variety of data collected. It also looks forward to the application of promising technological developments to mapping of the facial surface topography.

The amount of new information available is considerable. The data bank of North American Caucasians has been extended to include the youngest age groups, and anthropometric characteristics have been determined in three subgroups of Caucasians: North Americans, Germans, and Czechs. Age-related changes in the head and face in North American Caucasians are presented through a large number of linear and angular measurements. In addition, anthropometric norms have been developed for selected age groups of African-Americans and Chinese. Asymmetries of the face have been analyzed in young adult white, black, and Chinese subjects of both sexes, paying attention to the significant differences in anthropometry of the head and face of North American Caucasians, African-Americans, and Singapore Chinese in this age range.

Anthropometry has also been used to demonstrate quantitatively the differences in the degree of attractiveness in young adult North American Caucasian males and females. Chapters are devoted to assisting in the selection of measurements in clinical practice, avoiding errors in measurement, and interpretating findings and presenting results. Another chapter explores the possibilities offered by the use of computer-generated three-dimensional images of the head and face, technology that heralds the advent of a new era in anthropometry. The book describes the application of surface measurements in both medical and nonmedical fields, presenting a total of over 10,000 normative data, about 8,400 for anthropometric measurements and 1,600 for asymmetries.

Ostensibly, this book if for medical clinicians and researchers—plastic surgeons, maxillo-facial surgeons, otolaryngologists, orthodontists, geneticists, dysmorphologists, toxicologists, psychiatrists, anatomists, dermatologists, and pediatricians—but the audience goes well beyond the boundaries of medicine. In fact, the book is for anyone with an interest in dimensions of the head and face. Physical anthropologists, police detectives, forensic identification professionals, artists, and manufacturers of helmets and eyeglasses will also find the book useful.

Leslie G. Farkas, M.D.

Anthropometry of the Head and Face

Second Edition

Chapter 1

The Population Sample

Leslie G. Farkas

Surface measurement norms for head, face, orbits, nose, lips, mouth, and ears had been formulated from examinations of 2326 healthy Caucasian North American residents, with ages ranging from newborn to young adult. Incomplete records relating to Chinese and African-American populations have now been supplemented by examinations of 235 children and young Chinese adults (in Singapore and Canada) and 132 young black adults in the United States (Table 1–1).

In Caucasians, the cranial and facial norms were established in 20 age groups from birth to 18 years of age. The age of the young adults was in range of 19–25 years. The sample was almost equally divided between the sexes. Collection of the norms from individuals living in Western (Alberta), Central (Ontario), and an Eastern Province (Quebec) of Canada ensured a fair representation of various ethnic groups of Caucasians.

The craniofacial anthropometric norms in Chinese population were obtained from children 6, 12, and 18 years of age (1). The data about the face in young adulthood were also obtained from a group of Chinese residents of Toronto. The group of Chinese examined are ethnically very close to the Cantonese and Hong Kong Chinese, settled in North America.

Young adult blacks were represented by a group of African-Americans and blacks born in the Caribbean Islands, all university students in Buffalo, New York.

The Caucasian norms between 6 and 18 years of age were collected between 1973 and 1976 (2). The norms of children below 6 years of age and in young adults were established between 1983 and 1986. The Chinese and African-American norms were collected in the years 1987–1989.

TABLE 1-1. *Racial origin, age, and sex of the population sample*

Race	Age group	Sex[a]	Number of individuals measured	Location	Measured by
Caucasian	Birth–3 years	M	107	Calgary, Alberta, Canada	T. A. Hreczko
		F	101		
	4–18 years	M	714	Toronto, Ontario, Canada	L. G. Farkas
		F	718	Montreal, Quebec, Canada	
	19–25 years	M	275	Toronto, Ontario, Canada	L. G. Farkas
		F	411		
Subtotal			2326 (1096 males and 1230 females)		
Chinese	6, 12, 18 years	M	117	Singapore, Asia	L. G. Farkas
	19–25 years	F	118	Toronto, Ontario, Canada	
Subtotal			235		
African-American	19–25 years	M	66	Buffalo, New York	L. G. Farkas
		F	66	Toronto, Ontario, Canada	
Subtotal			132		
Total			2693 (1279 males and 1414 females)		

[a] M, male; F, female.

REFERENCES

1. Farkas LG, Ngim RCK, Lee ST. The fourth dimension of the face: A preliminary report of growth potential in the face of the Chinese population of Singapore. *Ann Acad Med Singapore* 1988;17:319–327.
2. Farkas LG. *Anthropometry of the head and face in medicine.* New York: Elsevier, 1981.

Chapter 2

Examination

Leslie G. Farkas

VISUAL ASSESSMENT (ANTHROPOSCOPY)

A well-organized system of observations renders work easier, more rapid, and more accurate.

Aleš Hrdlička
1920

Anthroposcopy (from the greek *anthropos,* "human," and *skopein,* "examine") means judging the body's build by inspection. Visual assessment, one of the oldest methods of examination still used in medicine today, is not reliable because it is highly subjective. We recorded 57 qualitative signs in the craniofacial complex (Table 2–1). Most of these signs cannot be expressed numerically, because their determination quantitatively would be difficult. Some signs were listed because they express the visual impression of the measurement (e.g., acute nasolabial angle).

TABLE 2–1. *Anthroposcopic (qualitative) signs in the craniofacial complex*

Region	Number of signs
Head	2
Face	6
Orbits	13
Nose	14
Lips and mouth	9
Ears	13
Total	57

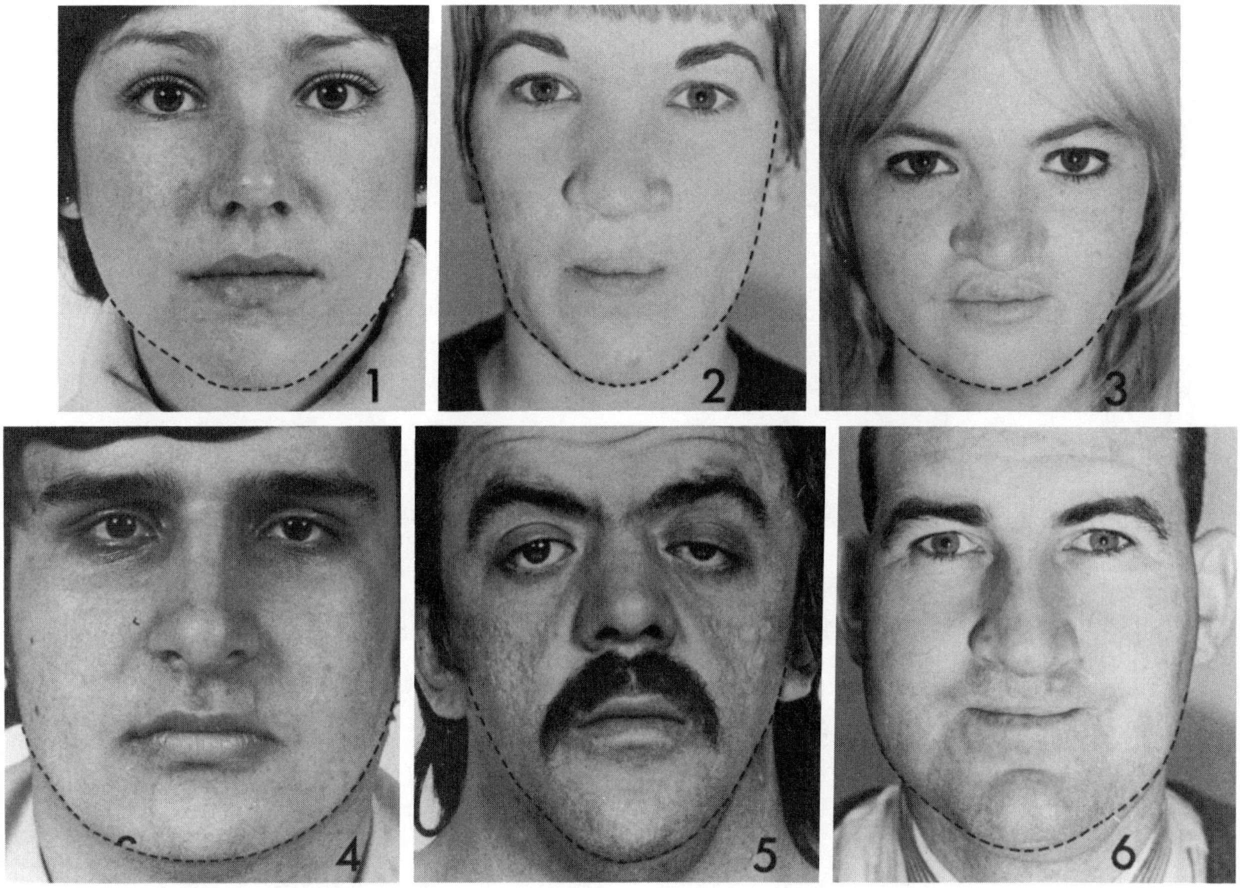

FIG. 2–4. General face shape: (**1**) proportionate in width and height; (**2**) long–narrow; (**3**) short–wide; (**4**) square; (**5**) triangular; (**6**) trapezoid.

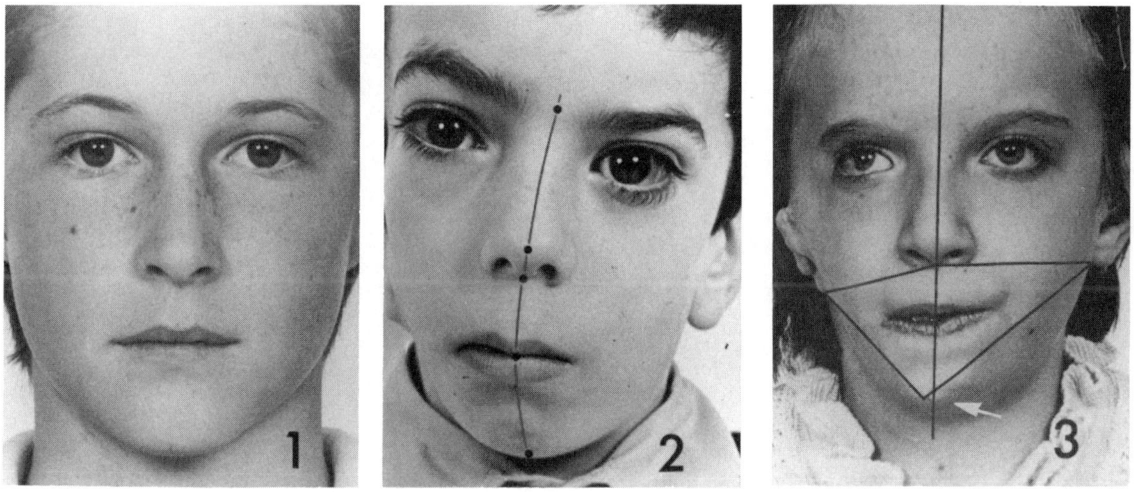

FIG. 2–5. Facial midaxis quality: (**1**) in midline; (**2**) concave; (**3**) with dislocated chin. (From ref. 12, with permission.)

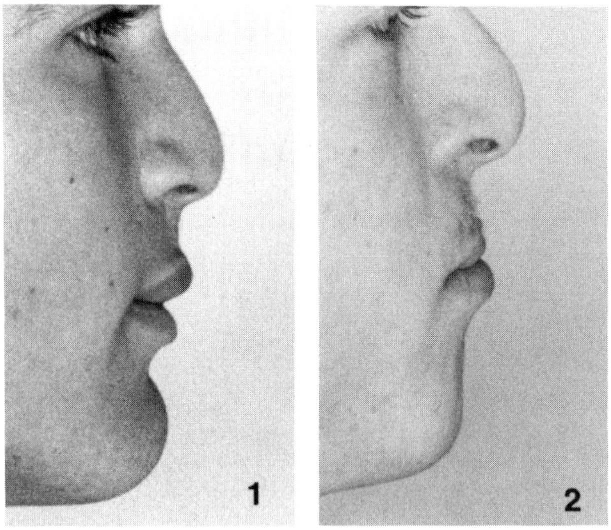

FIG. 2-6. Chin contour: (**1**) indented; (**2**) flat.

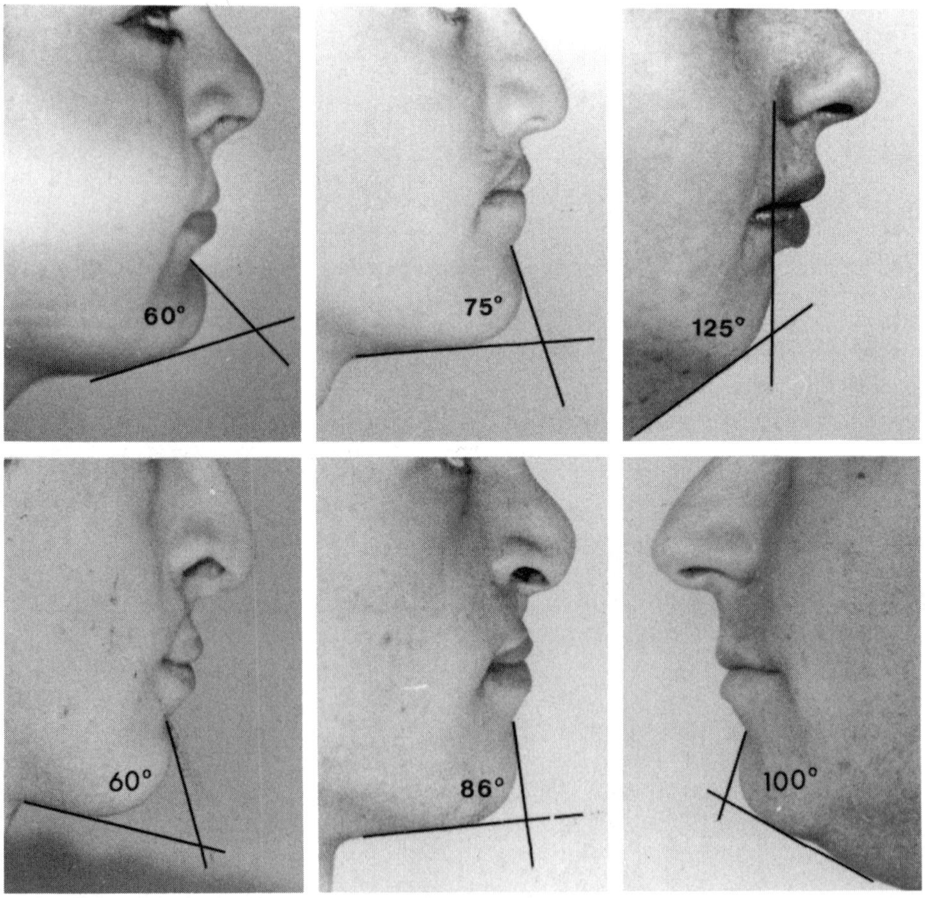

FIG. 2-7. Mentocervical angle. **Upper line:** Females. **Lower line:** Males. Qualities: 60 degrees indicates subnormal (acute) angles; 75 degrees and 86 degrees are regarded as optimal (close to the respective means); 125 degrees and 100 degrees are supernormal (obtuse).

Orbits

1. Relative positions of orbitale landmarks: level, uneven level (rt-lt).
2. Relative positions of the FHs: level, uneven level (rt-lt).
3. Direction of each eye fissure: normal, mongoloid, antimongoloid, other (Fig. 2–8).
4. Epicanthus of each eye (Fig. 2–9): minor, moderate, and marked. The marked epicanthus covers the endocanthion. In subjects who have marked epicanthi bilaterally, only the width of the area between the edges of the skin folds, and not the true intercanthal width, can be assessed (1).
5. Defect of each eyelid: ptosis, coloboma, hypoplasia, hyperplasia, other (Fig. 2–10).
6. Function of each eyelid: normal, ptotic, lagophthalmic.
7. Each eyebrow: normal, uneven relative to other eyebrow, defective, deformed in shape (Fig. 2–11).
8. Each supraorbital rim: normal, protruding, recessed, of uneven level, other.
9. Position of orbits: level, lower (rt-lt), deeper (rt-lt) (Fig. 2–12).
10. Position of eye fissures: level, one lower (rt-lt) (Fig. 2–12).
11. Position of endocanthions: level, one lower (rt-lt) (see Fig. 2–12).
12. Bulbous: normal in size, exophthalmos, microphthalmos, anophthalmos (rt-lt) (Fig. 2–13).
13. Eyelashes: normal, partly deficient, deficient (rt-lt).

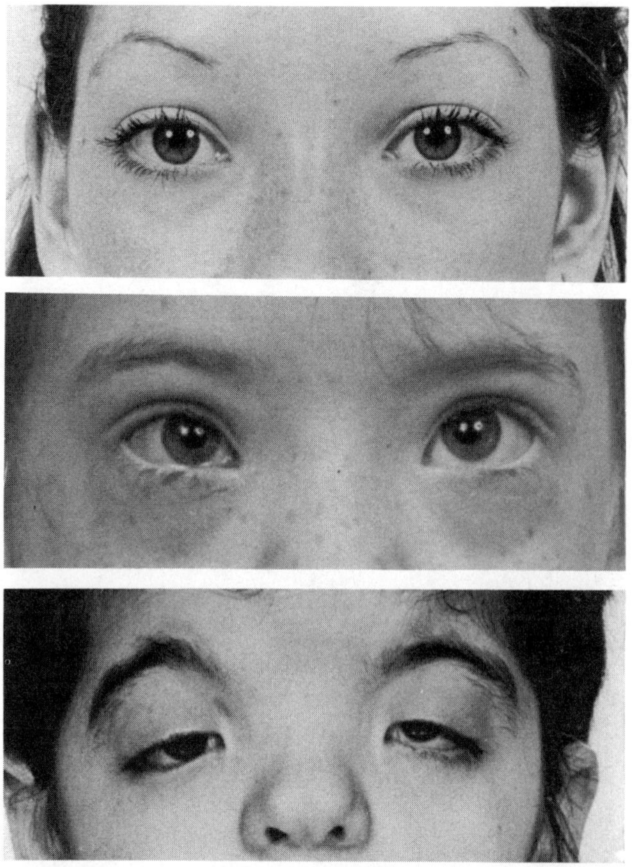

FIG. 2–8. Eye fissure direction. **Upper line:** Normal. **Middle:** Mongoloid. **Lower line:** Antimongoloid.

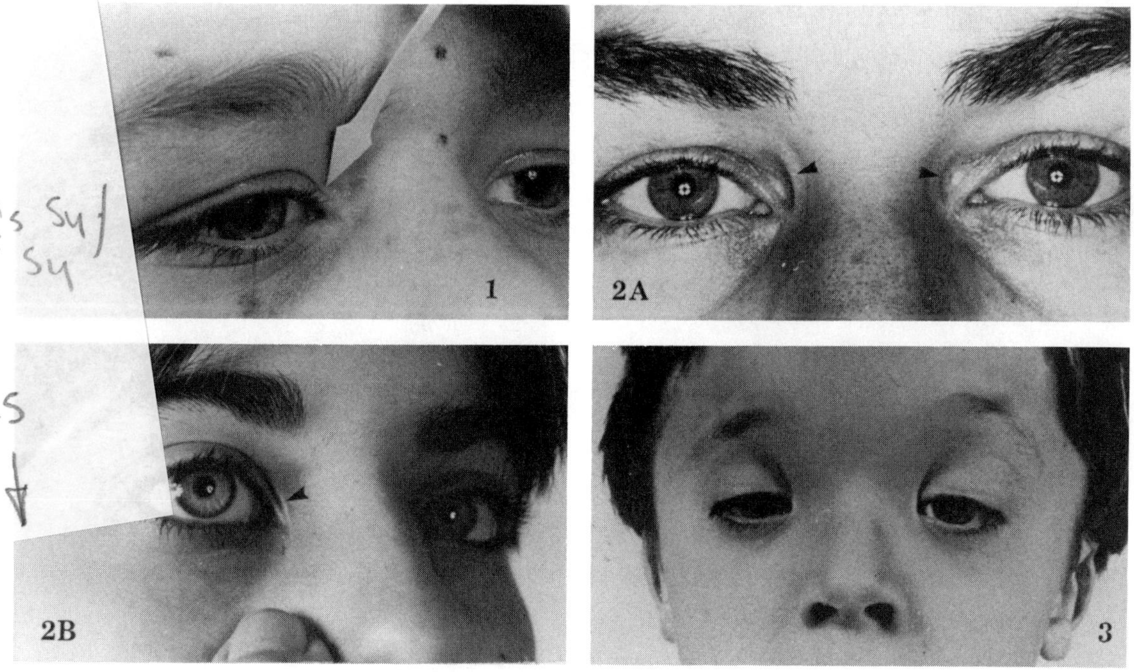

FIG. 2–9. Epicanthus: (**1**) Minor; (**2A, 2B**) moderate; (**3**) marked.

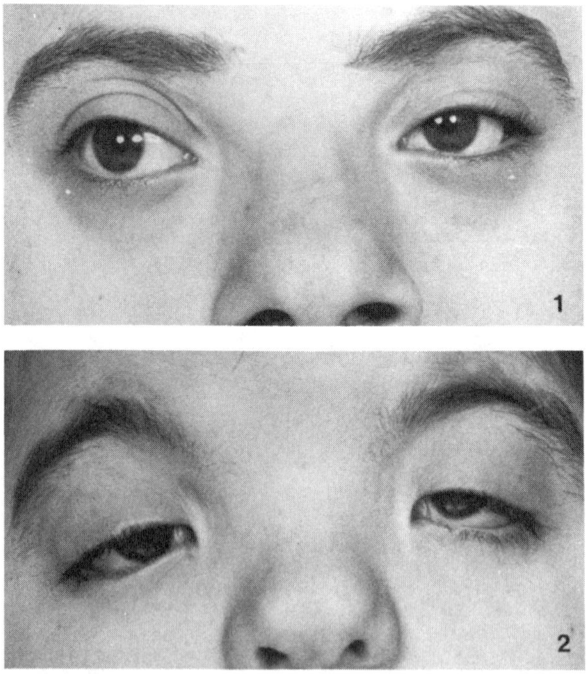

FIG. 2–10. Eyelid defect: (**1**) unilateral ptosis (lt); (**2**) bilateral ptosis.

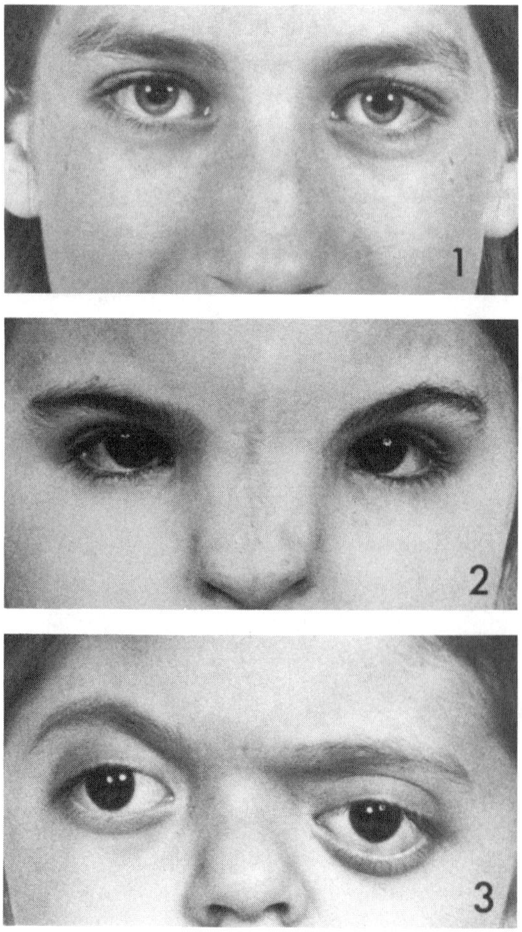

FIG. 2–12. Position of the orbits: (**1**) in level; (**2**) right lower; (**3**) left deeper.

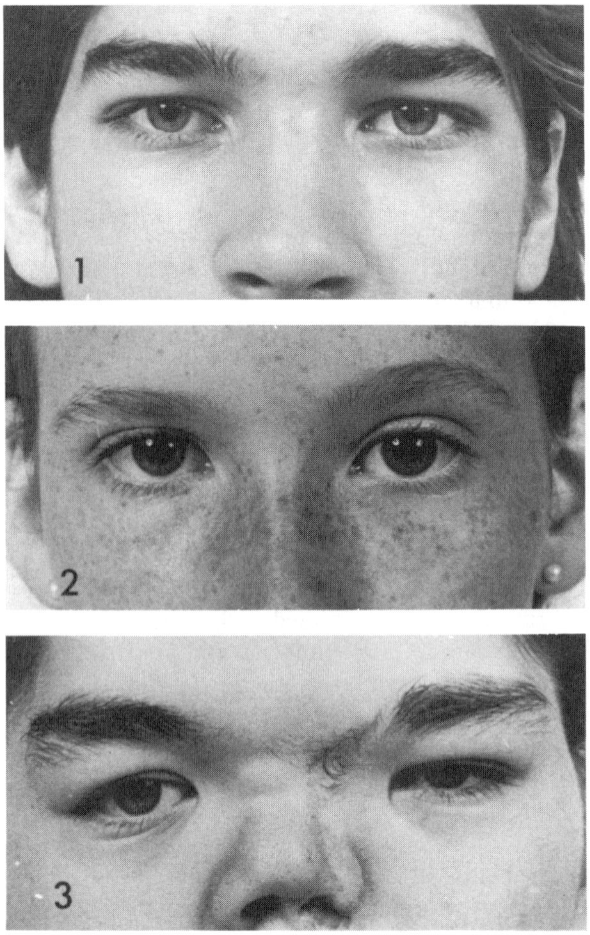

FIG. 2–11. Eyebrow quality: (**1**) normal and at the same level; (**2**) right brow lower located; (**3**) left brow defective.

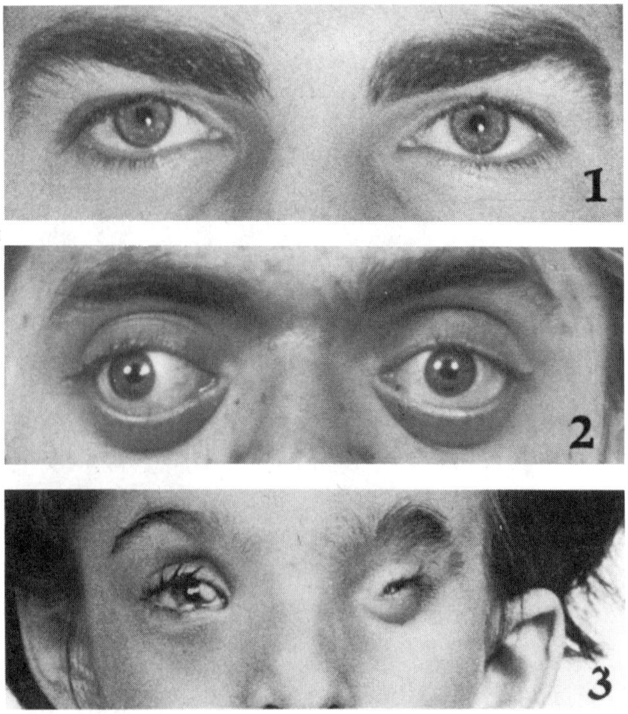

FIG. 2–13. Bulbous: (**1**) normal; (**2**) exophthalmos; (**3**) microphthalmos (lt).

Repaired cleft palate

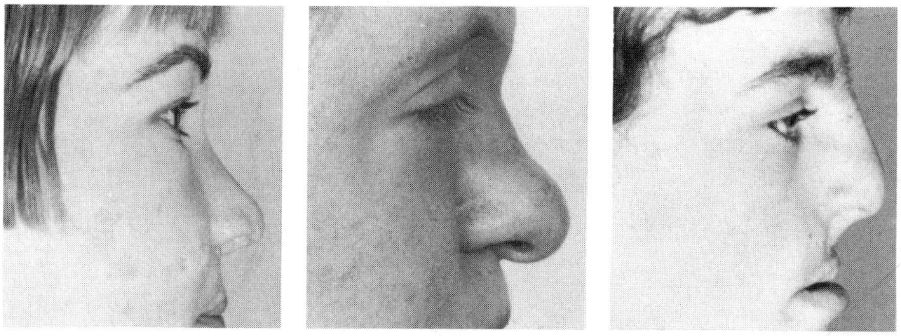

FIG. 2–14. Nasal root depth. **Left:** Shallow. **Middle:** Medium. **Right:** Deep.

Nose

1. Nasal root depth: shallow, medium, deep (Fig. 2–14).
2. Nasal root width: medium, narrow, wide (Fig. 2–15).
3. Nasal bridge contour: straight, hump, saddle (Fig. 2–16).
4. Nasal tip: medium, wide, pointed, flat, bifid, other (Fig. 2–17).
5. Shape of each nasal ala: normal, flat, angled, other (Fig. 2–18).
6. Defect of each nasal ala: cleft, coloboma, hypoplasia, aplasia, hypertrophy, other.
7. Type and size of nostrils: symmetrical, asymmetric type, asymmetric size, asymmetric type and size (Fig. 2–19).
8. Septum: no dislocation, dislocated (rt-lt) (Fig. 2–20).
9. Alar base configuration: short-curved, full-curved, straight + thin + short, straight + thin + long (rt-lt) (Fig. 2–21).
10. Vertical dislocation of the labial insertion of the alar base (sbal): lower, higher (rt-lt) (Fig. 2–22).
11. Sagittal dislocation of the facial insertion of the alar base (ac): deeper-located, protruding on one side (rt-lt).
12. Nasofrontal angle shape: normal, deep, flat (Fig. 2–23).
13. Nasofrontal angle degree: medium, acute, obtuse (see Fig. 2–23).
14. Nasolabial angle size: medium, acute, obtuse, other (Fig. 2–24).

APERT'S SYN.

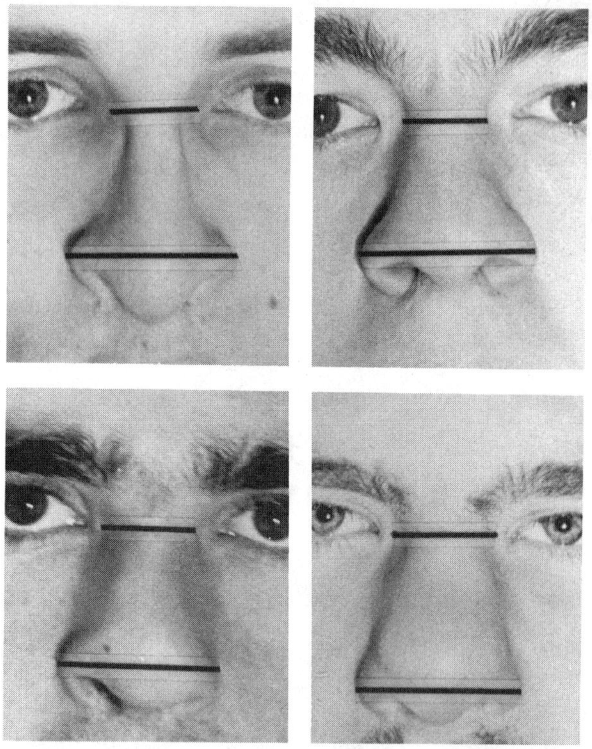

FIG. 2–15. Nasal root width. Variations from narrow-normal (**upper line left**) to medium-normal (**upper line right**) and abnormally wide nasal roots (**lower line**).

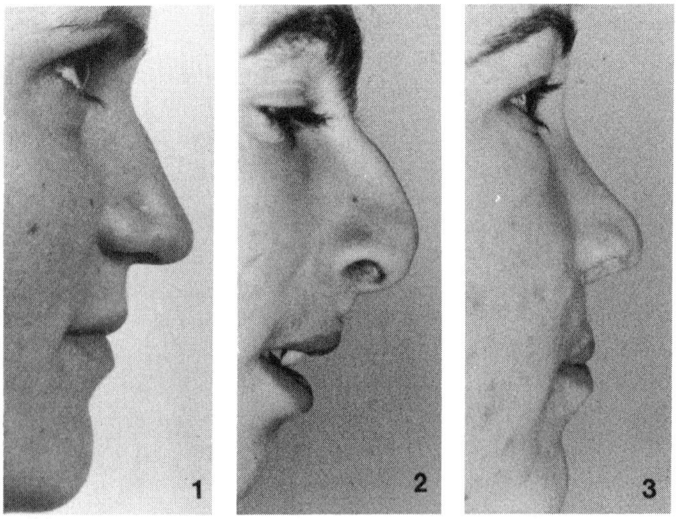

FIG. 2–16. Nasal bridge contour: (1) straight; (2) with hump; (3) with saddle.

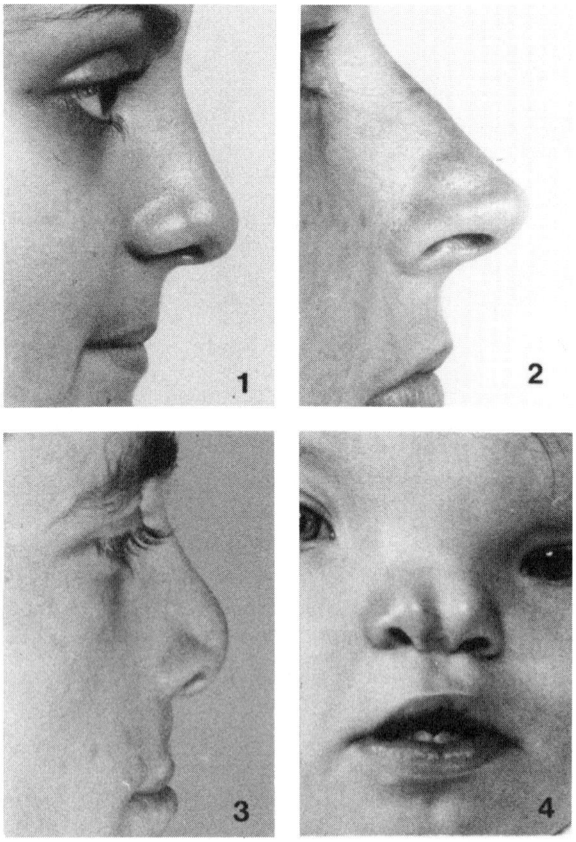

FIG. 2–17. Nasal tip quality: (1) medium; (2) pointed; (3) flat; (4) bifid.

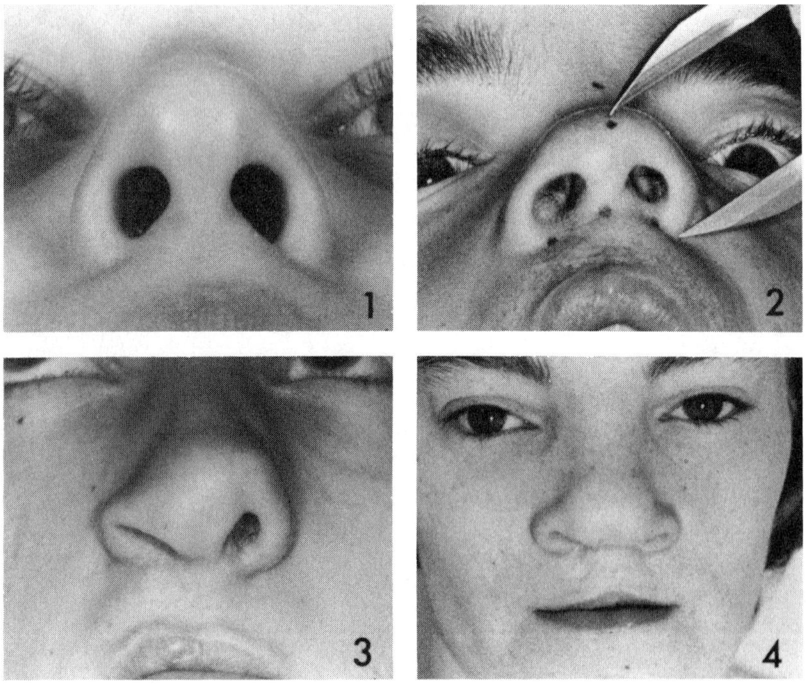

FIG. 2–18. Ala shape: (1) normal; (2) right flatter; (3) right-angled; (4) bilaterally angled.

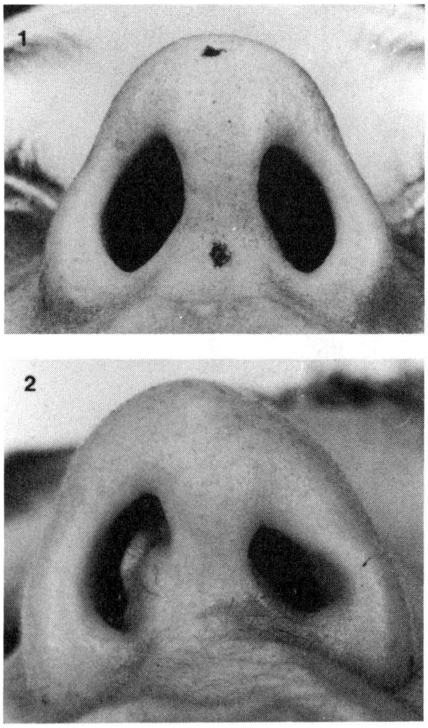

FIG. 2–19. Nostril shape quality: (1) symmetrical in shape (type); (2) asymmetrical in shape (type) and size.

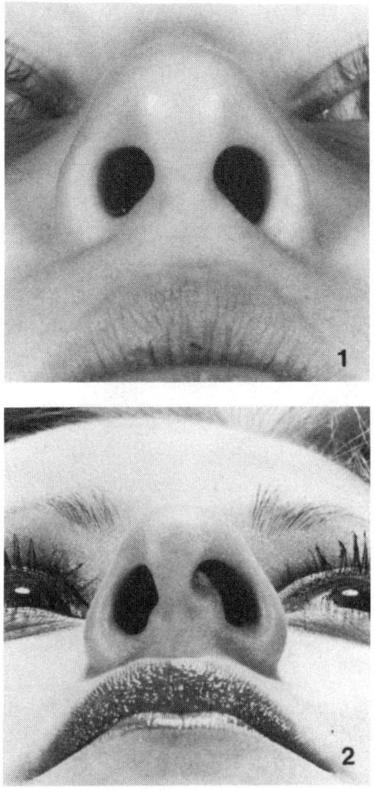

FIG. 2–20. Septum position. (**1**) in midaxis; (**2**) deviated to right.

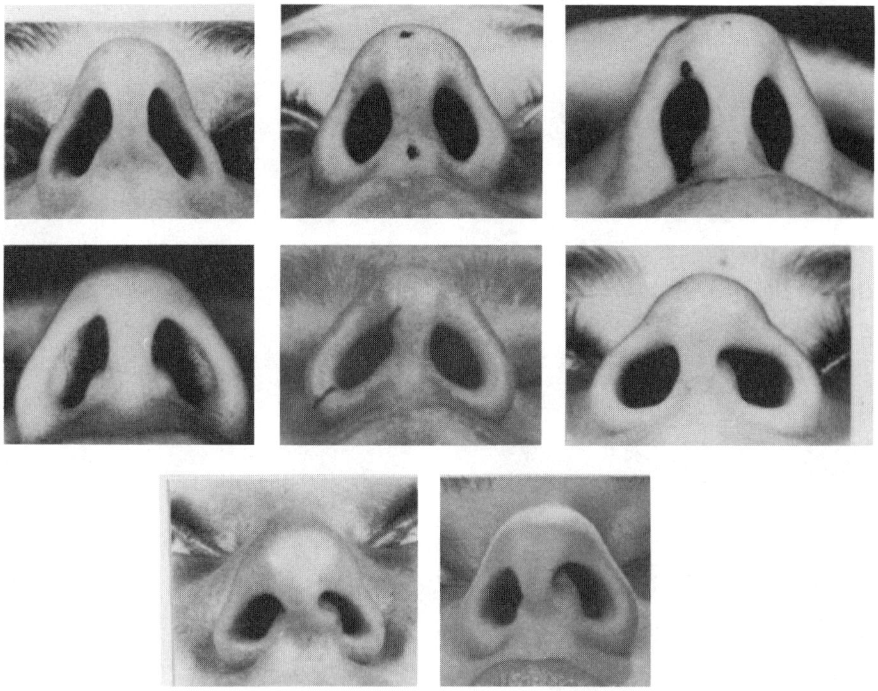

FIG. 2–21. Caucasian alar base configuration. **Upper line:** Variations of the alar base shape in type I nostrils; the base is slightly curved or straight. **Middle and lower lines:** Type II and III nostrils; variations between the slightly curved short base and noses with curved longer bases. (From ref. 37, with permission.)

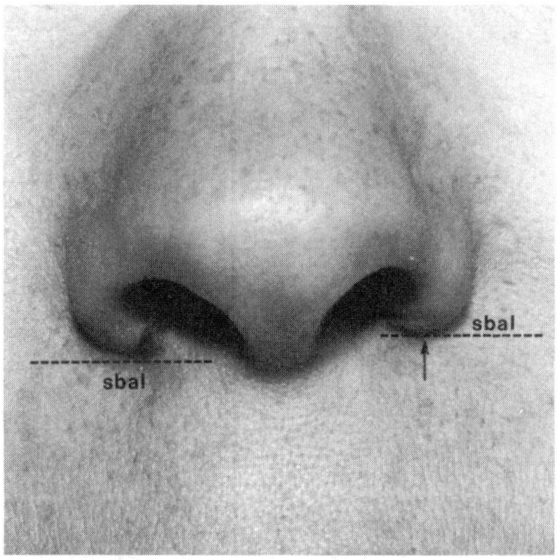

FIG. 2–22. Uneven vertical position of the labial alar bases (sbal) in a healthy normal subject.

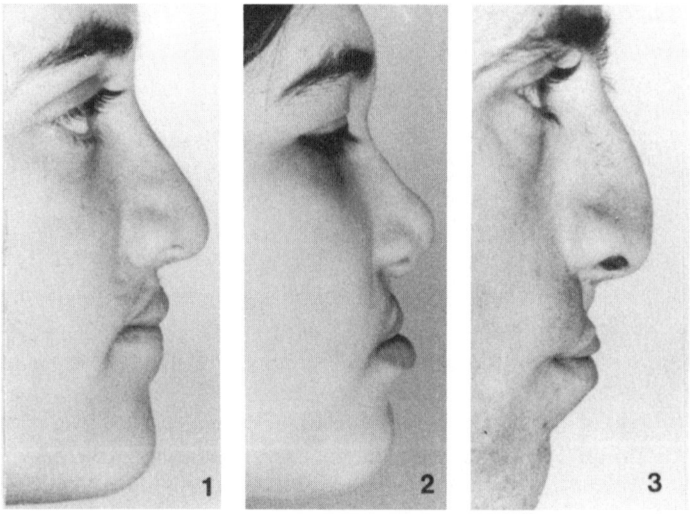

FIG. 2–23. Nasofrontal angle shape: (**1**) normal; (**2**) deep (acute); (**3**) flat (obtuse).

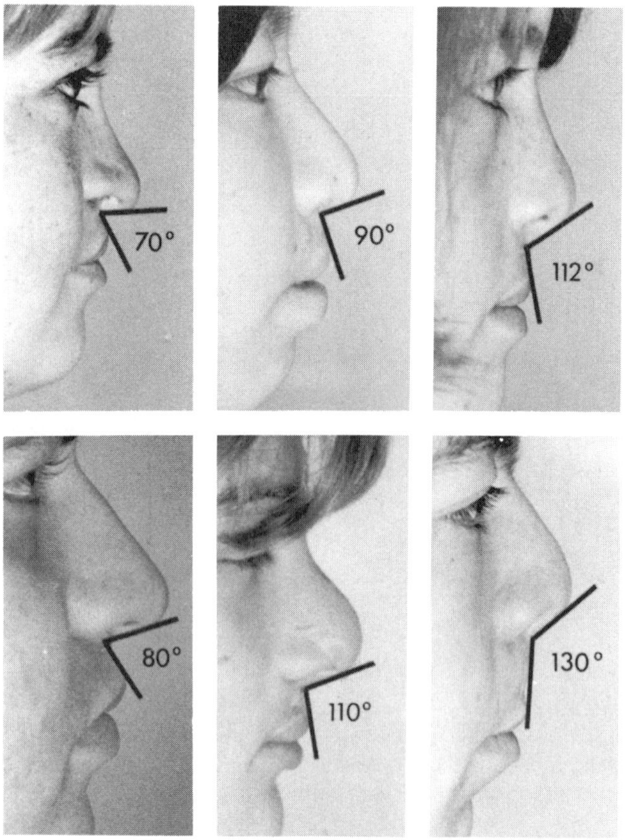

FIG. 2–24. Nasolabial angle size. **Upper line:** Females. **Lower line:** Males. Qualities: acute, 70 degrees and 80 degrees (in females it is subnormal, in males it is borderline-small); medium, 90 degrees and 110 degrees; obtuse, 112 degrees and 130 degrees (supernormal).

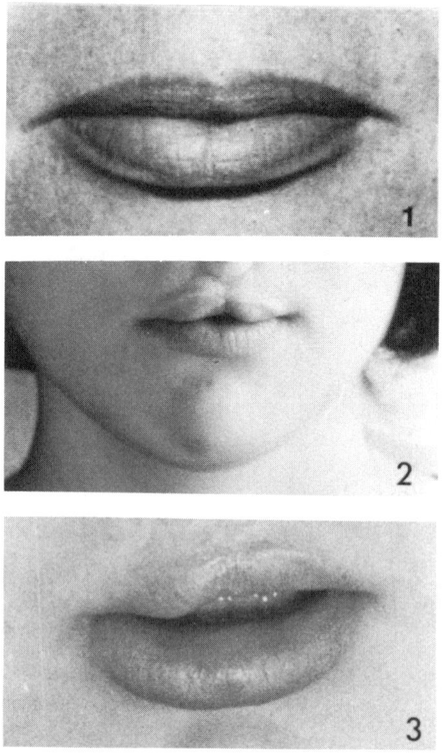

FIG. 2–25. Upper vermilion shape: (**1**) symmetrically formed; (**2**) notched in the middle; (**3**) unilaterally extended.

Lips and Mouth

1. Vermilion line of the upper lip: regular, irregular.
2. Shape of vermilion of the upper lip: symmetrically formed and normal, diminished, notched, extended, other (Fig. 2–25).
3. Defect of the upper lip: congenital scar, cleft, other.
4. Defect of the lower lip: fistula, cleft, other.
5. Relationship between the lateral heights of the upper lip: symmetrical, one side smaller due to lower location of the alar base (sbal) or higher location of vermilion line (ls′), one side larger due to higher location of the alar base (sbal) or lower location of the vermilion line (ls′).
6. Location of the labial fissure: in midportion of the face, dislocated to rt-lt.
7. Direction of the labial fissure: horizontal, one corner lower (rt-lt) (Fig. 2–26).
8. Defect of the labial fissure: asymmetric halves, pre- or postoperative macrostoma, other (Fig. 2–27).
9. Relationship between the upper and lower vermilion heights: proportionate, protruding lower lip vermilion (Fig. 2–28).

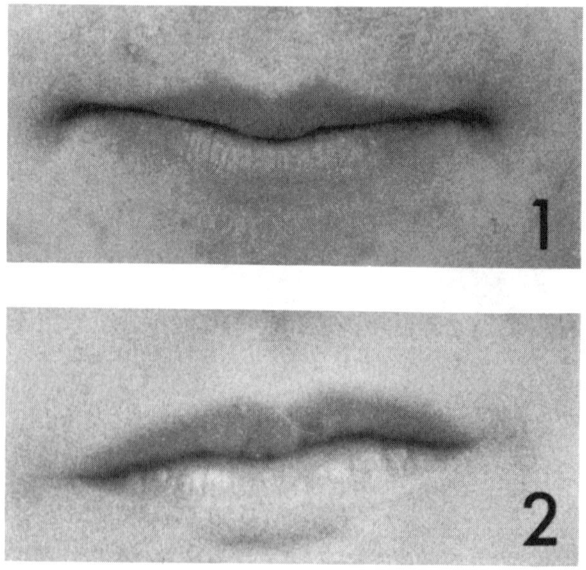

FIG. 2–26. Labial fissure position: (**1**) horizontal, (**2**) oblique.

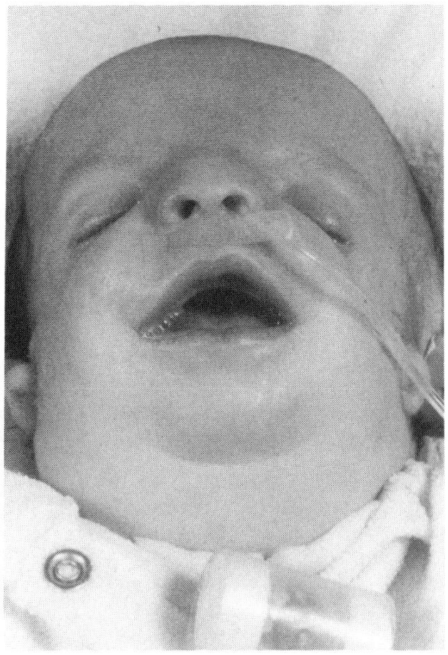

FIG. 2–27. Preoperative right macrostoma.

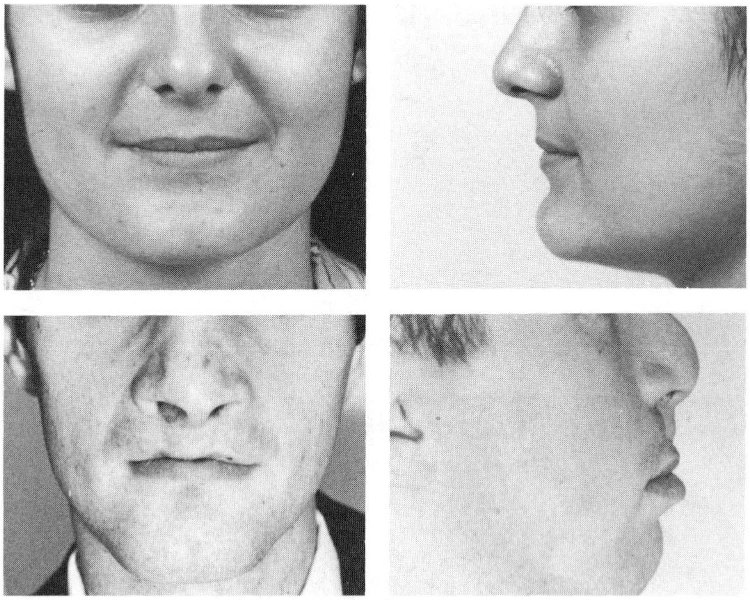

FIG. 2–28. Mutual relation between the vermilions. **Upper line:** Harmonious (proportionate). **Lower line:** Greatly reduced height of the upper lip vermilion with relatively protruding lower lip in adult patients with cleft lip palate operated upon in early childhood.

Ears

1. Each preauricular region: normal, with skin appendages, with fistula(e), with appendage(s) and fistula(e) (Fig. 2–29).
2. Each tragus: proportionate to the size of the ear, small, aplastic, large, bifid.
3. Each external auditory meatus: normal in size, narrow, atretic.
4. Position of the tragus in relation to external auditory meatus: At right ear the porion is *below* the level of tragion; at left ear the porion and the tragion landmarks are on the *same level* (Fig. 2–30).
5. Shape of each ear: symmetrical, asymmetrical in contour and/or anterior relief.
6. Anterior surface of each ear: helix covering scapha, normally rolled helix, flat helix, wide helix, Macacus-type ear, and helix impression on the upper half of the ear (Figs. 2–31 and 2–32).
7. Each ear-lobe attachment: free, attached (Fig. 2–33).
8. Developmental defects of each auricle: none, slight hypoplasia and shape deformity (mild microtia), true hypoplasia (moderate microtia), aplasia (severe microtia) (2), coloboma, hypertrophy, other.
9. Position of the ear canal horizontally: symmetrical in relation to the facial midline, horizontally one side is closer (anterior dislocation) (3) (Fig. 2–34).
10. Position of the ear canal vertically: symmetrical level between the two ears, one ear canal lower (3) (Fig. 2–35).
11. Upper edge of the ear in relation to orbital level: above eyebrow, at eyebrow tail, at upper eyelid level, at exocanthion level (Fig. 2–36).
12. Lower edge of each ear in relation to nasalabial level:

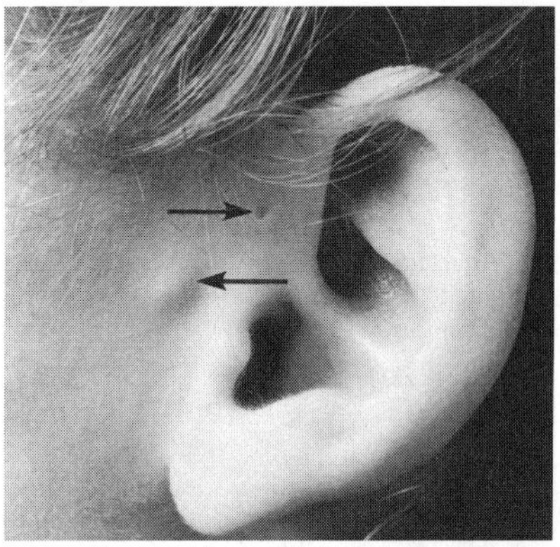

FIG. 2–29. Preauricular fistula and skin tag (*arrows*) in a healthy young boy.

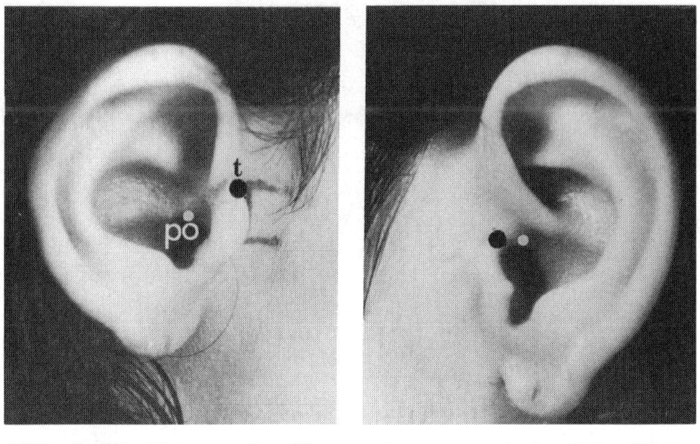

FIG. 2–30. Ear canal orifice and tragus relationship. **Right ear:** The porion (po) is below the level of the tragion (t). **Left ear:** The po and t are on the same level.

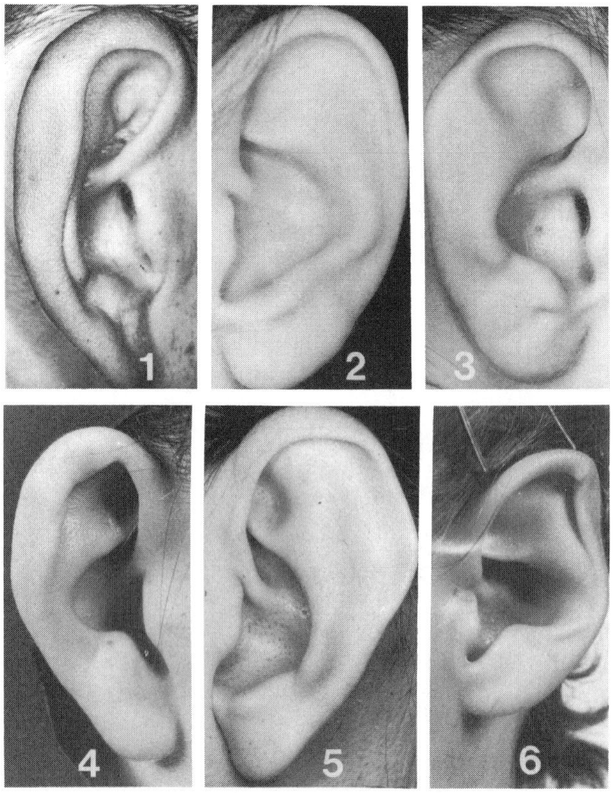

FIG. 2–31. Anterior surface qualities of the helix: (1) wide helix covering the scapha; (2) normally rolled helix; (3) flat helix; (4) wide helix; (5) Macacus-type ear; (6) helix impression in the upper half.

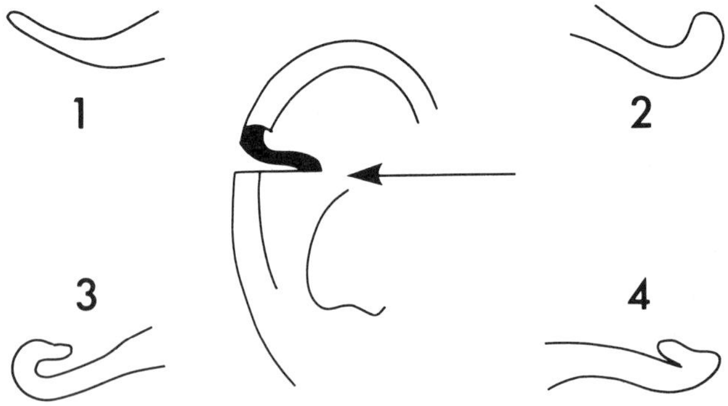

FIG. 2–32. Schematic representation of a cross section of the helix, showing: (1) flat helix; (2) normally rolled helix; (3) wide helix; (4) wide helix covering scapha.

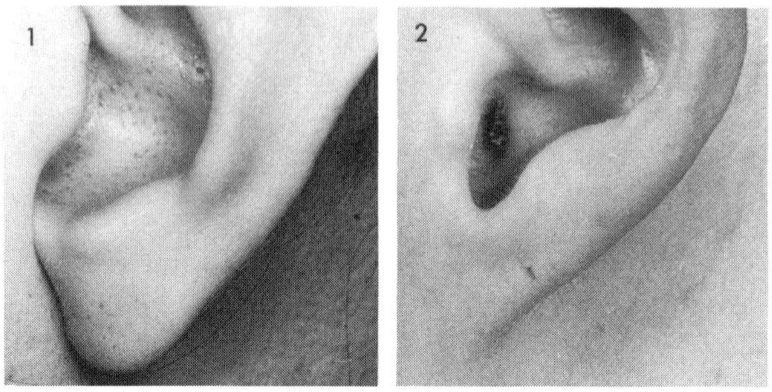

FIG. 2–33. Ear-lobe attachment: (1) normally developed free ear lobe; (2) attached hypoplastic ear lobe.

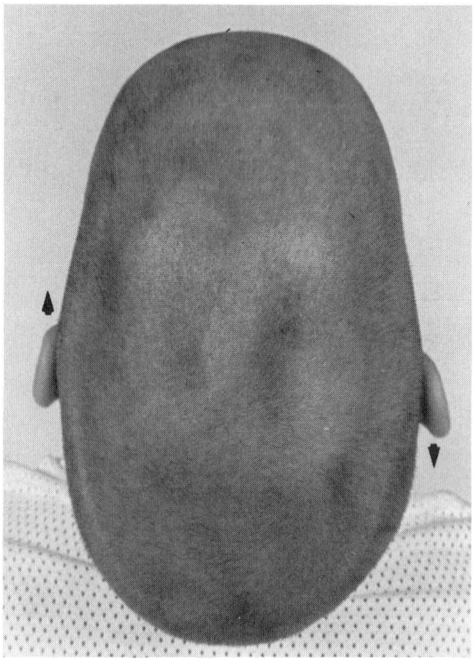

FIG. 2–34. Horizontal dislocation of the ear canal. The left ear is located closer to the facial midline.

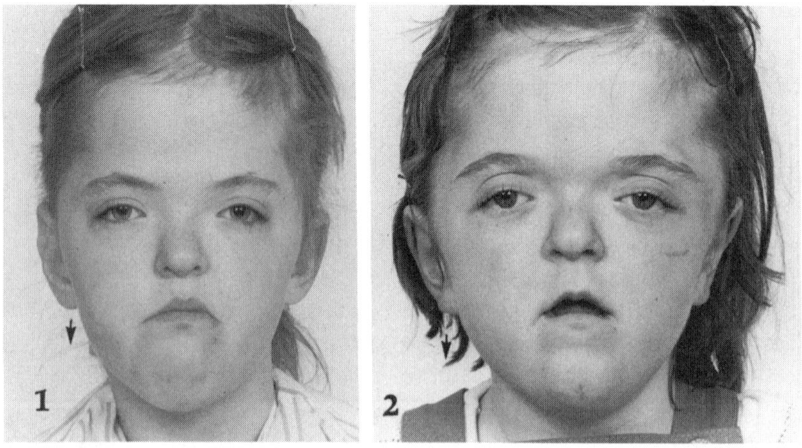

FIG. 2–35. Vertical dislocation of the ear canal. 1: The larger right ear (62 mm long) appears lower located than the shorter left ear (58 mm long), but the ear canals are approximately on the same level. 2: the right ear canal is lower located, but the ears are of virtually the same length (right 55 mm, left 56 mm).

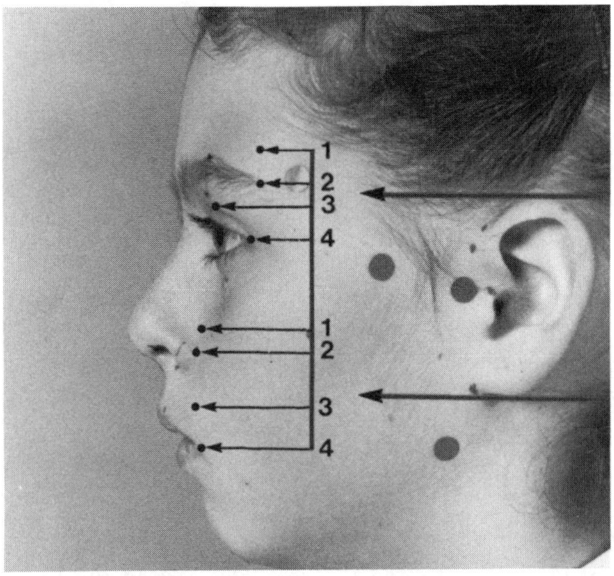

FIG. 2–36. Position of the auricle is determined by the relationship of its edges to other facial features in rest position of the head. *The upper edge of the ear* may be: (**1**) above the eyebrow; (**2**) level with the eyebrow tail; (**3**) level with the upper lid; (**4**) level with the exocanthion. *The lower ear edge* may be: (**1**) above the alar crest; (**2**) level with the alar crest; (**3**) level with the upper lip; (**4**) level with the cheilion.

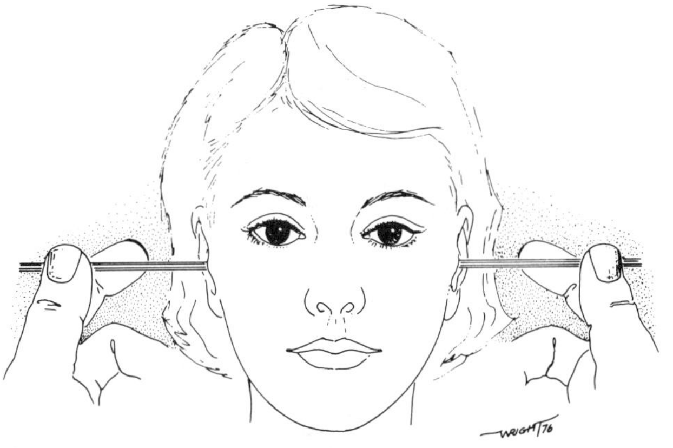

FIG. 2–37. Assessment of the relative positions of the tragions by holding two rods that extend horizontally from the tragions. The facial profile line must be vertical.

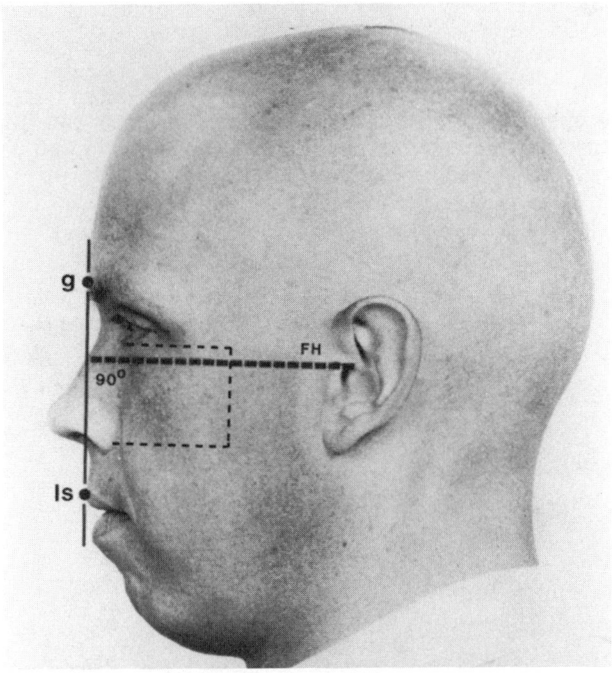

FIG. 2–38. Determination of the ear canal level according to Leiber's original proposal (4). The area of the normal ear canal location is outlined by the *fine dotted lines* on the face between the lower lid and the nasal ala. Leiber's profile line touches the glabella (g) and the most protrusive point (labiale superius, ls) of the upper lip. In this subject the Leiber line is vertical and the ear canal level is identical with the FH, putting the ear canal level to the upper third of the normal zone.

above alar crest, at alar crest level, at upper lip level, at cheilion level (see Fig. 2–36).

13. Relative positions of the tragions: on same level, one tragion lower (Fig. 2–37).

The following examination method for determination of the position of ears based on the level of the ear canal on the head [proposed by Leiber (4)] cannot be qualified as purely subjective or as completely objective. It determines the level of the ear canals in relation to the midface. Each canal level is given by the location of the point on the face at which a straight line drawn from the porion meets the special profile line (glabella–upper lip) at 90 degrees (Fig. 2–38). Leiber's profile line touches the glabella and the most protruding point of the upper lip (ls). The area of the face between the free margin of the lower eyelid and the upper edge of the nasal ala (normal area) must be crossed by the line drawn from the ear canal if it is to be considered at normal level. In a high-set ear this line meets the profile line above the normal area, and in a low-set ear it meets the profile line below the normal area.

In the *Farkas modifications* (5) of the Leiber proposal the level of the ear canal in relation to the midface is determined using a special device (6) (Fig. 2–39).

The normal area of the midface outlined by the lower lid and nasal ala is divided into three equal horizontal segments, which allows more accurate identification of the ear-canal level within the normal area (Fig. 2–40). The perpendicular (T-shaped) arm of the instrument is held along the special facial profile line (g-ls) and is moved up and down until the horizontal arm touches the tragion. The ear is considered to be in the normal location if the axial line of the horizontal branch lies within the area outline by Leiber (Fig. 2–41). The examination is carried out with the head in the rest position and the facial profile line in the vertical.

The test results are reliable only if the special profile line of Leiber is normal (Fig. 2–42a). Receding profile lines (Fig. 2–42c) incline to produce low-set ears, whereas markedly protruding lines (Fig. 2–42b) tend to shift the level of the ear canal into the upper third of the normal space or above the edge of the lower lid (set high) (7).

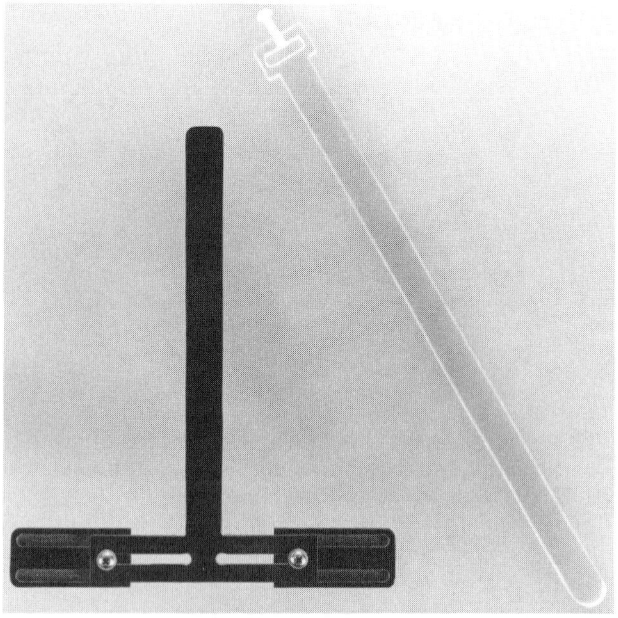

FIG. 2–39. Instrument designed by Farkas (5,6). The T-shaped part of the tool is placed along the Leiber line. The transparent branch attached to the T-shaped part at 90 degrees indicates the position of the ear canal.

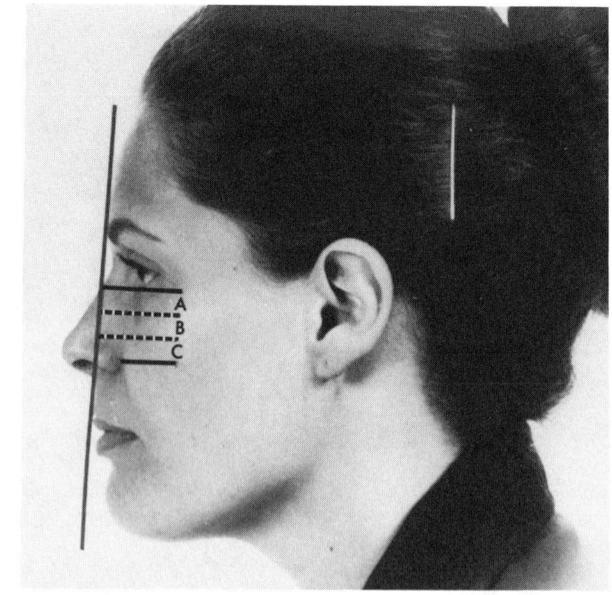

FIG. 2–40. The Farkas modification (6) of Leiber's proposal. The normal area is divided into three equal horizontal segments (**A–C**).

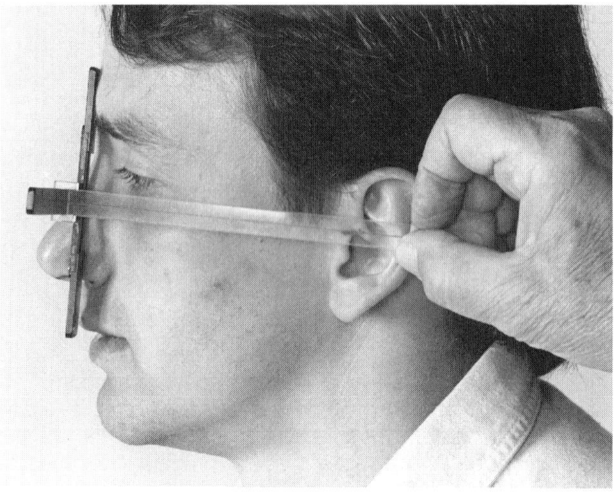

FIG. 2–41. Determination of the ear canal level by the Farkas modification (7,8). The ear canal in this adult subject is in a high normal position.

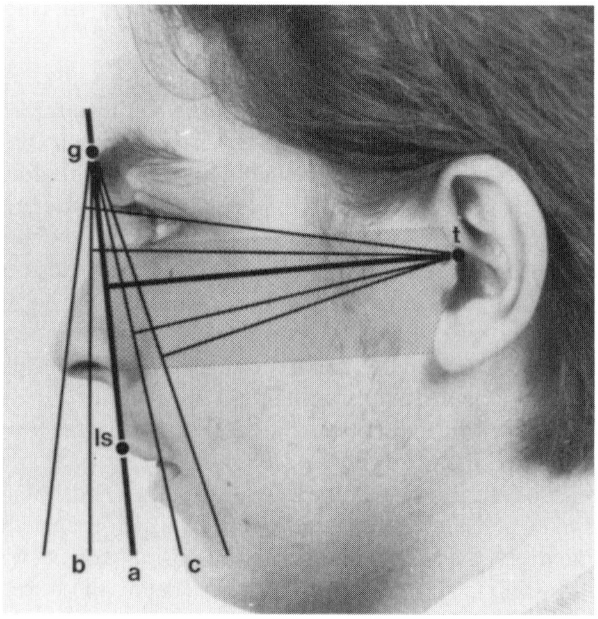

FIG. 2–42. The dependence of the Leiber test on the degree of line inclination (g-ls). **a:** (normal position indicated by the *solid line*): The ear canal is located in the middle third of the normal zone. **b:** Protruding profile line shifts the level of the ear canal upwards. **c:** Receding profile line shifts the level of the ear canal downwards. (From ref. 2, with permission.)

QUANTITATIVE ASSESSMENT (ANTHROPOMETRY)

Measurements, to be strictly comparable,
must be taken in a strictly defined way
and from or between the same anatomical points.

Aleš Hrdlička
1920

Anthropometry (from the Greek *anthropos,* "human," and *metron,* "measure") is the biological science of measuring the size, weight, and proportions of the human body.

Measuring Tools and Techniques

The standard instruments (e.g., sliding and spreading calipers) that are used in physical anthropology are made of metal, and the metric tape is made of fabric and has a millimeter scale (Fig. 2–43). Plastic measuring tapes on the market today are not ideal for measuring surface (tangential) distances, especially short ones (e.g., length of the alae, slopes of the nasal root, arcs of the vermilions). These tapes are not flexible enough to adhere to the skin surface and follow its changing reliefs.

The sliding caliper measures the linear projective distances between two landmarks in the same plane or in neighboring planes (e.g., eye fissure length, height of the

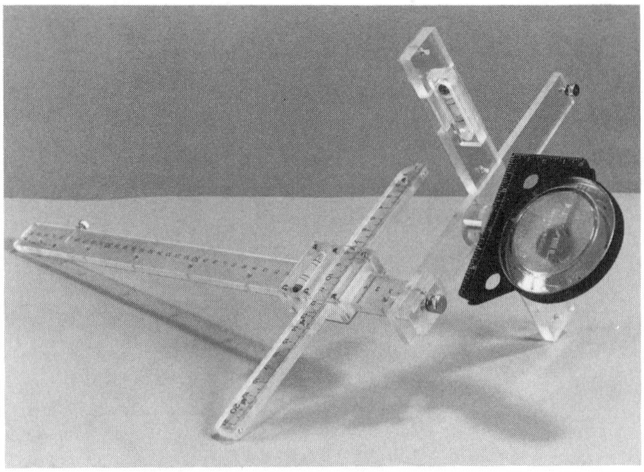

FIG. 2–44. Large double sliding caliper with levels. Used when measuring projective distances involving the vertex (v) or the opisthocranion (op), in FH position of the head.

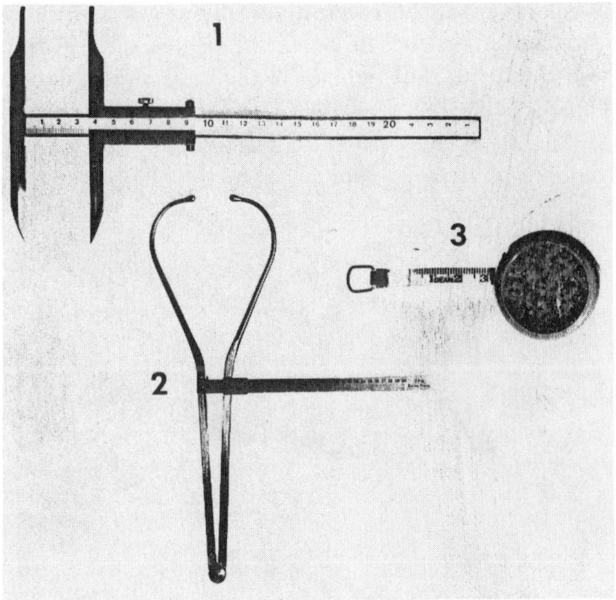

FIG. 2–43. Standard anthropometric instruments: (**1**) sliding caliper; (**2**) spreading caliper; (**3**) soft measuring tape.

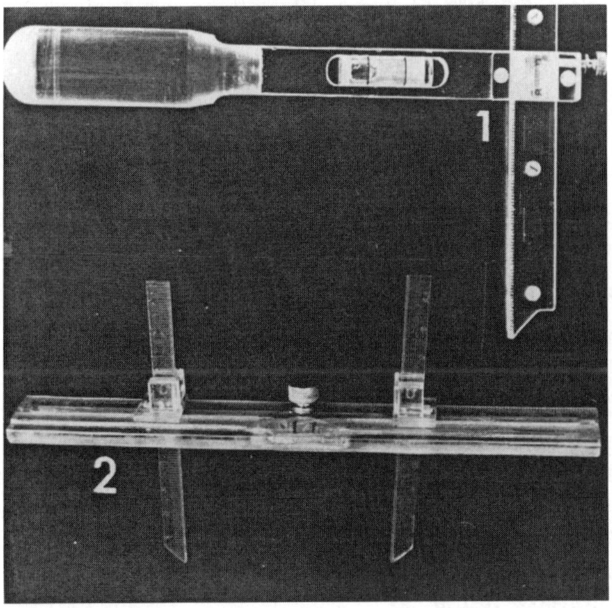

FIG. 2–45. Sliding caliper with level: (**1**) modified for depth measurement with a single branch to measure the nasal root and protrusion of the nasal tip; (**2**) modified with double branches and level for measuring depth differences between two landmarks (e.g., rt en and lt en).

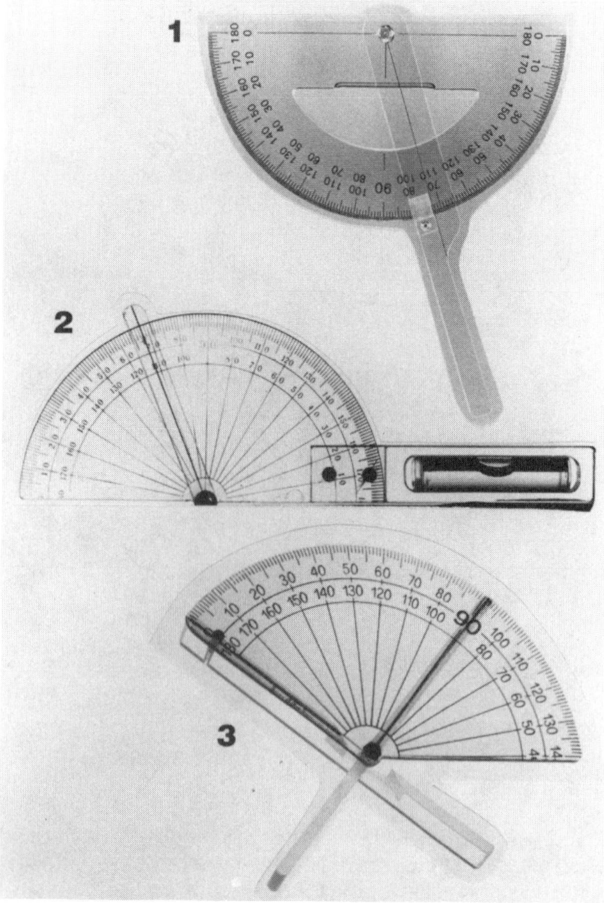

FIG. 2–46. Measuring tools with various modifications of transparent protractor designed by Farkas (8). **1:** The nose deviation protractor for measuring the degree of the nasal bridge and columella deviation. **2:** The nostril inclination protractor with level for measuring the nostril's longitudinal axis inclination. **3:** The nasal root and alar-slope angle meter used for determining the angle between the slopes of the nasal root and the slopes of the nasal alae.

nose, mouth width). The spreading caliper is used when the projective linear distance has to be determined between distant surfaces and various planes (e.g., length of the head; width of the head, forehead, and skull base; width of the face and mandible; depth measurements of the face). The soft metric tape is used for determining the tangential linear distances taken along the skin surface between two landmarks (e.g., the supraorbital, maxillary, and mandibular arcs of the face; circumference of the head; surface length of the nasal ala; upper and lower vermilion arcs of the lips). During the last two decades the following instruments were developed in our Craniofacial Measurement Laboratory: the *large double sliding caliper with levels* (Fig. 2–44) is used when measuring projective distances involving the vertex (v) and the opisthocranion (op) landmarks of the head which require standard position (FH) of the head (8). The *modified sliding caliper with level* [Fig. 2–45(1)] determines the sagittally directed distances [depth of the nasal root (en-se sag), protrusion of the nasal tip (prn) sagittally from

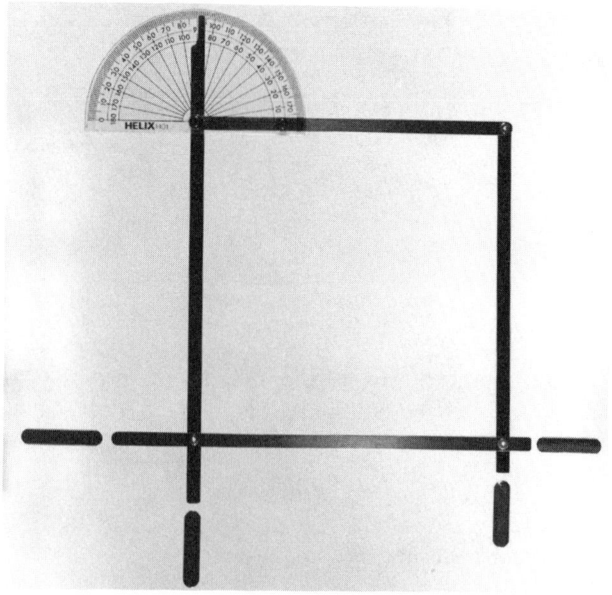

FIG. 2–47. Multipurpose facial angle meter (9). The pointed portion of the instrument is for measuring nasofrontal, nasolabial, and labiomental angles. The large and small forked portions of the instrument are for determining the glabellonasal, nasal tip, and mentocervical angles. By using extension pieces, both child and adult measurements can be taken. (From ref. 9, reprinted with permission.)

the facial insertion of the nasal ala (ac)]. *The modified sliding caliper with double mobile arms and level* is for measuring the sagittal level difference between two points (e.g., en & en, en & ex, sci & or, ac & ac) [Fig. 2–45(2)] (8).

Measuring tools with various modifications of a transparent protractor were designed by Farkas: *the nose deviation protractor* measures the degree of deviation from the midaxis of the face in the nasal bridge and in the columella [Fig. 2–46(1)] (8). *The nostril inclination*

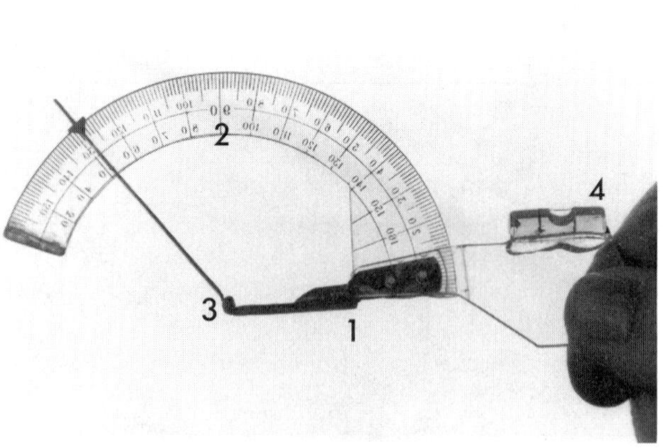

FIG. 2–48. The Singapore angle meter with level. This was designed by S. T. Lee (Department of Plastic Surgery, Singapore General Hospital) as modified by Farkas. It is used for measuring the inclination of the longitudinal axis of the eye fissure and nostril from the horizontal, as well as for measuring the nasolabial and nasofrontal angles. (**1**) Handle, (**2**) scale, (**3**) pointer, (**4**) level.

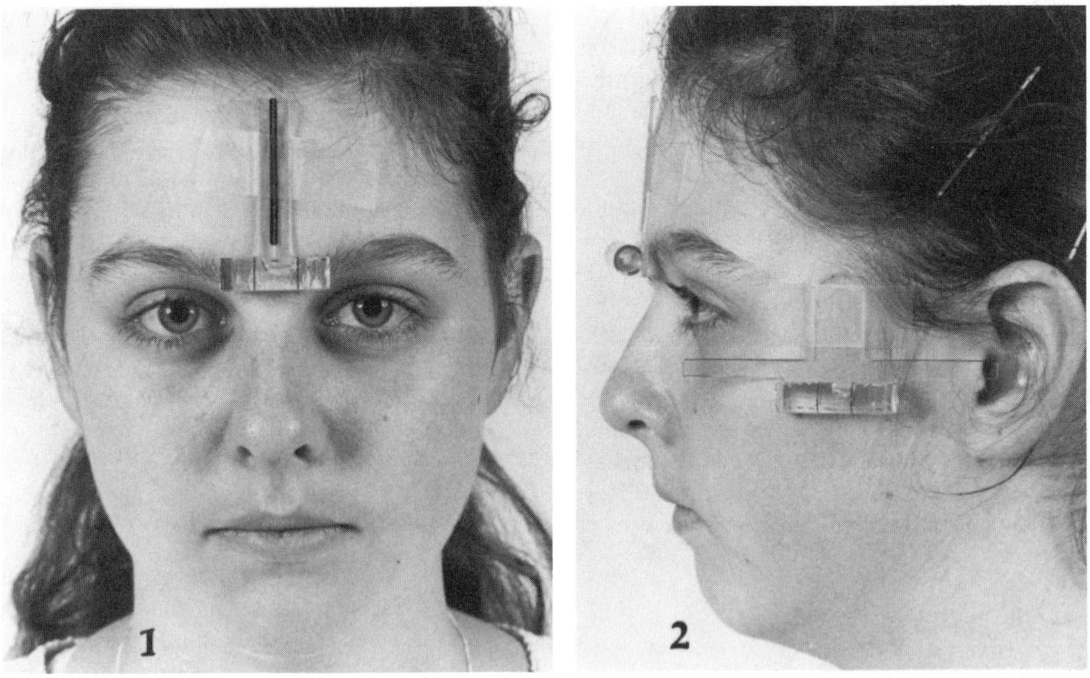

FIG. 2–49. Head position checking devices with level. **1:** The smaller instrument is taped to the forehead and helps to maintain the face in vertical. (For defining the midaxis of the face see page 20). **2:** The larger instrument's horizontal body is aligned with the line connecting the tragion point with the orbitale landmark. The level indicates the horizontal. When taped to the skin, the instrument helps to orient the head into FH and maintain this position if required.

protractor with level [Fig. 2–46(2)] determines the inclination of the nostril's longitudinal axis from the horizontal. It can be used also for measuring the inclination of the eye fissure's longitudinal axis (en-ex line) from the horizontal.

The nasal root and alar-slope angle meter [Fig. 2–46(3)] measures the angle created by the surfaces of the slopes of the nasal root and the nasal alae.

The multipurpose facial angle meter (Fig. 2–47) was developed recently for measuring six angles in the facial profile line (9).

The Singapore angle meter with level (Fig. 2–48) designed by S. T. Lee (Department of Plastic Surgery, Singapore General Hospital) and modified by Farkas is used for measuring the inclination of the longitudinal axis of the eye fissure and nostril from the horizontal, as well as for measuring the nasolabial and nasofrontal angles.

The head position checking devices with level (Fig. 2–49) help to maintain the head in standard position (FH) or to maintain the midline of the face in vertical (10).

The commercial level-finder, the *Dasco Pro Angle Finder plus level with a magnetic base* (Dasco Pro, Inc., Rockford, Illinois) proved to be a simple and practical tool for measuring any inclination on the surface of the head and face. Inclinations of short surfaces (columella, nasal bridge, upper lip, lower lip, chin) are measured with the short edge of the magnetic base. A metal ruler attached to the magnetic base makes it possible to measure the inclination of longer surfaces (e.g., forehead, upper face, lower face, general profile line, ear axis) (Fig. 2–50).

A general rule when measuring between two *soft landmarks* (e.g., pronasale on the nasal tip, alare on the nasal ala, curvature point of the facial insertion of ala, cheilion

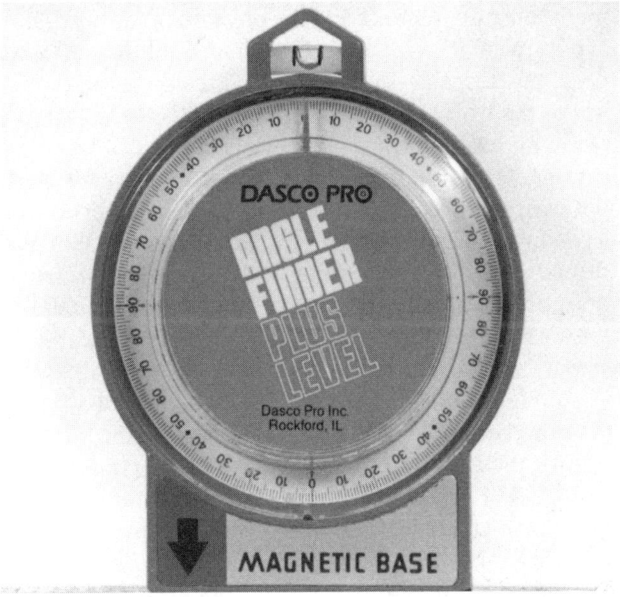

FIG. 2–50. The commercial *Angle Finder plus level with a magnetic base* (Dasco Pro, Inc., Rockford, Illinois) can be used for measuring the inclination of any surface of the head and face. Metal rulers of various lengths attached to the magnetic base increase its short edge to the required length.

point of the mouth, the auricular points) is that the hard tips of the sliding caliper touch, but do not press on, the skin surface. Similarly, when facial arcs are being measured, the tape must not be pressed into the soft tissues. On the other hand, the blunt pointers of the spreading caliper are pressed against the bony surface when measuring between *bony landmarks* (e.g., menton on the chin, gonion on the mandible, zygion on the face, glabella, frontotemporale on the forehead, eurion, opisthocranion, vertex on the head, nasion on the nose). When measuring the circumference, the length, or the widths of the head (or forehead, face, or mandible), the examiner must be certain that the metric tape (measuring the circumference) or the tips of the caliper are sufficiently pressed against the bone surface in the skull to eliminate the effect of thick hair cover and, for forehead and face, the varying thickness of the subcutaneous tissue.

Positioning of the Subject

For measurement, the subjects were seated in a dental chair with the head resting on the head support. The examiner is standing or sitting in front of the subject. The head of the examiner must be level with the head of the subject. The location of the landmark on the top of the head (vertex) and the validity of inclinations from the vertical or the horizontal are significantly influenced by the position of the head. We took readings at *rest* and *standard* positions of the head, according to the requirements for each measurement.

Rest position of the head is determined by the subject's own feeling of the natural head balance (11). Determination of the angle sizes and most of the linear measurements are not influenced by the position of the head. Full exposure of the soft nose (ala shape, columella, nasal floor, nostrils, alar–slope angle) and the upper and lower lips in the frontal plane is facilitated if the head is in the reclining position. In our experience, most of the measurements of the orbits are most easily obtained when the patient's body is in a recumbent position. The subject's eyes should be gazing straight up to the ceiling or should be closed, as required. Measurements to assess deviation from the facial midline or to determine relative positions of two symmetrical points are taken while the subject's profile line is in the vertical.

Standard orientation of the head was achieved by positioning the head in the FH. In this position the line connecting the orbitale (or) and the porion (po) points is horizontal (Fig. 2–1). When the subject is recumbent, the FH becomes vertical. Standard position of the head is required in projective measurements taken from the vertex point on the top of the head and all inclinations. If necessary, an assistant gently holds the subject's head in position. The rest position of the head is identical with the position in FH if the level of the ear canals and lower orbital rims is the same. In healthy Caucasian population subjects, asymmetrically located ear canals are indicated by uneven vertical levels of tragions in 13.4% (160 of 1197) between 6 and 18 years of age (8). The asymmetrical position increases to 79.4% (54 of 68) in patients with various forms of craniosynostosis and in 90.6% (126 of 139) of patients with facial microsomia (12).

Because of the possible ear canal level asymmetry, in repeated measurements the FH must be established always on the same side. In healthy persons in the rest position, the inclination of the line connecting the orbitale and the porion (or tragion) is about 5 degrees higher, on average, than it is in the FH (8). In healthy young adults the rest position of the head below the FH line is seldom found (*unpublished data*).

Facial Midline (Midaxis)

The facial midline is defined by three anatomical points: the nasion (root of the nose), the subnasale (base of the columella), and the gnathion, or menton (lower edge of the mandible). In a normal face the profile is oriented to the vertical by horizontal positioning of paired symmetrical features (e.g., the upper border of the eyebrows, the endocanthions, the rims of the lower eyelids, the insertion points of the alae, direction of the labial fissure or of the commissures of the labial fissure). At least one of these paired features must be horizontal in order to align the profile in the vertical. A commercial angle finder (Fig. 2–50) can be used to confirm that the required position has been achieved.

Two simple instruments designed in our laboratory are fixed to the skin of the midforehead and temporal region with two strips of adhesive tape. They help to maintain the facial profile line in the vertical and the head in FH during the examination (Fig. 2–49) (10).

Landmarks

For the reliability of linear or angular measurements, a knowledge of the precise location of the landmarks on the surface of the head and face is essential. Between these orientation points are measured the linear and tangential (surface) distances. They also define the FH. In this study, classical landmarks (13) and some new measuring points were used (6,14–16). To ease orientation and ensure uniformity in anthropometric terminology, the landmarks are named according to Greek or Latin anatomical terminology. Anyone who works with anthropometry must become familiar with these terms and their abbreviations.

Abbreviations (symbols) of the landmarks are used for marking the measurement, using the short abbreviation (n-gn) instead of full names of landmarks (nasion–gnathion) or the name of the measurement (face height). The symbols of the anthopometric landmarks are marked with lowercase letters (e.g., n denotes a nasal point on the surface), opposite to the symbols in roentgenocephalometry (in which the nasal point on the skeleton is marked with capital N). The use of the internationally accepted anthropometric symbols, without any individual modifications, is a *sine qua non* of easy understanding of papers based on anthropometry.

Landmarks that are used in craniofacial anthropometry and cephalometry may have the same name but not the same anatomical location (e.g., porion). The anthropometric landmarks are differentiated by the adjective "bony" or "osseous" when they are situated on the surface of the underlying bone and by "soft" when located on the actual skin surface.

Even on a "normal" face, accurate identification of landmarks requires some experience. Some points are well-defined on the skin surface and are easy to find (e.g., endocanthion, tragion), whereas identification of landmarks that are placed on bony prominences underlying the skin (e.g., gnathion, gonion) may cause difficulties for a beginner. Some measuring points are described only generally (e.g., opisthocranion), or their location depends on the position of the head (e.g., vertex).

To avoid errors in locating landmarks that are used for more than one measurement (e.g., nasion, tragion), landmarks should be marked on the skin with ink. This is particularly important for landmarks that are above those bony points that are covered by soft tissues (e.g., gnathion, gonion).

In measurements of the head and face, 47 landmarks are used: 6 on the head, 6 on the face, 8 on the orbits, 11 on the nose, 6 on the lips and mouth, and 10 on the ears.

Head

Vertex (v) is the highest point of the head when the head is oriented in the FH. The vertex is not identical to the bregma, the bony landmark in the middle of the top of the skull, where the coronal and sagittal sutures cross (Fig. 2–51).

Glabella (g) is the most prominent midline point between the eyebrows and is identical to the bony glabella on the frontal bone (Fig. 2–52).

Opisthocranion (op) is the point situated in the occipital region of the head and is most distant from the glabella; that is, it is the most posterior point of the line of greatest head length (17). It is close to the midline on the posterior rim of the foramen magnum. The location of this landmark depends on the shape of the os occipitale (Fig. 2–51).

Eurion (eu) is the most prominent lateral point on each side of the skull in the area of the parietal and temporal bones (Fig. 2–52).

Frontotemporale (ft) is the point on each side of the forehead, laterally from the elevation of the linea temporalis. This location was chosen due to difficulties of finding the linea temporalis in children with disfigurements

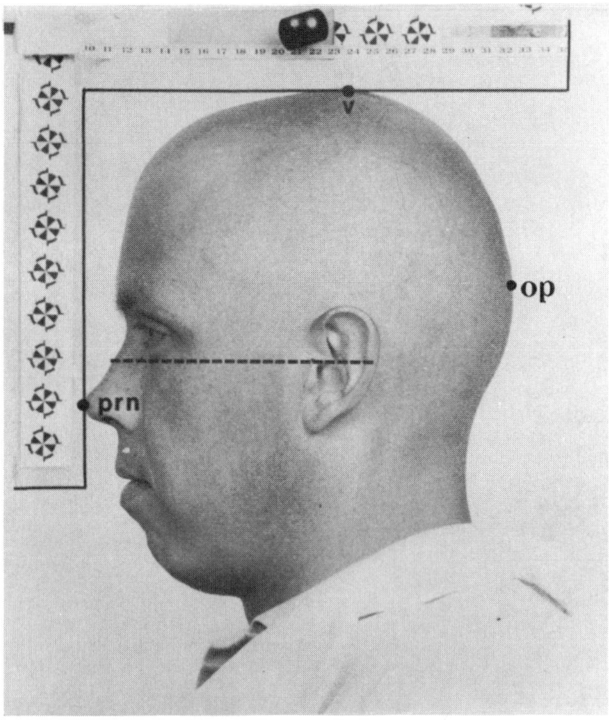

FIG. 2–51. In the FH position the vertex (v) is the highest point and the opisthocranion (op) is the most posterior landmark of the head.

of the head. The position of this landmark approximately corresponds with the level of the terminal points of the tail of the eyebrow (Fig. 2–52).

Trichion (tr) is the point on the hairline in the midline of the forehead. In early childhood, identification of this landmark may be difficult because of an irregular or indistinguishable hairline. It cannot be determined on a balding head (Fig. 2–52).

Face

Zygion (zy) is the most lateral point of each zygomatic arch and is identified by trial measurement, not by anatomical relationship (18). It is identical to the bony zygion of the malar bones (Fig. 2–52).

Gonion (go) is the most lateral point on the mandibular angle close to the bony gonion. It is identified by palpation. If the angle is flat or if there is rich soft-tissue cover, determination of this point is very difficult (Fig. 2–52).

Sublabiale (sl) determines the lower border of the lower lip or the upper border of the chin (16) (Fig. 2–52). It corresponds with the mentolabial ridge of anatomists (19), a point in the midline of the nasolabial sulcus (si) (20), or the point marked as inferior labial point (21), supramentale (22), submental point (23), or labiomentale (24). The identification of this landmark on the lower face with deep and indented mental sulcus is easy (16). On the shallow ridges and flat surfaces of the reced-

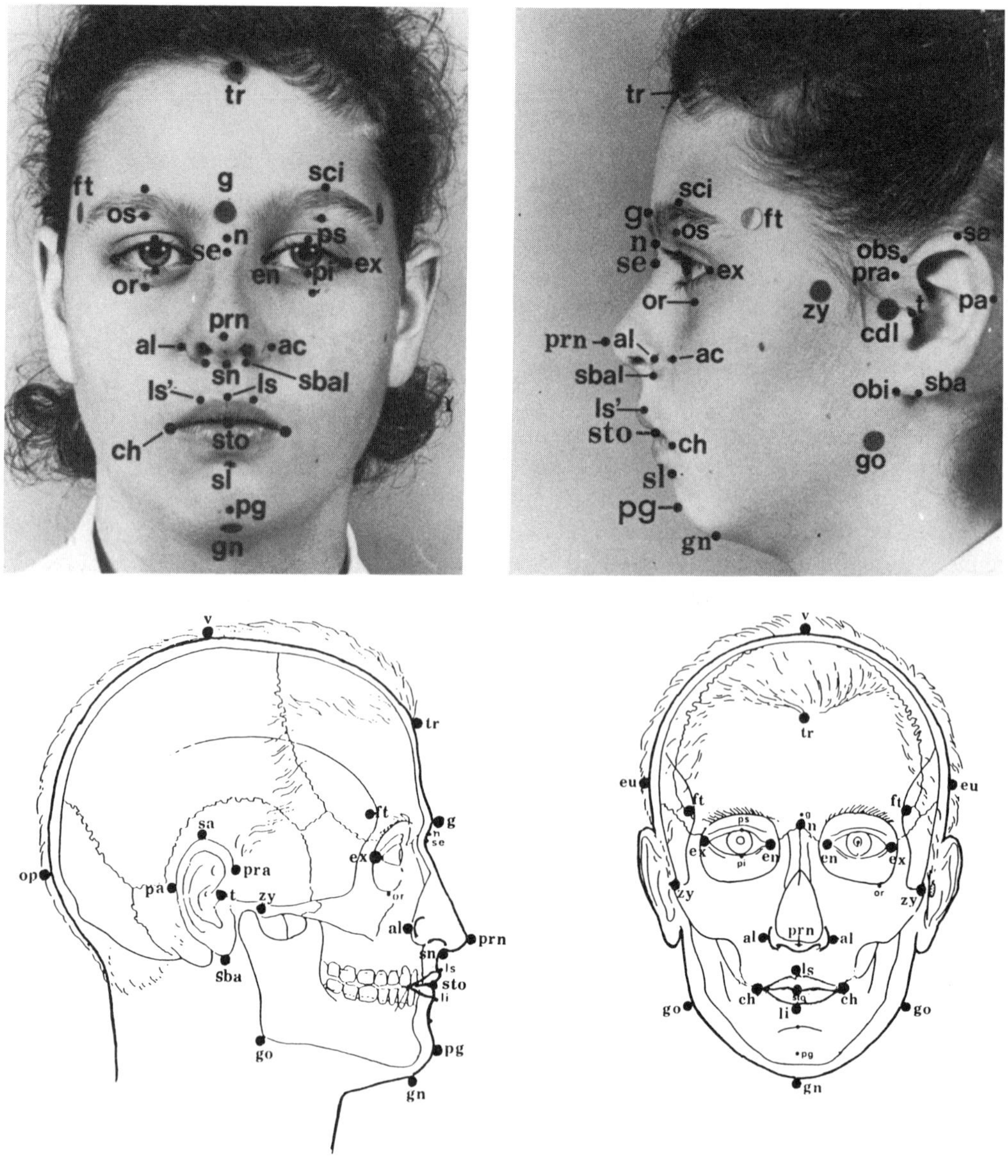

FIG. 2–52. Craniofacial surface landmarks of the head and face in frontal and lateral aspects (**upper figures**). The landmarks were marked on the skin of the subject *before* taking the pictures. The schematic drawings of the head and face (**lower figures**) show the surface landmarks in relation to the underlying craniofacial skeleton. (Modified from ref. 63, with permission.)

ing chins the landmark on the surface was determined with the help of the level of the bottom of the lower lip, by intraoral examination (Fig. 2–53).

Pogonion (pg) is the most anterior midpoint of the chin, located on the skin surface in front of the identical bony landmark of the mandible (25) (Fig. 2–52).

Menton (or gnathion) (gn) is the lowest median landmark on the lower border of the mandible. It is identified by palpation and is identical to the bony gnathion (26). This landmark is the lowest point used in measuring facial height (Fig. 2–52).

Condylion laterale (cdl) is the most lateral point on the surface of the condyle of the mandible. It is identified by palpation at each temporomandibular joint when the jaw is open (18) (Fig. 2–52).

Orbits

Endocanthion (en) is the point at the inner commissure of the eye fissure. The soft endocanthion is located lateral to the bony landmark (MO) that is used in cephalometry (27,28) (Fig. 2–52).

Exocanthion (or ectocanthion) (ex) is the point at the outer commissure of the eye fissure. The soft exocanthion is slightly medial to the bony exocanthion (LO) (Fig. 2–52) (28).

Center point of the pupil (p) is determined when the head is in the rest position and the eye is looking straight forward. Identification is easiest when the patient is reclining, the eye fissures are horizontal, and the eyes are gazing straight upward.

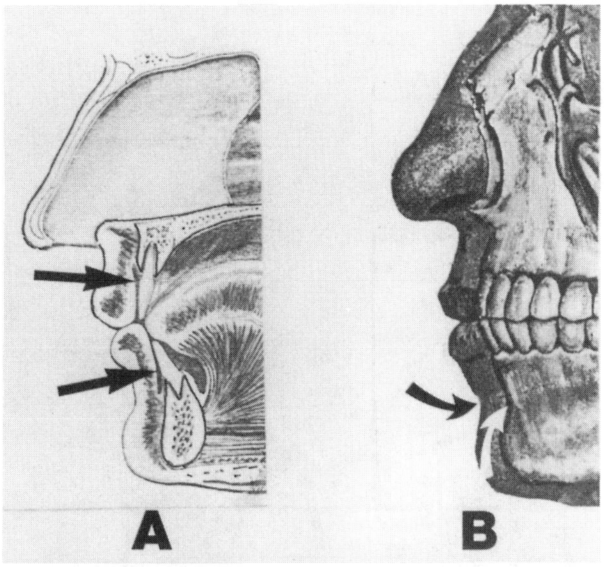

FIG. 2–53. Relationship between the anatomical position of the mentolabial ridge on the skin surface and the bony contour of the underlying mandible. (From ref. 16, with permission.)

Orbitale (or) is the lowest point on the lower margin of each orbit. It is identified by palpation and is identical to the bony orbitale (Fig. 2–52).

Palpebrale superius (ps) is the highest point in the midportion of the free margin of each upper eyelid (Fig. 2–52) (8).

Palpebrale inferius (pi) is the lowest point in the midportion of the free margin of each lower eyelid (Fig. 2–52) (8).

Orbitale superius (os) in young adults is the highest point on the lower border of the eyebrow, close to the highest bony point of the upper margin of each orbit (19) (Fig. 2–52), where the bony supraorbitale landmark (sor) is located (17).

Superciliare (sci) is the highest point on the upper borderline in the midportion of each eyebrow (Fig. 2–52) (8). Note that when the eyebrows have been cosmetically treated (plucked), this point cannot be located.

Nose

Nasion (n) is the point in the midline of both the nasal root and the nasofrontal suture. The slight ridge on which it is situated can be felt by the observer's fingernail. This point always is above the line that connects the two inner canthi (29) (Fig. 2–52). The soft nasion and the bony nasion are identical (30).

Sellion (subnasion) (se) is the deepest landmark located on the bottom of the nasofrontal angle (Fig. 2–52) (24), marked also as "m" (median) (31). The point usually occurs somewhere between the levels of the supratarsal fold and eyelash (32).

Maxillofrontale (mf) is at the base of the nasal root medially from each endocanthion, close to the bony

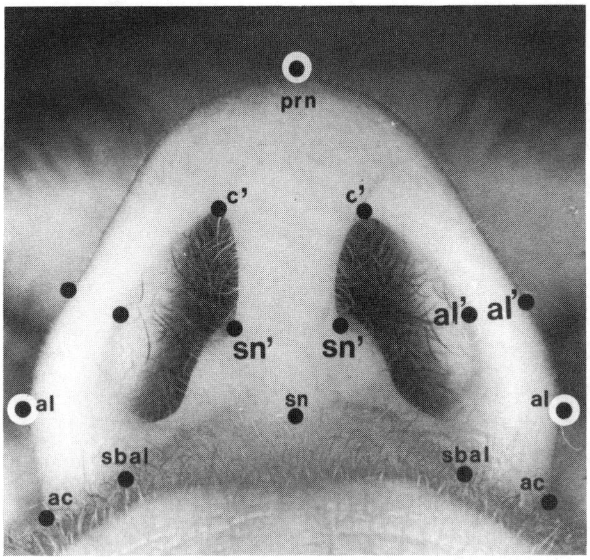

FIG. 2–54. Base view of the nose with surface landmarks.

maxillofrontale of the medial margin of each orbit, where the maxillofrontal and nasofrontal sutures meet (Fig. 2–52) (8).

Alare (al) is the most lateral point on each alar contour (Fig. 2–52).

Pronasale (prn) is the most protruded point of the apex nasi, identified in lateral view of the rest position of the head. This point is difficult to determine if the nasal tip is flat. In the case of the bifid nose, the more protruding tip is chosen for prn (Fig. 2–52).

Subnasale (sn) is the midpoint of the angle at the columella base where the lower border of the nasal septum and the surface of the upper lip meet (33) (Fig. 2–52). This point is not identical to the bony subnasion, or nasospinale (ns), which is "the midpoint of the anterior margin of the apertura piriformis at the base of the spina nasalis anterior" (26). The landmark is identified in base view of the nose, or from the side. Where nasolabial an-

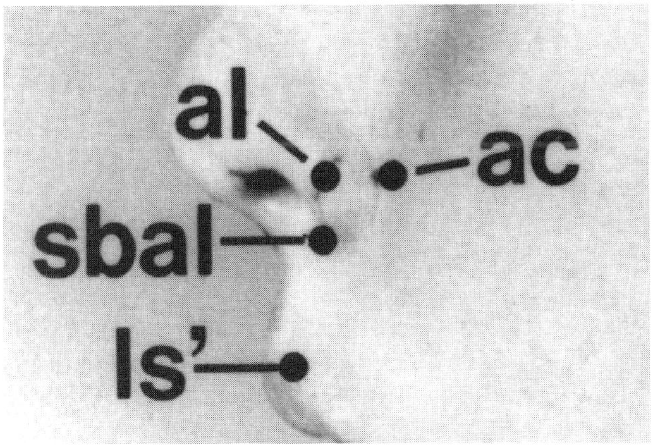

FIG. 2–55. Three landmarks of the ala: *al* (alare) on the most lateral point of the ala; *ac* (alar curvature point) indicating the facial insertion of the alar base, *sbal* (subalare) showing the labial insertion of the alar base. (From ref. 34, with permission.)

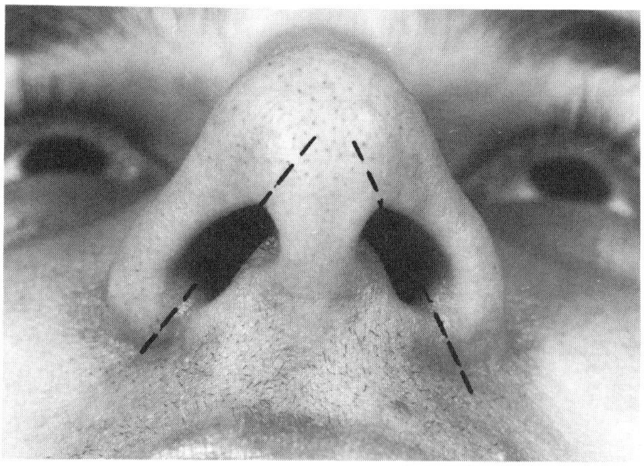

FIG. 2–56. Terminal points of the nostril axis.

gles are found with curved contours, locating the point can be difficult.

Subalare (sbal) is the point at the lower limit of each alar base, where the alar base disappears into the skin of the upper lip (14). The landmarks indicate the *labial insertion* of the alar base (Figs. 2–52 and 2–54).

Alar curvature (or alar crest) point (ac) is the most lateral point in the curved base line of each ala, indicating the *facial insertion* of the nasal wingbase (8,34) (Figs. 2–52, 2–54, and 2–55).

Highest point of the columella [*or columella breakpoint of Daniel (32)*] (c′) is the point on each columella crest, level with the top of the corresponding nostril (Fig. 2–54) (8).

Alare′ (al′) is the marking level at the midportion of the alae (al′-al′) where the thickness of each ala is measured (Fig. 2–54) (8,35).

Subnasale′ (sn′) indicates the midpoint of the columella crest where the thickness of the columella (sn′-sn′) is measured (Fig. 2–54) (8). When measuring the width of the nostril floor between the subalare (sbal) and the crest of the columella, the sn′ point designates the columella crest at the bottom line (36).

Terminal points of the nostril axis are the highest and the lowest spots of each nostril (Fig. 2–56) (37). Classification of the nostril types is based on inclination of the longitudinal axis of the nostrils from the horizontal.

Lips and Mouth

Crista philtri landmark (cph) is the point on each elevated margin of the philtrum just above the vermilion line (Fig. 2–57).

Labiale (or labrale) superius (ls) is the midpoint of the upper vermilion line (Figs. 2–52 and 2–58).

Labiale superius′ (ls′) laterally from the midpoint (ls) is also located on the upper vermilion line vertically beneath the right and left subalare (sbal) points (Figs. 2–52 and 2–58).

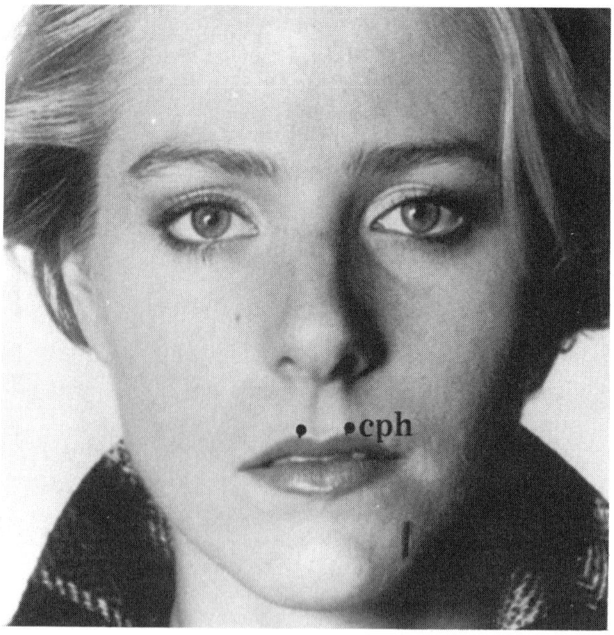

FIG. 2–57. Landmarks (cph) on the elevated margins of the philtrum.

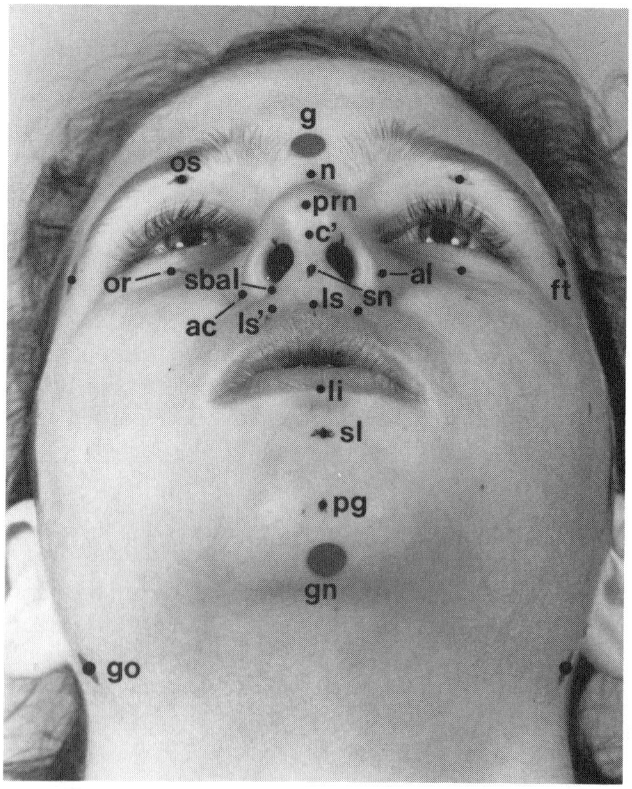

FIG. 2–58. Base view of the face: landmarks on the lips (ls, ls′, li, sl) marked on the skin *before* taking the picture.

Labiale (or labrale) inferius (li) is the midpoint of the lower vermilion line (Figs. 2–52 and 2–58).

Stomion (sto) is the imaginary point at the crossing of the vertical facial midline and the horizontal labial fissure between gently closed lips, with teeth shut in the natural position (Fig. 2–52).

Cheilion (ch) is the point located at each labial commissure (Fig. 2–52).

Ears

Superaurale (sa) is the highest point on the free margin of the auricle (Fig. 2–52).

Subaurale (sba) is the lowest point on the free margin of the ear lobe (Fig. 2–52).

Preaurale (pra) is the most anterior point of the ear, located just in front of the helix attachment to the head (Fig. 2–52).

Postaurale (pa) is the most posterior point on the free margin of the ear (Fig. 2–52).

Otobasion superius (obs) is the point of attachment of the helix in the temporal region (Fig. 2–52). It determines the upper border of the ear insertion.

Otobasion inferius (obi) is the point of attachment of the ear lobe to the cheek (Fig. 2–52). It determines the lower border of the ear insertion.

Porion (soft) (po) is the highest point on the upper margin of the cutaneous auditory meatus (Fig. 2–30). According to Ashley-Montagu (38), this point is a few millimeters medial to the bony porion. Hrdlička's (29) and Gosman's (18) "auriculo" corresponds with the soft porion.

Tragion (t) is the notch on the upper margin of the tragus (Fig. 2–52).

Points establishing the medial longitudinal axis: The uppermost point is determined by halving the upper portion of the ear (i.e., the portion above the upper insertion point of the ear); the lowest point is the middle of the free border of the ear lobe (Fig. 2–59) (6).

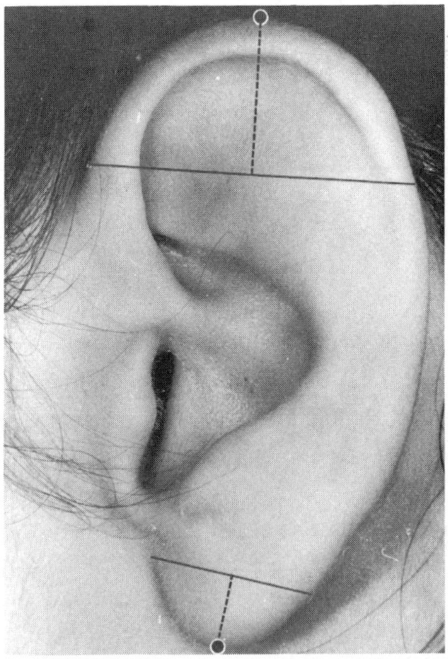

FIG. 2–59. Determination of the terminal points of the medial longitudinal axis of the auricle.

Measuring

Most of the head, face, and ear measurements are performed according to classical methods of physical anthropology (26,29,39–42). Some have been modified (6,8,14,15,43,44). Others were newly developed (5,6, 8,16,32,34,45–52). A number of measurements that were developed by the author are described here for the first time. The individual measurements can be *linear* or *angular*. The linear measurements are *projective,* in which the shortest distance is determined between two landmarks (e.g., intercanthal width = en-en), or *tangential,* in which the measurement is taken along the skin surface by a soft measuring tape between the two terminal landmarks (e.g., maxillary arc = t-sn-t). According to the direction, the measurement can be horizontal or vertical (oblique), providing metric information in two dimensions. Projective *depth* measurements belong to the third dimension showing the degree of the protrusion of the nasal root, sagittal level difference between the inner (endocanthion) and outer commissure (exocanthion) of the eye fissure, the degree of the protrusion of the upper orbital rim over the lower orbital rim, and so on. Lateral projective depth measurements of the face taken between the tragion of the ear and the glabella, nasion, subnasale, and gnathion in the facial midline provide certain information about level difference between the parallel frontal planes (placed at the tragion point of the ear and planes at the individual midpoints of the face) but do not define the true depth of the face.

Angular measurements record inclinations or angles. In *inclination,* the angular position of a line or surface is determined from the horizontal or vertical [head maintained in standard position (FH)]. Measurements of the inclination of the eye fissure, nostril axis, and labial fissure line require only vertical direction of the facial midline. *Angles* provide information about the mutual relationship between two intersecting lines or planes; for example, the nasal tip angle is formed by the contours of the nasal bridge and the columella, and the labiomental angle is defined by the surface of the lower lip and plane following the upper portion of the chin.

The linear and angular measurements can be single or paired. *Single measurements* assess the major dimensions of the head and face, and they define central areas of the face (e.g., length of the nose, width of the mouth). *Paired measurements* register the differences in size, lo-

TABLE 2–2. *Anthropometric measurements in the six regions of the craniofacial complex*

Region	Measurement[a] Single Linear, *N*	Single Angular, *N*	Paired Linear, *N*	Paired Angular, *N*	Subtotal
Head	12	1	2		15
Face	15	7	9		31
Orbits	2		17	2	21
Nose	8	11	7	2	28
Lips and mouth	10	4	4		18
Ears			17	2	19
Total	47	23	56	6	132

[a] Linear measurements are projective or tangential. Angular measurements include both inclinations and angles.

cation, and inclination of paired features or the two sides of the same organ (e.g., the right and the left nasal ala, or both halves of the labial fissure) or the head and face.

Each measurement is represented by an abbreviation consisting of a capital letter and an arabic number. The letter indicates whether the measurement is single (S) or paired (P). In the population norms the individual craniofacial regions are indicated by roman numbers (see Appendixes A–C).

The craniofacial measurements are listed separately in each region of the craniofacial complex (Table 2–2). Seventy of the 132 measurements (53%) are single and 62 (47%) are paired. The total number of linear (projective and tangential) measurements is 103 (78% of 132), and the angular ones (inclinations and angles) comprise 29 (22% of 132). The 62 paired measurements represent 124 examinations, and thus the total number of measurements which can be taken from one subject's head and face increases to 194. Anthropometric examination should be completed by recording the height and weight of the subject.

Although data about the asymmetries are provided by the paired measurements in the various regions of the craniofacial complex, seven additional examinations offer *quantitative data* (in millimeters) about vertical, sagittal, and horizontal aberrations of some points in the orbital, facial, nasal, labial, and aural regions from their normal position. The extent of asymmetry between the positions of the soft-tissue orbits in the frontal plane is given by the level differences of the eyebrows (superciliare points right and left, sci-sci) and/or inner commissures (endocanthion landmarks right and left, en-en) of the eye fissures. In the case of the soft nose, the asymmetrical position of the ear bases is proved by establishing the difference between the levels of the labial insertions of the alar bases (subalare point right and left, sbal-sbal). The extent of deviation of the scoliothic facial midaxis from the vertical can be shown by measuring the distance between the laterally dislocated chin point (gnathion) and the assumed vertical midaxis of the face.

Measurements in Six Regions of the Craniofacial Complex

Head

Single Linear Measurements (S)

Projective

- Horizontal

Head position: rest
Instrument: spreading caliper

S-1 Width of the head (diamètre transversal maximum, Kopfbreite) (eu-eu) is the distance between the eurions (Fig. 2–60).

S-2 Width of the forehead (ft-ft) is not the classical small width of the forehead measured between the temporal lines. The distance is measured between the points (Fig. 2–60) located laterally from the temporal lines (see page 54).

S-3 Skull-base width (bitragion diameter, diamètre biauriculaire, Kopfbreite über dem Tragus) (t-t) is the distance between the tragions (Fig. 2–60).

- Perpendicular

Head position: FH
Instrument: large double sliding caliper with levels

S-4 Height of the calvarium (v-tr) is the distance between the vertex and the trichion [Fig. 2–61(1)].

S-5 Anterior height of head [also forehead height (56), Vordere Hirnkopfhöhe of Knussmann (52)] (v-n) is the distance between the vertex and the nasion [Fig. 2–61(2)].

S-6 Special height of the head (v-en) is the distance between the vertex and the endocanthion level [Fig. 2–61(3)].

S-7 Height of the head and nose [Scheitel-Nasen-Höhe of Knussmann (52)] (v-sn) is the distance between the vertex and the subnasale [Fig. 2–61(4)].

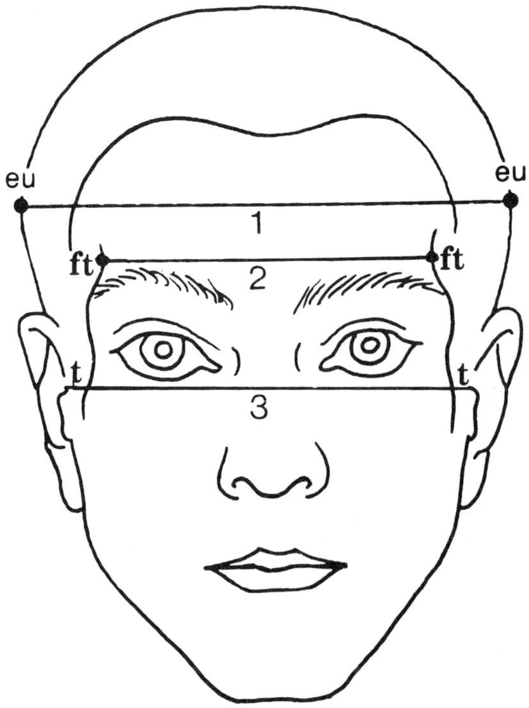

FIG. 2–60. Horizontal measurements of the head: (1) width of the head (eu-eu); (2) width of the forehead (ft-ft); (3) the skull-base width (t-t).

S-8 Combined height of the head and face (diamètre vertical total de la tête, craniofacial height, Ganze Kopfhöhe) (v-gn) is the distance between the vertex and the gnathion [Fig. 2–61(5)].

Note: The flat, fixed horizontal arm of the caliper is pressed to the vertex, the body of the caliper (with scale) is vertical, and the tip of its mobile arm points to the tr, n, en, and sn points or is pressed to the bony gnathion (Fig. 2–62).

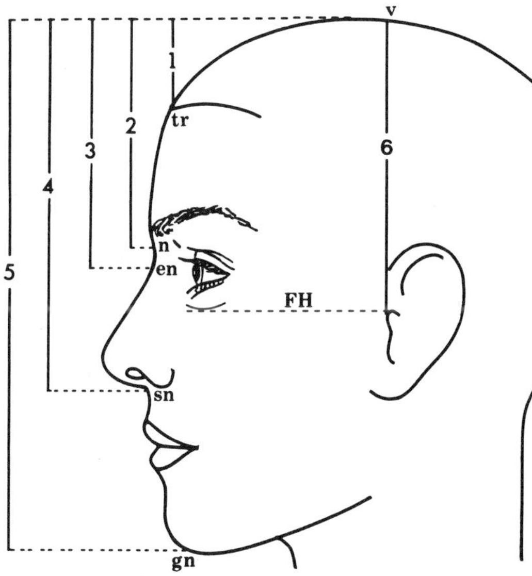

FIG. 2–61. Perpendicular measurements of the head: (1) v-tr; (2) v-n; (3) v-en; (4) v-sn; (5) v-gn; (6) v-t.

Head position: rest
Instrument: sliding caliper

S-9 Height of forehead I [Supraorbitale Stirnhöhe of Knussmann (52)] (tr-g) is the distance between the trichion and the glabella [Fig. 2–63(1)].

S-10 Height of forehead II (longueur frontale, Directe Stirnhöhe) (tr-n) is the distance between the trichion and the nasion [Fig. 2–63(2)].

• Sagittal
Head position: rest
Instrument: spreading caliper

S-11 Length of head (longueur maximum de la tête, Kopftiefe or maximum glabello-occipital diameter) (29) (g-op) is the distance between the glabella and the opisthocranion [Fig. 2–64(1)].

Tangential

Head position: rest
Instrument: soft measuring tape

S-12 Circumference of the head (occipitofrontal circumference, périmètre de la tête, Kopfumfang) is measured in the horizontal plane around the

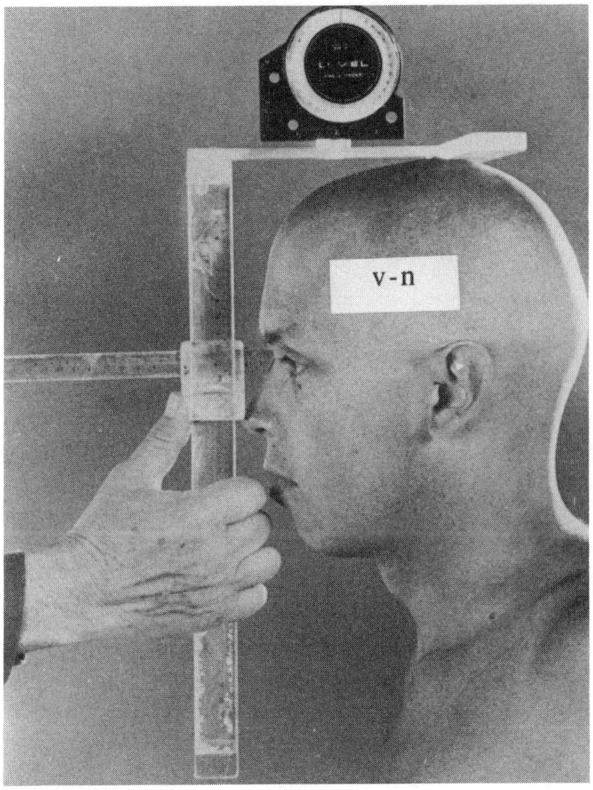

FIG. 2–62. Measuring the anterior height of the head (v-n) with the large double sliding caliper with levels.

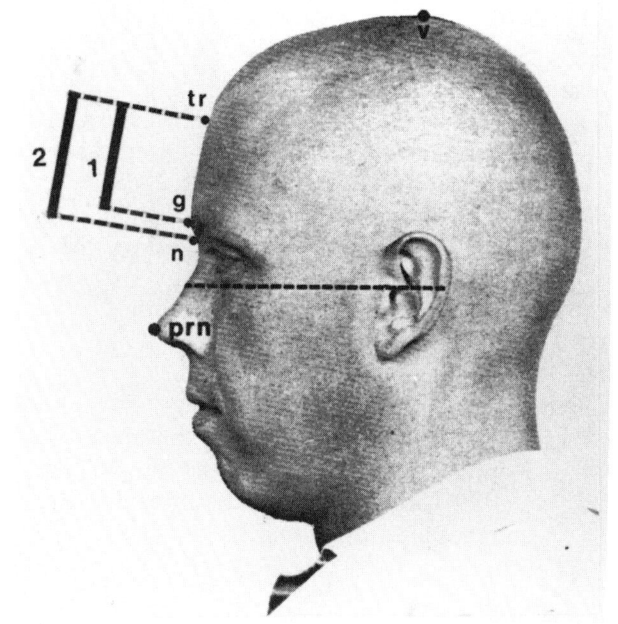

FIG. 2–63. Height measurement of the forehead: (**1**) forehead height I (tr-g); (**2**) forehead height II (tr-n).

head, through the glabella and opisthocranion. The tape must be held tightly around the head [Fig. 2–64(2)].

Single Angular Measurement (S)

Inclination

Head position: FH
Instrument: commercial angle meter

S-13 Inclination of the anterior surface of the forehead from the vertical is determined along the general trichion–glabella line (tr-g) (8) (Fig. 2–65). Zero inclination indicates that the surface is vertical. Inclination in positive degrees indicates a protrusion from the vertical. In dorsally tilted foreheads, where the surface of the forehead is behind the vertical, the inclination registers negative degrees.

Paired Linear Measurements (P)

Projective

• Perpendicular
Head position: FH
Instrument: large double sliding caliper with levels

P-14 Auricular height of the head (38) or the relation of each ear-canal level to the vertex (v-po) is the distance between the vertex and each porion. The fixed arm of the caliper rests horizontally on the top of the head, and the mobile arm is moved to the level of the edge of the ear canal (Fig. 2–66).

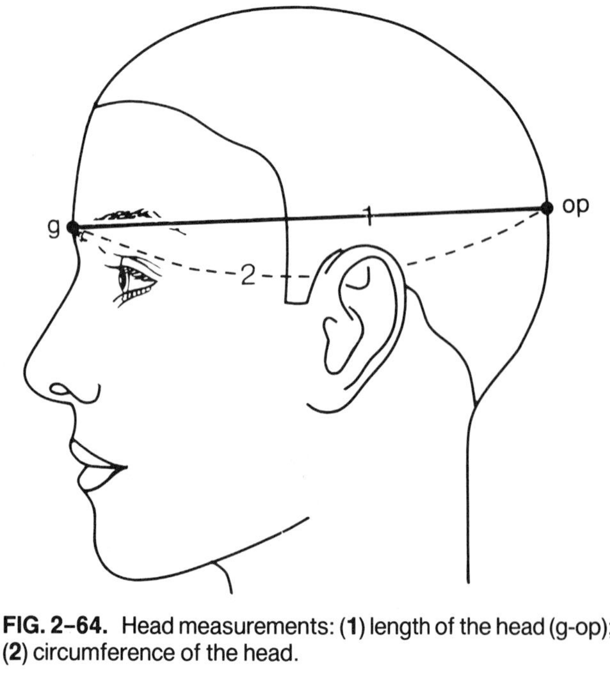

FIG. 2–64. Head measurements: (**1**) length of the head (g-op); (**2**) circumference of the head.

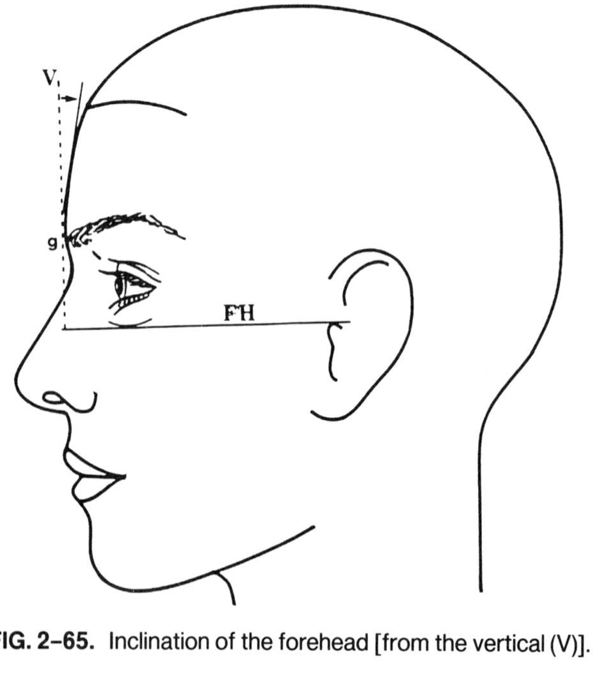

FIG. 2–65. Inclination of the forehead [from the vertical (V)].

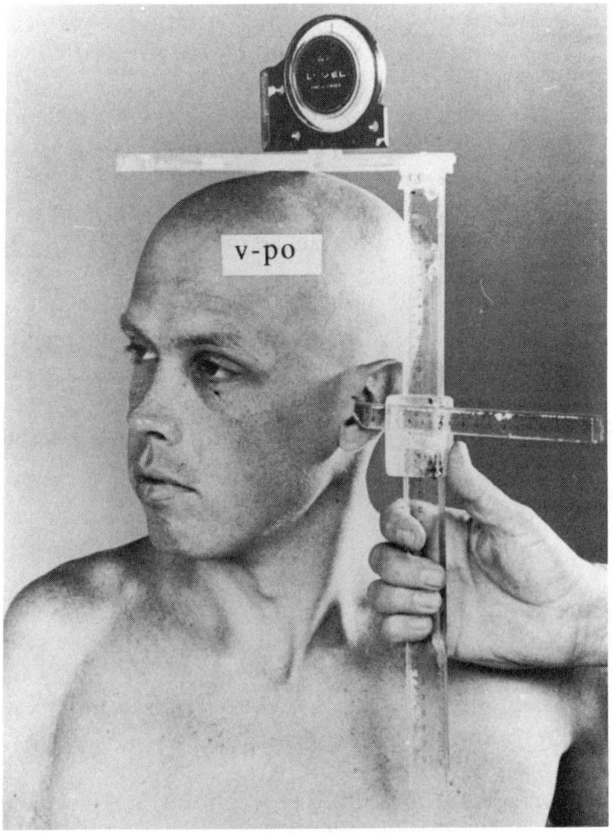

FIG. 2-66. Measuring the auricular height of the head (v-po) with the large double sliding caliper with levels.

The measurement is shown on the perpendicular body of the caliper. Because the location of the bony basion cannot be determined in a living person, this measurement substitutes for the true skull height (26).

P-15 Distance between the vertex and each tragion (v-t) is measured in a way that is similar to P-14 (v-po) [Fig. 2–61(6)].

Face

Single Linear Measurements (S)

Projective

- Horizontal

Head position: rest

Instrument: spreading caliper

S-1 Width of the face (bizygion diameter, upper face width, maximum interzygomatic breadth, or distance bizygomatique) (zy-zy) is measured at the widest part of the face, between the zygions (Fig. 2–67). The tips of the calipers are pressed over the zygomatic arch until the maximum breadth is determined.

S-2 Width of the mandible (bigonial diameter, width of the lower face, largeur mandibulaire or Unter-

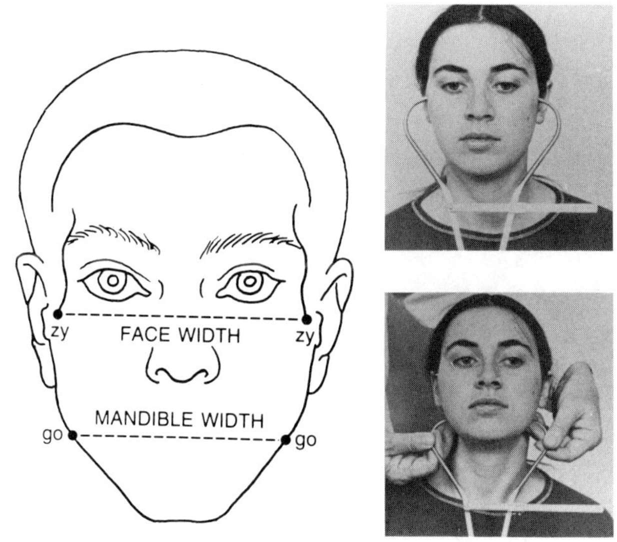

FIG. 2-67. Horizontal measurements of the face using the spreading caliper: face width (zy-zy); mandible width (go-go).

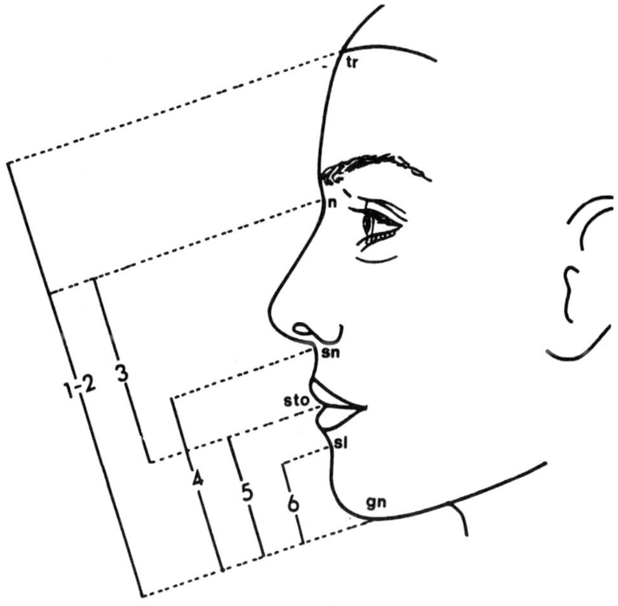

FIG. 2-68. Perpendicular measurements of the face: (**1**) tr-gn; (**2**) n-gn; (**3**) n-sto; (**4**) sn-gn; (**5**) sto-gn; (**6**) sl-gn.

kieferwinkelbreite) (go-go) is the distance between the gonions (Fig. 2–67). (Because the thickness of the soft-tissue cover over the mandibular angles varies with individuals, the pointers of the caliper must be firmly pressed against the bony surfaces.)

• Perpendicular
Head position: rest
Instrument: sliding caliper

S-3 Physiognomical height of the face (longueur total du visage, Physiognomische Ganzgesichtshöhe) (tr-gn) is the distance between the trichion and the gnathion [Fig. 2–68(1)].

S-4 Height of the face (morphological face height, hauteur nasomentonnière, Nasomentale Gesichtshöhe) (n-gn) is the distance between the nasion and the gnathion [Fig. 2–68(2)].

S-5 Height of the upper face (physiognomical upper face height, diamètre nasiobuccal or Mittelgesichtshöhe) (n-sto) is the distance between the nasion and the stomion [Fig. 2–68(3)].

S-6 Height of the lower face (total jaw height, longueur spino-mentonnière or Untergesichtshöhe) (sn-gn) is the distance between the subnasale and the gnathion [Fig. 2–68(4)].

S-7 Height of the mandible (height of the lower third of the face or Direkte Untergesichtshöhe) (sto-gn) is the distance between the stomion and the gnathion [Fig. 2–68(5)].

S-8 Height of the chin is the distance between the sublabiale and the gnathion (Kinnhöhe) (sl-gn) (16) [Fig. 2–68(6)].

S-9 Height of the upper profile (tr-prn) is the distance between the trichion and the pronasale [Fig. 2–69(1)].

S-10 Height of the lower profile (prn-gn) is the distance between the pronasale and the gnathion [Fig. 2–69(2)].

S-11 Lower half of the craniofacial height (en-gn) is in the neoclassical two-section facial profile canon (v-en and en-gn) measured between the endocanthion and the gnathion (47) [Fig. 2–69(3)].

S-12 Distance between the glabella and the subnasale (g-sn) is the third quarter of the neoclassical four-section facial profile canon (v-tr = tr-g = g-sn = sn-gn) (48) [Fig. 2–69(4)].

Tangential

Facial-arc measurements provide data about the soft-tissue contours in the supraorbital, maxillary, and mandibular regions.
Head position: rest
Instrument: soft measuring tape. During measurement the tape gently touches the skin surface. The tragi

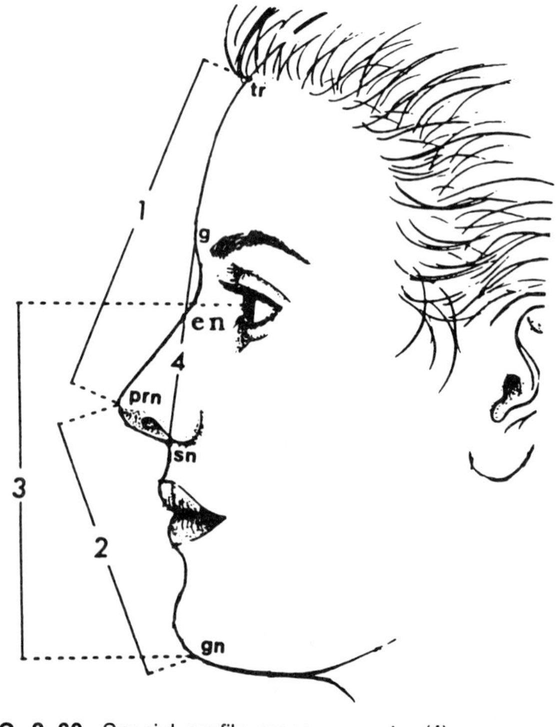

FIG. 2–69. Special profile measurements: (**1**) upper profile height (tr-prn); (**2**) lower profile height (prn-gn); (**3**) lower half of the craniofacial height (en-gn); (**4**) the glabella–subnasale distance (g-sn).

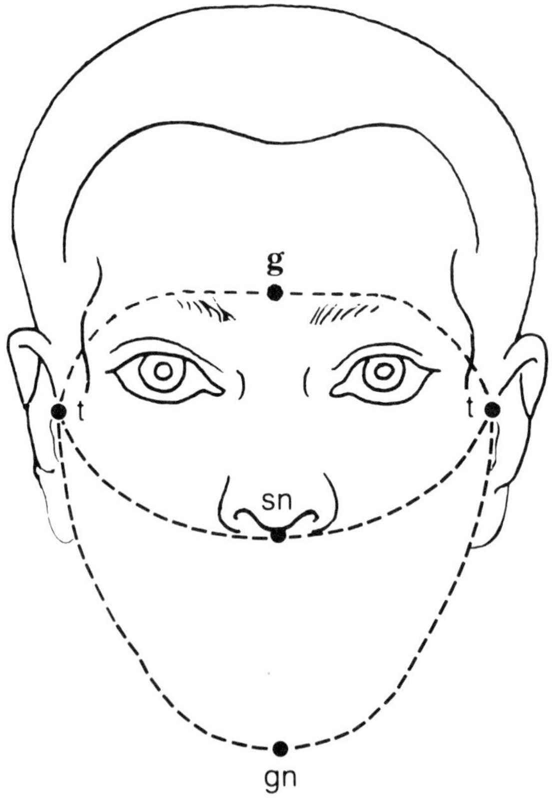

FIG. 2–70. Arc measurements of the face schematically: the supraorbital arc (t-g-t); the maxillary arc (t-sn-t); the mandibular arc (t-gn-t).

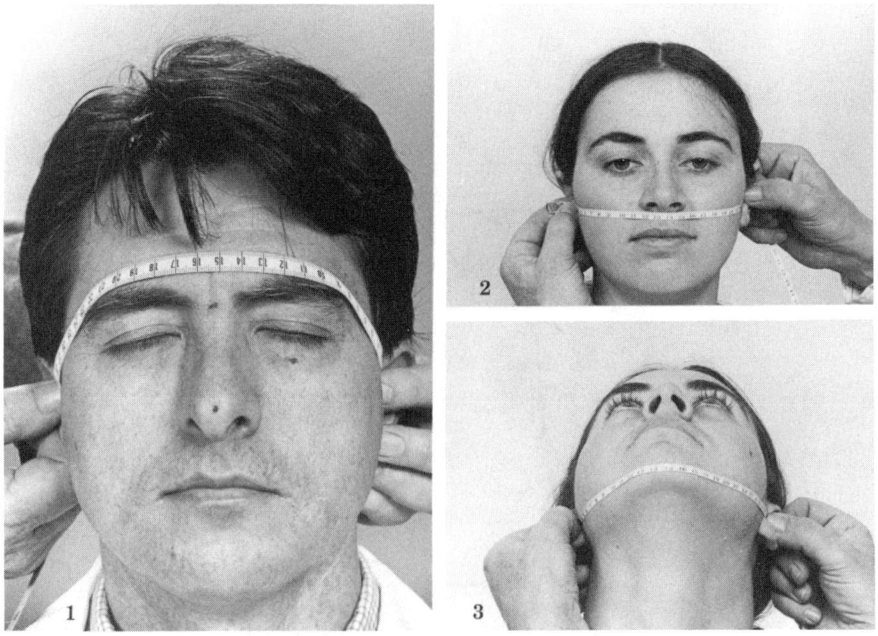

FIG. 2–71. Facial arcs measured by soft tape: (**1**) the supraorbital arc; (**2**) the maxillary arc; (**3**) the mandibular arc.

should not be depressed when the ends of the tape are held at the tragions.

S-13 Supraorbital arc (t-g-t) is measured between the tragions, following a curved line along the upper orbital rims around the glabella [Figs. 2–70 and 2–71(1)].

S-14 Maxillary arc (subnasal arc or bitragion–subnasale arc) (t-sn-t) is measured between tragions, following a mildly curved line across the cheeks around the subnasale. The zero mark of the tape is held to the right tragion with the left hand; the right hand holds the other end of the tape at the left tragion [Figs. 2–70 and 2–71(2)].

S-15 Mandibular arc (submandibular arc or bitragion–menton arc) (t-gn-t) is measured along a very curved line on the lower edge of the mandible. The tape is held gently at the right tragion with the left hand, crosses the chin below the menton, and rises to the left tragion [Figs. 2–70 and 2–71(3)].

Single Angular Measurements (S)

Inclinations

Head position: FH
Instrument: commercial angle meter

S-16 Inclination of the upper face profile from the vertical (g-sn) is measured on the profile contour between the glabella and subnasale (Fig. 2–72).

S-17 Inclination of the Leiber line from the vertical (g-ls) is measured between the glabella and the labiale superius (Fig. 2–73).

S-18 Inclination of the lower face profile from the vertical (sn-pg) is measured on the line connecting the subnasale with the pogonion (Fig. 2–74).

S-19 Inclination of the mandible from the vertical (li-pg) is measured on the line connecting the labiale inferius with the pogonion (Fig. 2–75).

S-20 Inclination of the chin from the vertical is measured along the upper surface of the chin, from the sublabiale downwards (53) (Fig. 2–76).

S-21 Inclination of the general profile line from the vertical (g-pg) is measured on the line connecting the glabella with the pogonion (53) (Fig. 2–77).

Note: Zero inclination indicates that the profile line is vertical. Inclination in positive degrees indicates protrusion from the vertical. Inclination of facial profile lines receding from the vertical are marked in negative degrees.

Angle

Head position: rest
Instrument: multipurpose facial angle meter

S-22 Mentocervical angle is formed by the upper contour of the chin and the surface beneath the mandible (51) (Fig. 2–78).

Paired Linear Measurements (P)

Projective

• Perpendicular
Head position: rest
Instrument: spreading caliper

P-23 Height of the mandibular ramus [Hintere Ge-

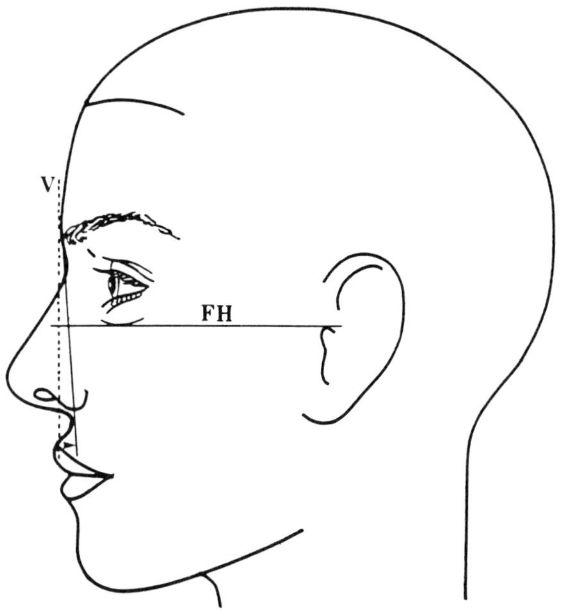

FIG. 2–72. Inclination of the upper face profile line from the vertical (V, indicated by the *dotted line*).

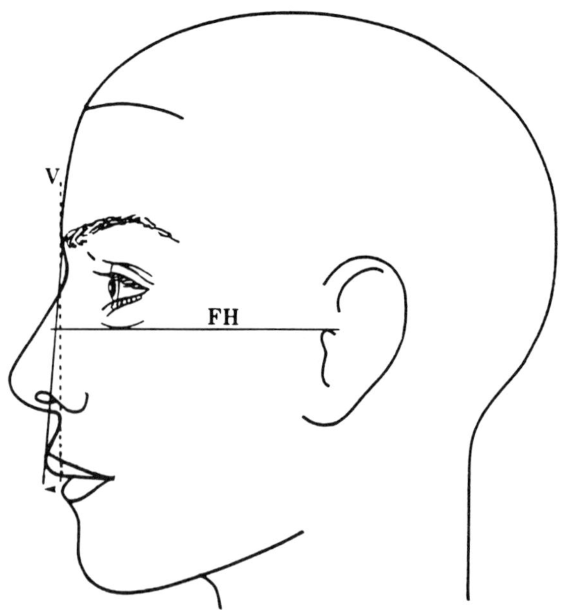

FIG. 2–73. Inclination of the Leiber line from the vertical (V).

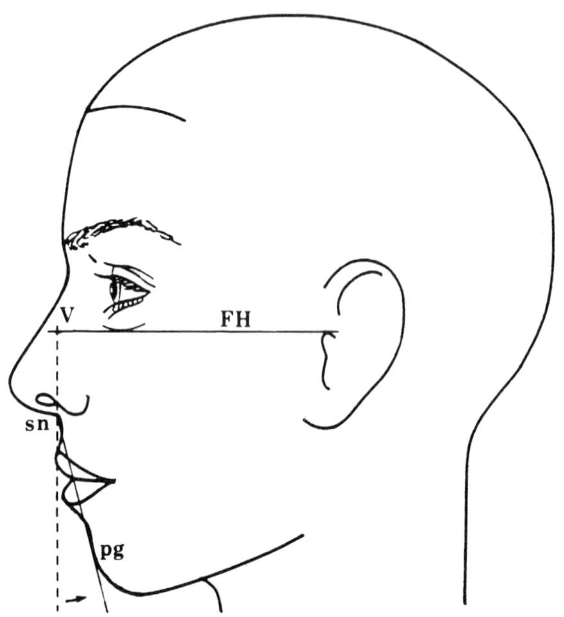

FIG. 2–74. Inclination of the lower face profile line from the vertical (V).

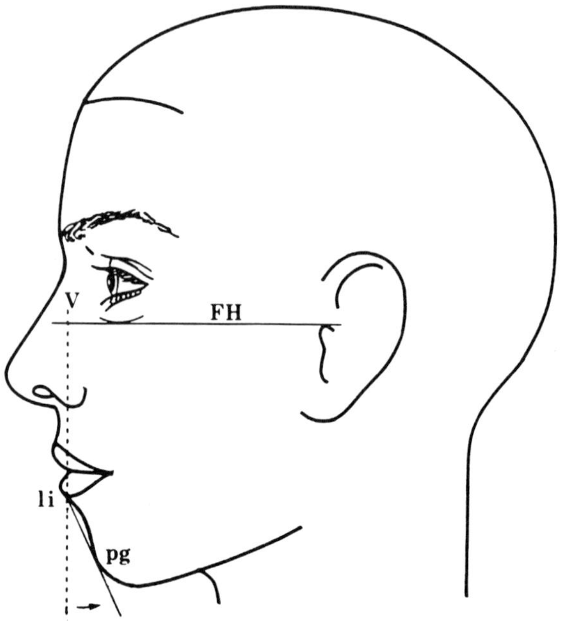

FIG. 2–75. Inclination of the mandible from the vertical (V).

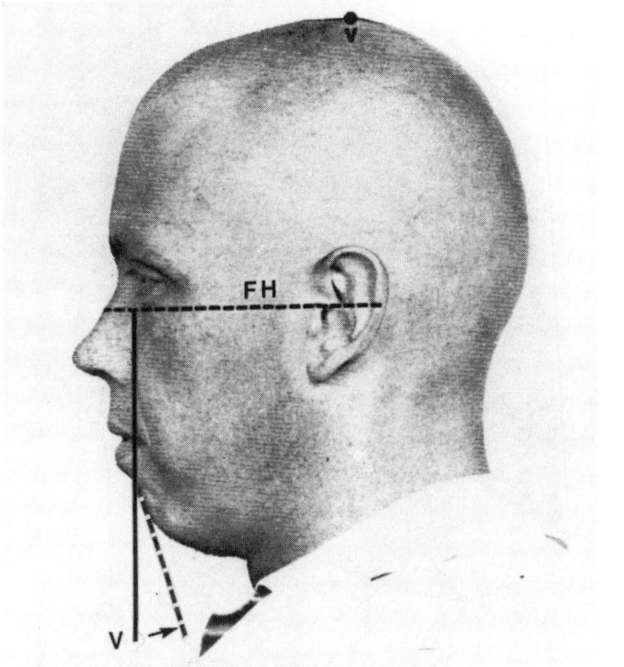

FIG. 2-76. Inclination of the chin from the vertical (V).

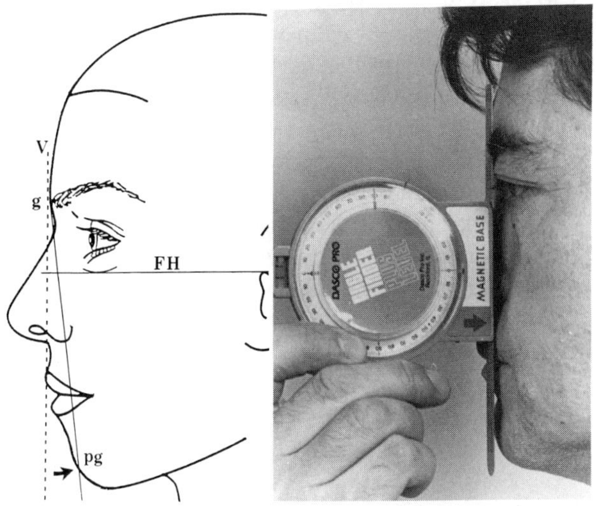

FIG. 2-77. Inclination of the general profile line from the vertical (V) (schematically) and measuring the inclination by the commercial Angle Finder (Dasco Pro).

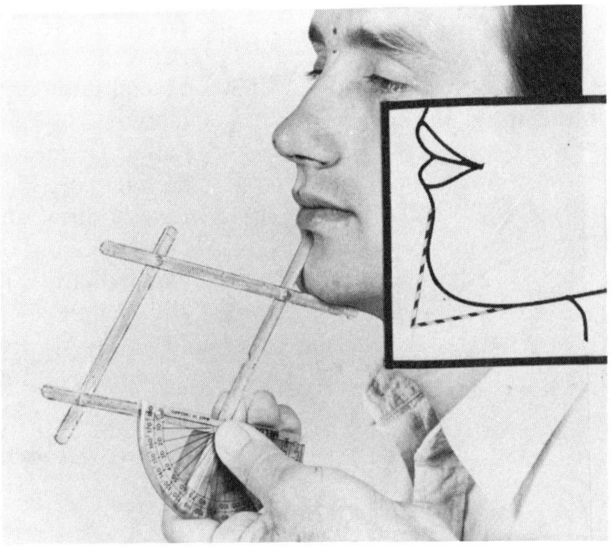

FIG. 2-78. Measuring the mentocervical angle with the multipurpose facial angle meter.

sichtshöhe of Knussmann (24)] (go-cdl, rt & lt) is measured between the gonion and the condylion laterale (11) [Fig. 2–79(1)].

Lateral oblique measurements indicate the depth of the face.

Head position: rest

Instrument: spreading caliper

P-24 Tragion–glabellar depth or depth of the upper third of the face is measured between the tragion and the glabella (t-g) (Fig. 2–80).

P-25 Tragion–nasion depth or depth of the upper third of the face is measured between the tragion and the nasion (t-n) (Fig. 2–80).

P-26 Tragion–subnasale depth or depth of the middle third (maxillary region) of the face is measured between the tragion and subnasale (t-sn) (Fig. 2–80).

P-27 Tragion–gnathion depth or depth of the lower third (mandibular region) of the face is measured between the tragion and the gnathion (t-gn) (Fig. 2–80).

P-28 Gonion–gnathion depth or depth of the mandible (Untergesichtstiefe) (go-gn) is measured between the gonion and the gnathion (8) [Fig. 2–79(2)].

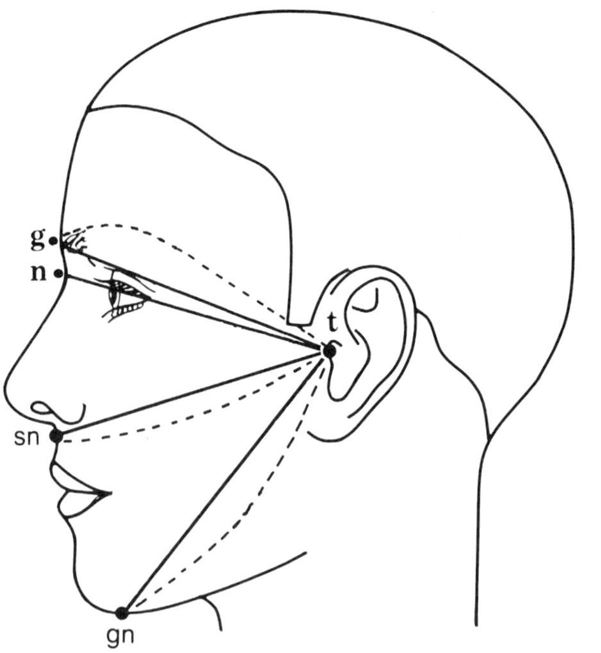

FIG. 2–80. Depth measurements of the face: the tragion–glabellar depth (t-g); the tragion–nasion depth (t-n); the tragion–subnasale depth (t-sn); the tragion–gnathion depth (t-gn). The surface half-arcs of the face are indicated by *dotted lines.*

Tangential

Head position: rest

Instrument: soft measuring tape

P-29 Lateral surface half-arc in the upper third of the face (t-g surf) is measured along each eyebrow, between the tragion and the glabella (Fig. 2–80).

P-30 Lateral surface half-arc in the middle third (maxillary region) of the face (t-sn surf) or maxillary half-arc is measured between the subnasale and each tragion (Fig. 2–80).

P-31 Lateral surface half-arc in the lower third (mandibular region) of the face (t-gn surf) or mandibular half-arc is measured between the gnathion and each tragion (Fig. 2–80).

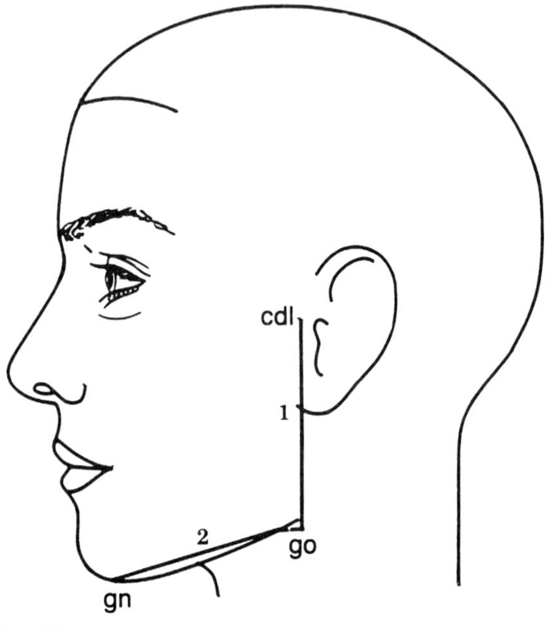

FIG. 2–79. Measurements of the mandible: (**1**) height of the ramus (go-cdl); (**2**) depth of the body (go-gn).

Orbits

Single Linear Measurements (S)

Projective

• Horizontal

Head position: rest

Instrument: sliding caliper

S-1 Intercanthal width [interocular diameter, intercanthal distance, inner canthal distance, endocanthal distance, interorbital distance, Obere Nasenbreite of Martin (39), largeur interoculaire or Interokularbreite] (en-en) is the distance between the endocanthions (Fig. 2–81). The measurement does not correspond to the bony-interorbital distance between the two dacrion landmarks (27,40),

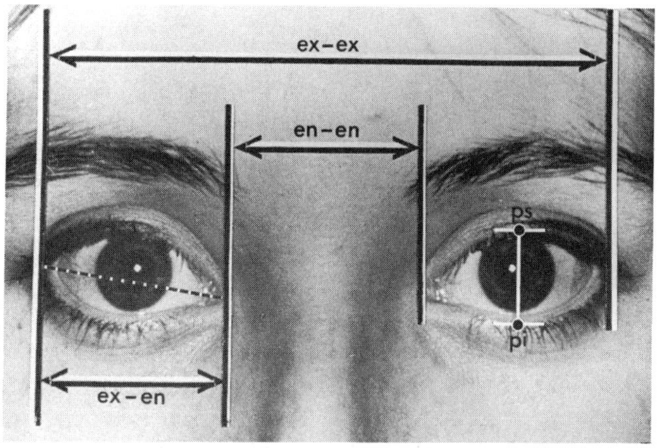

FIG. 2-81. Basic measurements of the soft orbits: the intercanthal width (en-en); the biocular width (ex-ex); the eye fissure length (ex-en rt & lt); the height of the eye fissure (ps-pi rt & lt); inclination of the eye fissure (inclination of the line connecting the en and ex points from the horizontal). (From ref. 64, with permission.)

which are medial to the endocanthions at the junction of the nasal process of the frontal bone, the frontal process of the maxilla, and the lacrimal bone where the medial canthal ligament is attached.

S-2 Biocular width (biocular diameter, external biocular breadth, outer canthal distance, largeur bioculaire externe or Biokularbreite) (ex-ex) is the distance between the exocanthions (Fig. 2–81). The biocular width is shorter than the external orbital breadth [diamètre orbitaire of Martin (39)] measured between the outer bony edges of the orbits at the level of the exocanthions, which is similar to the biorbital diameter of the skull (39).

Paired Linear Measurements (P)

Projective

• Horizontal
Head position: rest
Instrument: sliding caliper

P-3 Length of the eye fissure (longueur palpebrale, palpebral fissure length or Direkte Lidspaltenbreite) (ex-en) is the distance between the endocanthion and the exocanthion of each eye (Fig. 2–81).

P-4 Endocanthion–facial midline (projective) distance (en-se) is measured between each endocanthion and the midpoint of the nasal bridge (sellion) at the level of the eye fissure (Fig. 2–82).

Head position: recumbent
Instrument: sliding caliper

P-5 Pupil–facial midline distance (se) is measured between the center of each pupil (p) and the facial midline at the level of the pupil (sellion) (Fig. 2–83).

Lateral oblique measurements indicate the position of the orbits from the ear and mandible.
Head position: rest
Instrument: spreading caliper

P-6 Orbito-aural distance (ex-obs) is measured between the exocanthion and the otobasion superius on each side [Fig. 2–84(1)].

P-7 Orbito-tragion distance (ex-t) is measured between the exocanthion and the tragion on each side [Fig. 2–84(2)].

P-8 Orbito-gonial distance (ex-go) is measured between the exocanthion and the gonion on each side [Fig. 2–84(3)].

P-9 Orbito-glabellar distance [Augentiefe of Schwidetzky (54)] (ex-g) is measured between each exocanthion and the glabella [Fig. 2–84(4)].

• Perpendicular
Head position: recumbent
Instrument: sliding caliper

P-10 Height of the orbit [Orbitalhöhe of Martin and Saller (13)] (or-os) is the distance between the orbitale and the orbitale superius of each eye (Fig. 2–85).

P-11 Combined height of the orbit and the eyebrow (or-sci) is the distance between the orbitale and the superciliare of each eyebrow (Fig. 2–85).

P-12 Height of the eye fissure (Lidspaltenhöhe) (ps-pi) is the distance between the free edges of each eyelid (Fig. 2–85).

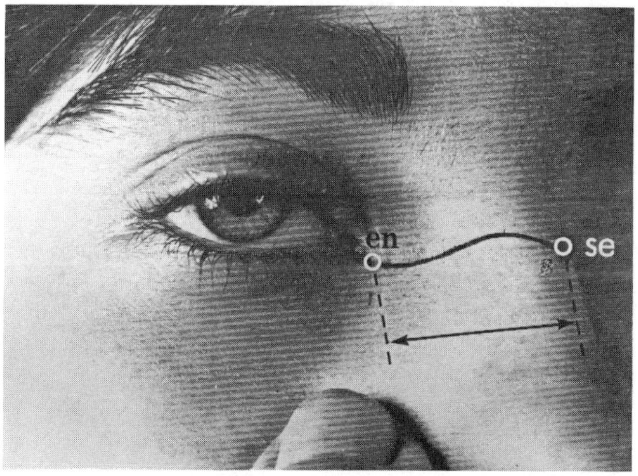

FIG. 2-82. Position of the soft orbit in relation to the nasal root midpoint: projective distance (en-se); tangential distance (en-se surf).

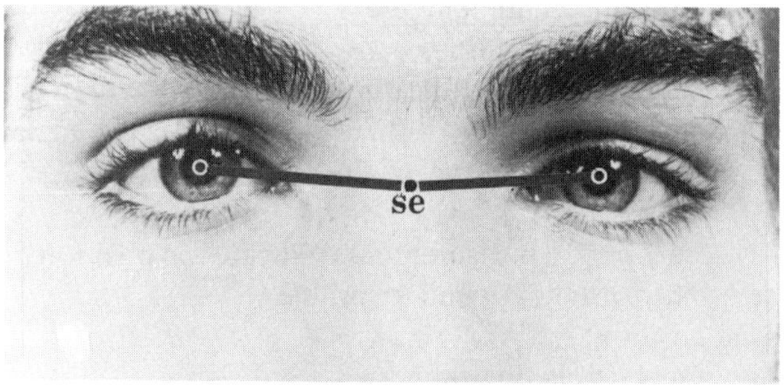

FIG. 2–83. Pupil–nasal root midline distance (p-se).

P-13 Height of the upper lid (os-ps) is the distance between the orbitale superius and the palpebrale superius of each eye (Fig. 2–85).

P-14 Pupil–upper lid height is the distance between the pupil center and the orbitale superius of each eye (p-os) (Fig. 2–85).

P-15 Height of the lower lid (pi-or) is the distance between the palpebrale inferius and the orbitale of each eye (Fig. 2–85).

P-16 Pupil–lower lid height is the distance between the pupil center and the orbitale of each eye (p-or) (Fig. 2–85).

• Sagittal

Head position: FH and recumbent, with the midfacial plane in the vertical

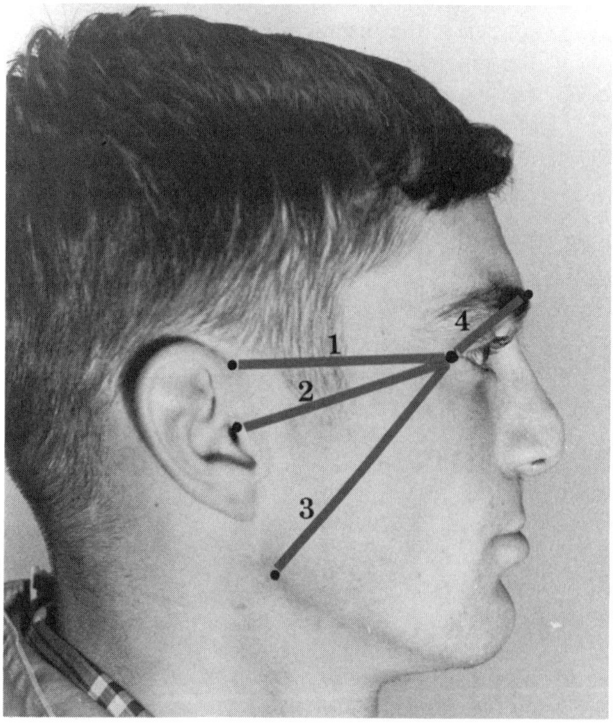

FIG. 2–84. Position of the soft orbits in relation to the ear: (**1**) ex-obs; (**2**) ex-t to the mandible; (**3**) ex-go to the gonion; (**4**) ex-g to the glabella.

Instrument: modified sliding caliper with two mobile arms and level

P-17 Difference between the levels of the upper and the lower orbital rims (os′/or′) in the sagittal direction on each side. The mobile arms of the caliper are resting on the rims of the orbit. The body of the caliper is lifted up to the horizontal, indicated by the level (L). The sagittal distance between the two orbital rims is calculated from the scales on the arms (Fig. 2–86).

P-18 Difference between the levels of the endocanthion and the exocanthion (en/ex) of each orbit in the sagittal direction. One mobile arm is pushed to the bone surface close to the endocanthion, and the other is pushed to the bony rim behind the exocanthion. The body of the caliper is lifted up to the horizontal indicated by the level (L). The sagittal level difference is calculated from the scales on the arms (Fig. 2–86).

Tangential

• Horizontal

Head position: rest and recumbent

Instrument: soft measuring tape

P-19 Endocanthion–facial midline surface distance (en-se surf) is measured between each endocanthion and the midpoint of the nasal bridge (sellion) at the level of the eye fissure (Fig. 2–82).

Paired Angular Measurements (P)

Inclinations

• Horizontal

Head position: FH, with the facial profile in the vertical

Instrument: commercial angle meter

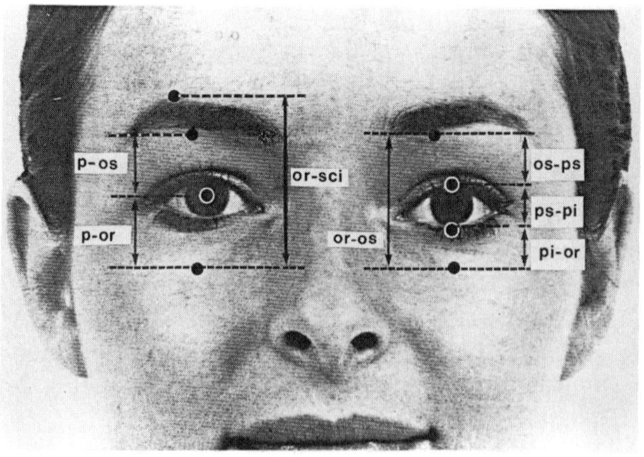

FIG. 2–85. Perpendicular measurements of the soft orbits: height of the orbit (or-os); height of the orbit and eyebrow (or-sci); eye fissure height (ps-pi); heights of the upper lid (os-ps) and lower lid (pi-or); pupil–upper lid distance (p-os); pupil–lower lid distance (p-or).

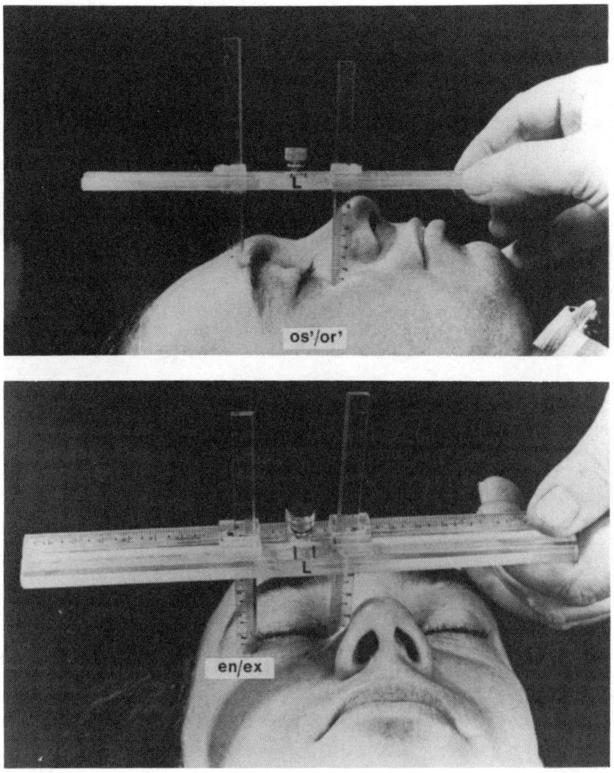

FIG. 2–86. Sagittal orbital measurements. **Upper half:** Measuring the difference in level between the upper and lower orbital rims (os′/or′). **Lower half:** Measuring the difference in level between the commissures of the eye fissure (en/ex).

P-20 Inclination of the line connecting the endocanthion and exocanthion of the eye fissure (en-ex line) from the horizontal. The straight side of the angle meter follows the line between the commissures of each eye (Figs. 2–81, 2–87, and 2–88).

When measuring the eye fissure, the patient's gaze should be directly forward; if the patient is recumbent, his or her gaze should be directly towards the ceiling. In the normally opened eye, the exocanthion located at the lateral commissure of the eye fissure may not be fully discernible (in adult males) because of the slight drooping of the upper lid edge. In such circumstances, the patient should be requested to look upwards (without moving the head) while the length of the eye fissure or its inclination is measured.

- Perpendicular

Head position: FH and recumbent, with the midfacial plane in the vertical

Instrument: commercial angle meter

P-21 Inclination of the line placed over the upper and lower orbital rims (os′/or′) (Fig. 2–89). The straight side of the angle meter is pressed to the bony edges in the middle portion of the orbit on each side.

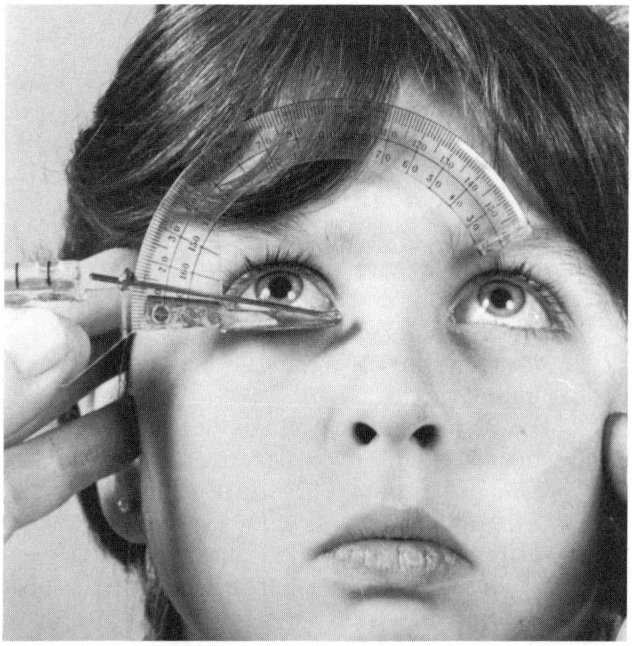

FIG. 2–87. Measuring the inclination from the horizontal of the eye fissure using the Singapore angle meter with level. During measurement the eyes look upward.

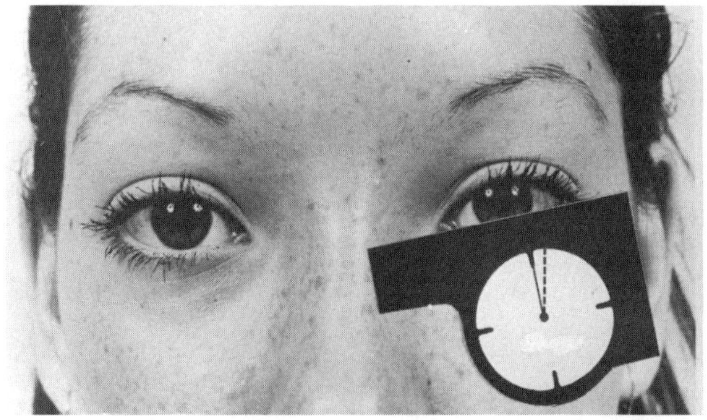

FIG. 2–88. Measuring the inclination of the eye fissure axis with a commercial angle meter. The facial midline must be vertical.

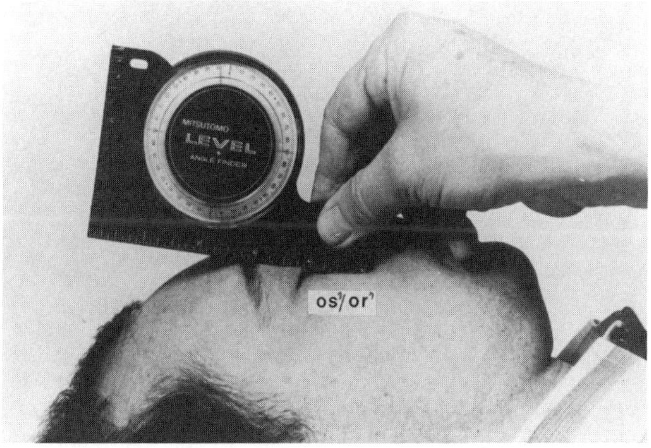

FIG. 2–89. Measuring the inclination along the line connecting the upper and lower orbital rims (os′/or′) with a commercial angle meter. The inclination from the horizontal is measured while the subject is in a recumbent position.

Nose

Single Linear Measurements (S)

Projective

• Horizontal

Head position: rest

Instrument: sliding caliper

S-1 Width of the nasal root (mf-mf) (8) is the diameter of the root above the base line at the level of the eye fissures, anatomically between the maxillofrontalia (Fig. 2–90).

S-2 Width of the nose [width of the soft nose, bi-alar diameter, largeur du nez, interalar distance or Untere Nasebreite of Martin (39), or morphological width of the nose of Farkas et al. (34)] (al-al) is the distance between the most lateral points on the alae [Fig. 2–91(1)]. The arms of the caliper only touch the skin.

S-3 Anatomical width of the nose (34) is the width between the facial insertion points of the alar base [Nasenansatzbreite of Knussmann (24)] (ac-ac) and is measured between the most lateral points in the curved base lines of alae (8) oriented towards the face [Fig. 2–91(2)].

S-4 Width between the labial insertions of the alar base (sbal-sbal) is the distance between the points

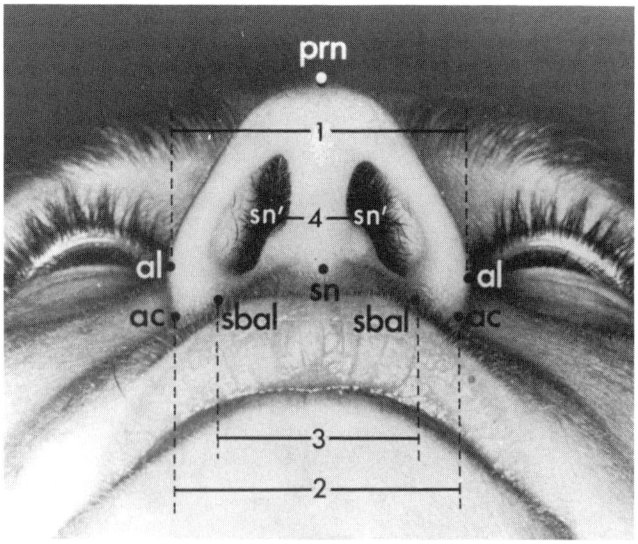

FIG. 2–91. Horizontal measurements of the soft nose: (**1**) nose width (al-al); (**2**) width between the facial insertions of alar bases (ac-ac); (**3**) width between the labial insertions of alar bases (sbal-sbal); (**4**) width of the columella (sn′-sn′).

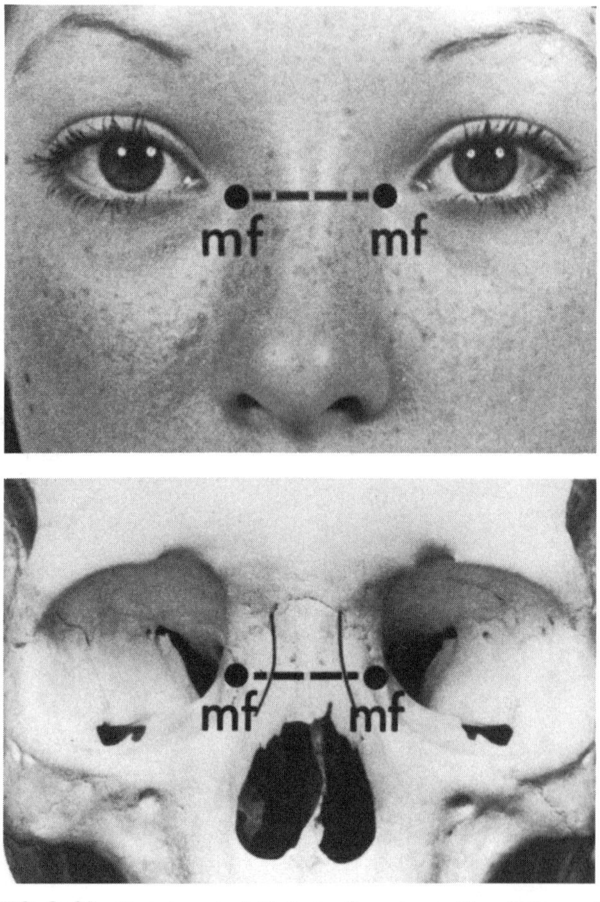

FIG. 2–90. Nasal root width (mf-mf) measured by sliding caliper. **Upper figure:** The blunt edges of the instrument are pressed to the bony walls of the nasal root, approximately at the level of the eye fissures. **Lower figure:** Width measurements of the root include a portion of the frontal process of maxilla.

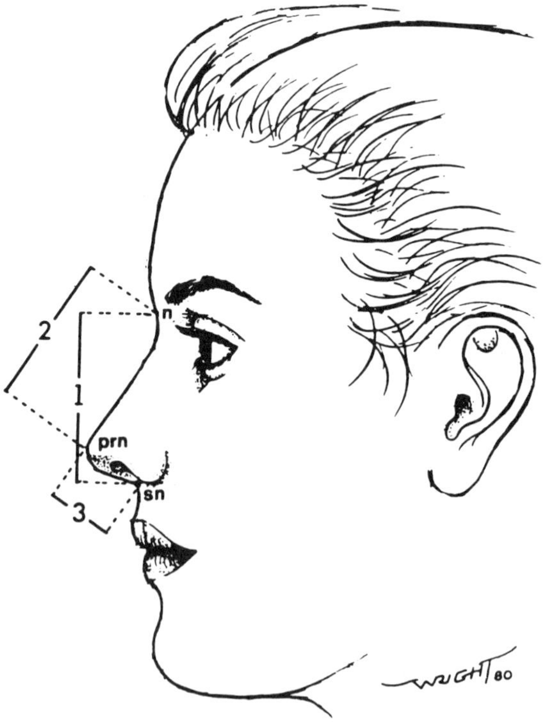

FIG. 2–92. Perpendicular and sagittal measurements of the nose: (**1**) nose height (n-sn); (**2**) bridge length (n-prn); (**3**) nasal tip protrusion (sn-prn).

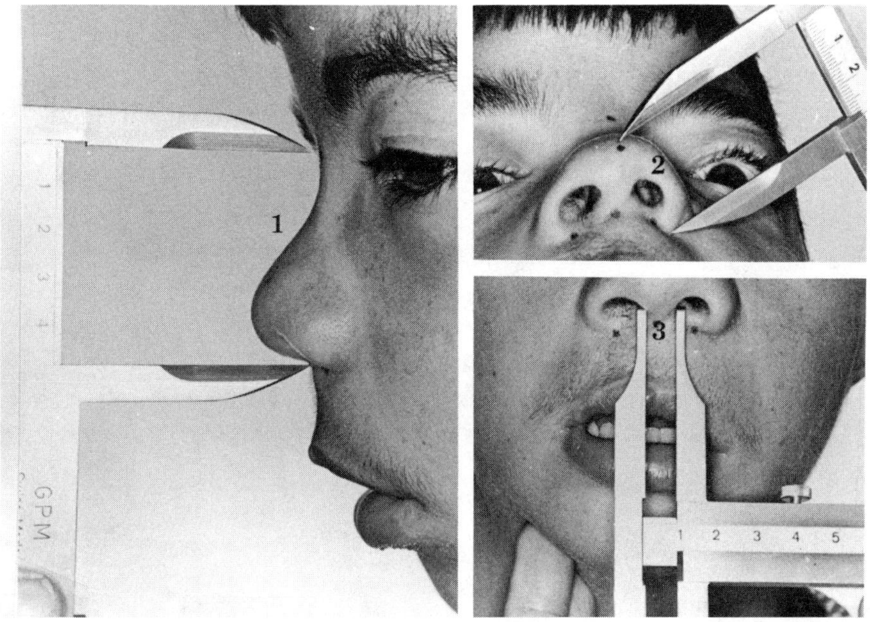

FIG. 2-93. Measuring with the sliding caliper: (**1**) the height of the nose (n-sn); (**2**) the length of the ala (ac-prn); (**3**) the width of the columella (sn′-sn′).

where the alar base disappears into the skin of the upper lip [Fig. 2–91(3)].

S-5 Width of the columella (8) [Nasenseptumbreite of Knussmann (24)] (sn′-sn′) is measured in its midportion with the arms of the caliper only touching the skin [Figs. 2–91(4) and 2–93(3)].

• Perpendicular
Head position: rest
Instrument: sliding caliper

S-6 Height of the nose [hauteur du nez, Nasenhöhe or upper facial height] (n-sn) is the distance between the nasion and the subnasale [Figs. 2–92(1) and 2–93(1)].

S-7 Length of the nasal bridge (Nasenrückenlänge) (n-prn) is the distance between the nasion and the pronasale [Fig. 2–92(2)].

• Sagittal
Head position: reclined head
Instrument: sliding caliper

S-8 Nasal protrusion [nasal tip protrusion, nasal depth, Nasenelevation of Martin (39), saillie de la base du nez, or Nasenbodentiefe (52)] (sn-prn) is the distance between the subnasale and the pronasale [Fig. 2–92(3)].

Single Angular Measurements (S)

Inclinations

• Perpendicular
Head position: FH, with the facial profile in the vertical
Instrument: commercial angle meter

S-9 Inclination of the nasal bridge from the vertical (8) [Nasenrückenwinkel (52)] is measured with the straight side of the angle meter resting on the nasal bridge (Fig. 2–94). In a nose with hump or saddle deformity the inclination of the general bridge-line is estimated.

S-10 Inclination of the columella [also nasal tip rota-

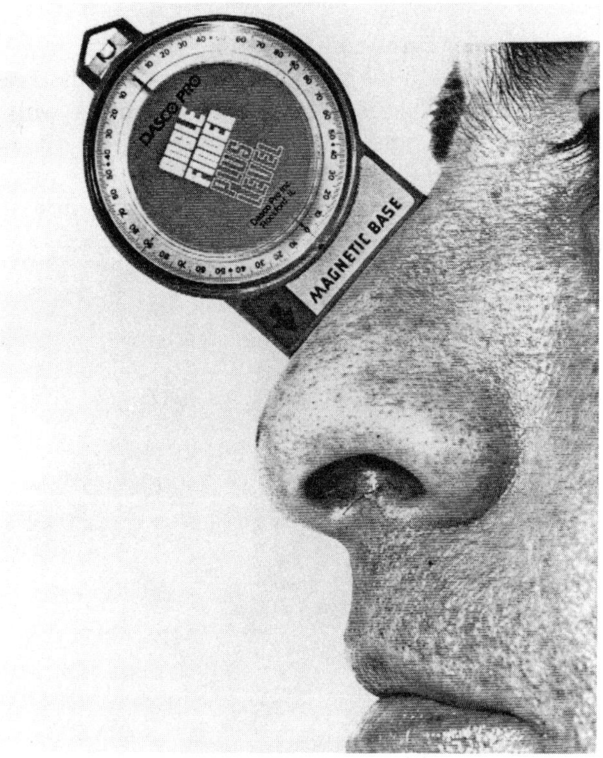

FIG. 2–94. Measuring the inclination of the nasal bridge from the vertical using the commercial Angle Finder (Dasco Pro).

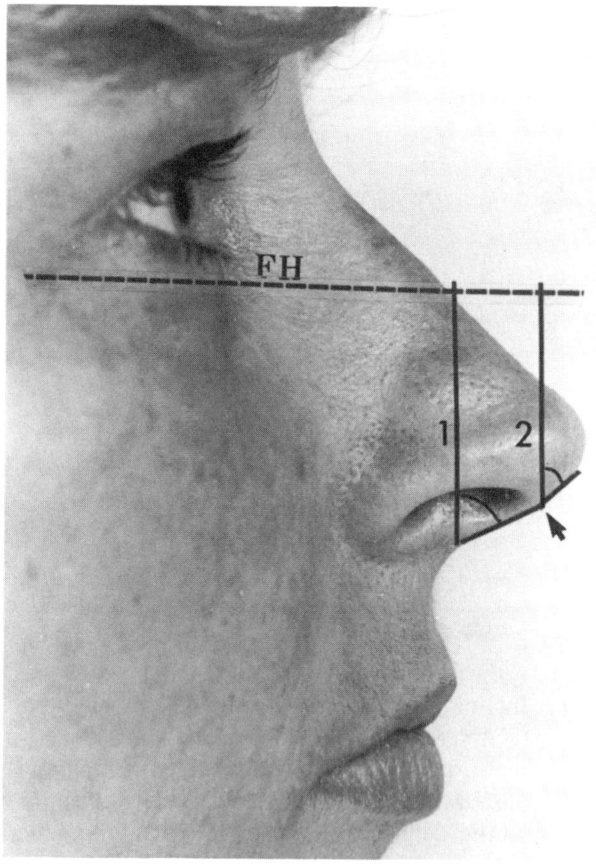

FIG. 2–95. **1:** Inclination of the columella from the vertical. **2:** Inclination of the tip portion of the nose from the vertical. The *arrow* points to the breakpoint of the columella.

tion of Bernstein (55)] from the vertical is measured with the shorter, straight side of the angle meter [Fig. 2–95(1)]. In the columella with a curved contour the inclination of the general columella-line is examined (34).

S-11 Inclination of the tip portion of the nose from the vertical is measured along the line extended anteriorly from the columella breakpoint towards the nasal tip [Fig. 2–95(2)] (32). The nasal tip inclination can be measured only on the flat surface of the subapical portion of the nose.

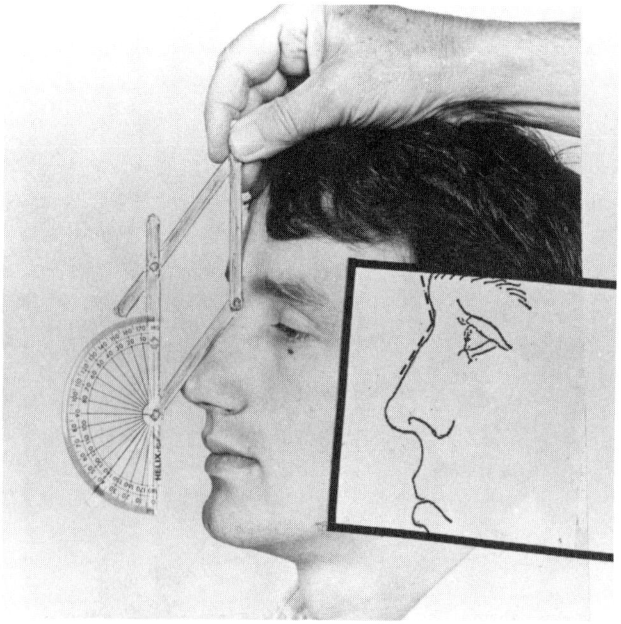

FIG. 2–97. Measuring the nasofrontal angle with the multipurpose facial angle meter.

Angles

Head position: rest
Instrument: multipurpose facial angle meter

S-12 Glabellonasal angle is measured above the nasal root between the contours of the anterior surface of the forehead below and above the glabella (Fig. 2–96) (51).

S-13 Nasofrontal angle is measured between the proximal nasal bridge contour and the anterior surface of the forehead below the glabella (Fig. 2–97).

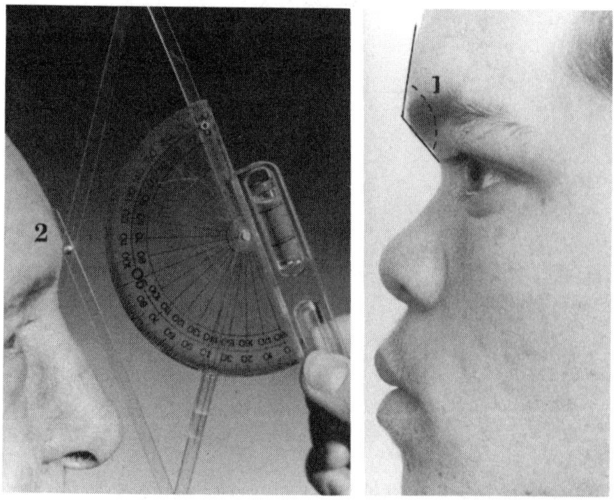

FIG. 2–96. Measuring the glabellonasal angle (**1**) with the multipurpose facial angle meter (**2**).

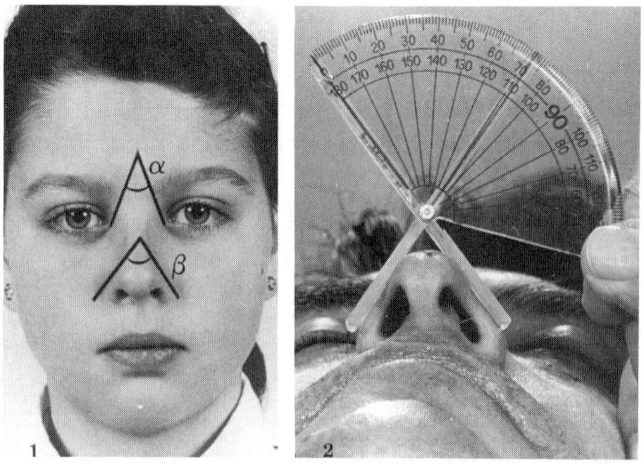

FIG. 2–98. Nasal root–slope and alar–slope angles. **1:** Schematic drawing of the slope contours in the two areas of the nose (α and β). **2:** Measuring the alar–slope angle with a special angle meter.

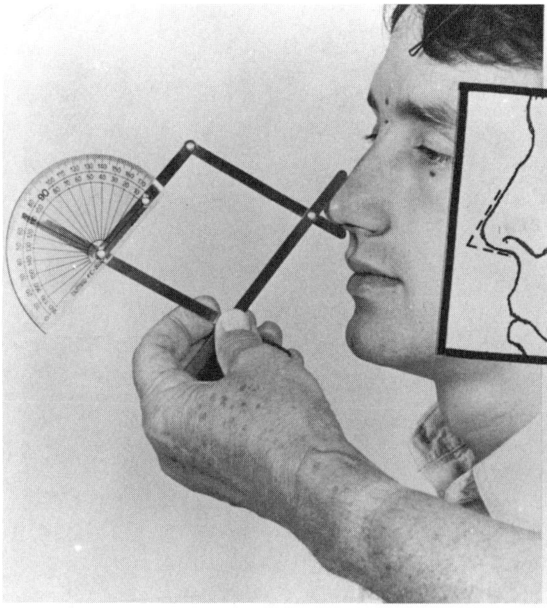

FIG. 2-99. Measuring the nasal tip angle with the multipurpose facial angle meter.

S-14 Nasal root–slope angle is measured between the inclined sides of the root at the level of the eye fissure [Fig. 2–98(1α)].

S-15 Nasal tip angle [Joseph's septodorsal angle or Septodorsalwinkel (56)] is formed by the lines following the general direction of the columella and the nasal bridge (Fig. 2–99) (35).

S-16 Alar–slope angle is created by the lines following the tilted surfaces of the alae [Fig. 2–98(1β) and 2–98(2)] (34).

S-17 Nasolabial angle (septolabial angle, columella–labial angle, or labial–columellar angle) is measured between the surfaces of the columella and of the upper-lip skin (Fig. 2–100).

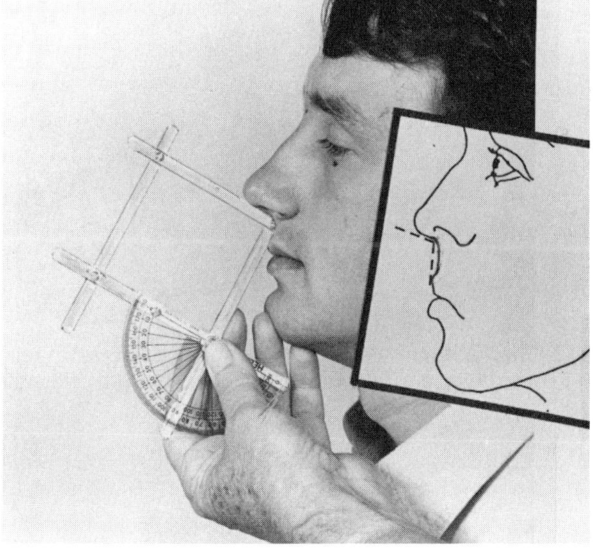

FIG. 2-100. Measuring the nasolabial angle by multipurpose facial angle meter.

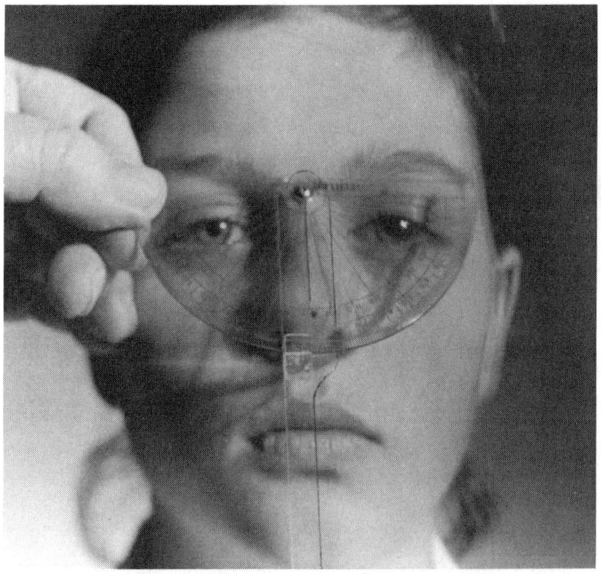

FIG. 2-101. Measuring the deviation of the nasal bridge from the midaxis of the face.

Deviations

Head position: rest, with the facial midline in the vertical (nasal bridge); reclined, with the profile line in the vertical (columella)

Instrument: nose deviation protractor

S-18 Deviation of the bridge from the midline of the face. The zero mark of the special protractor is placed on the nasion, and the mobile arm is moved to the nasal bridge (Fig. 2–101). The deviation is read on the scale of the protractor. The direction of the deviation is to the right or left.

S-19 Deviation of the columella from the perpendicular. The zero mark of the protractor is on the subnasale, and the mobile arm follows the direction of the columella axis (Fig. 2–102).

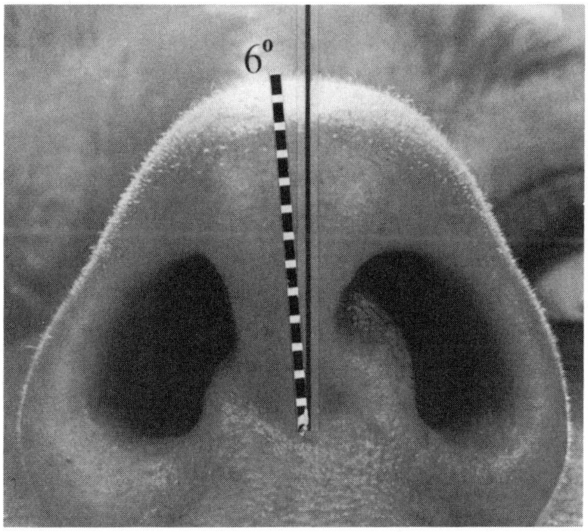

FIG. 2-102. Measuring the deviation of the columella from the vertical.

Paired Linear Measurements (P)

Projective

• Horizontal (oblique)
Head position: reclined
Instrument: sliding caliper

P-20 Width of the nostril floor (sbal-sn) is the distance between the subnasale and each subalare (Fig. 2–103).

P-21 Thickness of the ala (al′-al′) (8) is measured in the midportion of each ala (Fig. 2–103).

P-22 The position of the facial insertion points of the alae in relation to the midpoint of columella base (ac-sn) is given by the projective distance between the ac landmark and subnasale on each side (Fig. 2–104).

P-23 Length of the ala (Nasenflügellänge) (ac-prn) is the distance between each facial insertion point of the alar base and the pronasale [Fig. 2–93(2)].

• Depth Measurement
Head position: reclined
Instrument: sliding caliper

P-24 Length of the columella (sn-c′) is measured along the crest of the columella between levels of the subnasale and the top (breakpoint) of the columella at the level of the tip of each nostril (Fig. 2–105) (15).

• Sagittal
Head position: recumbent, with the midfacial plane in the vertical
Instrument: modified sliding caliper with one mobile arm and a level

P-25 Depth of the nasal root (8) (protrusion of the nasal root, depth of the endocanthion or Nasenwurzeltiefe) (en-se sag) is measured between each endocanthion and the sellion. The body of the caliper is resting horizontal on the nasal bridge with the help of the level. The mobile arm (MA) with scale is moved downward until the tip gently touches the skin close to the inner commissure (en) of the eye fissure. The depth (d) is shown on the scale of the mobile arm (Fig. 2–106).

Tangential

• Horizontal
Head position: rest or reclined
Instrument: soft measuring tape

P-26 Surface length of the ala (ac-prn surf) is measured between the facial insertion of the alar base and the pronasale, on both sides (Fig. 2–104).

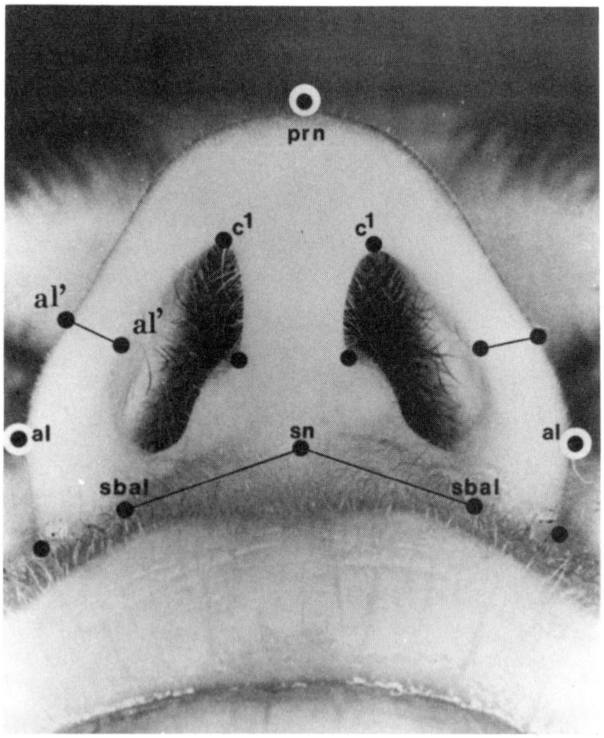

FIG. 2–103. Paired linear measurements of the soft nose: the nostril floor width (sbal-sn rt & lt); the ala thickness (al′-al′ rt & lt).

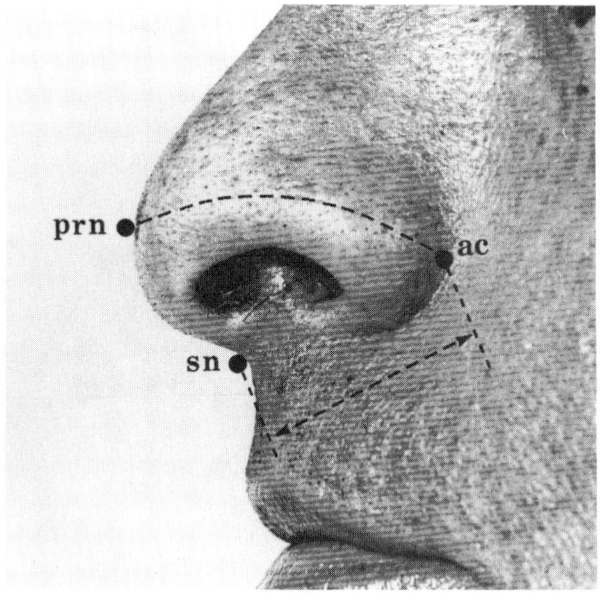

FIG. 2–104. Position of the facial alar base point (ac) in relation to the subnasale (sn), and the tangential length of the ala (ac-prn surf).

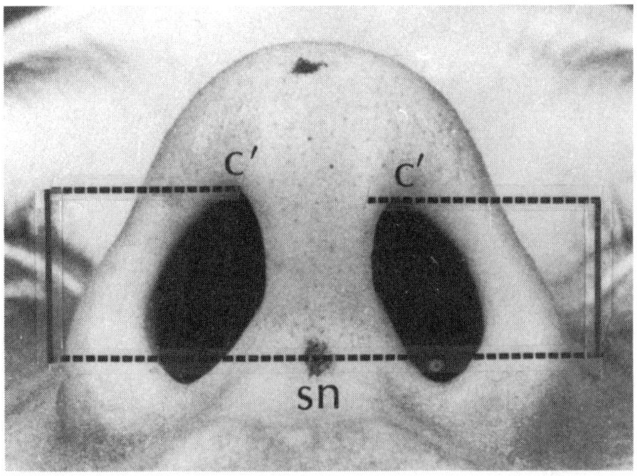

FIG. 2–105. Length of the columella (sn-c') is measured along the crest on both sides. (From ref. 12, with permission.)

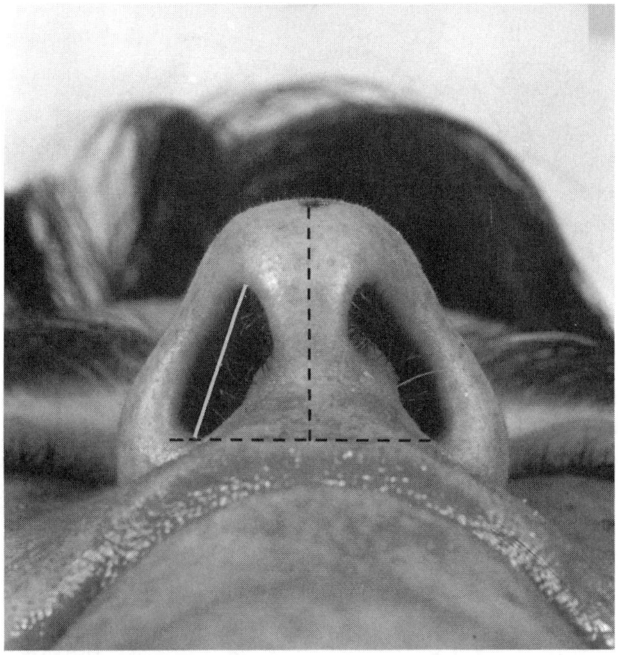

FIG. 2–107. Inclination of the longitudinal nostril axis measured from the horizontal.

Paired Angular Measurements (P)

Inclinations

• Horizontal

Head position: reclined, with the midfacial plane in the vertical

Instrument: nostril inclination protractor with level

P-27 Inclination of the longitudinal nostril axis (37) is measured from the horizontal along the line connecting the highest and lowest spots of each nostril (Fig. 2–107). In healthy Caucasians the direction of the longitudinal axis is convergent. In some other races, divergent nostril axis can be seen. The straight edge of the protractor is kept horizontal with the help of the level at the level of the nostril floor with its zero mark on the lowest terminal point of the axis. The pointer moved to the highest terminal point of the axis indicates the degree of the inclination on the scale of the protractor (Fig. 2–108).

P-28 Seven nostril types have been categorized by reference to degree of inclination (37,44) (Fig. 2–109).

Facial position: reclined, with the midfacial plane in the vertical

Instrument: commercial angle meter

P-29 Inclination of the alar–slope line is measured from the horizontal (al-prn). The straight side of the angle meter follows the flat surface of the ala on each side (Fig. 2–110).

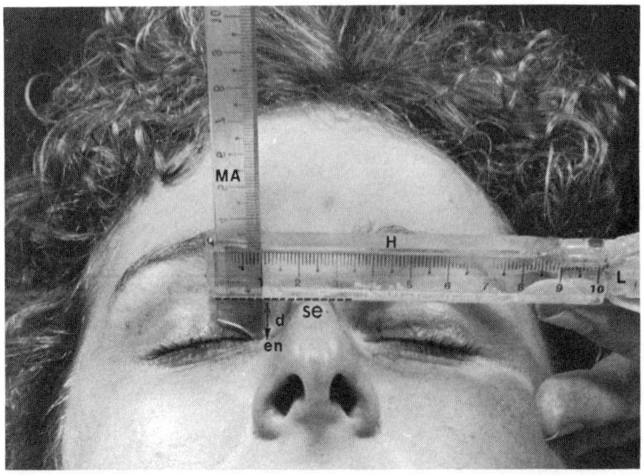

FIG. 2–106. Measuring the depth of the nasal root with the modified sliding caliper with one mobile arm and a level. Instrument: MA, mobile arm of the caliper; L, level; H, horizontally held body of the caliper. Landmarks: en, endocanthion; se, sellion; d = sagittal distance between the en and se points, depth of the nasal root.

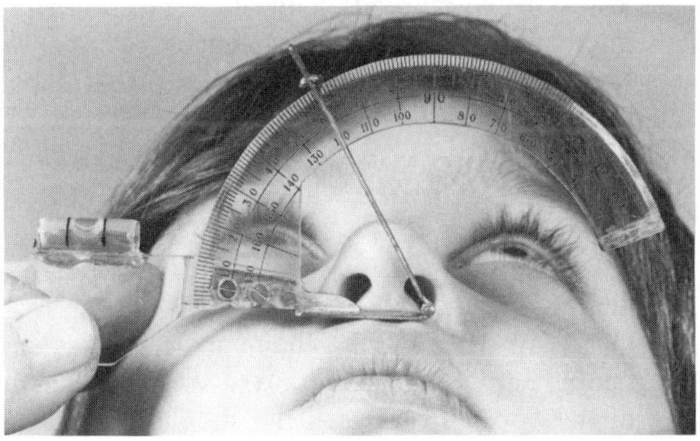

FIG. 2–108. Measuring with the Singapore angle meter: nostril inclination.

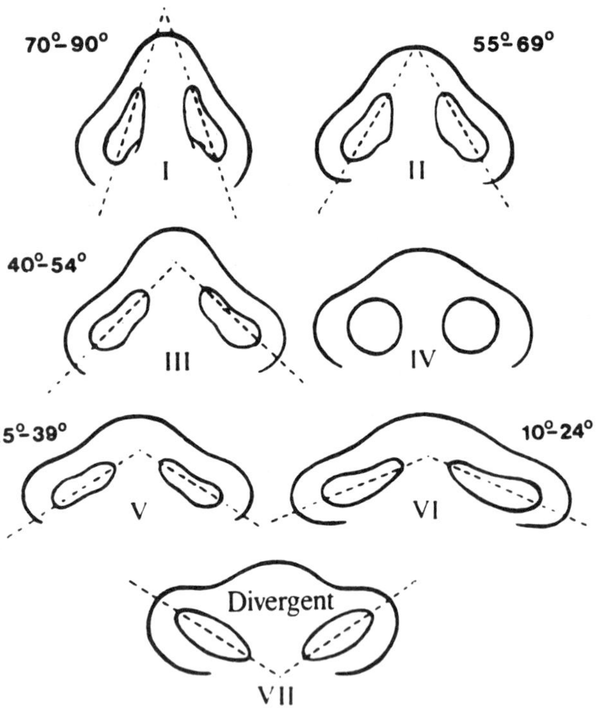

FIG. 2-109. Nostril types. Of the seven quantitatively determined nostril types, the nostril axes are convergent in six cases, with one being divergent. In North American Caucasians, types I to III are most frequently seen. In non-Caucasians, types IV to VII are most common. (From ref. 37, with permission.)

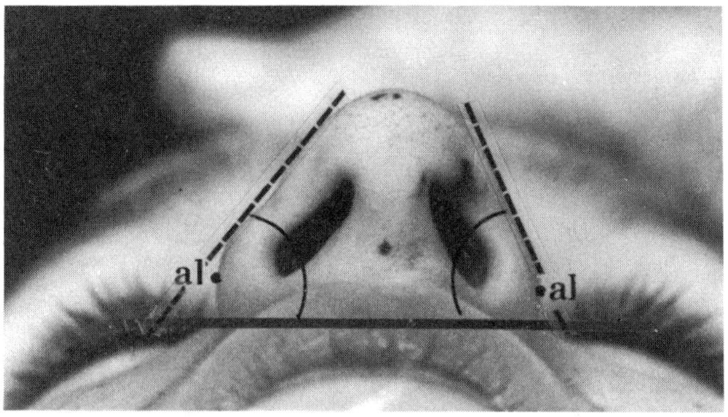

FIG. 2-110. Inclination of the alar-slope surface from the horizontal.

Lips and Mouth

Single Linear Measurements (S)

Projective

• Horizontal
Head position: rest
Instrument: sliding caliper

S-1 Width of the philtrum (cph-cph) is measured above the vermilion line, between the elevated edges of the line (crista philtri) (Fig. 2-111).

S-2 Width of the mouth (length of the labial fissure, intercommissural distance, longueur buccale, or Mundspaltenbreite) (ch-ch) is the distance between the cheilions of the closed mouth (Fig. 2-112).

• Perpendicular
Head position: rest
Instrument: sliding caliper

S-3 Height of the upper lip (hauteur de la lèvre superieure or Oberlippenhöhe) (sn-sto) is the height of the upper lip between the subnasale and the stomion [Fig. 2-113(1)].

S-4 Height of the cutaneous upper lip [Hautoberlippenhöhe, height of the skin portion of the upper lip, or length of the philtrum] (sn-ls) is the length of the cutaneous area between the subnasale and the labiale superius [Fig. 2-113(2)].

S-5 Vermilion height of the upper lip (ls-sto) is the thickness of the vermilion in the facial midline between the labiale superius and the stomion [Fig. 2-114(1)].

S-6 Vermilion height of the lower lip (sto-li) is the thickness of the vermilion in the facial midline between the stomion and the labiale inferius [Fig. 2-114(2)].

S-7 Height of the cutaneous lower lip (medial vertical lower lip length, Hautunterlippenhöhe) (li-sl) is the height of the skin portion of the lower lip between the labiale inferius and the sublabiale (16) [Fig. 2-113(3)].

S-8 Height of the lower lip (medial vertical height of the lower lip, Unterlippenhöhe) (sto-sl) is measured between the stomion and the sublabiale (16) [Fig. 2-113(4)].

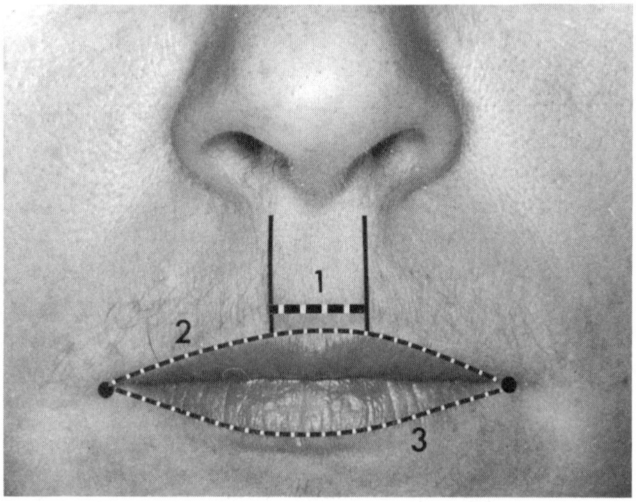

FIG. 2-111. 1: Width of the philtrum (cph-cph). 2: The upper vermilion arc (ch-ls-ch). 3: The lower vermilion arc (ch-li-ch).

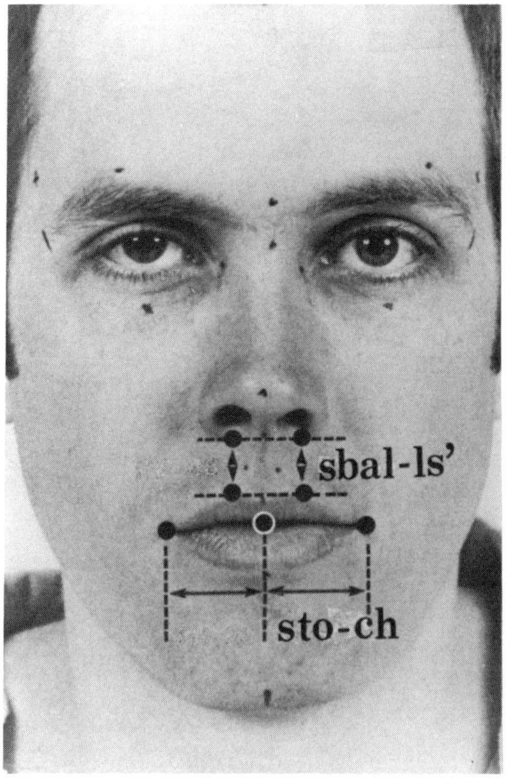

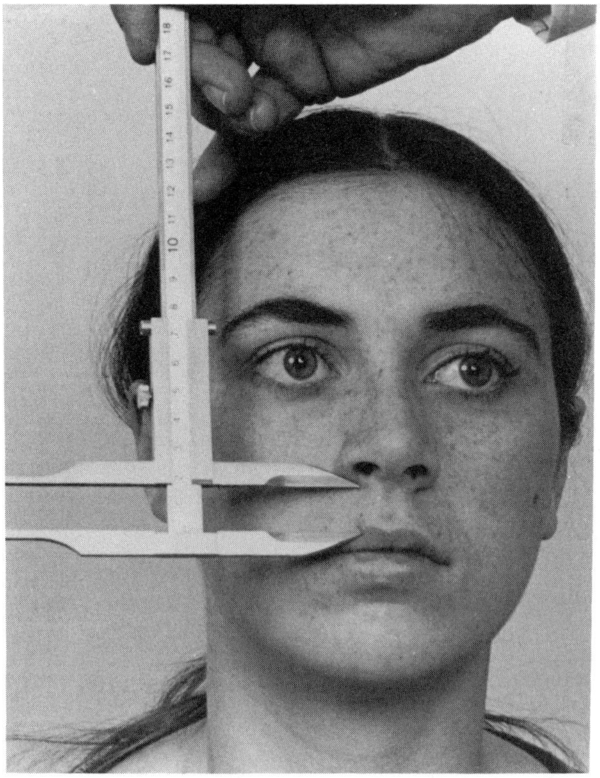

FIG. 2–112. Width of the mouth (ch-ch), half of the labial fissure (ch-sto), and the lateral upper lip height (sbal-ls'). Measuring the lateral upper lip height with the sliding caliper.

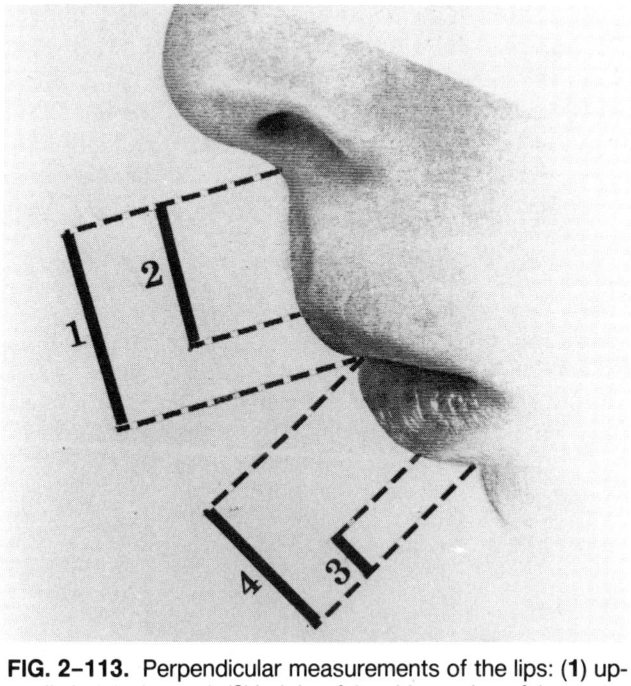

FIG. 2–113. Perpendicular measurements of the lips: (**1**) upper lip height (sn-sto); (**2**) height of the skin portion of the upper lip (sn-ls); (**3**) height of the skin portion of the lower lip (li-sl); (**4**) lower lip height (sto-sl).

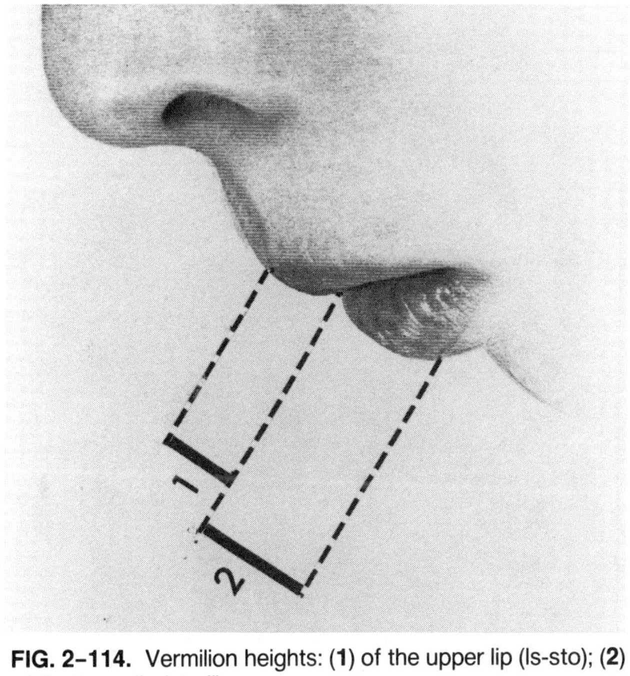

FIG. 2–114. Vermilion heights: (**1**) of the upper lip (ls-sto); (**2**) of the lower lip (sto-li).

Tangential

• Horizontal
Head position: rest
Instrument: soft measuring tape

S-9 Vermilion surface arc of the upper lip (upper vermilion arc) (ch-ls-ch). The soft measuring tape is following the vermilion line of the upper lip from one commissure of the labial fissure to the other, on gently closed lips (16) [Fig. 2–111(2)].

S-10 Vermilion surface arc of the lower lip (lower vermilion arc) (ch-li-ch). The soft measuring tape follows the vermilion line of the lower lip from one commissure of the labial fissure to the other, on gently closed lips (16) [Fig. 2–111(3)].

Single Angular Measurements (S)

Inclinations

• Horizontal
Head position: rest, with the midfacial plane in the vertical
Instrument: commercial angle meter

S-11 Inclination of the labial fissure from the horizontal (8) is measured along the plane between commissures of the mouth (ch-ch) (Fig. 2–115). If the labial fissure is oblique, the side of the lower commissure is recorded.

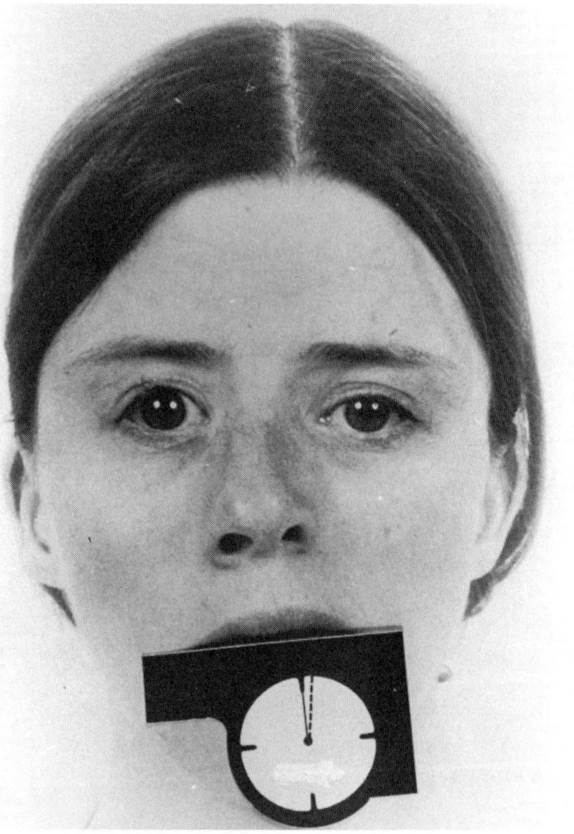

FIG. 2–115. Measuring the inclination of the labial fissure from the horizontal using a commercial angle meter.

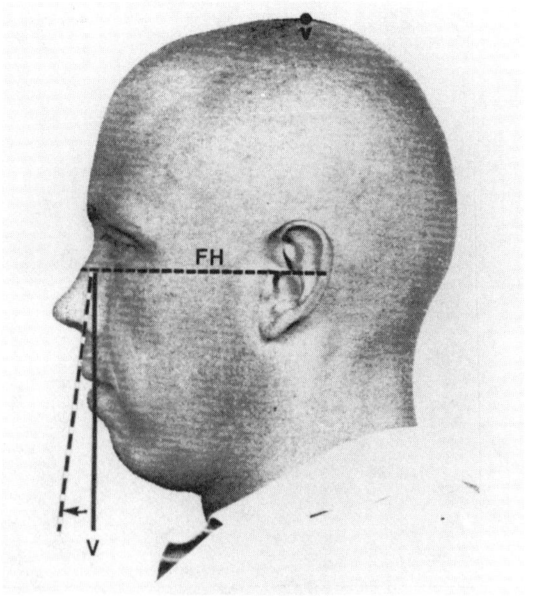

FIG. 2–116. Measuring the inclination of the upper lip from the vertical (V).

• Perpendicular
Head position: FH, with midface plane in the vertical
Instrument: commercial angle meter

S-12 Inclination of the upper lip (8) is the general direction of the upper lip skin surface in relation to the vertical, approximately along the line between the subnasale and labiale superius (Fig. 2–116).

S-13 Inclination of the lower lip (16) is the general direction of the lower lip skin surface in relation to the vertical, approximately along the line between the labiale inferius and sublabiale (Fig. 2–117).

Angle

Head position: rest
Instrument: multipurpose facial angle meter

S-14 Labiomental angle is formed by the skin surface of the lower lip and the surface contour of the chin above the pogonion landmark (Fig. 2–118) (52).

Paired Linear Measurements (P)

Projective

• Horizontal
Head position: rest
Instrument: spreading caliper

P-15 Halves of the labial fissure length (ch-sto right and sto-ch left) (8) are measured between each chei-

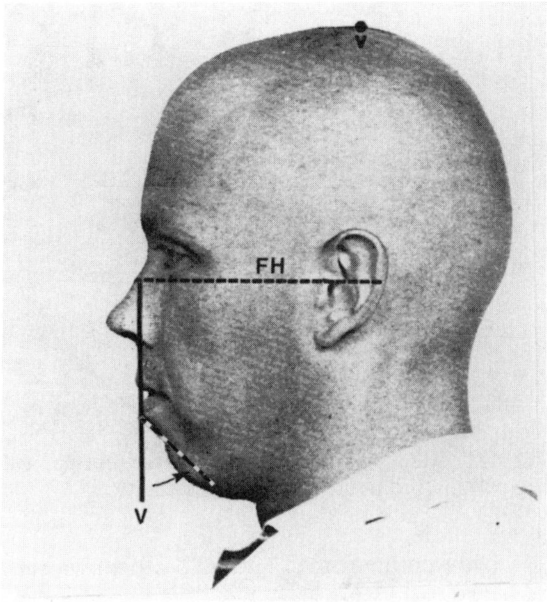

FIG. 2–117. Measuring the inclination of the lower lip from the vertical (V).

lion and the stomion, between gently closed lips (Fig. 2–112).

• Perpendicular
Head position: rest
Instrument: sliding caliper
P-16 Lateral upper lip height (lateral vertical upper lip length) (sbal-ls′) is the cutaneous lip height below each alar base, between the subalare and the point on the vermilion line marked as ls′ (Fig. 2–112), vertical from the subalare (46).

• Depth Measurement
Head position: rest
Instrument: spreading caliper
P-17 Labiotragial distance (ch-t) is measured between the cheilion and the tragion on each side [Fig. 2–119(1)] (8).

Tangential

• Lateral Arc Measurement
Head position: rest
Instrument: soft measuring tape
P-18 Lateral labiotragial arc (ch-t surf) is measured between the cheilion and the tragion on each side [Fig. 2–119(2)] (8).

Ears

Paired Linear Measurements (P)

Projective

• Horizontal
Head position: rest
Instrument: sliding caliper

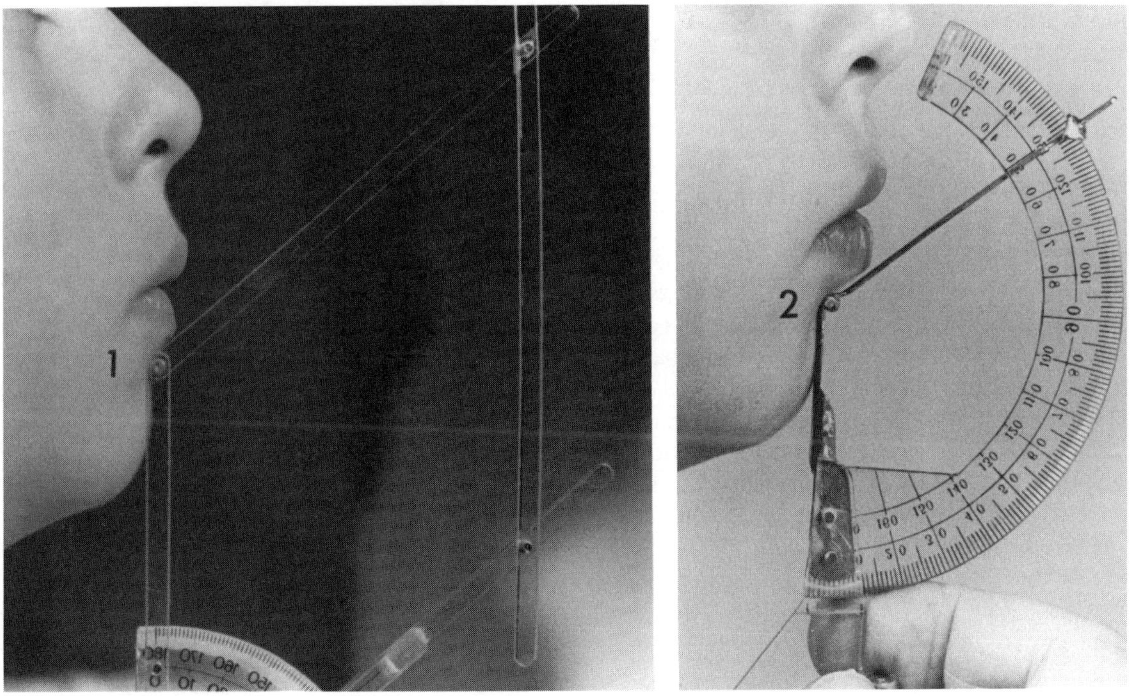

FIG. 2–118. Measuring the labiomental angle: (**1**) with the multipurpose facial angle meter; (**2**) with the Singapore angle meter.

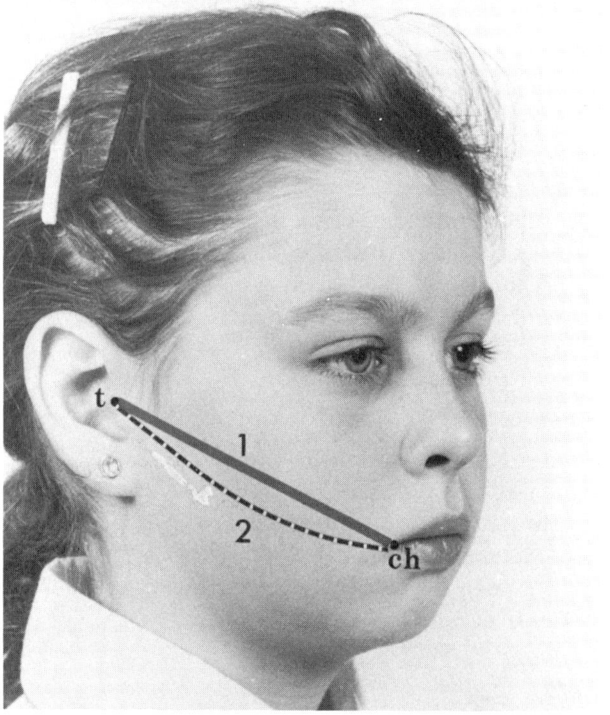

FIG. 2–119. Establishing the position of the mouth in relation to the ear canal: (1) projectively by measuring the ch-t distance; (2) tangentially by measuring the same distance along the skin surface (ch-t surf).

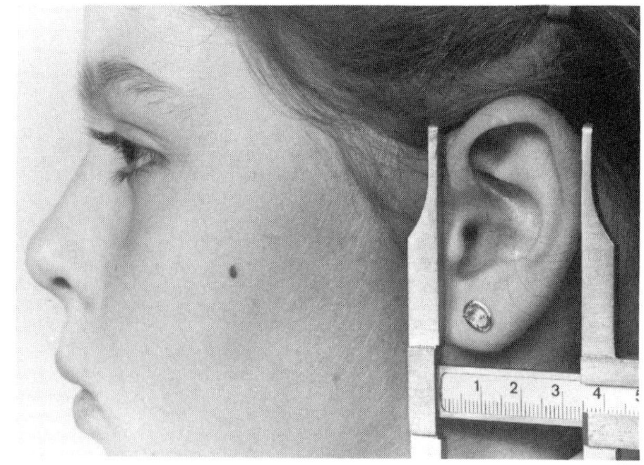

FIG. 2–120. Measuring the width of the ear (pra-pa) with the sliding caliper. (From ref. 12, with permission.)

P-1 Width of the auricle (physiognomic ear breadth, largeur de l'oreille, Ohrbreite) (pra-pa) is the maximum width between the preaurale and the postaurale of each ear (Fig. 2–120).

• Perpendicular
Head position: rest
Instrument: sliding caliper

P-2 Height of the tragus is measured between its base and the tip on each side (Fig. 2–121) (8).

P-3 Length of the auricle (physiognomic ear length, longueur maxima de l'oreille, or Ohrlänge) (sa-sba) is the maximum length of the long axis of the pinna between the superaurale and the subaurale of each ear (Fig. 2–122).

P-4 Morphological width of the ear (26) (width of the ear insertion to the head) (obs-obi) is the width of the attachment of the ear to the head, measured between the otobasion superius and the otobasion inferius of each ear (Fig. 2–122).

Position of the ear in relation to the facial midaxis landmarks (or additional projective depth measurements of the face) (45).
Head position: rest
Instrument: spreading caliper

P-5 Upper naso-aural distance (n-obs) is measured between the nasion and each otobasion superius [Fig. 2–123(1)].

P-6 Lower naso-aural distance (n-obi) is measured between the nasion and each otobasion inferius [Fig. 2–123(1)].

P-7 Upper subnasale–aural distance (sn-obs) is measured between the subnasale and each otobasion superius [Fig. 2–123(2)].

P-8 Lower subnasale–aural distance (sn-obi) is measured between the subnasale and each otobasion inferius [Fig. 2–123(2)].

P-9 Upper gnathion–aural distance (gn-obs) is measured between the gnathion and each otobasion superius [Fig. 2–123(3)].

P-10 Lower gnathion-aural distance (gn-obi) is measured between the gnathion and each otobasion inferius [Fig. 2–123(3)].

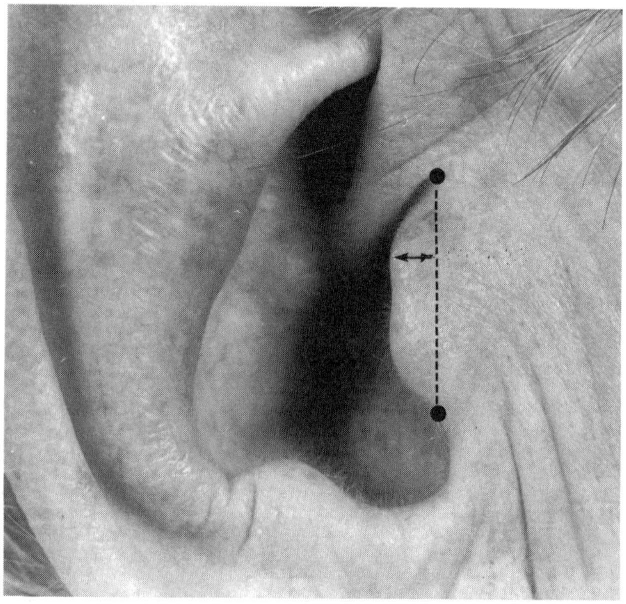

FIG. 2–121. Height of the tragus.

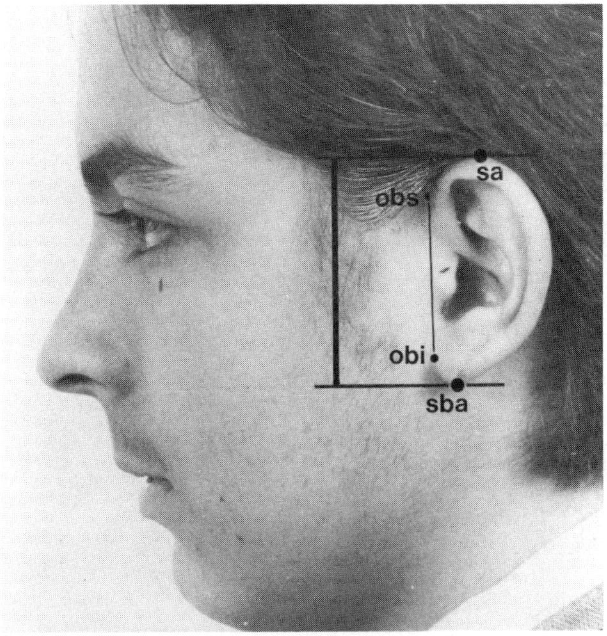

FIG. 2-122. Perpendicular measurements of the auricle: length of the ear (sa-sba); width of the ear insertion to the head (obs-obi).

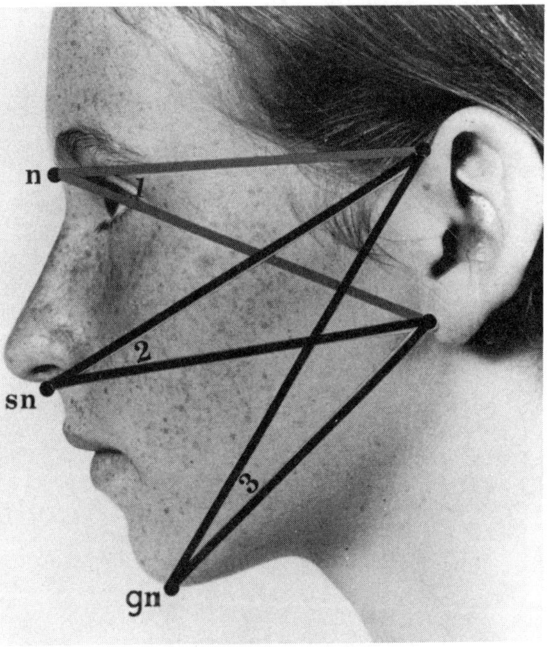

FIG. 2-123. Position of the ear in relation to the facial midline landmarks: (**1**) to the nasion (n-obs and n-obi); (**2**) to the subnasale (sn-obs and sn-obi); (**3**) to the gnathion (gn-obs and gn-obi).

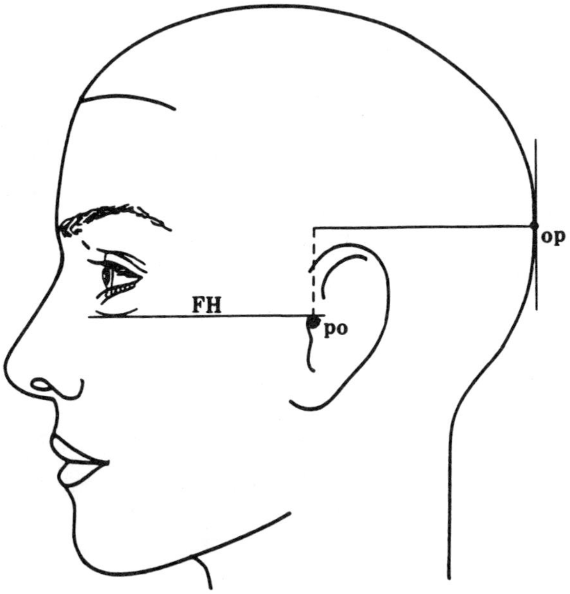

FIG. 2-124. Position of the ear canal in relation to the opisthocranion point (the most posterior point of the head in FH position).

- Sagittal

Head position: FH, with midface plane in the vertical, in recumbent position

Instrument: large double sliding caliper with levels

P-11 Occipito-aural distance (op-po) (8) is measured between the opisthocranion and the porion (Fig. 2-124). The most posterior point of the head is resting on the horizontal arm of the instrument. The mobile rider is moved upwards to the level of the ear canal. The exact distance between the ear canal and the most posterior point of the head (opisthocranion) is obtained when the mobile horizontal branch touches the upper edge of the ear canal. The distance is shown on the scale of the fixed perpendicular arm (Fig. 2-125).

Tangential

- Lateral Arc Measurement (8)

Head position: rest

Instrument: soft measuring tape

P-12 Upper naso-aural surface distance (n-obs surf) is measured between the nasion and each otobasion superius [Fig. 2-126(1)].

P-13 Lower naso-aural surface distance (n-obi surf) is measured between the nasion and each otobasion inferius [Fig. 2-126(1)].

P-14 Upper subnasale-aural surface distance (sn-obs surf) is measured between the subnasale and each otobasion superius [Fig. 2-126(2)].

P-15 Lower subnasale-aural surface distance (sn-obi surf) is measured between the subnasale and each otobasion inferius [Fig. 2-126(2)].

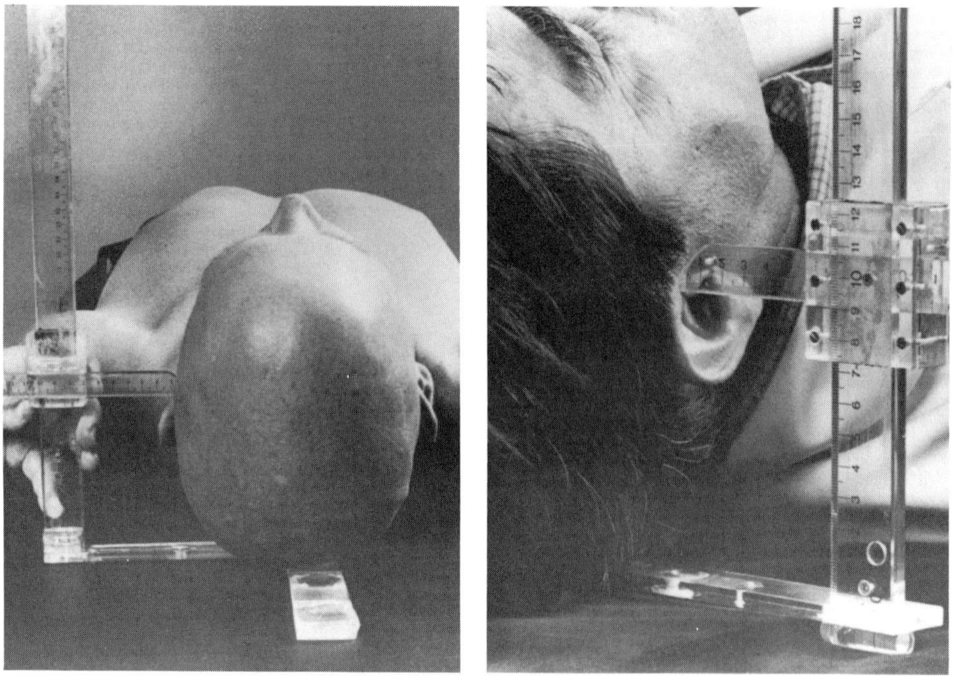

FIG. 2–125. Measuring the sagittal distance between the ear canal (porion) and the opisthocranion landmarks with the large double sliding caliper with levels. **Left:** The measurement is shown from behind the head. **Right:** The measurement seen from the side.

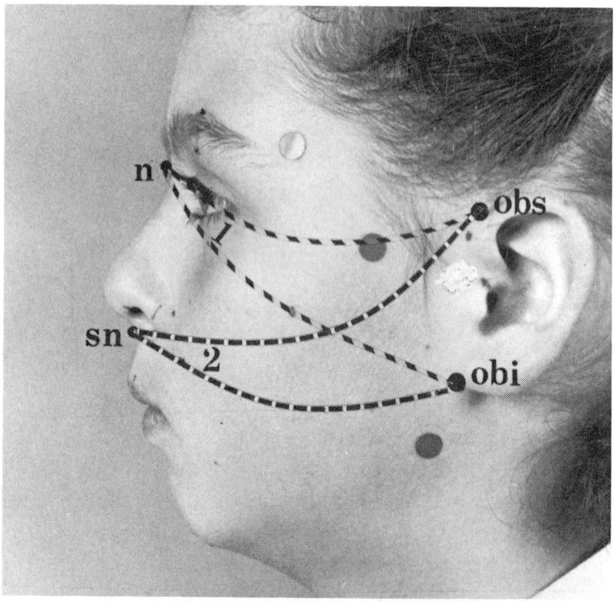

FIG. 2–126. Position of the ear in relation to the nasion and subnasale points in midline of the face by taking arc measurements: (**1**) from the nasion point to the upper (obs) and lower (obi) insertion points of the ear; (**2**) from the subnasale to the upper (obs) and lower (obi) insertions of the ear.

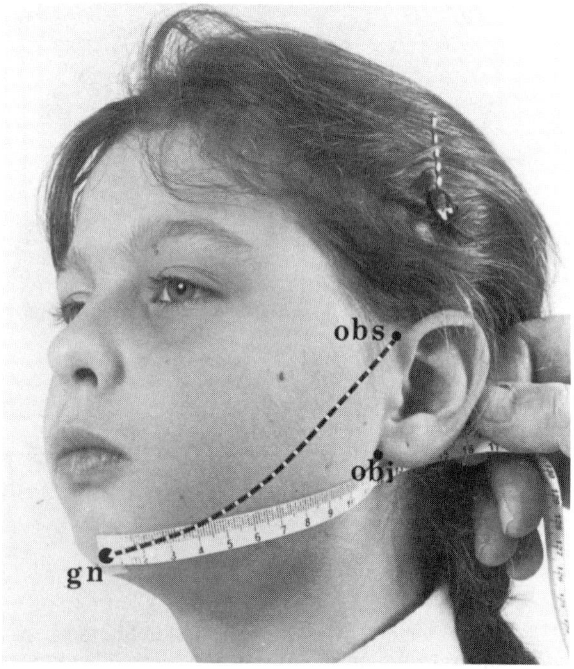

FIG. 2–127. Position of the ear in relation to the chin point by taking arc measurements from the gnathion to the upper (obs) and lower (obi) insertion points of the ear.

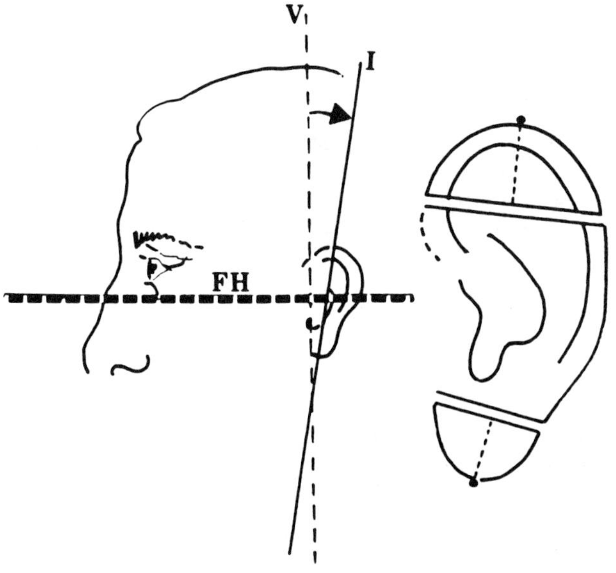

FIG. 2–128. Inclination of the medial longitudinal axis of the auricle. Determination of the terminal points of the axis (see also Fig. 2-59). I, inclination; V, vertical. (From ref. 6, with permission.)

P-16 Upper gnathion–aural surface distance (gn-obs surf) is measured between the gnathion and each otobasion superius (Fig. 2–127).

P-17 Lower gnathion–aural surface distance (gn-obi surf) is measured between the gnathion and each otobasion inferius (Fig. 2–127).

Paired Angular Measurements (P)

Inclination

• Perpendicular

Head position: FH, with the facial profile in the vertical

Instrument: commercial angle meter

P-18 Inclination of the medial longitudinal axis of the ear (ear rotation) from the vertical is measured by placing the long side of the angle meter along the line connecting the two most remote points of the medial axis (6) (Fig. 2–128).

Angle

• Ear protrusion

Head position: rest

Instrument: multipurpose facial angle meter, or transparent protractor

P-19 Protrusion of each ear (57) is the angle between the posterior aspect of the pinna and the mastoid plane. The V-shaped portion of the Singapore angle meter or the multipurpose angle meter is placed between the ear and the head in the area of the greatest protrusion. The tip of the V-shaped branches is touching the deepest spot of the retroauricular space. The angle of the protrusion is shown on the scale of the protractor (Fig. 2–129).

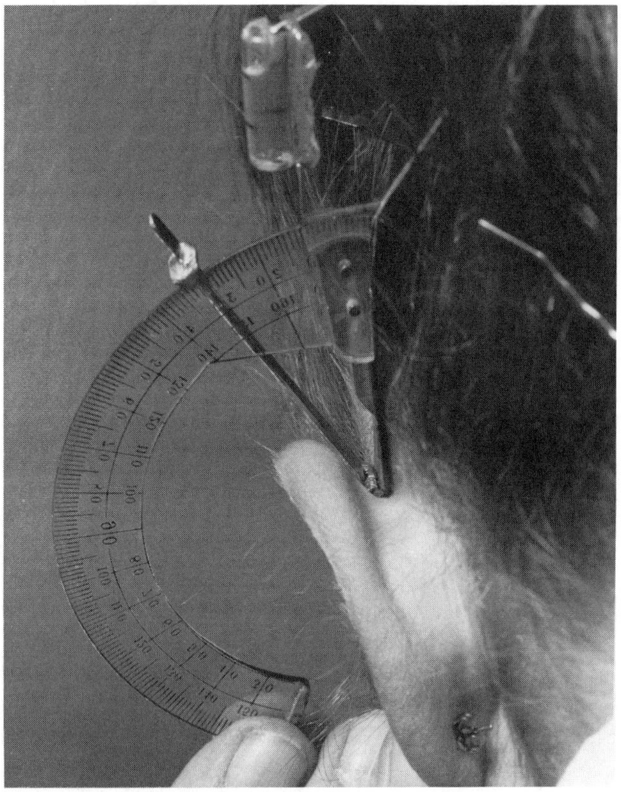

FIG. 2–129. Measuring the ear protrusion with the Singapore angle meter. (From ref. 2, with permission.)

Additional Measurements

Level Difference in the Frontal Plane of the Orbits

Head position: rest, with midface plane in the vertical

Instrument: sliding caliper and commercial angle finder with magnetic base (Dasco)

1. Level difference between the right and left superciliare landmarks of the brows (Fig. 2–130). A metal

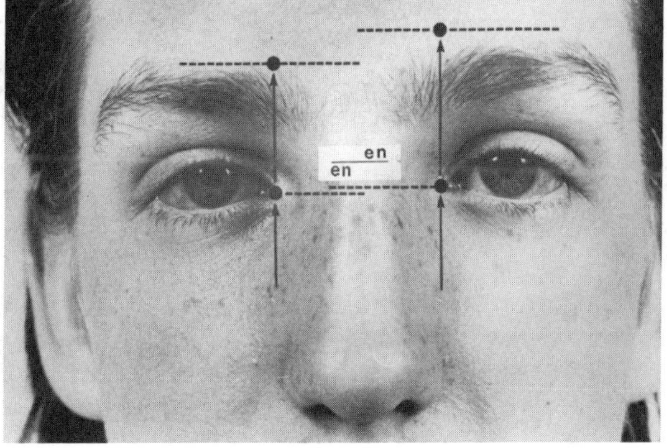

FIG. 2–130. Uneven levels of the eyebrows and between the position of the endocathions.

ruler attached to the magnetic base is held in the horizontal at the level of the lower located landmark. The level difference in the landmarks is measured by the caliper.

2. Level difference between the right and left endocanthion landmarks (Fig. 2–130).

Sagittal Level Difference in the Orbits

Head position: recumbent with midface plane in the vertical

Instrument: The modified sliding caliper with double mobile arms and level

3. Level difference between the position of endocanthions sagittally. The mobile arms are moved medially to the nasal root, and then pushed down gently to the skin surface close to the endocanthion landmarks. The horizontal position of the instrument's body is restored with the help of the level (L). The depth difference in the endocanthions is shown on the scale of the mobile arms (Fig. 2–131). The examination can be carried out also by the modified sliding caliper with level [see Fig. 2–45(1)] with one mobile arm, but the measurement must be repeated on both sides.

Level Difference in the Frontal Plane of the Soft Nose

Head position: recumbent with midface plane in the vertical

Instrument: sliding caliper and the commercial Angle Finder with magnetic base

4. Level difference between the alar bases of the nose. (For technique of the measurement see the frontal dislocation of the superciliare landmarks.) (Fig. 2–130).

Sagittal Level Difference in the Soft Nose

Head position: recumbent with midface plane in the vertical

Instrument: The modified sliding caliper with double mobile arms and level

5. Level difference between the depths of the alar bases (subalare rt and lt) (Fig. 2–132). (For technique of the measurement see "Sagittal Level Difference in the Orbits," above.)

Chin Point Dislocation to the Side

Head position: rest, with midface plane in the vertical

Instrument: sliding caliper

6. Dislocation of the gnathion point to the right or left from the vertical facial midline. A ruler is placed in continuation of the midface line preserved in the midaxis of the face to reach the chin point. The distance between the dislocated chin point and the ruler is measured by sliding caliper [Fig. 2–5(3)]. At routine examination of the face from the front, a smaller dislocation of the chin cannot be noticed. Examination of the direction of the midaxis of the face in recumbent position of the patient from behind the head is more sensitive for visual discovering of irregularities in the shape of the facial midline.

Horizontal Ear Canal Dislocation on One Side

Head position: rest, recumbent with midface plane in the vertical

Instrument: spreading caliper

7. Dislocation of the ear canal on one side (toward the facial midline) can be easily discovered when mea-

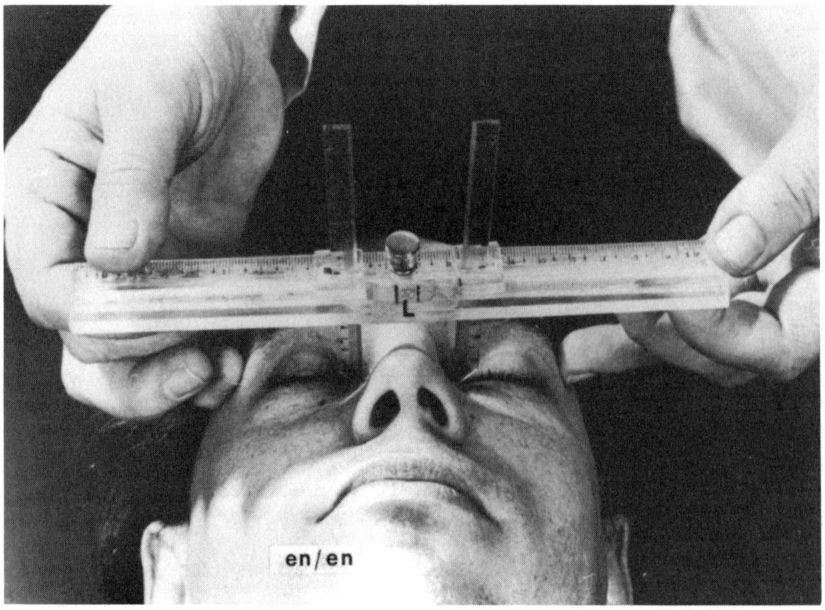

FIG. 2–131. Measuring the level difference between the endocathions, sagittally.

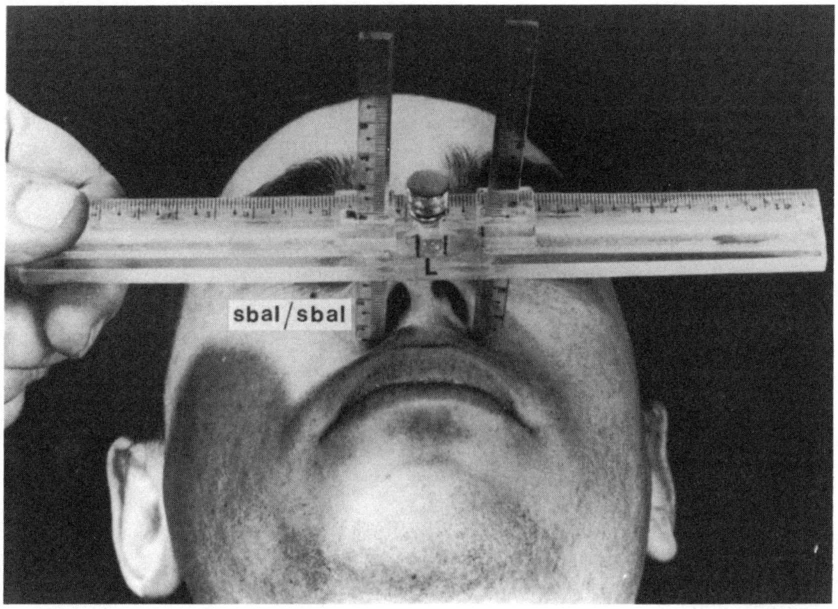

FIG. 2–132. Measuring the difference in level between the alar bases in a sagittal direction.

suring the width of the skull base (t-t) with the spreading caliper. It is important to pay attention to the position of the scale of the caliper in relation to the frontal plane of the face (Fig. 2–133). Symmetrical location of the ear canals in relation to the midaxis of the face is indicated by parallel position of the caliper's scale to the frontal plane of the face. Ear canal dislocation is present if the scale forms an angle with the frontal plane of the face. The extent of dislocation depends on the size of the angle. Although the examination does not determine the exact degree of the dislocation, the successive paired projective distances between the ear canals (or tragions) and the subnasale point in the facial midline

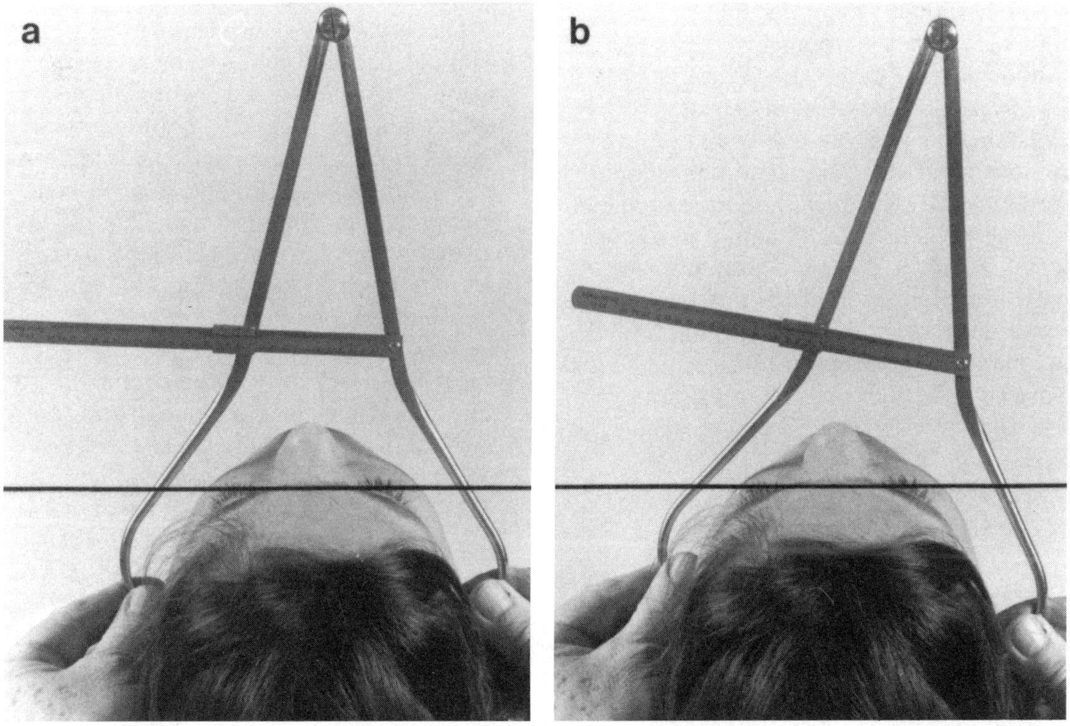

FIG. 2–133. Determination of the ear canal positions in relation to the facial midline using the spreading caliper. (**a**) Scale of the caliper parallel with the frontal plane of the face indicates symmetrical location of the ear canals. (**b**) Oblique position of the scale to the facial plane designates asymmetrical position of the ear canals. The ear canal on the side of the elevated branch of the caliper is located closer to the facial midline. (From ref. 2, with permission.)

will provide an objective proof of it. In paired depth measurements of the face the asymmetrical location of the ear canals and/or the irregular directions of the facial midline are responsible for the facial asymmetry in most of the cases.

Body Weight and Height

Weight was recorded in kilograms, taken on a portable scale (SECA, Germany) with subjects lightly clad and without shoes.

Height was taken at erect stance (barefoot) in centimeters using a medical metric scale (58).

CONCLUSIONS

The anthropometric landmarks and measurements of the head and face have generally been established in healthy subjects. The anthropometric terms used here are identical to the classical ones in *Anthropologie,* edited by Knussmann in 1988 (24), which is regarded as the fourth edition of Martin's anthropological monographs. Over the years some of the classical measuring methods have been modified and new surface landmarks and/or measuring techniques introduced because of the need to determine the extent of disfigurements seen in patients with facial syndromes. My own experience, acquired over 25 years of measuring more than 1000 patients with facial defects, from infants to young adults, prompted me to modify certain anthropometric examination methods. Although some of the new measurements do not precisely determine the extent of defects, and location of new landmarks may cause difficulties, these innovations in technique were required to express quantitatively morphological signs previously described only anthroposcopically. Goldstein (59), author of a longitudinal postnatal growth study of the head and face, declared that "extreme precision of the physical sciences (e.g., gross anatomy) is scarcely possible" but added that "the maximum precision possible must be achieved." In practice, however, the meaning of "maximum" and "possible" depends heavily on the actual morphological conditions.

Landmarks

The precise determination of surface landmarks may cause problems, even in healthy persons. Thus, Davenport (60) acknowledged the difficulties in measuring the interpupillary distance in his longitudinal postnatal growth study of the head and face in North American Caucasians, as follows: "It seemed to be impracticable to measure the distance between the center of the pupils directly, especially in babies." Instead, he measured "the distance between the outer angles of the eyes; also the distance between inner angles, taking half the sum of these two measurements for pupillary distance." In a hyperteloric orbit, this technique may not produce the expected result; nevertheless, Davenport's approach is a good example of the continuing effort to overcome technical problems.

In clinical anthropometry, changes in surface measurements are justified if they fulfill their purpose. An example is the measurement of forehead width. The frontotemporal landmark is located on the most medial point of the curved, laterally concave linea temporalis, a few millimeters above the tails of the eyebrow [Knussmann (52)]. Davenport (60) noted that errors could occur when measuring the growth of minimum frontal width (ft-ft) because "the landmarks are not easy to locate in the relatively undeveloped head." The same difficulties are encountered in patients with facial syndromes, especially those associated with cranial disfigurements caused by craniosynostosis. I tried to overcome them by modifying the classical "minimal frontal width" measurement. In my modification, the blunt pointers of the spreading caliper are pressed firmly to the medial wall of the bony deepenings *lateral* to the temporal line right and left. Firm pressure is necessary because the deepenings vary in depth and are covered by skin, subcutis, and muscle. This modification has been criticized recently (61), but in my experience classical measuring techniques sometimes require change to meet current needs in corrective surgery.

The quantitative determination of the depth and width of the nasal root also presents a problem. Knussmann (24) rightly noted that the depth measurement I introduced (see page 42) determines the depth between the levels of the sellion and endocanthion and not the depth of the nasal root alone. In my experience, the nasal root protrusion contributes much more to the measurement than the difference between the baseline of the nasal root and level of the endocanthion. Thus, the depth measurement is influenced much more by the changes in the depth of the nasal root than by the possible changes in level at the inner commissure of the eye fissures. In Knussmann's monograph (24), the measurement of the nasal root width (*Nasenwurzelbreite, Obere Nasenbreite*) is identical to the intercanthal width (en-en). Davenport (60) admitted the existence of the nasal root as a morphologically independent feature within the intercanthal width; in periods of no growth in the nasal root, the distance between the inner eye angles increased and vice versa. Identification of the landmark at the base of the nasal root (maxillofrontale, mf) is not easy (see page 38). Visually, the baseline of the root is a few millimeters medially from the endocanthion. The need to measure the width of the nasal root became evident when the strikingly wide nasal width did not change after correction of the hyperteloric orbit. Analysis of 54 patients with moderate and severe degrees of hypertelorism showed that the nasal root was abnormally wide in 63% (34 of 54), borderline-wide (mean + 2 SD) in 16.7% (9 of 54), and optimal (mean ± 1 SD) in 20.3% (11 of 54). These findings indicate that the degree of reduction in width of the en-en distance and in nasal root width can be calculated simultaneously (62).

Measurements

Generally, the schematic drawings showing the linear and angular measurements of the head and face serve only as visual guides. Linear measurements not requiring FH, especially those in the facial profile, are taken with the sliding caliper placed in various positions, depending on the inclination of the surface. The figure (Fig. 2-113) showing these measurements illustrates the technique of measurement adjusted to the actual changes in facial contours. A standard schematic drawing cannot show all variations in position of the instrument. Much simpler are the illustrations visualizing the vertical linear measurements of the head and face taken in the standard (FH) position (Fig. 2-61).

According to Weiner and Lourie (42), "the methods, even where these are described in great detail, demand a thorough acquaintance with their underlying principles, as well as the technical details and use of the instruments." The instructions required to become familiar with particular techniques can be provided in special short-term courses.

REFERENCES

1. Farkas LG, Cheung G. Orbital measurements in the presence of epicanthi in healthy North American Caucasians. *Ophthalmologica* 1979;179:309–315.
2. Farkas LG. Anthropometry of the normal and defective ear. *Clin Plast Surg* 1990;17:213–221.
3. Farkas LG. Otoplastic architecture. In: Davis J, ed. *Aesthetic and reconstructive otoplasty.* New York: Springer-Verlag, 1987;13–52.
4. Leiber B. Ohrmuscheldystopie, Ohrmuscheldysplasie und Ohrmuschelmissbildung-Klinische Wertung und Bedeutung als Symptom. *Arch Klin Exp Ohren Nasen Kehlkopfheilkd* 1972;202:51–84.
5. Farkas LG. Vertical location of the ear, assessed by the Leiber test, in healthy North American Caucasians 6–19 years of age. *Arch Otorhinolaryngol* 1978;220:9–13.
6. Farkas LG. Anthropometry of normal and anomalous ears. *Clin Plast Surg* 1978;5:401–412.
7. Farkas LG. Anthropometry of the normal and defective ear. *Clin Plast Surg* 1990;17:213–221.
8. Farkas LG. *Anthropometry of the head and face in medicine.* New York: Elsevier, 1981.
9. Venkatadri G, Farkas LG, Kooiman J. Multipurpose anthropometric facial anglemeter. *Plast Reconstr Surg* 1992;90:507–510.
10. Farkas LG, Deutsch CK. Two new instruments to identify the standard positions of the head and face during anthropometry. *Plast Reconstr Surg* 1982;69:879–880.
11. Solow B, Tallgren A. Postural changes in craniocervical relationships. *Tandlaegebladet* 1971;75:1247–1257.
12. Farkas LG. Basic anthropometric measurements and proportions in various regions of the craniofacial complex. In: Brodsky L, Ritter-Schmidt DH, Holt L, eds. *Craniofacial anomalies: an interdisciplinary approach.* St Louis: Mosby–Year Book, 1992;41–57.
13. Martin R, Saller K. *Lehrbuch der Anthropologie in Systematischer Darstellung.* Stuttgart: Gustav Fischer Verlag, 1957.
14. Hajniš K, Farkas LG. Proposición del examen anthropológico de fissuras de labio (labio leparino) de maxilar y de paladar. *Rev Latinoam Cirur Plast* 1964;8:194–201.
15. Farkas LG, Lindsay WK. The columella in cleft lip and palate anomaly. In: Hueston LT, ed. *Transactions of the fifth international congress of plastic and reconstructive surgery.* Melbourne: Butterworth, 1971;373–381.
16. Farkas LG, Katic MJ, Hreczko TA, Deutsch C, Munro IR. Anthropometric proportions in the upper lip–lower lip–chin area of the lower face in young white adults. *Am J Orthod* 1984;86:52–60.
17. Wood Jones FW. Measurements and landmarks in physical anthropology. *Bernice P. Bishop Museum Bull (Honolulu)* 1929;63: 3–9.
18. Gosman SD. Anthropometric method of facial analysis in orthodontics. *Am J Orthod* 1950;36:749–762.
19. Zrzavy J. *Anatomy for artists* [in Czech.]. Prague: Health Publishing, 1957.
20. Legan HL, Burston CJ. Soft tissue cephalometric analysis for orthognathic surgery. *J Oral Surg* 1980;38:744–751.
21. Burston CJ. The integumental profile. *Am J Orthod* 1958;44:1–25.
22. Peck H, Peck S. A concept of facial esthetics. *Angle Orthod* 1970;40:284–318.
23. Berkowitz S, Cuzz J. Biostereometric analysis of surgically corrected abnormal faces. *Am J Orthod* 1977;72:526–538.
24. Knussmann R, ed. *Anthropologie. Handbuch der vergleichenden Biologie des Menschen.* Stuttgart: Gustav Fischer Verlag, 1988.
25. Scott JH. *Dento-facial development and growth.* Oxford: Pergamon. 1967.
26. Godycki M. *Basic anthropometry* [in Polish]. Warsaw: State Publishing House, 1956.
27. Munro IR, Das SK. Improving results in orbital hypertelorism correction. *Ann Plast Surg* 1979;2:499–507.
28. Farkas LG, Ross RB, Posnick JC, Indech GD. Orbital measurements in 63 hyperteloric patients. Differences between the anthropometric and cephalometric findings. *J Craniomaxillofac Surg* 1989;17:249–254.
29. Hrdlička A. *Anthropometry.* Philadelphia: The Wistar Institute of Anatomy and Biology, 1920.
30. Ashley-Montagu MF. Location of the nasion in the living. *Am J Phys Anthropol* 1935;20:81–93.
31. Hajniš K, Farkas LG. Anthropological record for congenital developmental defects of the face (especially clefts). *Acta Chir Plast* 1969;11:261–267.
32. Daniel RK, Farkas LG. Rhinoplasty: image and reality. *Clin Plast Surg* 1988;15:1–10.
33. Howells WW. The designation of the principal anthropometric landmarks on the head and skull. *Am J Phys Anthropol* 1937;22: 477–494.
34. Farkas LG, Hajniš K, Posnick JC. Anthropometric and anthroposcopic findings of the nasal and facial region in cleft patients before and after primary lip and palate repair. *Cleft Palate Craniofac J* 1993;30:1–12.
35. Farkas LG, Kolar JC, Munro IR. Geography of the nose: a morphometric study. *Aesthetic Plast Surg J* 1986;10:191–223.
36. Farkas LG. Anthropometry of the face in cleft patients. In: Bardach J, Morris HL, eds. *Multidisciplinary management of cleft lip and palate.* Philadelphia: WB Saunders, 1990;474–482.
37. Farkas LG, Hreczko T, Deutsch CK. Objective assessment of standard nostril types—a morphometric study. *Ann Plast Surg* 1983;11:381–389.
38. Ashley-Montagu MF. Location of the porion in the living. *Am J Phys Anthropol* 1939;25:281–295.
39. Martin R. *Lehrbuch der Anthropologie in Systematischer Darstellung mit besonderen Berücksichtigung der anthropologischen Methoden,* 2nd ed. Jena: Gustav Fischer Verlag, 1928.
40. Günther H. Konstitutionelle Anomalien des Augenabstandes und der Intreorbitalbreite. *Virchows Arch* [*Pathol Anat*] 1933;290:373–384.
41. Martin R, Saller K, eds. *Lehrbuch der Anthropologie,* vol 3. Stuttgart: Gustav Fischer Verlag, 1962.
42. Weiner JS, Lourie JA, eds. *Human biology: a guide to field methods.* Oxford: Blackwell Scientific Publishing, 1969.

43. Lindsay WK, Farkas LG. The use of anthropometry in assessing the cleft-lip nose. *Plast Reconstr Surg* 1972;49:268–293.
44. Farkas LG, Lindsay WK. Morphology of the adult face following repair of unilateral cleft lip and palate in childhood. *Plast Reconstr Surg* 1973;52:652–655.
45. Figalová P, Farkas LG. Localisation of auricle by means of anthropometric methods. *Acta Chir Plast* 1968;10:7–14.
46. Farkas LG, Lindsay WK. Morphology of the adult face following repair of bilateral cleft lip and palate in childhood. *Plast Reconstr Surg* 1971;47:25–32.
47. Farkas LG, Katic MJ, Kolar JC, Munro IR. The adult facial profile: relationships between the inclinations of its segments. *Dtsch Z Mund Kiefer Gesichts Chir* 1984;8:182–186.
48. Farkas LG, Hreczko TA, Kolar JC, Munro IR. Vertical and horizontal proportions of the face in young adult North American Caucasians: revision of neoclassical canons. *Plast Reconstr Surg* 1985;75:328–337.
49. Farkas LG, Kolar JC. Anthropometric guidelines in cranio-orbital surgery. *Clin Plast Surg* 1987;14:1–16.
50. Farkas LG, Kolar JC. Anthropometrics and art in the aesthetics of women's faces. *Clin Plast Surg* 1987;14:599–616.
51. Farkas LG, Ngim RCK, Lee ST. The fourth dimension of the face: a preliminary report of growth potential in the face of the Chinese population of Singapore. *Ann Acad Med Singapore* 1988;17:319–327.
52. Knussmann R. Zur Methode nach Integument und nach Morphologischen Merkmalen des Kopfes. *HOMO* 1961;12:193–217.
53. Farkas LG, Sohm P, Kolar JC, Katic MJ, Munro IR. Inclinations of the facial profile: art versus reality. *Plast Reconstr Surg* 1985;75: 509–519.
54. Schwidetzky I. Die metrisch-morphologischen Merkmale und der falische Typus. In: Schwidetzky I, Walter H, eds. *Untersuchungen zur anthropologischen gliederung westfalens.* Münster: Aschendorff, 1967.
55. Bernstein L. Esthetics in rhinoplasty. *Otolaryngol Clin North Am* 1975;8:705–715.
56. Joseph J. *Nasenplastik und sonstige Gesichtsplastik.* Leipzig: C Kabitzsch, 1931.
57. Farkas LG. Growth of normal and reconstructed auricle. In: Tanzer RC, Edgerton MT, eds. *Symposium on reconstruction of the auricle.* St Louis: CV Mosby, 1974;24–31.
58. Farkas LG, Wood MM, Height and weight in Caucasian school children in Central Canada. *Can J Public Health* 1982;73:328–334.
59. Goldstein MS. Changes in dimensions and form of the face and head with age. *Am J Phys Anthropol* 1936;22:37–89.
60. Davenport CB. Post-natal development of the head. *Sci Monthly* 1941;52:197–202.
61. Kolar JC. Methods in anthropometric studies, [Letter]. *Cleft Palate Craniofac J* 1993;30:429–431.
62. Farkas LG, Hreczko T, Katic M. Anthropometric growth patterns in the nasal root and intercanthal widths of a North American Caucasian population. *Anthropologie* 1993;in press.
63. Farkas LG, Posnick JC, Hreczko T. Anthrometry of the head and face in 95 Down syndrome patients. In: Epstein CJ, ed. *The morphogenesis of Down syndrome.* New York: Wiley–Liss, 1991;55.
64. Hreczko T, Farkas LG, Katic M. Clinical significance of age-related changes of the palpebral fissures between age 2 and 18 years in healthy Caucasians. *Acta Chir Plast* 1990;32:194–204.

Chapter 3

Sources of Error in Anthropometry and Anthroposcopy

Leslie G. Farkas

With proper instruments and proper instructions, and unyelding sense of honesty, the worker in anthropometry must develop a habit of minute care and accuracy, until these become automatic.

Aleš Hrdlička
1920

Sources of error in anthropometry of the normal face have been often discussed (1–6). The commonest sources of error are improper identification of landmarks, inadequate use of measuring equipment, and improper measuring technique.

Even if the "extreme precision of the physical sciences is scarcely possible, nor actually necessary, in biological research, especially with reference to observations in gross anatomy" (1), the maximum precision possible must be achieved. Errors in measurement can be greatly reduced by marking the position of landmarks on the skin with a skin-marking pen. We recommend that the measurements be taken at least twice and the average recorded; this is particularly important if the examination requires special positioning of the head and face. The quality of the surface measurements can greatly improve with practice.

ANTHROPOMETRY

Improper Identification of Landmarks

Soft landmarks on the skin surface (e.g., cheilion, labiale superius, or inferius pronasale) or on the edge of the soft facial features (e.g., subalare, alare, the main aural and palpebral landmarks) are readily identifiable in a healthy person's face. However, they are difficult to locate on a deformed face (e.g., pronasale on a flat nasal tip, or subalare on a disfigured alar base). It is difficult to locate the tragion in a poorly developed tragus. Marked epicanthus prevents the use of the endocanthion for measurements.

Bony landmarks used in surface measurements are easily identified by palpation when they are covered only by a thin layer of tissue (e.g., orbitale), although some (e.g., vertex) can be located only when the head is placed in the standard [Frankfort' horizontal (FH)] position. Bony landmarks covered by rich subcutaneous tissue (e.g., gonion), or muscle (e.g., modified frontotemporale) may not be adequately identifiable. Irregular bone surface of the skull, present in many facial syndromes, hampers or precludes accurate identification of the main landmarks of the head.

In congenitally or traumatically deformed faces, the appropriate location of some landmarks may be identifiable by reference to adjacent features. If the landmark is missing, an effort must be made to replace it with new reference points whenever possible. If one part of a paired landmark is missing, the same substitute point must be used for measurements on both sides.

Head

Vertex (*v*) landmark located on the top of the head can be identified only in the standard (FH) position of the head. In a longitudinal follow-up, the same side of the face must be used for establishing the FH. Following orbit surgery, the position of the orbitale point on the lower orbital rim can be changed, thereby affecting the position of the FH. In order to avoid this problem, at the time of the preoperative examination the normal or less deformed side of the face should be chosen for the FH.

Eurion (*eu-eu*) landmarks, which are the most prominent lateral points on each side of the skull of the parietal and temporal bones, may be difficult to locate on a head with an irregular bone surface.

Glabella (*g*) identification is difficult if the area above the nasal root and/or between the eyebrows is flat or even depressed. The landmark can be replaced by a point in midaxis of the face at the level of the upper edge of the eyebrows (if they are thin) or at the midlevel of thick eyebrows.

Frontotemporale (*ft*) points cannot be located if the forehead forms a semicircular curve in the horizontal direction with no palpable flatness laterally, or in a forehead with unilateral protrusion on one side and flatness on the opposite side. In such cases the landmarks can be replaced by points located above the tail of the eyebrows. Forehead disfigurements are frequent in patients with craniosynostosis.

Trichion (*tr*) is the landmark located on the hairline in the midaxis of the face, easy to identify if the hairline is symmetrically curved on both sides and clearly discernible visually. The determination of the location of the landmark is difficult in early childhood when the hairline is indistinct and in adulthood at the first signs of baldness. In facial anomalies the hairline can be irregular or asymmetrical on the sides. Because reliable identification of this landmark can be difficult, the measurement of the forehead height (tr-n) or calvarium height (v-tr) must be taken with due care.

Face

Zygion (*zy*) landmarks can be easily located with a spreading caliper on a well-developed zygomatic arch, at the most lateral points of the arch right and left. Flatness of the arches in first and second branchial syndromes can make the identification of this landmark difficult.

Gonion (*go*) is one of the most difficult anthropometric points to identify. The shape of the angulation formed by the body and ramus of mandible cannot always be determined with absolute certainty by palpation. The identification of the landmarks by palpation is difficult if the angles are covered with mobile thick skin, or if the angles are flat. It is almost impossible to locate the landmarks on a hypoplastic mandible.

Pogonion (*pg*) located at the most protruding point of the soft chin contour is influenced by corresponding formation of the mandible. On a receding and flat chin contour the pogonion landmark cannot be located. In a small number of patients with hypoplastic mandible a "false pogonial protrusion" was found on the chin formed by accumulation of soft subcutaneous tissue only. In these cases the measurement of the inclination of the general profile line (glabella–pogonion line) did not show the true inclination of the profile.

Gnathion (*gn*) palpated in midaxis of the face on the lower border of the mandible should be identified without difficulty, even on a markedly receding lower jaw.

Orbits

Endocanthion (*en*) cannot be located if covered by a *marked* epicanthal skin fold. Our study showed that in healthy Caucasians the mean width of the skin fold covering the endocanthion was 1.5 mm (7).

Exocanthion (*ex*) landmark, even in a normally open eye, cannot always be seen clearly due to a small portion of the upper lid overlapping the outer commissure of the eye fissure. This happens more often in males than in females, and it occurs more frequently in adults than in young children. When measuring the length of the eye fissure or its inclination, it is recommended that the patient be asked to look upwards.

Superciliare (*sci*) Any measurement involving the landmark on the highest point of the eyebrow's upper edge should be carried out on the forehead in a rest position. The smallest contraction of the frontal muscle moves the level of the eyebrows upwards. The landmark cannot be located on trimmed eyebrows.

Palpebrale superius (*ps*) *and palpebrale inferius* (*pi*), in normal conditions, are found in the midaxis of the upper and lower lid, on the rim of the lids. In congenitally deformed or defective lids after injury, the landmarks of the palpebral rims are shifted to the area of the largest distance between the open palpebral rims.

Nose

Nasion (*n*) landmark identification requires a certain amount of experience when the location is carried out by "palpation" of the nasofrontal suture line. After injury of the nasal bones or in noses with bony or silastic implants after corrective surgical measures, the nasion cannot be defined by palpation. A very deep nasofrontal angle may also produce difficulties in definition of the nasion by palpation, especially in very young children. If the nasion point cannot be located, the substitute point will be at the crossing of the nasal root midline and the horizontal that is tangential to the highest points on the superior palpebral sulci (8). Goldstein (1) found that the accuracy of the reading was satisfactory when the physician's applicator was held horizontally at the height of the line that connects the folds in the upper eyelids as the patient's eyes looked straight forward and a marker was placed in the middle, where the applicator crossed the nasal root.

Sellion (*se*), the midpoint [also called "median" (m)] of the nasal root at the level of the eye fissures, is mistakenly regarded as nasion (6). Anatomical studies showed that the nasion landmark in young adults is located higher than the sellion by a mean of 4.9 mm (9).

Subnasale (*sn*) landmark is located in the midpoint of the base line of the columella. Its identification is easy if a sharp angle at the columella/upper lip junction is present. The landmark can be more easily located from the side than from the front of the face. The level of the labial insertion points of alae (subalare right and left) should not be used for determination of the position of the subnasale point. The landmark can be on the same level as the subalare points but also higher or lower than the alar bases (Fig. 3–1). If the nasolabial junction has a curved shape, the subnasale point would be on the bottom of the curve. In such cases Hrdlička (10) recommended that one should "press slightly the branch of the instrument in the line of the septum, until the requisite point is reached." The proper identification of the subnasale point is very important when measuring the height of the nose and the height of the upper lip.

Subnasale' (*sn'*) is an auxiliary landmark which is used when the measurement is taken from the *edges* of the columella, not from the midpoint at its base (sn). Thus, the landmark is used when measuring the width of the columella in its midportion (sn'-sn'). Because the columella is sometimes much wider above its base, the width measurement at that level would not give the real information about the actual size. The same landmark (sn'), but located on the edge of columella base, is used when measuring the real width (sn'-sbal) of the nostril floor between the columella (sn') and the labial insertion of the alar base (sbal). This measurement gives much more precise information about the differences between the nostril floor widths [especially in cleft lips (11)] than does the

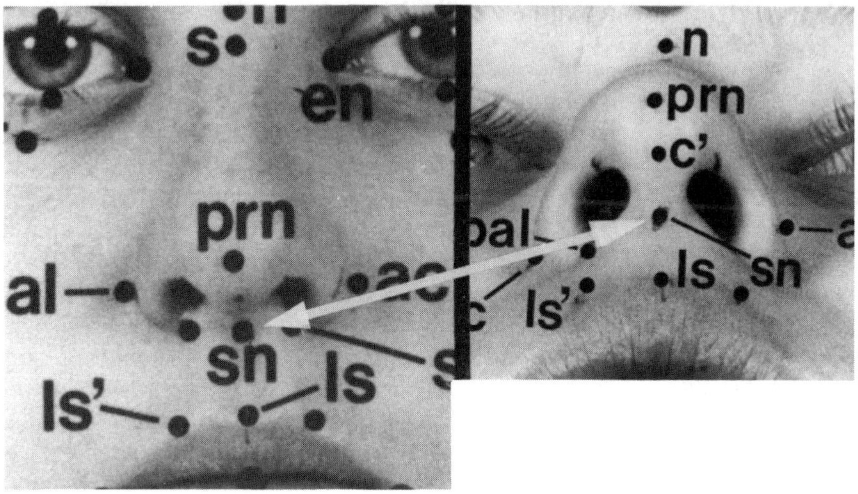

FIG. 3–1. Location of the subnasale point (sn) influenced by the position of the head. The location of the sn point was determined *before photographing* the face in the rest position from the front (**left figure**). In this position the sn is almost level with the labial insertions of the alae (sbal). The **right figure** shows the changes which occur when the head is uptilted to show the base view of the nose. The sn point suddenly appears above the labial insertion points of the alae. This clearly indicates that marking the subnasale point on the print (after photographing) may not register the real position of the landmark.

conventional width measurement, which incorporates half of the columella width at its bottom (sbal-sn).

Pronasale (*prn*) point is located with the head in the rest position. When viewed from the side, this point is easily marked. However, there are great variations of head rest positions in healthy subjects, and even more so in patients with facial anomalies. On a pointed nasal tip the identification of the pronasale point is easy. Very difficult is the determination of prn point on a flat or large curved surface. The columella-length–nasal-tip protrusion index [(sn-c′ × 100)/sn-prn] (12) may offer some help to determine the approximate location of the nasal tip, provided that the columella length was normal. In a bifid nose with uneven tip heights, the nasal protrusion (sn-prn) is measured from the columella base to both the highest and lowest tip points. The measurement closer to the normal will determine the definitive tip of the nose. In the case of a cleft lip nose with a flat nasal tip of uneven contour levels (oblique) the protrusion is measured from the subnasale to the most protruding tip point only.

Alare (*al*) points are more accurately defined, and therefore the width measurements of the soft nose are also more precise if the head is reclined rather than in the rest position.

Alar curvature landmark (*ac*) is the most lateral point on the curved base line between the ala and the face indicating the facial insertion of alae. The points can be seen only from the side of the face. Marking the landmarks on the skin with a pen makes the width measurement between the facial insertion points of alae accurate, provided that the measurement was taken in the reclined position of the head (see Chapter 2, Fig. 2–55).

Maxillofrontale (*mf*) landmarks are approximately at the base of the protruding nasal root separated from the inner commissures of the eye fissures by a narrow skin portion, but located at a higher plane. Before taking the measurement with the sliding caliper, examination of the shape of the slopes of nasal root by palpation helps to locate the landmarks. Such examination is essential on a seemingly flat nasal root which may often be caused only by a thick skin cover.

Lips and Mouth

Stomion (*sto*) is the midpoint of the labial fissure between gently closed lips. If the mouth is dislocated from its central position in the face, the stomion landmark is located laterally from the midaxis of the face. Consequently, the halves of the labial fissure (ch-sto rt, sto-ch lt) become asymmetrical in length. In a dislocated mouth the midpoint of the philtrum [labiale superius (ls)]—usually dislocated as well—is used as the orientation point for determination of the location of the stomion.

Labiale superius laterale (*ls′*) landmark is located laterally from the midpoint [labiale superius (ls)] of the upper vermilion line and vertically below the labial insertion point of the alar base (subalare) on the right and left side. In healthy subjects the right and left ls′ points should be on the same level. In patients with cleft lip after surgery, the lateral labiale superius landmarks are very often on various levels. The asymmetrical position of these landmarks can be one of the causes of asymmetry between the lateral heights of the upper lip.

Cheilion (*ch*) points are found at the right and left commissures of the labial fissure. Precise location can become difficult—especially in older patients—because of a fine shallow ridge in the skin, laterally from the commissures. The process of finding the points can be assisted by examination of the labial fissure corners. This will require the assistance of the patient who will need to open the mouth for repeated inspection, as necessary. During measurement of the mouth width, the lip must be gently closed. In some facial syndromes (e.g., lateral facial dysplasia) the mouth can be dislocated to the side from its central position in face. The deformity is usually associated with (a) ipsolateral deviation of philtral crests, which may cause dislocation of the labiale superius (ls) landmark from its midaxial position, and (b) an oblique direction of the labial fissure. Measurements would reveal asymmetry between the halves of the labial fissure (ch-sto rt and sto-ch lt) and vertical level difference between the cheilion landmarks.

Ears

Preaurale (*pa*) point is located in front of the helix attachment to the head. It cannot be seen on the skin surface. The landmark is palpated in front of the cartilage buried beneath the skin.

Subaurale (*sba*) is the lowest point on the free margin of the ear lobe, easy to define on well-developed and free ear lobes. In an ear with a hypoplastic ear lobe attached to the head, the lowest point (sba) of the ear is at the point of the attachment.

Otobasion inferius (*obi*) is located at the insertion of the ear to the cheek distally. It is not clearly visible if the ear lobe is well-developed; therefore, marking the precise location of the landmark with ink on the skin is strongly recommended when the measurement between the lower insertion of the ear and the chin point is taken. In an ear with hypoplastic ear lobe attached to the cheek (Fig. 3–2), the location of the otobasion inferius point (obi) is identical with the subaurale (sba). A unilaterally attached ear lobe can lead to asymmetry in three directions. First, it affects the length of the ear: The ear length may be smaller or larger than that on the opposite side, depending on the size of the lobule (well-developed or hypoplastic). Visually, that may give the false impression of vertical dislocation of the ear due to the uneven level of the ear lobes. Second, the projective measurements between the lower insertions of the ears and the land-

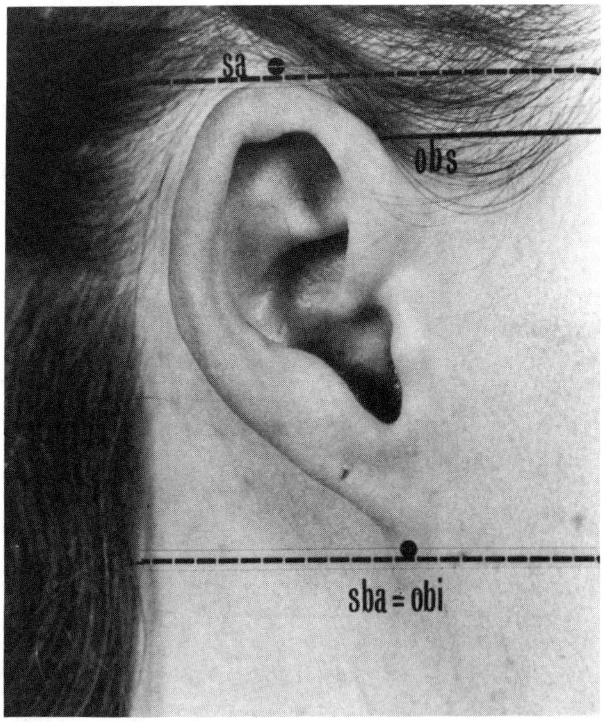

FIG. 3–2. Hypoplastic ear lobe attached to the face. The lower edge of the auricle [subalare (sba)] and the lower insertion of the ear [otobasion inferius (obi)] share the same spot.

marks located in the midaxis of the face would be asymmetrical. The greatest difference would be seen in the measurement between the lower attachment of the ears and the chin point (gnathion). The ear with lobule attachment may be closer or further from the chin point, depending on the size of the lobule. The third asymmetry will occur in the inclination of the longitudinal axis of the auricle caused by the change of the lower terminal point of the axis, from the midpoint of the ear lobe to the point at the lobe attachment.

Porion (*po*) landmark is located on the highest point of the upper rim of the external meatus. In routine anthropometric praxis we use the landmark only when measuring the auricular height of the head [vertex-porion (v-po)] or at determination of the position of the ear canal in relation to the most posterior point of the head [opistocranion-porion (op-po)].

If the external auditory meatus is atretic, the missing porion landmark can be replaced by the tragion, if present. Both horizontal and oblique measurements between the porion and the facial midpoints are slightly longer than those between the tragion and the facial midpoints, but the biporion diameter differs minimally from the bitragion diameter (2).

Tragion (*t*) landmark, the "classical" surrogate for the porion point, is positioned on the same level as the upper rim of the external meatus for the great majority of the healthy population. In some congenital anomalies of the face the tragion point may be shifted proximally or distally from the ear canal level (see Chapter 2, Fig. 2–30). There are great variations in the size and shape of the tragus, ranging from the large or bifid formation to the hypoplastic or almost flat tragus. Missing porion and tragion points can be replaced by a landmark at the temporomandibular joint [condylion laterale (cdl)], but this increases the likelihood of inaccurate measurements. The diameter between two condylions must be determined from points projected on the skin; if the spreading caliper is pressed on the condylions, the diameter will be short by a mean of 5 mm. Standard measurement of facial depth in the mandibular region (gn-t) is slightly shorter when the condylion (gn-cdl) is substituted for the tragion. Test measurements in 35 young adult men and 35 young adult women showed that in each sex the average bicondylion diameter was 9 mm shorter than the bitragion diameter.

Ear damage may extend to the temporomandibular joint, obliterating the above landmarks (po, t, cdl) and necessitating substitution of the mastoidal landmark (ms). The classic mastoidal point is on the most protruding spot of the mastoid process, level with the center of the auditory meatus (10). Testing the bimastoidal diameter in the 70 young adults mentioned above revealed that the bitragion diameter was greater than the bimastoidal diameter by an average of 9.8 mm in men and 8.5 mm in women (5). In patients with severe microtia the mastoid is flat and even depressed, making the determination of the landmark problematic.

Problems with Measuring Tools

Standard measurements require the use of standard instruments. Errors in measurement may occur when improper measuring tools are used or when standard instruments are used improperly. The scales of anthropometric measuring tools are millimeters or degrees. Normally, measurements with fractions are rounded up to the next higher figure.

It is evident that accurate measurement requires correct use of the instrument and knowledge of the peculiarities of the landmark. The examiner must be familiar with (a) the areas in which the tip of the instrument used should be pressed to the bony surface to obtain correct measurements and (b) the areas where the instrument barely touches the skin surface at measurement.

Various instruments permit different degrees of accuracy. The most accurate is the sliding caliper: The pointed arms ensure that errors in measurement between well-defined and clearly visible points should not exceed a fraction of a millimeter. With a spreading caliper, errors in measurement may be up to 3 mm, depending on the region and the examiner's experience. The most difficult for a beginner is the use of a soft measuring tape. The tape must follow the contours of the surface being measured. Pressing the tape into the tissue shortens the measurement. Conversely, when measuring the circumference of the head, the tape must be tightened around the head to eliminate the effect of the hair.

Measuring tape made of cloth in contrast to plastic tape has the great advantage that it can precisely follow the variously curved skin surface of the face, lips, and nose; thus, the measurements are more precise. The use of a transparent ruler by amateur anthropometrists is still very much favored. The ruler can give correct measurements if used between two points located on the same plane, with no projective obstacle between the landmarks. The use of a ruler for measuring a projective distance between two points located in the same plane (e.g., en-en) but separated by a protruding formation (e.g., in this case, by the nasal root) or on a curved surface relief (e.g., the curved protrusion of the lips between the cheilion landmarks when measuring the width of the mouth) offers questionable results. Relatively short projective distances should be measured with a sliding caliper.

The soft measuring tape is reserved for measuring surface arcs (tangential measurements), but it would be a poor substitute for the sliding caliper. The usefulness of the spreading caliper lies in its ability to (a) obtain data about projective distances (diameters) between two distant points of the head or face (e.g., eu-eu, g-op, zy-zy, go-go) or (b) measure the depth of the face (e.g., g-t, n-t, sn-t, gn-t). However, it is definitely less accurate than the sliding caliper, the pointed branches of which give more precise readings than the blunt edges of the spreading caliper, even when measuring between well-defined soft landmarks (e.g., en-en, ch-ch). Again, the blunt edges of the spreading caliper allow a gentle pressure to be applied when measurements between bony landmarks are being taken (e.g., ft-ft, go-go).

When measuring the angles in the facial profile line, the multipurpose facial angle meter developed in our laboratory (13) offers more precise values than does a transparent protractor (see Chapter 2).

The use of improper instruments in anthropometrically oriented studies may produce conflicting results, making the worth of scientific information doubtful.

Improper Measuring Technique

Errors can result from failure to fix the head in the required position during measurement or by improperly taking the measurements.

Head Position

Because the subject's head tends to return to the rest position during examination, the standard head position (FH), if required, must be checked before each measurement. Failure to maintain the appropriate position of the head may result in errors of 5–10 degrees of inclination. The FH is about 5 degrees lower than the rest position in healthy normal persons, but the difference is noticeably greater in patients who have a cranial or facial anomaly (5).

Positioning of the head in the FH during measurement may appear unnatural or be very uncomfortable for a patient whose face is markedly deformed. In such cases, the examiner can assess *inclinations* with the patient's head in the rest position but must record (in degrees) the differences between resting and FH positions. Some projective *linear measurements* (e.g., v-gn, v-po, op-po) must always be taken with the head in the FH.

In a patient with asymmetric FHs, the difference between the two horizontals must be assessed with an angle meter, and measurements requiring the standard position of the head have to be taken from the FH closer to the head's resting position. In longitudinal studies, whichever FH is used must be marked in the chart, and the same FH must be used throughout the follow-up. In our experience, the left FH is often closer to the rest position than the right FH.

Some measurements (e.g., those involving the vertex and opisthocranion landmarks, and all inclinations) require positioning of the head in two planes simultaneously: in FH and with the profile in the vertical. In an uncooperative subject, it is difficult to achieve reliable measurements without an experienced assistant to hold the head in the required position.

Facial Midline (Vertical Orientation of the Facial Profile)

Vertical positioning of the facial profile line is easy in a healthy face which has the midfacial landmarks (nasion, subnasale, gnathion) in one vertical line. However, identification is difficult if the facial midpoints are not located in the vertical axis of the face. The facial midline is the borderline between the right and left half of the face; errors can be made in judging facial symmetry if the midline of the face is not properly determined. The nasion point is the most stable of the three main facial midpoints and can be used in most faces with a curved profile line. Any dislocation of the main facial midpoints from the established midline must be recorded in the chart. If only one facial midpoint is available, vertical orientation of the face requires the help of symmetrical paired landmarks (e.g., endocanthions) or facial features (e.g., nasal alae, commissures of the mouth). When the midline is vertical, it is possible to determine the deviation of features from the vertical and the inclination from the horizontal.

If the ideal facial midline cannot be determined because of severe axial facial deformity, for measurements requiring this midline, an arbitrary, visually acceptable line is chosen and maintained in the vertical with the help of horizontal paired facial features.

Problems When Executing the Measurements

Head

The measurement of the head circumference should be taken in the plane defined by the glabella and opisthocranion. Thus, forehead protrusion and/or increased head width above the glabella–opisthocranion level may not affect the size of the circumference. Additional circumference measurements can be taken in this area if required.

Similarly, in addition to the true head length taken between the glabella and the opisthocranion, an increased length between the area of the forehead protrusion and the most posterior head point should also be recorded. If the frontal surface of the forehead is concave or convex, the inclination of the forehead is determined from a substituting general line. If the forehead has two distinctively different inclinations (usually an anteriorly tilted portion in the lower half and a posteriorly tilted surface in the upper half), both inclinations can be recorded in the chart.

Difficulties measuring the conventional width of the forehead (ft-ft) experienced by Davenport (14) in his longitudinal cranial growth study in a healthy population may cause certain problems also in children with various forms of craniosynostosis (see Chapter 2, section entitled "Conclusions"), even when using the modified method.

Face

The heights of the upper face (n-sto) and mandible (sto-gn) become smaller when taken while the patient's mouth is open. Likewise, an open mouth (while crying or smiling) increases the negative inclination of the mandible and the height of the lower face (sn-gn).

Orbits

The eye of the examiner should be level with the eyes of the examinee, with head of the subject in the rest position, when length of eye fissures (or their height or inclination) is measured. Measurements involving the exocanthion landmark (biocular width, eye fissure length, inclination of the eye fissure) will be more precise when taken with upward-looking eyes. When measuring the inclination of the eye fissures, the facial profile line must be kept vertical. Vertical measurements of the orbits, involving the superciliare, must be taken in the natural position of the eyebrows. In a wrinkled forehead the eyebrows move upwards, causing a false increase in the combined eyebrow and orbit height.

Nose

Where the nasal bridge has an uneven surface, the length is determined by measuring the projective distance between the nasion and pronasale. For measuring the inclination on a concave or convex bridge surface, the measurement is taken from a general line. Similarly, a general line should replace the concave or convex contours of the columella when its length or inclination is ascertained. Downward dislocation of one of the labial insertion points (sbal) of the alar base can be established directly by measuring the actual difference between the two subalare levels. Indirectly, the same result can be obtained by measuring the lateral heights of the upper lip (sbal-ls′), provided that the lateral labiale superius (ls′) points of the upper vermilion line are on the same level.

Lips and Mouth

When examining the lips and mouth, closed lips are a sine qua non for correct measuring. With open mouth the heights of the upper (sn-sto) and lower (sto-sl) lips are reduced and the projective horizontal distance between the commissures of the mouth (ch-ch) is shortened. The heights of the upper and lower lip vermilion may seem larger due to the poorly discernible borderline of the oral mucosa. Gently closed lips are also essential for correct measurement of the inclinations of the Leiber line (g-ls), the mandible (li-pg), and both the upper and lower lips. The degree of inclination of an oblique labial fissure cannot be established if the mouth is open. The correctness of the height measurements of the upper and lower lips is not influenced by concave or convex lip contours because projective measurements are taken between the subnasale and stomion, respectively, between the stomion and sublabiale. Ascertainment of the inclination on concave or convex lip contours is carried out from a line following their general contours.

Ears

It is essential to determine the location of the landmarks of the ear, especially that of the tragion, otobasion inferius, supraaurale and subaurale, because not only the sizing of the ears depends on them but also the depth measurements of the face and the position of the auricle in relation to the facial midline and to the chin. Equally important is to establish the mutual position of the ears, both vertically and horizontally, in relation to the landmarks in the midaxis of the face.

Asymmetries in the depth measurements of the face are caused mainly by asymmetrical positioning of the ear canals on the head. Difficulties are rising when the abnormally low position of the ear should be determined. While the diagnosis of an extremely low-set ear does not create problems, the mild or moderate degrees of abnormality cannot be proven by one single surface measurement. Simultaneous examination of the ear canal position on the head, both anthropometrically and cephalometrically, may solve this problem.

Comments on the Methods of Indirectly Obtained Linear Measurements

The basic idea of the indirectly obtained measurement is that with the help of two measurements a third can be calculated. This formula was followed when the height of the mandible (sto-gn) was calculated by subtracting the height of the upper face (n-sto) from the height of the face (n-gn) (14).

In the orbits, the width of the eye fissures were obtained by taking half of the difference between the biocular (ex-ex) and intercanthal (en-en) width measurements (14). The interpupillary distance (p-p) by the same investigator was calculated by taking half of the sum of the intercanthal (en-en) and biocular (ex-ex) widths.

The described methods would be acceptable only in case that the landmarks of the measurements were located at the same plane. In congenitally damaged faces we can hardly find nasion, stomion, and gnathion points at the same frontal plane. Similarly, in the orbits it would be difficult to find one patient with endocanthions and exocanthions located at the same frontal plane. An additional problem is that the symmetry or asymmetry of the length of the eye fissures could not be detected.

Investigators measuring the upper or lower lip height must be cautioned not to use the indirect method for obtaining a missing component in the lips: The landmarks in the upper lip—namely, the subnasale, the labiale superius, and the stomion—are not at the same frontal plane, similar to what is reported for the stomion, the labiale inferius, and the sublabiale points of the lower lip. Thus, from two known segments the size of the third one cannot be calculated in either lip.

ANTHROPOSCOPY

Successive visual and quantitative evaluation of patients has revealed the subjectivity of anthroposcopy. Impressions based on visual perception of the quality of the craniofacial regions require anthropometric confirmation. Errors in anthroposcopy are greater than those of anthropometry. Visual evaluation alone does not provide enough information about the extent of a defect. Errors result from the method of examination (e.g., incorrect positioning of the patient's head, wrong angle of view of the examiner's eyes), poor lighting, or the examiner's failings (e.g., poor aesthetic judgment, little talent for geometry, insufficient medical experience).

The main problems in anthroposcopy are caused by the three-dimensional nature of the face and especially by the impulse to make estimates based on subconscious comparisons of the sizes of various facial features (Fig. 3-3) (15).

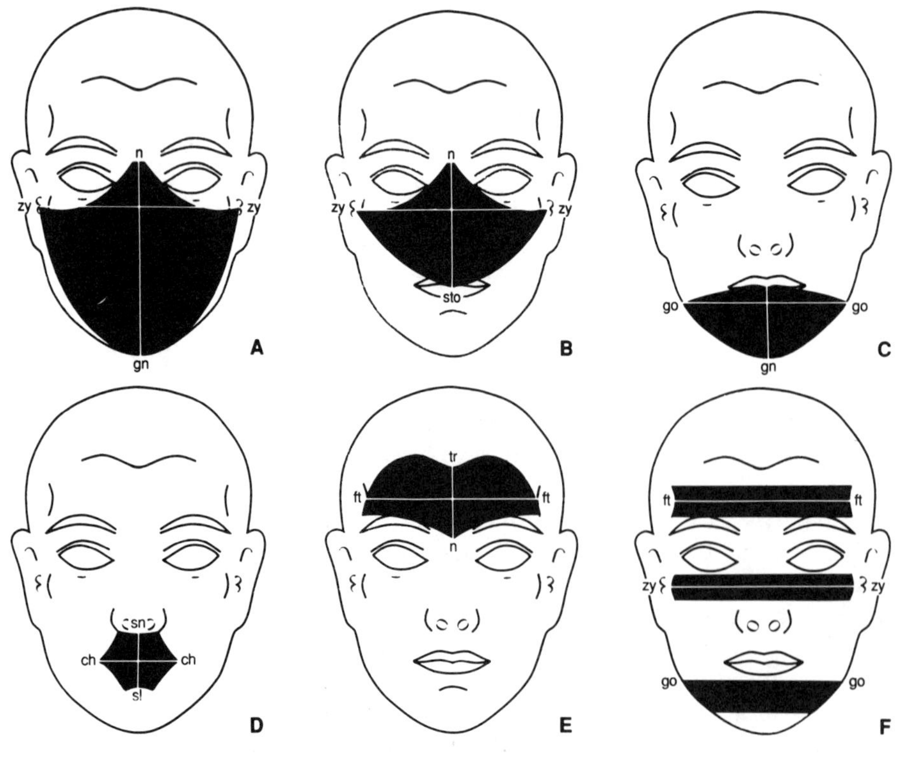

FIG. 3–3. When visually estimating the size or individual areas of the face from the front, our judgment would be subjected to the quality of relationship between the width and height in each of the areas. (From ref. 15, with permission.)

Problem of Various Planes

In the *frontal view* the projective distances appear short if their terminal landmarks are located on different planes (e.g., one point is closer and the other more distant to the examiner's eyes). This applies for both horizontal and perpendicular linear distances.

In the *lateral view* our visual judgment of the individual profile segments is greatly influenced by the inclination of the segment as well as by the quality of the angles separating the individual sections of the profile. A vertical surface makes the forehead high, compared to the forehead of the same size but tilted dorsally. A flat nasal bridge appears longer than a protruded one (Fig. 3–4). The nose appears longer if its nasofrontal angle is obtuse, whereas it appears shorter in the presence of a deep, acute nasofrontal angle. A columella inclination greater than 90 degrees makes the nose larger, although only the nasal bridge (n-prn) may be increased and not the height of the nose (n-sn) (Fig. 3–5). A small inclination of the columella associated with a large nasolabial angle creates the impression of a "small" nose, although only the length of the bridge of the nose and not the height of the nose might has been decreased in size.

Visual assessment of *depth* (assessed in the *sagittal* direction) is significantly influenced by the effects of the light on the three-dimensional surface relief of the face. Dark areas appear narrow and/or deep, whereas lighter spots appear wide and/or shallow (e.g., in the nasal root area) (Fig. 3–6).

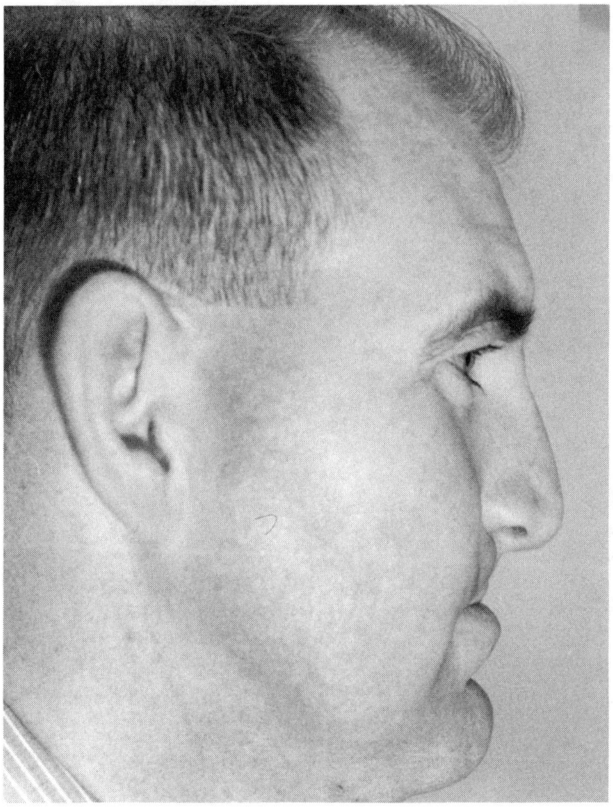

FIG. 3–4. A severely subnormal inclination (−18 degrees) of the nasal bridge in an adult face. The flat nasal bridge creates the visual impression of a long nasal bridge.

Anthroposcopic Illusion in the Face and Head

Head

From the front, our impression of the width of the head (forehead) is influenced by the width of the face: In the presence of a narrow face, even the normal forehead appears wide. The visual judgment of the height of the head (forehead) depends on (a) the position of the head during the inspection and (b) the inclination of the anterior surface of the forehead. A vertical or protruding forehead looks high. In a greatly protruding forehead the supraorbital rims appear depressed. Posterior tilt of the forehead reduces its projected height—for a viewer from the front. The seemingly low position of the ears can produce the impression of a high head.

Face

Most of the mistakes made in visual judgment occur during frontal examination of the face. Of the two main horizontal diameters—the face width (zy-zy) and the width of the mandible (go-go)—only the extremely narrow or strikingly wide faces would be judged correctly (as abnormalities). In *frontal view,* the vertical midline distances (n-sto, sn-gn, sto-gn) can be mistakenly defined by anthroposcopy, caused by the various inclinations of the individual sections of the facial profile. Thus in the dish-like face, the concave upper face may appear vertically shorter (n-sto) than in reality. Visually, the lower face height (sn-gn) would appear small in the presence of a receding mandible. In contrast, a protruding mandible would markedly contribute to the impression of a long face. Inspection of the facial profile *from the side* of the face can correct some of these false frontal impressions. The estimate of the size of the individual section of the facial profile is influenced also by the size of the neighboring part: The size of the face is judged in relation to the height of the head (or forehead), the lower face is compared to the entire face, and chin is compared to the lower face (15) (Fig. 3–7).

Especially in the face, the results of the visual judgment are markedly influenced by the position of the head of the examinee. In a high rest position, even a subnormal (flat) nose or face inclination may appear close to normal.

Orbits

Anthroposcopy of the soft-tissue orbits is influenced by the size and shape of the nasal root, the sagittal level

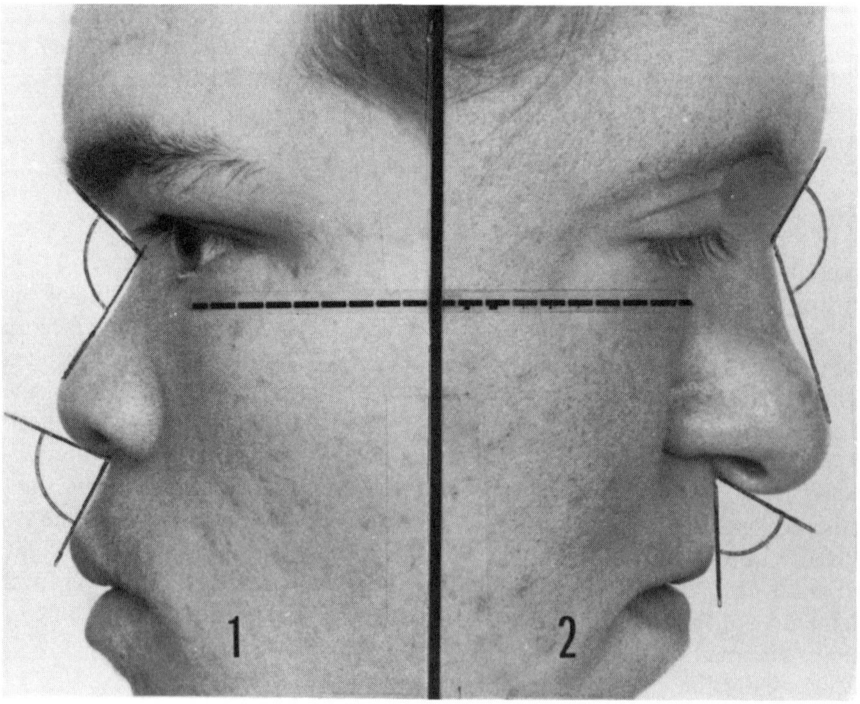

FIG. 3–5. The influence of the angular measurements (angles and inclinations) on the quality of visual perception of the nose. **1:** The nasofrontal angle is 100 degrees, bridge inclination is 40 degrees, columella inclination is 65 degrees, and the nasolabial angle is 86 degrees. **2:** The nasofrontal angle is 130 degrees, inclination of the bridge is 20 degrees, columella inclination is 120 degrees, and the nasolabial angle is 60 degrees. The nose on the right side appears larger because of the smaller nasal bridge inclination, combined with a larger columella inclination and a smaller nasolabial angle.

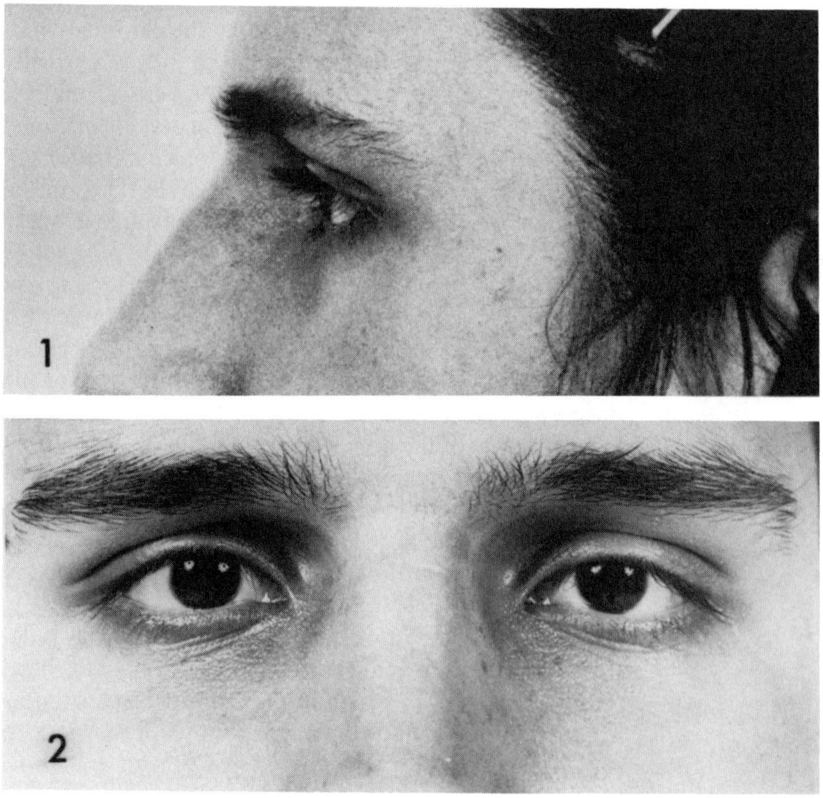

FIG. 3–6. This healthy young man has a borderline-large nasal root depth (**1:** en-se sag 19 mm) and a borderline-large intercanthal width (**2:** en-en 38 mm). Visually, in the presence of the seemingly deep nasal root the space between the orbits appears relatively narrow.

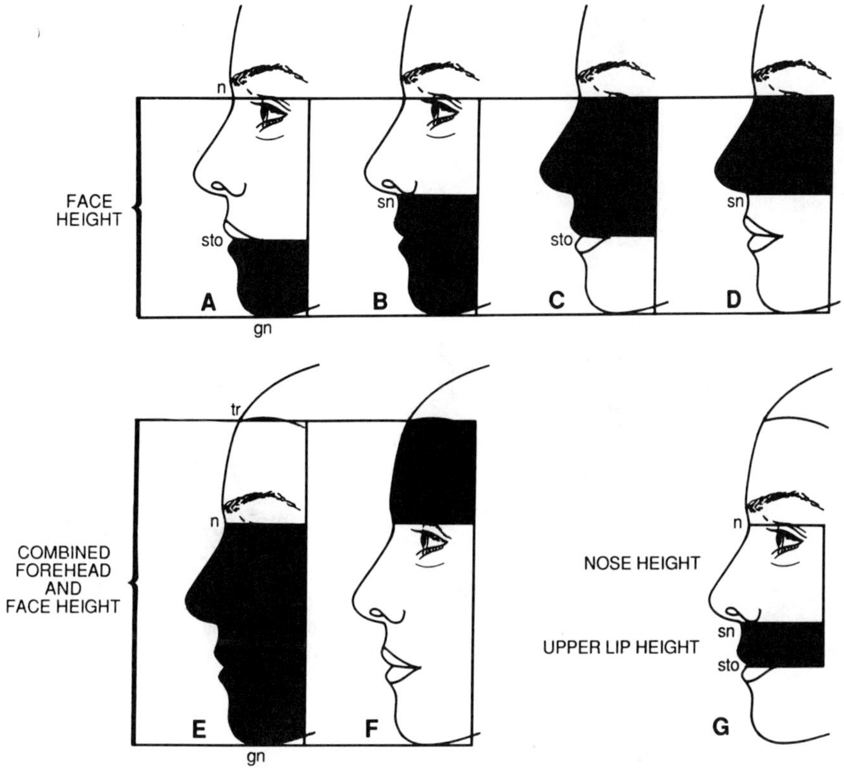

FIG. 3–7. Main portions of the head and face seen from the side. In visual examination, judgment about size is influenced by the size of the adjacent areas. (From ref. 15, with permission.)

difference between the inner (en) and outer (ex) commissures, and the inclination of the eye fissures (16). Additional problems arise when a skin fold (epicanthus) blocks the view of one or both endocanthions.

A hardly discernible elevation of the nasal root greatly contributes to the illusion of a large intercanthal width (en-en). In contrast, on the sides of a well-protruding nasal root, shady spots at the base of the slopes create the impression of a small intercanthal width. Near a wide interorbital space the length of the eye fissures appears small. The appearance of short eye fissures can also be created by marked inclination of the eye fissures (17). In our experience a horizontal eye fissure direction can cause the impression of antimongoloid inclination. Major skin folds (epicanthi) covering the endocanthions, seen often in the presence of shallow nasal roots, share partly the responsibility for the impression of an abnormally large intercanthal width. These skin folds do contribute to certain shortening of the eye fissures, as found in 28 of 94 patients with Down syndrome, but without making them abnormal (18). Our studies showed that in abnormally short eye fissures combined with marked epicanthi the surgical correction of the skin folds would not restore the normality (*unpublished data*). Epicanthi covering the inner commissures of the eye fissures greatly influence the inclination of their longitudinal axis by a shift of the medial point of the axis from the endocanthion to a lower positioned point at the crossing of the epicanthal skin fold with the lower lid rim (18).

The visual impression of the eye fissure lengths is strongly influenced by the differences between the anterior–posterior (sagittal) levels of the endocanthions (en-en) and exocanthions (ex-ex). An increased sagittal distance between the two lines creates the visual illusion of short eye fissures (16).

Nose

Visual assessment of the height of the nose (n-sn) in lateral view is influenced by two profile angles: proximally by the nasofrontal and distally by the nasolabial angles, as well as by inclination and length of the nasal bridge (n-prn). A flat nasofrontal angle (Fig. 3–8) combined with a deep nasal root (en-se sag) would show a "long" nose, especially in the presence of an acute nasolabial angle. In contrast, the nose between a widely open nasolabial and a deep nasofrontal angle would appear "short," more so in combination with a great nasal bridge inclination. In describing the size of the nose, the height of the nose (n-sn) is often confused with the length of its bridge (n-prn) (6). Thus, a nose with an overhanging tip portion is proclaimed to be "long." Errors in visual judgment can also occur when the estimate of the size of the nose is based on its relationship to the size of the surrounding facial features. The seemingly large nose in patients with Treacher Collins syndrome may be judged as abnormal, but has proved to be the only

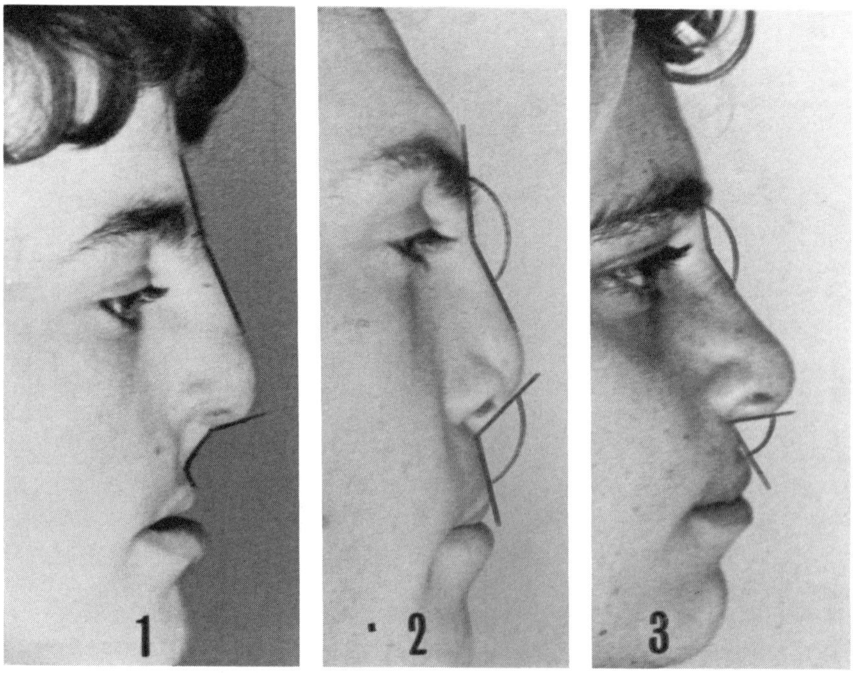

FIG. 3–8. A variety of abnormally flat nasofrontal angles (170-160-150 deg) associated with (1) abnormally wide open nasolabial angle (140 deg), (2) borderline large size angle (120 deg), and (3) a subnormal acute angle (60 deg). The visual judgment of the bridge length is greatly influenced by the size of the nasolabial angle.

normal feature in the facial profile when measured (19) (Fig. 3–9).

The inclination of the nasal bridge alone can be a factor influencing our visual judgment: The nasal bridge appears longer in a flat nose than in a nose with great protrusion.

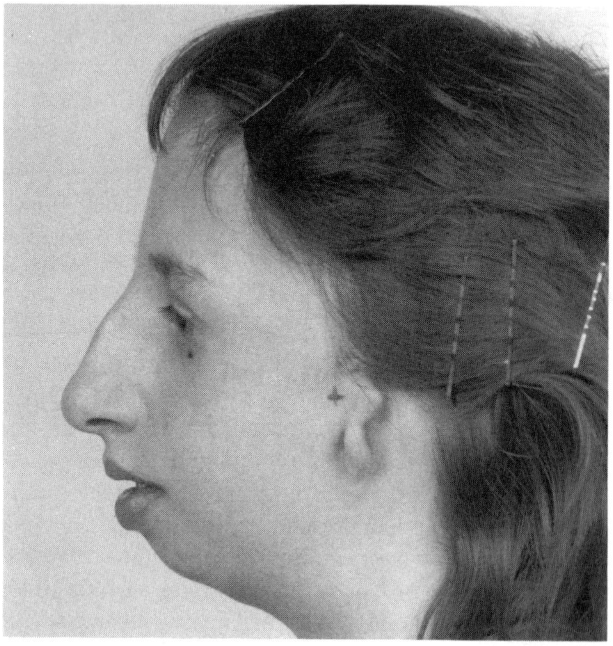

FIG. 3–9. A young girl with Treacher Collins syndrome. Anthroposcopically the nose appears large, but anthropometrically it is of optimal size.

Lips and Mouth

From the frontal view, the visual impression of the height of the upper lip depends on the size of the nasolabial angle and the position of its skin surface. In the presence of a curved columella–upper lip junction (nasolabial angle) the upper lip appears short. A protruding or receding lip contour projection shows a seemingly short upper lip. Vertical upper lip surface combined with an obtuse nasolabial angle makes the lip "longer." Likewise, at the vertical position or with a small inclination the lower lip appears high, accentuated even more by a flat labiomental angle. Greater protrusion of the lower lip emphasizes the height of the vermilion, while the skin portion of the lip is scarcely discernible.

Ears

During visual examination a primary requirement is to maintain an identical level between the eyes of the examiner and those of the examinee. When the position of the ears on the head or the ear sizes are judged, the profile line of the face must be kept in the vertical. Shortcomings of visual evaluation may lead to erroneous estimates of the size or location of the auricles and failure to see the connections between the displaced ears and asymmetries in depth measurements of the face.

Ear Size

In the frontal view, the differences between the upper levels (at the tip of the ears) and the lower levels (at the free margin of the lobules) can give some indication of the length of the ears. The judgment of auricular size by relating the ear's upper and lower edges to the adjacent eyebrow and the nasal ala levels had proved to be crude and unreliable, differing in sexes and subject to change with age (20).

The old belief which suggests that nose height can be used as a guide in estimating ear size is equally unreliable (21). A narrow ear can appear longer than it really is. An auricle with greater inclination is seen as shorter. A characteristic which further complicates size comparison of ear and face height should be noted, because it can give the false impression of the ear being disproportionately small (18). Studies show that ear length in early childhood is more developed than length of face [at ages 1 year and 18 years, both sexes: 76.4% and 67.8% mean values, respectively (22)], and therefore appears disproportionately large. At 1 year of age the ear occupies more than half the face height, whereas at 18 years it occupies less than half of it (20).

The width of the auricle, as seen from the front, may give some information about the degree of ear protrusion from the head. From the side, a flat anterior relief creates the illusion of a wide ear, whereas a deep concha with a well-protruding antihelix, emphasized by a deep scapha, makes the auricle appear narrower.

Inclination of the Longitudinal Axis

Visually only the extreme directions of the longitudinal axis can be noticed, even if the head is in the rest position. Inclination of the ears should not be judged according to the inclination of the nasal bridge, as claimed by the neoclassical canon. For the great majority of a healthy population the inclination of the nasal bridge is larger than that of the ear (21).

Locating the Ears on the Head

Determination of the mutual position of the ear canals on the head in vertical directions, as well as in relation to the midaxis of the face, is essential for appreciation of the facial depth measurements at the orbital (t-g, and t-ex), maxillary (t-sn and t-ch), and mandibular (t-gn) regions. However, visual determination of the symmetry or asymmetry in the horizontal or vertical position of the ear canals (or tragions) is highly unreliable. What we see is the difference in the levels of ear lobes. It may be caused by asymmetry between the ear canal levels, but also by different ear lengths, sometimes resulting from unilateral attachment of the ear lobe to the cheek. A lower located auricle is not necessarily low-set. An uneven level of tragion was found in 13.4% (160 of 1197) of healthy North American Caucasians between 6 and 18 years of age, whereas low-set ears were seen only in 1% (14 of 1432) (5). Asymmetrical positioning of the auricle was observed with much greater frequency in patients with facial syndromes. Thus, in Down syndrome, ear level asymmetry was reported in 22.3% (21 of 94) (18), but none of the ears were low-set. In various forms of craniosynostosis the percentage of ear level asymmetry increased to 79.4% (54 of 68) and in hemifacial microsomia patients reached 90.6% (126 of 139) (23). Examination of ear location on the head in North American Caucasians with the help of the Leiber test (24) revealed age-related level changes within the normal area between the lower lid and nasal ala (25): In about two-thirds of 6- to 9-year-old males and females the most common location was in the lower zone (with ear canal level close to the nasal ala level). With age (in both sexes), the location "moved up", making the middle zone dominant in young adults. Visual examination showed neither dystopia nor asymmetry in ear locations in these healthy subjects. However, by visual examination the position of the ear(s) in the lower zone of the normal area in patients with facial syndromes may easily give the impression of a low-set ear in the presence of a small or severely receding

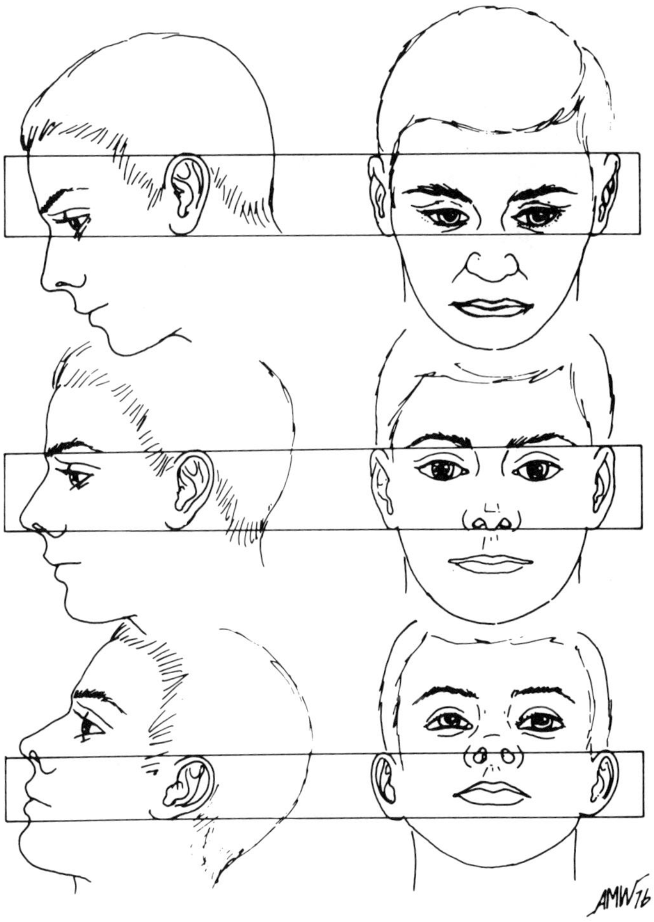

FIG. 3-10. Visual impression of ear location dependent on position of the head. The **middle drawing** shows the head in standard (FH) position.

chin, high head, large inclination of the auricle, hypoplasia of the nose, highly positioned eyebrows, and an uptilted head (26) (Fig. 3–10).

Anthroposcopic determination of abnormal ear positions can be accepted only if a *microtic ear with atretic ear canal* is seen at a very low level of the face.

REFERENCES

1. Goldstein MS. Changes in dimensions and form of the face and head with age. *Am J Phys Anthropol* 1936;22:37–89.
2. Godycki M. *Basic anthropometry* [in Polish]. Warsaw: State Publishing House, 1956.
3. Alexander M, Laubach LL. *Anthropometry of the human ear (a photogrammetric study of USAF flight personnel)*. Report AMRL-TR-67-203. OH: US Air Force Aerospace Medical Research Laboratories, Wright-Patterson Air Force Base, January 1968, 1–28.
4. Jamison PL, Zegura SL. A univariate and multivariate examination of measurement error in anthropometry. *Am J Phys Anthropol* 1974;40:197–203.
5. Farkas LG. *Anthropometry of the head and face in medicine.* New York: Elsevier, 1981.
6. Knussmann R. ed. *Anthropologie. Handbuch der vergleichenden Biologie des Menschen.* Stuttgart: Gustav Fischer Verlag, 1988.
7. Farkas LG, Cheung G. Orbital measurements in the presence of epicanthi in healthy North American Caucasians. *Ophthalmologica* 1979;179:309–315.
8. Ashley-Montagu MF. Location of the nasion in the living. *Am J Phys Anthropol* 1935;20:81–93.
9. Daniel RK, Farkas LG. Rhinoplasty: image and reality. *Clin Plast Surg* 1988;15:1–10.
10. Hrdlička A. *Anthropometry.* Philadelphia: Wistar Institute of Anatomy and Biology, 1920.
11. Farkas LG. Anthropometry of the face in cleft patients. In: Bardach J, Morris HL, eds. *Multidisciplinary management of cleft lip and palate.* Philadelphia: WB Saunders, 1990:474–482.
12. Farkas LG, Munro IR., eds. *Anthropometric facial proportions in medicine.* Springfield, IL: Charles C Thomas, 1987.
13. Venkatadri G, Farkas LG, Kooiman J. Multipurpose anthropometric facial anglemeter. *Plast Reconstr Surg* 1992;90:507–510.
14. Davenport CB. *Post-natal development of the head.* Sci Monthly 1941;52:197–202.
15. Farkas LG, Kolar JC. Anthropometrics and art in the aesthetics of women's faces. *Clin Plast Surg* 1987;14:599–616.
16. Farkas LG, Posnick JC, Winemaker MJ. Orbital protrusion index in Treacher–Collins syndrome: a tool for determining the degree of soft-tissue damage. *Dtsch Z Mund Kiefer Gesichts Chir* 1989;13: 429–432.
17. Farkas LG, Munro IR, Kolar JC. Abnormal measurements and disproportions in the face of Down's syndrome patients: preliminary report of an anthropometric study. *Plast Reconstr Surg* 1985;75:159–167.
18. Farkas LG, Posnick JC, Hreczko T. Anthropometry of the head and face in 95 Down syndrome patients. In: Epstein CJ, ed. *The morphogenesis of Down syndrome.* New York: Wiley–Liss, 1991;53–97.
19. Farkas LG, Posnick JD. Detailed morphometry of the nose in patients with Treacher Collins syndrome. *Ann Plast Surg* 1989;22: 211–219.
20. Farkas LG. Otoplastic architecture. In: Davis J, ed. *Aesthetic and reconstructive otoplasty.* New York: Springer-Verlag. 1987;13–52.
21. Farkas LG, Hreczko TA, Kolar JC, Munro IR. Vertical and horizontal proportions of the face in young adult North American Caucasians: revision of the neoclassical canons. *Plast Reconstr Surg* 1985;75:328–337.
22. Farkas LG, Posnick JC, Hreczko TM. Anthropometric growth study of the ear. *Cleft Palate Craniofac J* 1992;29:324–329.
23. Farkas LG. Basic anthropometric measurements and proportions in various regions of the craniofacial complex. In: Brodsky L, Hue Ritter-Schmidt D, Holt L, eds. *Craniofacial anomalies. An interdisciplinary approach.* St Louis: Mosby–Year Book, 1992;41–57.
24. Leiber B. Ohrmuscheldystopie, Ohrmuscheldysplasie und Ohrmuschelmissbildung-Klinische Wertung und Bedeutung als Symptom. *Arch Klin Exp Ohren Nasen Kehlk* 1972;202:51–84 (see Chapter 2 in ref. 4, cited herein).
25. Farkas LG. Vertical location of the ear, assessed by the Leiber test, in healthy North American Caucasians 6–19 years of age. *Arch Otorhinolaryngol* 1978;220:9–13.
26. Farkas LG. Anthropometry of the normal and defective ear. *Clin Plast Surg* 1990;17:213–221.

Chapter 4

Anthropometry of the Head and Face in Clinical Practice

Leslie G. Farkas

The selection of the measurements for a particular piece of study is not as difficult as might seem, once we are well conscious of the exact aims of the study to be undertaken.

Aleš Hrdlička
1920

In clinical practice, the determination of morphological aberration from the normal is the primary goal of all those dealing with patients with congenital anomalies of the head and face. Craniofacial anthropometry provides a quantitative dimension to the dysmorphologist's observational skills (1) in the identification of anomaly and its proper classification in the system of facial syndromes. The surgeon's goal—in addition to the precise determination of location of the defect—is the detailed assessment of the extent of the defective areas, which is the basic requirement for planning the treatment. Anthropometry is one of the methods serving this purpose. The main factors which determine the quality of examination are (a) the experience of the examiner in selecting the proper measurements and instruments and (b) his or her skills in obtaining reliable data about the defective area. In small children, an additional and unknown factor is the degree of cooperation available during the examination.

LANDMARKS

Precise location of the individual anatomical points on the surface of the head and face is essential in anthropometry (2). The landmarks determine the length of the projective and tangential linear distances and the size of the angular measurements (angles and inclinations). An inexperienced examiner is advised to mark the location of the landmarks on the skin with water-soluble ink. A missing point or a landmark difficult to identify can be replaced by a supplementary point (see Chapter 3). The missing or dislocated landmark(s) must be recorded in the anthropometric chart.

SELECTION OF MEASUREMENTS

Introductory anthroposcopy may help to detect the location of a striking defect in the face. However, judgment can be easily distorted by visual illusions.

Basic Measurements

A set of 28 linear and 10 angular measurements of the head and face with 19 proportion indices (Table 4–1) proved to be valuable in our praxis when screening the individual regions of the craniofacial complex for morphological defects. After locating the major disfigurement(s), the region is subjected to a thorough analysis using all the measurements available. It is strongly recommended that other regions be screened for possible morphological defects, even if visually they do not show deviations from the normal.

EVALUATION OF FINDINGS

Anthroposcopy, Anthropometry

Anthroposcopy, the oldest examination method in medicine, cannot be excluded from the everyday praxis, but its quality must be greatly improved. The eye can be trained to observe more accurately by the systematic use of anthropometry. Anthroposcopic analysis of the facial

TABLE 4–1. *Basic measurements and proportion indices*

Region	Measurement	Proportion index
Head	Head width (eu-eu) Head length (g-op) Head height (v-n) Craniofacial height (v-gn)	eu-eu × 100/g-op v-n × 100/v-gn
Face	Face width (zy-zy) Mandible width (go-go) Face height (n-gn) Upper face height (n-sto) Mandible height (sto-gn) Supraorbital arc (t-g-t) Maxillary arc (t-sn-t) Mandibular arc (t-gn-t) Maxillary depth (t-sn rt & lt) Mandibular depth (t-gn rt & lt) Upper face inclination Mandible inclination	n-gn × 100/zy-zy sto-gn × 100/go-go n-sto × 100/n-gn sto-gn × 100/n-gn go-go × 100/zy-zy t-sn × 100/t-gn
Orbits	Intercanthal width (en-en) Biocular width (ex-ex) Eye fissure length (ex-en rt & lt) Eye fissure height (ps-pi rt & lt) Eye fissue inclination rt & lt	en-en × 100/ex-ex ps-pi × 100/ex-en rt & lt
Nose	Nose width (al-al) Nose height (n-sn) Alar base width (ac-ac) Tip protrusion (sn-prn) Bridge inclination Columella inclination Nasofrontal and nasolabial angles	al-al × 100/n-sn sn-prn × 100/al-al n-sn × 100/n-gn al-al × 100/zy-zy
Lips and mouth	Upper lip height (sn-sto) Mouth width (ch-ch) Lower lip height (sto-sl) Upper lip inclination Lower lip inclination Labiomental angle	sn-sto × 100/ch-ch sto-sl × 100/sn-sto ch-ch × 100/zy-zy
Ears	Ear width (pra-pa rt & lt) Ear length (sa-sba rt & lt) Lower gnathion-aural distance (gn-obi rt & lt)	pra-pa × 100/sa-sba rt & lt sa-sba × 100/n-gn rt & lt

Note: The measurement of the body height is recommended.

morphology alone can be erroneous. It can proclaim the wrong measurement as the basic cause of a disproportion. It fails to distinguish between disharmony and disproportion, as experienced by us, when the disharmoniously wide interorbital space was diagnosed as hypertelorism. Quantitative data of the surface anatomy of the head and face, with the possibility of defining the extent of defectiveness, is provided by anthropometry. For evaluation of anthropometric findings, valid population norms matching the race, sex, and age of the examinee are essential.

Normal, Abnormal

Determination of the normality or abnormality by measurements obtained from the patients requires a system of quantitative criteria.

The *normal range* of measurements is given by the mean ± two standard deviations (SD) (Fig. 4–1). Measurements within +1 SD and −1 SD of the mean are regarded as *optimal* (3).

A measurement of 2 SD below the mean is considered the smallest normal value, marked as *normal-small* or *borderline-small.* A finding of 2 SD above the mean is the largest normal value, referred to as *normal-large* or *borderline-large* (4,5).

Subnormal describes a measurement smaller than a mean − 2 SD, and *supernormal* refers to a measurement larger than a mean + 2 SD. In proportion indices, a value in the range of mean ± 1 SD of normal indicates proportionality—that is, *harmony.* Index values at the mean − 2 SD or mean + 2 SD levels designate *disharmony* (disharmoniously small, disharmoniously large). *Disproportions* are present if the index values are larger than the mean + 2 SD or smaller than the mean − 2 SD levels (supernormal, subnormal disproportions) (4).

Degree of Abnormality

In an abnormal finding the determination of the amount of defectiveness is essential. The degree of abnormality in a measurement or proportion index is established by calculating the difference between the finding and the appropriate terminal value of the normal range (e.g., mean − 2 SD or mean + 2 SD). The difference is expressed as a percentage of the terminal value: 0.1–2.9% indicates a *mild* abnormality (or disproportion), 3.0–9.9% indicates a *moderate* one, and 10.0% or more indicates a *severe* one (4).

Analysis of Proportion Qualities

Anthropometric analysis of harmony, disharmony, and disproportion offers objective information about the value of the participating measurements. Harmony can be produced not only by two optimal (e.g., in the range of the mean ± 1 SD) measurements, but also by two subnormal or two supernormal measurements with the same degree of defectiveness. Disharmony is most often caused by two normal measurements located at the extremities of the normal range, or by an optimal and a borderline (small or large) measurement. Disproportion occurs not only by the combination of a normal (optimal or borderline) and abnormal measurement, but also by two unequally defective measurements.

Interpreting the Borderline Values

Although the measurements positioned at the terminal points of the normal range (at mean − 2 SD or at mean + 2 SD) are still defined as normal, in longitudinal follow-up they may change and become optimal or abnormal. In our experience, during the postnatal de-

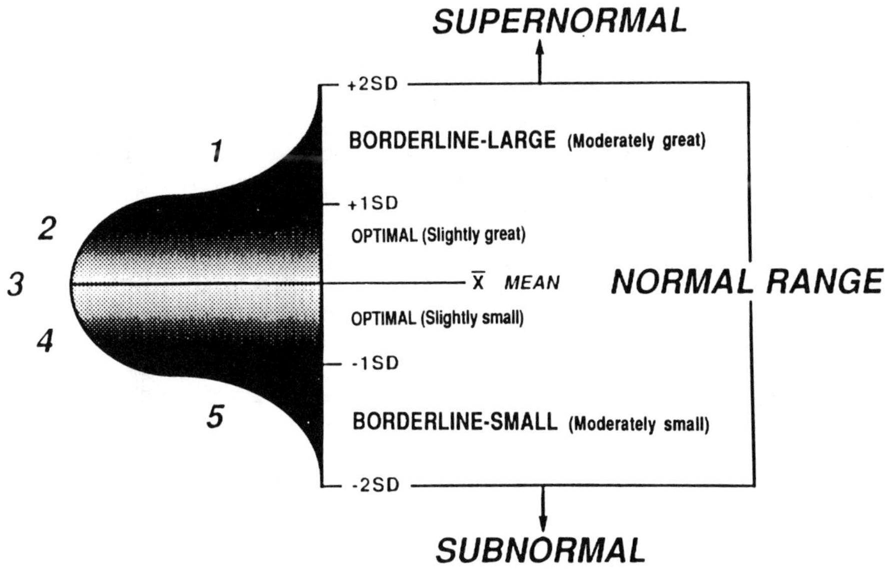

FIG. 4–1. Schematic representation of the normal range (mean ± 2 SD). There are five sections within the normal range: (1) borderline-large, (2) optimal-great, (3) mean value, (4) optimal-small, (5) borderline-small. (From ref. 3, with permission.)

velopment in congenitally damaged heads and faces, the borderline-short or borderline-long measurements showed a tendency to become abnormal in areas where in older ages the trend was abnormality. Therefore, such terminal measurement values can be regarded as potential candidates for becoming defective. In longitudinal studies of patients with facial syndromes, the terminal values found in stigmatized areas were reported in one group with the abnormal findings (5).

The Preliminary Anthropometric Examination

A one-time examination is the most frequent in clinical praxis. The quantitative methods of evaluation of the morphological changes permit the identification of minor defects (microforms), which are signs of great diagnostic value in dysmorphologic practice (6). The surgeon should appreciate the variety of data obtained by measurement, because it helps in understanding the linear and angular relationships involved and their influence on proportions. The identification of the main measurements responsible for disfigurement of the face is essential. Defects assessed by anthropometry may assist the surgeon in deciding the proper course of treatment.

The Longitudinal Preoperative Follow-up

This informs us about the postnatal development of the facial syndromes. Even a short period limited in the number of years provides valuable data about age-related changes (7). Our present knowledge about morphological changes in the head and face of patients with anomalies is based mostly on occasional observations. Each time the findings are compared with population norms. This comparison shows whether development has advanced or fallen behind. It will assist in deciding

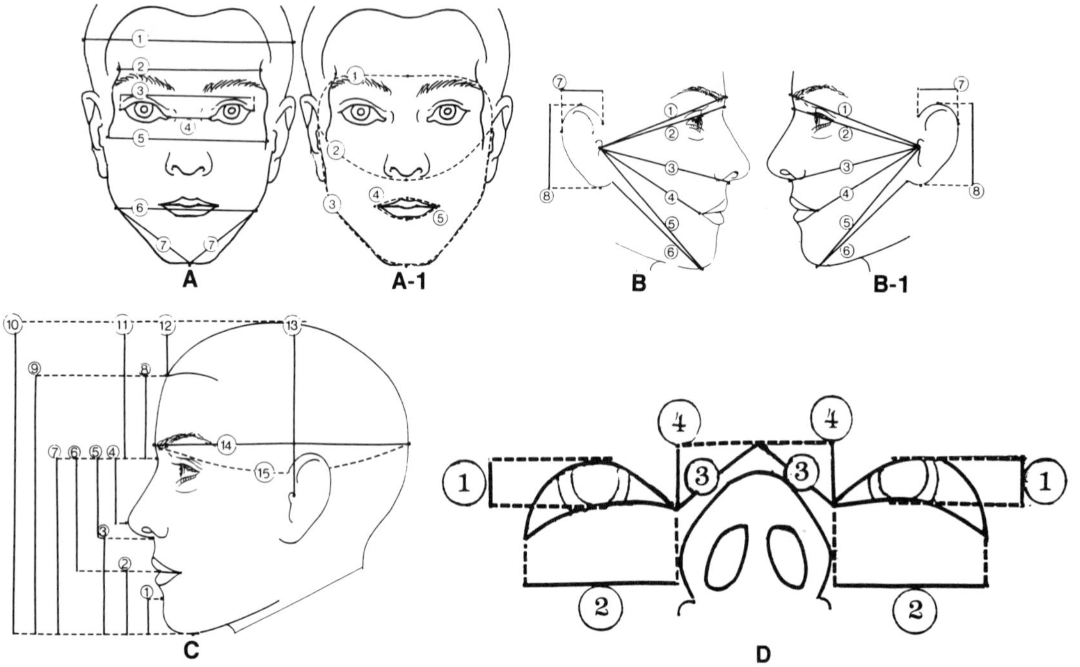

FIG. 4–2. Schematic drawings of the report form for the main craniofacial measurements. **A:** Horizontal linear measurements of the head, face, and orbits. (1) Head width (eu-eu); (2) forehead width (ft-ft); (3) biocular width (ex-ex); (4) intercanthal width (en-en); (5) face width (zy-zy); (6) mandible width (go-go); (7) mandible depth (gn-go). **A-1:** Arc-measurements of the face and lips. (1) Supraorbital arc (t-g-t); (2) maxillary arc (t-sn-t); (3) mandibular arc (t-gn-t); (4) upper vermilion arc (ch-ls-ch); (5) lower vermilion arc (ch-li-ch). **B, B-1:** Depth measurements of the face and the size of the ears (rt and lt). *Upper third face:* (1) supraorbital rim depth (t-g); (2) depth of the upper third of the face (t-n). *Middle third face:* (3) middle third face (maxillary) depth (t-sn); (4) labio-aural distance (ch-t). *Lower third face:* (5) lower third face (mandibular) depth (t-gn); (6) lower gnathion–aural distance (gn-obi). *Size of the ears:* (7) width (pra-pa); (8) length (sa-sba). **C:** Vertical measurements of the head, face, nose and the length and circumference of the head. *Face:* (1) chin height (sl-gn); (2) mandible height (sto-gn); (3) lower face height (sn-gn); (6) upper face height (n-sto); (7) face height (n-gn); (9) physiognomical face height (tr-gn). *Nose:* (4) nasal bridge length (n-prn); (5) nose height (n-sn). *Head:* (8) forehead height II (tr-n); (10) craniofacial height (v-gn); (11) anterior head height (v-n); (12) calvarium height (v-tr); (13) auricular head height (v-po); (14) head length (g-op); (15) head circumference. **D:** Measurements of the soft orbits and the nasal root (rt and lt). (1) Eye fissure height (ps-pi); (2) eye fissure length (ex-en); (3) endocanthion–facial midline distance (en-se); (4) nasal root depth (en-se sag).

whether to postpone or proceed with surgical intervention. The existence of the often discussed "self-reparatory" effort of nature can be demonstrated only in a longitudinal follow-up (6). A long-term analysis of development in which the examination is carried out in consecutive age groups but in various children with the same facial syndrome—the cross-sectional approach—is a modest alternative to the longitudinal follow-up. It is assumed, but not confirmed, that the rapid and slow growth periods and the maturation times determined in healthy children can be applied—ad analogiam—also in children with facial anomalies.

PLANNING SURGERY

The goal of the preoperative examination is to detect the defective elements responsible for disfigurement in the dense network of mutually bound relationships between the measurements, expressed by abnormal proportion index values. The produced failure can be morphological and/or functional. As soon as the regions with the most severe disproportions have been ascertained, the measurements composing them are subjected to a detailed analysis. The aim is to determine which of the measurements is the most suitable and surgically feasible to be changed in order to secure a good morphological and/or functional result. Changes in measurements can be calculated manually or by computer, guided by the appropriate proportion index formulae and the population norms. Facial defects must be corrected without producing disproportions in other areas. Restoration of harmony between measurements often helps to correct functional defects (e.g., improvement of vision by repairing a ptotic upper lid) (6).

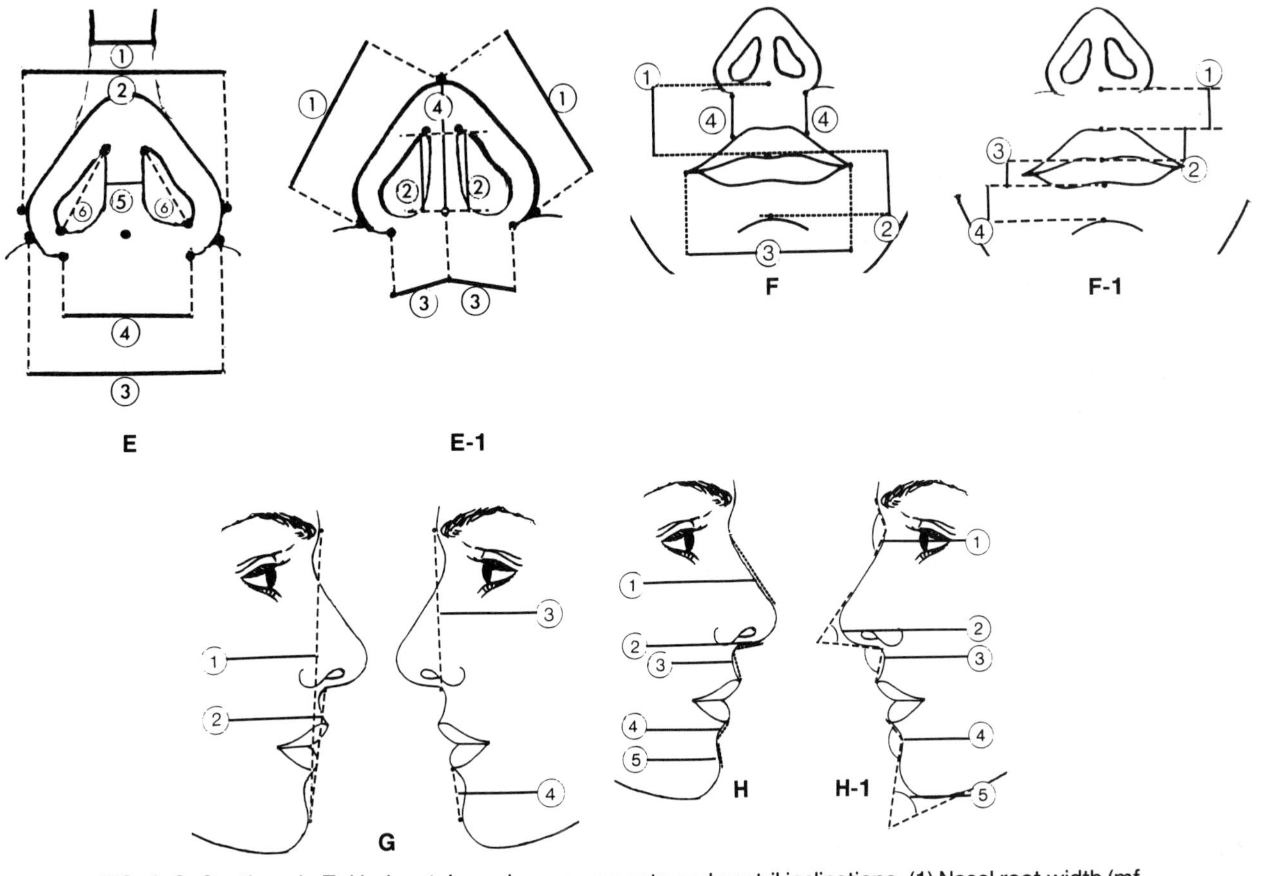

FIG. 4–2. *Continued.* **E:** Horizontal nasal measurements and nostril inclinations. (1) Nasal root width (mf-mf); (2) morphological nose width (al-al); (3) anatomical nose (alar base) width (ac-ac); (4) labial insertion width (sbal-sbal); (5) columella width (sn′-sn′); (6) nostril axis inclination (rt and lt). **E-1:** Soft nose measurements (rt and lt). (1) Ala length (ac-prn); (2) columella length (sn-c′); (3) nostril floor width (sbal-sn); (4) nasal tip protrusion (sn-prn). **F:** Main lip and mouth measurements. (1) Upper lip height (sn-sto); (2) lower lip height (sto-sl); (3) mouth width (ch-ch); (4) lateral upper lip heights (sbal-ls′) (rt and lt). **F-1:** Measurements of the skin portions and vermilions of the lips. *Upper lip:* (1) skin portion height (sn-ls); (2) vermilion height (ls-sto). *Lower lip:* (3) vermilion height (sto-li); (4) skin portion height (li-sl). **G:** Main facial profile line inclinations. (1) general profile line (g-pg); (2) lower face line (sn-pg); (3) upper face line (g-sn); (4) mandible (li-pg). **H:** Nose, lip, and chin inclinations. (1) Nasal bridge; (2) columella; (3) upper lip; (4) lower lip; (5) chin. **H-1:** Angles in the facial profile line. (1) Nasofrontal; (2) nasal tip; (3) nasolabial; (4) labiomental; (5) mentocervical.

Population norms for the inclinations and angles inserted between the inclination of the facial profile can be regarded only as general guidelines. Some help is offered by the discovery of a significant relationship between the following *basic* profile inclinations: (a) between the general profile line and the upper face, lower face, and mandible inclinations and (b) between the upper face and Leiber line, upper face and lower face, and mandible inclinations. Significantly related basic profile and *segmental* inclinations were found between the general profile line and the nasal bridge, upper face and nasal bridge, lower face and upper lip, mandible and lower lip, and mandible and chin inclinations (8). The individual inclinations and angles are creating a visually pleasing unit in an average healthy face. In surgical correction the final decision is still left to the aesthetical feelings of the surgeon. Although the defects of the profile line inclination are determined in FH of the head, the general rule should be that the corrections needed on the profile must be adjusted to the patient's rest position of the head (9). Analysis of the face in a group of young attractive North American Caucasian women showed that a vertical general profile line is not mandatory for making the profile attractive (3).

In surgical correction of the face the use of the neoclassical canons is still widely popular. However, an anthropometric study carried out in our laboratory demonstrated (10) that some canons do not reflect the reality (e.g., the three-section profile canon, the four-section facial profile canon), and others (two-section profile canon; the nasoaural, orbitonasal, orbital, naso-oral, and nasofacial proportion canons; and the nasoaural inclination canon) represent only one of the possible variations of the proportion qualities (11).

When planning a secondary corrective operation, the surgical intervention should avoid age periods in which accelerated growth was found in the healthy population (12). Around school age the growth generally slows down in face, which makes this period suitable for surgery. In corrective surgery when a change by operation is not urgent, the knowledge of maturation time ensures that the surgical correction would not hinder the growth (12).

Postoperative Assessment

By repeating the preoperative measurements after the operation, an objective picture is obtained about the effect of surgical treatment. The first postoperative assessment in children is carried out 1 year after surgery, whereas in adults it is carried out after 6 months. The manner of healing in adults, and both the healing and the extent of facial growth in children, may influence the findings. From the morphological point of view the operation is regarded as successful if the examination confirms a complete restoration of proportion normality. The result may be considered satisfactory if the degree of abnormality has been markedly reduced, best demonstrated by the drop of the percentage of defectiveness in measurements and proportions recorded prior to surgery (e.g., by a severe disproportion becoming a mild one).

Repeated Postoperative Examinations

These demonstrate the long-term effect of surgical treatment. A major factor affecting the outcome of primary surgery, especially in early childhood, is the growth of the face, which is an unknown factor in most facial syndromes. This process is best demonstrated by the changes of a fairly acceptable nose (i.e., after the primary cleft lip repair) into a "cleft lip nose" by school age. In adults, the stability of the results obtained by surgery can be determined by two to three examinations with 1-year intervals. In children, the final effect of surgical correction cannot be judged until the head and face reach full maturity (for maturation ages, see Chapter 6).

Recording the Findings

The *anthropometric chart* is an important part of the clinical documentation about the patient. It meets an immediate need, and it also contributes material for clinical research purposes. In the chart used in our laboratory, we record (a) personal notes about the patient and diagnosis, (b) date of examination, and (c) the quality of the cooperation during the examination. Space is reserved for the visual impressions about the most striking defects of the head and face. A section is devoted to the qualitative signs in all regions of the craniofacial complex. Most of the chart is kept for anthropometric findings. In each of the six regions (head, face, orbits, nose, lips and mouth, and ears) the complete set of linear and angular measurements is recorded with all level values of the normal range (−2 SD, −1 SD, mean, +1 SD, +2 SD). This makes a prompt evaluation of the findings possible. There is space for recording the degree of abnormality, calculated after examination.

Anthropometric charts are available for both sexes in all age groups between birth and 18 years of age. For evaluation of the findings, the North American Caucasian norms are used (see Appendix A).

REPORTING THE FINDINGS

The main findings of the anthropometric chart are gathered in a two-page document. On the front page the findings are summarized in each region separately. In the case of abnormal measurements, in addition to the absolute values found in patients, the mean values of the measurements are also given. We now also report the borderline measurements with their mean values. Optimal findings (measurements in range of mean ± 1 SD) are shown only in the most important measurements, if present. The aim of this short report is to give the surgeon information about the most defective area(s) of the head and face. The mean values help the surgeon to appreciate the extent of changes needed to restore normality. After surgery, the changes achieved by the operation are briefly described.

The reverse side of the document records the findings of importance. A set of simple schematic drawings shows the various regions of the head and face, with fine lines indicating the basic measurements. The measurements which were found to be subnormal are colored red, the supernormal measurements are colored blue, and those on the borderline level are colored yellow. Each measurement shown has its absolute value (in millimeters or degrees) added, while abnormal ones also show the degree of defectiveness (in percentage). By this arrangement, a glimpse of the drawings gives fast and reliable information about the location(s) and nature of the abnormality (Fig. 4–2). Repeated preoperative examinations in early childhood offer valuable data about the development which may influence the date of the primary surgical intervention. After the first operation, the changes in the colors of the measurements and/or in degrees of the previously abnormal findings are promptly perceptible. They are visualizing the effect of surgery. Repeated postoperated examinations are the way to obtain quantitative data about the stability of the morphological results of the surgical treatment, illustrated in the schematic drawings.

REFERENCES

1. Ward RE. Facial morphology as determined by anthropometry: keeping it simple. *J Craniofac Genet Dev Biol* 1989;9:45–60.
2. Farkas LG. *Anthropometry of the head and face in medicine.* New York: Elsevier. 1981.
3. Farkas LG, Kolar JC. Anthropometrics and art in the aesthetics of women's faces. *Clin Plast Surg* 1987;14:599–616.
4. Farkas LG. The proportion index. In: Farkas LG, Munro IR, eds. *Anthropometric facial proportions in medicine.* Springfield, IL: Charles C Thomas. 1987;5–8.
5. Farkas LG, Posnick JC, Hreczko T. Anthropometry of the head and face in 95 Down syndrome patients. In: Epstein, CJ, ed. The morphogenesis of Down syndrome. New York: Wiley-Liss, 1991; 53–97.
6. Farkas LG. Basic anthropometric measurements and proportions in various regions of the craniofacial complex. In: Brodsky L, Ritter-Schmidt DH, Holt L, eds. *Craniofacial anomalies: an interdisciplinary approach.* St Louis: Mosby–Year Book, 1992;41–57.
7. Feingold M. Difficulties encountered in the clinical description of the dysmorphic patient. *J Craniofac Genet Dev Biol* 1989;9:3–5.
8. Farkas LG, Katic MJ, Kolar JC, Munro IR. The adult facial profile: relationships between the inclinations of its segments. *Dtsch Z Mund Kiefer Gesichts Chir* 1984;8:182–186.
9. Farkas LG, Kolar JC, Munro IR. Geography of the nose: a morphometric study. *Aesth Plast Surg J* 1986;10:191–223.
10. Farkas LG, Hreczko TA, Kolar JC, Munro IR. Vertical and horizontal proportions of the face in young adult North American Caucasians: revision of neoclassical canons. *Plast Reconstr Surg* 1985;75:328–337.
11. Farkas LG, Munro IR. *Anthropometric facial proportions in medicine.* Springfield, IL: Charles C Thomas, 1987.
12. Farkas LG, Posnick JC. Growth and development of regional units in the head and face based on anthropometric measurements. *Cleft Palate Craniofac J* 1992;29:301–329.

Chapter 5

Photogrammetry of the Face

Leslie G. Farkas

Routine medical (nonstandardized) photography complements the narrative part of a patient's chart (1,2). Its usefulness is limited to indicating the location of the injury or showing disfigurement in detail. Standardized photographs as used in photogrammetry, on the other hand, may provide considerably more benefit.

Photogrammetry of the face (indirect anthropometry) is anthropometry adapted for quantification of surface features from standard (20%, 25%, 33%, 50%, or life-size) photographs. In direct anthropometry the measurements are taken directly from the subject's face, whereas in indirect anthropometry (photogrammetry) they are taken from the standard photographs. The use of standard photographic methods to produce prints of standard sizes and views allows photogrammetry to be scientific, accurate, and documentary.

In the last few decades, photogrammetry of the face has been widely used in medicine (3–9). However, assessment of the reliability of the indirect measurements obtained by this method has been limited (10–12) because of differing methods of evaluation and analysis of the mostly lateral-view photographs (13).

Our system of anthropometry involves a large number of direct facial measurements. In view of this, we were prompted to determine how many of these measurements could also be reliably obtained from photographs (i.e., indirectly) (14). Identification of a significant number of accurate measurements that could be obtained by photogrammetry would greatly reduce the time needed for direct measurements, especially in small and/or uncooperative children.

MATERIALS AND METHODS

The study group consisted of 36 healthy young North American Caucasians (18 men and 18 women). The standard landmarks were marked on the skin of each

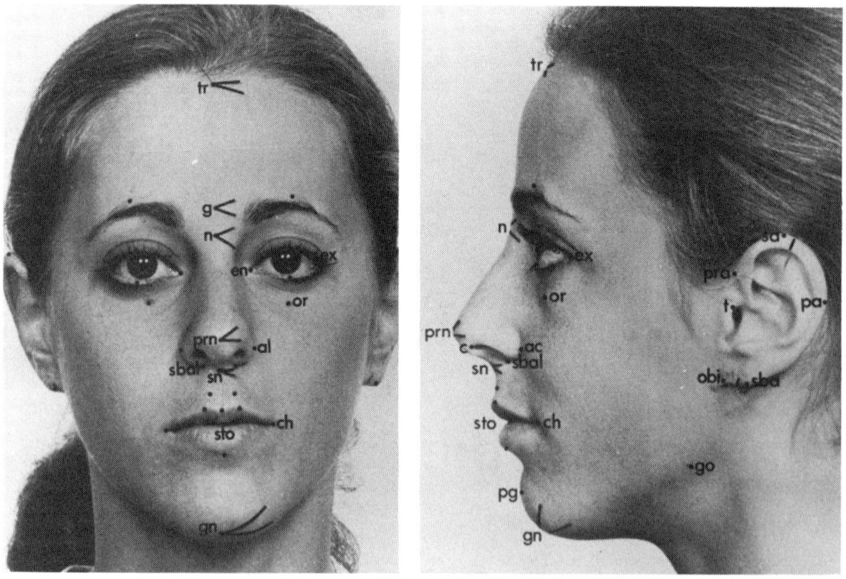

FIG. 5–1. Landmarks: The frontal print and the left lateral print of the face. (From ref. 14, with permission.)

with small ink dots (Fig. 5–1). Points in the facial midline (profile line landmarks: tr, g, n, prn, sn, gn) were marked by a horizontal V-shaped sign. The tip of this sign touched the landmark, and the wings extended to the left.

Prior to photographing, 100 measurements were taken from the head, face, and ears of the subjects (13) (Table 5–1). Following this, we took standard frontal and left lateral black-and-white photographs of our subjects. Life-size photographs were printed, and a second set of 60 measurements was taken from them. The two sets of measurements were then compared to test the validity of the second set of indirect measurements. The standard photographic technique developed at the Hospital for Sick Children was described earlier (14) (see also Chapter 15).

Measurements from the Prints

On the prints, linear distances between landmarks were measured by sliding caliper. Inclinations and angles

TABLE 5–1. *Photogrammetry: comparative results analysis of direct measurement/photo measurement*

Region	Measurements taken directly, *N*	Measurements taken from the prints	*N*
Head	12	ft-ft, t-t, tr-g, tr-n, forehead inclination	5
Face	23	zy-zy, go-go, tr-gn, n-gn, n-sto, sn-gn, sto-gn, left n-t, sn-t, gn-t, go-gn, inclinations of the upper face (g-sn) and lower face (sn-pg)	13
Orbits	16	en-en, ex-ex, ex-en rt & lt, ps-pi rt & lt, or-sci rt & lt, eye fissure inclination rt & lt	10
Nose	23	n-sn, n-prn, al-al, sn-prn, sbal-sn rt & lt, left ac-prn and sn-c′, nasal bridge inclination, nasal bridge deviation, nasofrontal and nasolabial angles	12
Lips and mouth	14	cph-cph, ch-ch, ch-sto rt & lt, sn-sto, sn-ls, ls-sto, sto-li, sbal-ls′ rt & lt, left ch-t, upper lip and labial fissure inclinations	13
Ears	12	sa-sba rt & lt, left pra-pa, gn-obi, gn-obs, obs-obi, ear inclination	7
Total	100		60

Region	Measurements taken directly, *N*	Measurements which cannot be taken from the prints	*N*
Head	12	eu-eu, v-n, v-gn, g-op, v-po rt & lt, head circumference	7
Face	23	t-sn-t, t-gn-t, t-sn surf rt & lt, t-gn rt & lt, right t-n, t-sn, t-gn, gn-go	10
Orbits	16	or-os rt & lt, ex-t rt & lt, ex-go rt & lt	6
Nose	23	mf-mf, en-se rt & lt, al′-al′ rt & lt, sn′-sn′, right ac-prn and sn-c′, ac-prn surf rt & lt, columella deviation	11
Lips and mouth	14	right ch-t	1
Ears	12	right pra-pa, gn-obi, gn-obs, obs-obi, ear inclination	5
Total	100		40

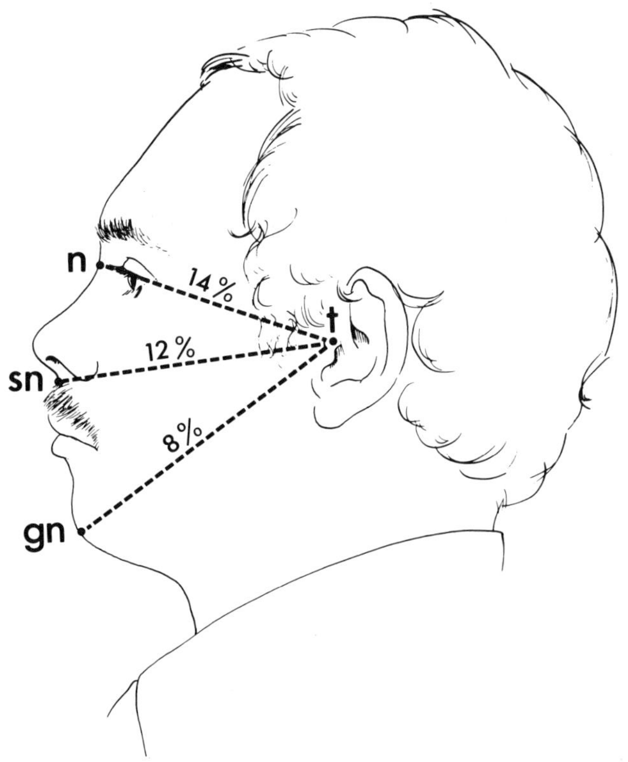

FIG. 5–2. Calculation of the extent of error in photogrammetry. Shortening of the lateral (horizontal and oblique) depth measurements in the upper, middle, and lower third of the face is expressed as a percentage of the measurements taken directly from the face of the subject. (From ref. 14, with permission.)

were determined with the help of a transparent protractor, and the degree of inclination was calculated from the horizontal or vertical as defined by the FH. Fractions of millimeters or degrees were rounded to the next highest figure.

Analysis of the Data

Each measurement obtained from the print was compared to the corresponding direct measurement, and the difference was registered. The differences in all 18 subjects of each sex were then averaged. A measurement was regarded as reliable only if the average difference between the indirect and direct measurement was not greater than 1 mm or 2 degrees. Otherwise, it was considered inaccurate. Inaccurate measurements were consistently longer or consistently shorter than the corresponding direct measurements, or a combination of the two, depending on the individual. In selected measurements the extent of error found in photogrammetry was expressed as a percentage of the direct measurement (Fig. 5–2).

Because the analysis of measurements taken separately in boys and girls showed the same trend, the averages of the differences were calculated jointly.

Measurements

Of a total of 100 measurements, we obtained 77 projective distances, 9 tangential (arc) measurements, and 14 angular measurements (10 inclinations and 4 angles) from each subject before photographing. From the frontal and left lateral photographs of each individual a total of 60 measurements could have been obtained: 62.3% of the projective distances (48 of 77), 85.7% of the angular measurements (12 of 14), but none of the 7 tangential arcs. Among the regions, the prints of the head allowed the smallest number of measurements [41.7% (5 of 12)], with the highest number [92.9% (13 of 14)] in the region of the lips and mouth. The two-dimensional character of the prints (one frontal and one lateral) and the land-

TABLE 5–2. *Photogrammetry: usable results*

Region	Measurement: Linear	Measurement: Angular	Subtotal *N*
Head	None	Forehead inclination from the vertical (Fig. 5–3D)	1
Face	Upper face height (n-sto) (Fig. 5–3A)	Upper face inclination from the vertical (g-sn); lower face inclination from the vertical (sn-pg) (Fig. 5–3D)	3
Orbits	Intercanthal width (en-en) (Fig. 5–3A); eye fissure height (ps-pi) (lateral view); orbit and brow height (or-sci) (Fig. 5–3C)	Eye fissure inclination from the horizontal rt & lt (Fig. 5–3B)	5
Nose	Nose height (n-sn); columella length (sn-c′) (Fig. 5–3C)	Nasal bridge inclination from the vertical (Fig. 5–3D); nasal bridge deviation from the facial midline (Fig. 5–3B)	4
Lips and mouth	Upper lip height (sn-sto); lower vermilion height (sto-li) (Fig. 5–3C); philtrum width (cph-cph); lateral upper lip heights (sbal-ls′) rt & lt (Fig. 5–3A)	Upper lip inclination from the vertical (Fig. 5–3D); labial fissure inclination from the horizontal (Fig. 5–3B)	7
Ears	None	None	
			Total: 20

marks not visible or covered by hair prevented taking 40 of the 100 measurements obtained directly from the subjects. The greatest reduction was in the head [58.3% (7 of 12)], whereas the least reduction was in the lips and mouth [7.1% (1 of 14)] (Table 5–1).

Accuracy of Indirect Measurements from the Prints

Table 5–2 shows the reliable measurements in each area of the head and face. Of the 60 indirect measurements, 20 (33.3%) were reliable, and seven of these were from the area of the lips and mouth. No accurate measurements of the ears were registered. The largest number (nine) of reliable measurements were of inclinations (Fig. 5–3A–D).

Inaccuracy in Indirect Measurements from the Prints

Of the 60 measurements taken from the prints, 40 were unreliable: Three (7.5% of 40) were consistently longer and 22 (55% of 40) consistently shorter than the identical direct measurements. The remaining 15 (57.5% of 40) were mixed, because they were longer in some individuals and shorter in others.

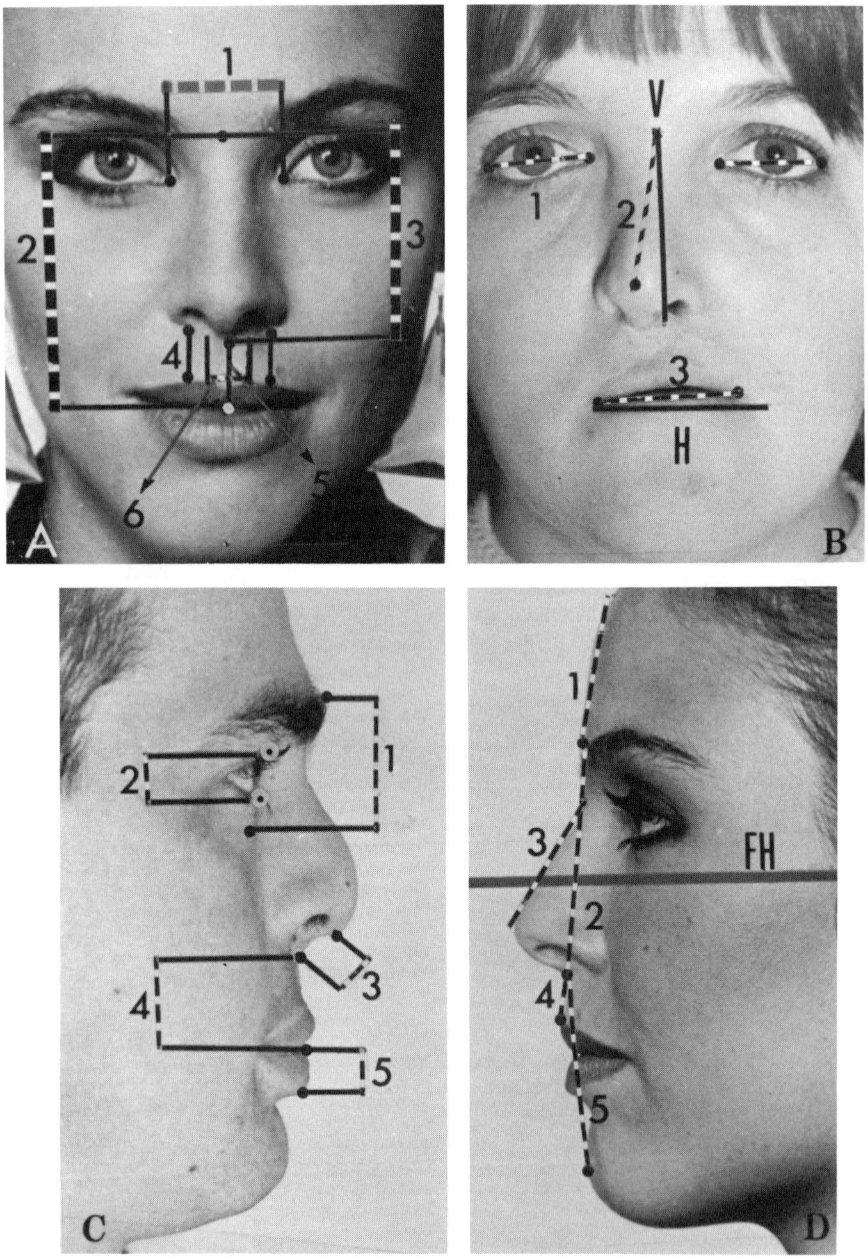

FIG. 5–3. Reliable measurements obtained by photogrammetry. Frontal view **A:** (1) en-en, (2) n-sto, (3) n-sn, (4) sbal-ls, (5) sn-sto, (6) cph-cph. Frontal view **B:** (1) en-ex inclination, (2) nasal bridge deviation from the vertical (V), (3) labial fissure inclination from the horizontal (H). Lateral view **C:** (1) or-sci, (2) ps-pi, (3) sn-c′, (4) sn-sto, (5) sto-li. Lateral view **D:** (1) forehead inclination, (2) upper face inclination, (3) nasal bridge inclination, (4) upper lip inclination, (5) lower face inclination.

Consistently Longer Measurements

Three measurements were consistently longer: the width of the face (zy-zy), by an average of 3.6 mm; the width of the nose (al-al), by 2.4 mm; and the width of the lower face (go-go), by 21.6 mm. All of these measurements were horizontal and were obtained from the frontal prints.

Consistently Shorter Measurements

Eleven consistently shorter measurements were taken from each print (Table 5–3); most of them were horizontal (15). The greatest shortenings were in the depth measurements of the upper face: 17.6 mm in t-n and 15.3 mm in t-sn on the profile print. On the frontal print, the greatest difference (5.5 mm) was in halves of the labial fissure (ch-sto rt and lt). In the semisagittal direction the depth of the nasal root was greatly reduced (7.2 and 7.4 mm).

Mixed Differences in Measurements

Fifteen measurements had mixed differences, depending on the individual facial characteristics of the subjects. Eleven of these measurements, mostly vertical, were taken from profile prints, and four (three horizontal) were obtained from frontal prints (Table 5–4). The greatest differences in measurements were in angles. The linear measurements differed little from the equivalent direct measurements. In general, vertical measurements showed smaller differences than did horizontal measurements.

TABLE 5–3. *Photogrammetry: consistently shorter measurements*

Print and type of measurement	Measurement	Average difference (mm)
Left lateral ($N = 11$)		
Horizontal	lt t-n	17.6
	lt t-sn	15.3
	lt t-gn	11.6
	lt go-gn	13.2
	lt ac-prn	3.5
	lt ch-t	5.6
	lt obs-gn	14.6
	lt obi-gn	9.6
Vertical	tr-n	2.0
	sn-gn	4.0
	sto-gn	3.9
Frontal ($N = 11$)		
Horizontal	ex-ex	2.0
	rt ex-en	2.3
	lt ex-en	2.4
	rt sbal-sn	1.7
	lt sbal-sn	2.5
	rt ch-sto	5.5
	lt ch-sto	5.1
Vertical	tr-g	3.5
	rt or-sci	1.9
Semisagittal	rt en-se	7.2
	lt en-se	7.4

SOURCES OF ERROR IN PHOTOGRAMMETRY

Identification of Profile Landmarks

The landmarks of the facial profile and forehead (n, prn, sn, ls, sto, pg, gn, tr, and g) are not always visible on

TABLE 5–4. *Photogrammetry: indirect measurements that are longer or shorter, depending on facial characteristics*

Print and type of measurement	Measurement	Average differences: Longer, Millimeters	Longer, Degrees	Shorter, Millimeters	Shorter, Degrees
Left lateral ($N = 11$)					
Horizontal	pra-pa	1.3	—	2.5	—
Vertical	tr-gn	2.5	—	4.3	—
	n-gn	2.8	—	2.8	—
	sa-sba	1.6	—	1.5	—
	obs-obi	1.5	—	2.6	—
	sn-prn	1.5	—	1.4	—
	n-prn	1.1	—	1.4	—
	ls-sto	1.3	—	1.3	—
Angle	Nasolabial	—	4.5	—	6.0
	Nasofrontal	—	5.8	—	4.5
Inclination	Ear axis	—	2.9	—	4.7
Frontal ($N = 4$)					
Horizontal	ft-ft	4.0	—	3.4	—
	t-t	3.4	—	2.9	—
	ch-ch	2.8	—	2.3	—
Vertical	ps-pi	1.2	—	1.4	—

the lateral print, even when they have been marked on the skin before photography. In profile, flattening along the axial line (root of the nose, tip of the nose, chin, etc.) and in some cases a slight axial depression (at the base of the columella, in the middle of the border of the upper vermilion, on the chin, etc.) make landmarks invisible on lateral prints. From the side, the glabella can be blocked by a bushy eyebrow and the trichion can be hidden by the hair. However, with the exception of the glabella, which is often covered by hair, our V-shaped markers are helpful in locating these landmarks.

Measurement from Bony Landmarks

Direct measurements between bony landmarks (e.g., zy-zy, gn-go) are taken by caliper, with the tips of the instrument pressed to the bony surface. Because this cannot be done on prints, errors occur even when landmarks are marked on the skin before photography. The errors might be much greater if the anatomical points were left unmarked.

Difficulty in Identifying Landmarks Located on the Edges or Contours of Anatomical Features

Some landmarks cannot be marked (e.g., inner or outer commissure of the eye fissure and the mouth, or upper and lower edges of the eye fissure), and others (the most lateral point of the ala, the highest and lowest points or the most anterior and posterior points of the auricle) are not always seen clearly because the anatomical feature bearing the landmark is not sharp enough on the print. This may result from the differing intensities of reflection in various areas of the face. Errors caused by this phenomenon are usually small.

Landmarks Covered by Hair or Hidden Behind Facial Features

Indirect measurement of the length, width, or height of the head produced significant errors unless the subject was bald. However, it may be possible to locate the head landmarks (v, eu, op) if a device similar to our headframe is used to position the head. The widest diameter of the head (eu-eu) can be determined from the frontal print if the perpendicular arms of the frame are pressed to the head at its greatest width during photography. The horizontal bar of this frame can be used to identify the vertex on the lateral-view print if it is lowered to the vertex (Fig. 5–4). The opisthocranion can be identified on the profile print by pressing the posterior arm of the frame to the posterior aspect of the head during photography.

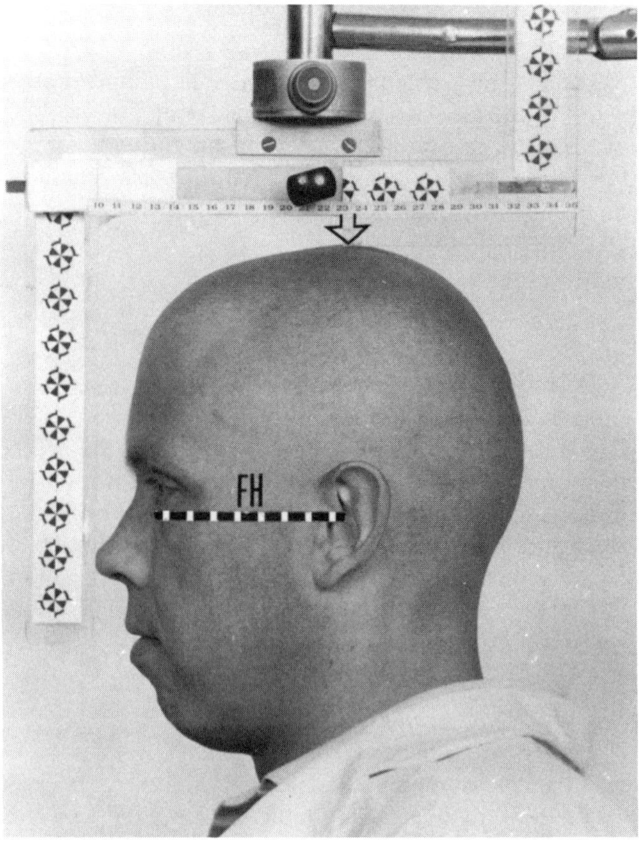

FIG. 5–4. Identifying the vertex point of the head by the lowered horizontal bar of the kephalostat (*arrow*).

The view of some landmarks is blocked by other features. For example, on the lateral print the porion landmark may be hidden behind the tragus, or the commissure of the labial fissure may be covered by a skin fold. On the frontal print the exocanthion landmark cannot be seen clearly in all people. In some, the edge of the upper lid overlaps the lateral commissure of the eye fissure. The small error eye fissure length (mean 2.4 mm) represents a mean 7.6% reduction in size of the eye fissure in a young adult, but it would be 9.3% in a 1-year-old child (Table 5–3).

Inaccuracies in Measurement of Angles

The visual assessment of any facial profile line consisting of six inclinations separated by six angles may cause difficulties, especially if the midportion of the face is flattened or depressed (Fig. 5–5). In our experience the quantitative evaluation of the angular measurements of the facial profile revealed that the findings differed by a range of ±5 degrees from those assessed visually (Table 5–4).

Errors Introduced by Photographic Distortion

Albrecht Dürer (15,16) believed that the face resembled a multifaceted formation consisting of a number of small geometric areas joined in various angles (Fig. 5–6). When photographing the face in the profile, one of these small areas is chosen for focusing. The degree of distortion depends on the differences in the level of the individual facets. In young men the tragion point was 70 mm from the focusing plane, which is at the level of the nasal tip (Fig. 5–7). The greatest effect was in shortening in the upper third of the face (Fig. 5–2).

Figure 5–8 shows the differences between various points of the facial profile and those of the focusing plane (orbitale) when the frontal print is being taken. In our study, the greatest distortion was observed when measuring the distance between the base of the columella (subnasale) and the tip of the nose (pronasale); this distance was 46% shorter on the print than on the subject because of the great level difference between the two landmarks. On the other hand, the measurement between the root of the nose (n) and the labial fissure (sto) proved to be accurate because of the similar relationship of the two landmarks to the focusing plane.

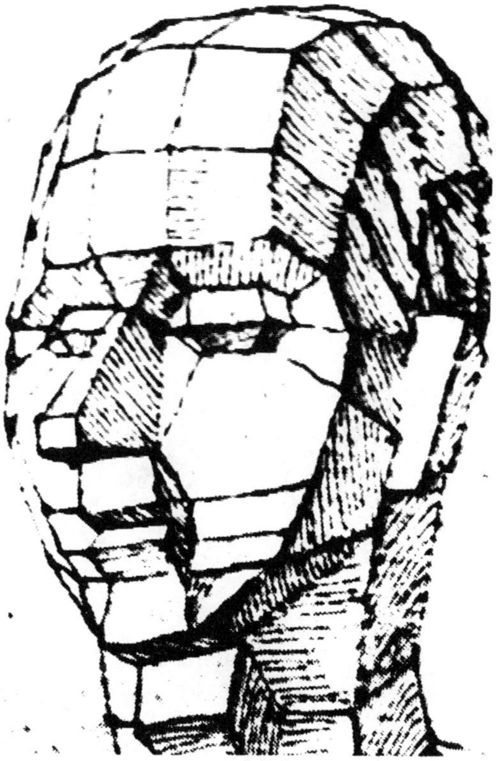

FIG. 5–6. Dürer's concept (1591) of the surface relief of the craniofacial complex.

Two-Dimensional Nature of the Print

The two-dimensional nature of the prints makes it impossible to measure the tangential arcs (supraorbital, t-g-t; maxillary, t-sn-t; mandibular, t-gn-t) or their halves (e.g., t-g surf rt and lt) or to perform the tangential measurements following the skin surface between the mouth (ch) and the ear (t) or along the nasal ala surface (ac-prn surf). Sagittal measurements, such as (a) the nasal root depth (en-se sag) and (b) level differences between the upper and lower orbital rims (or/os) or between the endocanthion and exocanthion (en/ex), cannot be taken at all from a frontal-view print. On the profile print the width measurement of the ear (pra-pa) depends on the degree of protrusion from the head: The greater the protrusion, the narrower the ear.

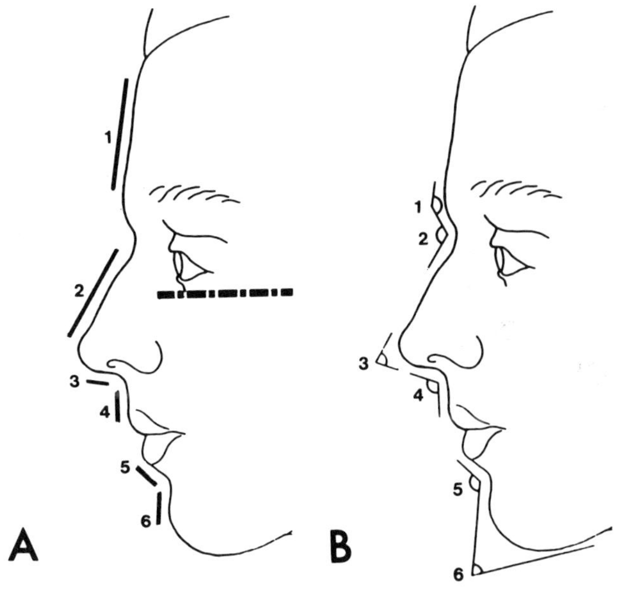

FIG. 5–5. A: Schematic drawing showing the head and face with the six inclinations: (1) forehead, (2) nasal bridge, (3) columella, (4) upper lip, (5) lower lip, (6) chin. **B:** Schematic drawing showing the head and face with the six angles: (1) glabellonasal, (2) nasofrontal, (3) nasal tip, (4) nasolabial, (5) labiomental, (6) mentocervical.

VALUABLE VIEWS IN PHOTOGRAMMETRY

Frontal-view prints supplied the most precise measurements of orbits, lips, and mouth. The profile prints were generally more useful, because they offered (a) accurate inclinations in the profile and (b) a number of vertical measurements which differed from the corresponding direct measurements by little more than 1 mm.

STUDY DATA IN THE LITERATURE

The three reports on the reliability of photogrammetry in comparison to anthropometry (10–12) mentioned

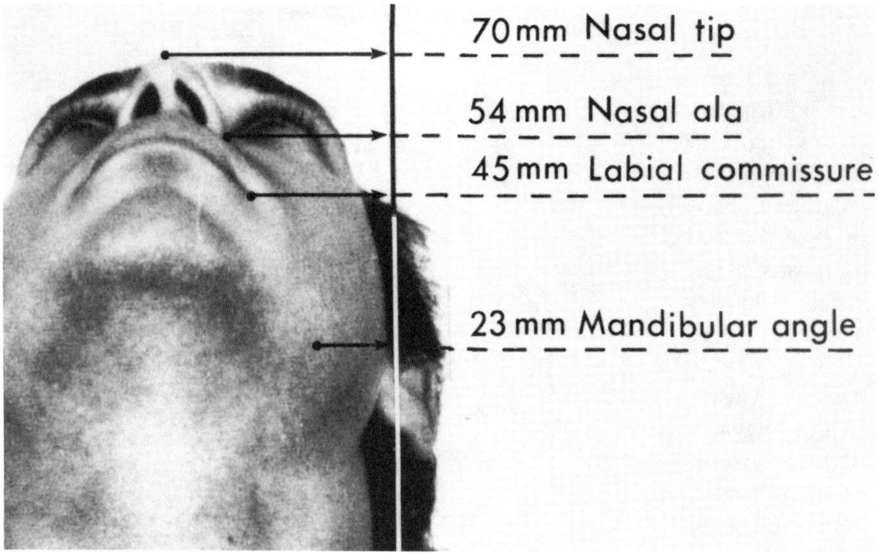

FIG. 5–7. Average distances of the pronasale (nasal tip), nasal ala, labial commissure, and the mandibular angle from the plane of the tragion (*perpendicular line*). The tragion is the point closest to the camera in lateral view.

only 16 of our 100 measurements of the head and face. A few of these 16 measurements showed the same trend as those in our study (Table 5–5). Some discrepancies between data from the literature and our results may be caused by differences in marking technique and head positioning (10,12). The sample of Gavan et al. (11) (two subjects) was too small for valid conclusions to be drawn, but this was the only study that expressed differences quantitatively.

PHOTOGRAMMETRY TODAY AND TOMORROW

Although standard photographs help to visualize the areas of marked changes in a patient's face, they cannot be considered adequate substitutes for the live face. If photogrammetry should become an equal partner of anthropometry, significant improvements are required in (a) the technique of taking photographs, (b) the method

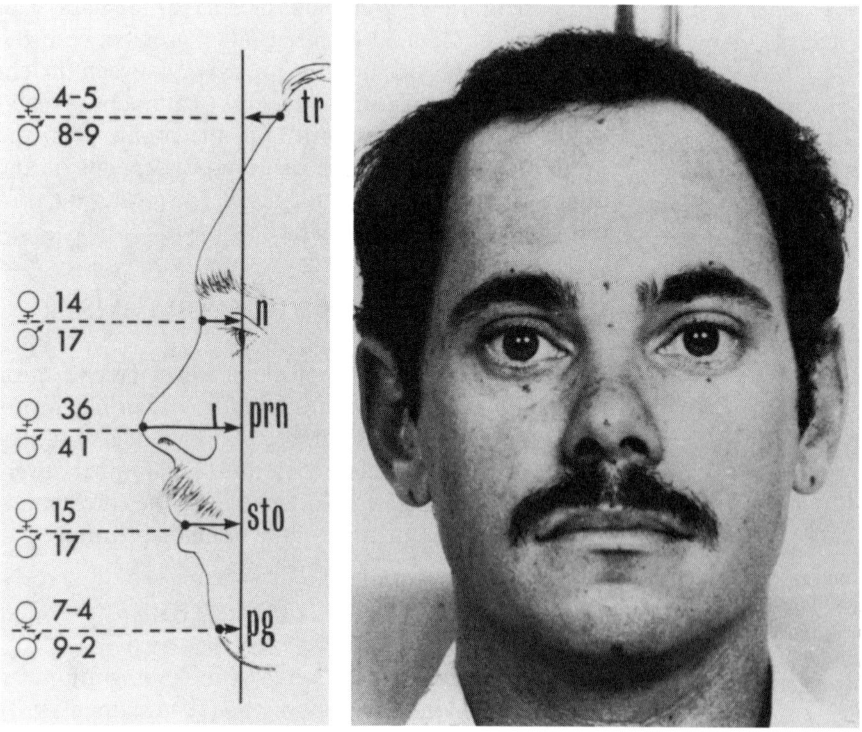

FIG. 5–8. Average distances (in millimeters) between the facial profile landmarks (trichion, tr; nasion, n; pronasale, prn; stomion, sto; pogonion, pg); and the focusing plane (*vertical line*). (From ref. 14, with permission.)

TABLE 5–5. *Comparison of photogrammetric data in the study and in the literature*

Area	Measurement	Tanner and Weiner (10), 1949 (*N* = 70)	Gavan et al. (11), 1952 (*N* = 2)	Fraser and Pashayan (12), 1970 (*N* = 50)	Our study (*N* = 36)
Head	v-po		Longer by 2 mm		
	g-op		Longer by 5 mm		Could not be taken from prints
	eu-eu		Longer by 18 mm		
	ft-ft	Longer	Longer by 12 mm		Longer by 4 mm or shorter by 5.9 mm
Face	zy-zy	Same		Correlated well	Longer by 3.6 mm
	go-go	Same		Correlated well	Longer by 21.6 mm
	n-gn		Longer by 18 mm	Correlated well	Longer or shorter by 2.8 mm
	gn-go			Correlated well	Shorter by 13.2 mm
Orbits	en-en			Correlated well	Same (±1 mm)
	ex-ex			Correlated well	Shorter by 2 mm
Nose	n-sn	Longer		Correlated well	Same (±1 mm)
	al-al	Longer	Longer by 4 mm	Correlated well	Longer by 2.4 mm
	n-prn			Correlated well	Longer by 1.1 mm or shorter by 1.4 mm
	sn-prn			Correlated well	Longer by 1.5 mm or shorter by 1.4 mm
Lips and mouth	ch-ch	Approximately same		Correlated well	Longer by 2.8 mm or shorter by 2.3 mm
	cph-cph	Approximately same		Correlated well	Same (±1 mm)

of identifying landmarks, and (c) measuring techniques. The requirements for these improvements are equally important for surgeons and dysmorphologists: The common goal is to determine *quantitatively* the deviation of the abnormal from the normal.

To produce the best results, the subject must be photographed with the landmarks marked on the skin and in the same standardized position of the head from the front and side, utilizing (a) Frankfurt horizontal for ensuring uniformity (2) and (b) reliable data about the inclinations in the facial profile line (13). Vertical measurements of the face showing mixed (longer or shorter) differences in comparison with the corresponding direct measurements (Table 5–4) might have been caused by the varying frontal position of the head facing the camera.

The usefulness of photogrammetry can be greatly increased by (a) developing new techniques for identification of all important landmarks, (b) introducing additional special views of the face and head, and (c) improving the photographing technique of the facial profile, making its contour sharp on the lateral print. By these changes, the number of reliable measurements taken from photographs of standard sizes would greatly increase. Sharp facial profile contours could eliminate the differences between the direct and indirect measurements of the nasolabial and nasofrontal angles. The contrasting facial line would make it possible to extend the measurements to all inclinations and angles of the profile, recently introduced (17–20). The now standardized enlargements of the photographs do not affect the values of these angular measurements (21).

It would be worthwhile to study those indirect measurements which consistently reveal small differences from corresponding direct measurements in large population samples with the intention to calculate "correction factors" for each measurement which would then convert into acceptable values. A similarly designed study in the most frequent facial syndromes would greatly increase the value of photogrammetry in congenital anomalies of the face.

The reliability of new apparatus for measuring the face (e.g., computer-assisted three-dimensional imaging) must be verified by consecutive direct anthropometry (22). For routine everyday praxis, such apparatus would be expensive for investigators not affiliated with large institutions. Thus, photogrammetry, hopefully in an improved version, will be available in the coming decades.

REFERENCES

1. Morello DC, Converse JM, Allen D. Making uniform photographic records in plastic surgery. *Plast Reconstr Surg* 1977;59:366–372.
2. Davidson TM. Photography in facial plastic and reconstructive surgery. *J Biol Photogr Assoc* 1979;47:59–67.
3. Stoner MM. A photometric analysis of the facial profile: a method assessing facial changes induced by orthodontic treatment. *Am J Orthod* 1955;41:453–469.
4. Neger M. A quantitative method of evaluation of the soft-tissue facial profile. *Am J Orthod* 1959;45:738–751.
5. Alexander M, Laubach LL. *Anthoropometry of the human ear (a*

photogrammetric study of USAF flight personnel). Report AMRL-TR-67-203. OH: US Air Force Aerospace Medical Research Laboratories, Wright-Patterson Air Force Base, January 1968;1–28.
6. Peck H, Peck S. A concept of facial esthetics. *Angle Orthod* 1970;40:284–318.
7. D'Ottaviano N, Baroudi R: Surgical and esthetic aspects of the facial profile. *Int J Oral Surg* 1974;3:243–246.
8. Sushner NI. A photographic study of the soft-tissue profile of the Negro population. *Am J Orthod* 1977;72:373–385.
9. Pech A, Cannoni M, Abdul S, Gitenet P, Zanaret M, Thomassin JM. Les aspects pratiques de la profiloplastie. *Rev Laryngol Otol Rhinol* 1978;99:39–46.
10. Tanner JM, Weiner JS. The reliability of the photogrammetric method of anthropometry with description of a miniature camera technique. *Am J Phys Anthropol* 1949;7:145–186.
11. Gavan JA, Washburn SL, Lewis PH. Photography: anthropometric tool. *Am J Phys Anthropol* 1952;10:331–353.
12. Fraser FC, Pashayan H. Relation of face shape to susceptibility to congenital cleft lip. A preliminary report. *J Med Genet* 1970;7: 112–117.
13. Farkas LG. *Anthropometry of the head and face in medicine.* New York: Elsevier, 1981.
14. Farkas LG, Bryson W, Klotz J. Is photogrammetry of the face reliable? *Plast Reconstr Surg* 1980;66:346–355.
15. Dürer A. *Della simmetria dei corpi humani.* Venetia: Presso D Nicolini, 1591.
16. Russel F. *World of Dürer.* New York: Time, 1967.
17. Farkas LG, Sohm P, Kolar JC, Katic MJ, Munro IR. Inclinations of the facial profile: art versus reality. *Plast Reconstr Surg* 1985;75: 509–519.
18. Farkas LG, Kolar JC, Munro IR. Geography of the nose: a morphometric study. *Aesthetic Plast Surg J* 1986;10:191–223.
19. Farkas LG, Ngim RCK, Lee ST. The fourth dimension of the face: a preliminary report of growth potential in the face of the Chinese population of Singapore. *Ann Acad Med Singapore* 1988;17:319–345.
20. Venkatadri G, Farkas LG, Kooiman J. Multipurpose anthropometric facial anglemeter. *Plast Reconstr Surg* 1992;90:507–510.
21. Hautvast J. Analysis of the human face by means of photogrammetric methods. *Anthropol Anz* 1971;33:39–47.
22. Farkas LG. Basic anthropometric measurements and proportions in various regions of the craniofacial complex. In: Brodsky L, Ritter-Schmidt DH, Holt L, eds. *Craniofacial anomalies: an interdisciplinary approach.* St Louis: Mosby–Year Book. 1992;41–57.

Chapter 6

Age-Related Changes in Selected Linear and Angular Measurements of the Craniofacial Complex in Healthy North American Caucasians

Leslie G. Farkas and Tania A. Hreczko

Surface measurements obtained through cross-sectional anthropometric studies of growing Caucasians (1–3) have proven useful in quantitative evaluation of postnatal development in head and face. Knowledge of the normal changes in soft-tissue cover of the head and face assists judgment of changes to skeletal morphology (4). Awareness of the nature of normal growth is essential if abnormal growth patterns are to be understood. The need to collect and record such information has been emphasized by many investigators and clinicians (3,5–17).

To demonstrate the main factors during postnatal development, the following data are reported separately in each region of the craniofacial complex: the degree of development of measurements at 1 year of age; total growth achieved between 1 and 18 years of age; intensiveness of growth after age 1 year; age at which maturation occurs in individual measurements.

DEFINITIONS

Development level at 1 year of age: a mean measurement expressed as a percentage of the mean value of the same measurement at age 18 (18).

Total growth increment (TGI) (or absolute total growth): difference (in millimeters or degrees) between the mean value of the measurement which is increasing between ages 1 and 18 years (19). Inclinations and angles during the postnatal development may show decreasing values.

Early growth increment (EGI) sums up the mean increments in linear measurements during the accelerated (rapid) growth periods between 1 and 6 years of age. Accelerated (rapid) growth periods observed between 6 and 16 years are marked as *late growth increments (LGI).*

Relative total increment (RTI) is TGI expressed as a percentage of the mean value of the measurement at age

1 year. RTI value is influenced by the degree of development at age 1 year and was introduced to help display the volume of increment between 1 and 18 years of age.

Relative annual increment (20) is the annual percentage increment (21), or growth percent per annum (22), showing the intensity of growth from year to year; it is calculated from the following formula: absolute annual growth (the actual difference between two age groups) × 100/mean value at the lower age group (22). An increment smaller than 2% is regarded as *below-average,* an increment of 2–3.9% is seen as *moderate,* an increment of 4–5.9% is referred to as *above-average,* and an increment of 6% or higher is regarded as *rapid* (2).

Age of full maturation was calculated using Blalock's method (23):

1. In measurements *increasing* with age, it is the age at which the mean value of the measurement *plus* two standard errors of the mean (SEM) reached the mean *minus* two standard errors of the mean (SEM) at age 18 years.
2. In measurements *decreasing* with age, it is the age at which the mean value of the finding *minus* two standard errors of the mean (SEM) reached the mean *plus* two standard errors of the mean (SEM) at age 18 years.

$$SEM = \text{standard deviation}/\sqrt{N}$$

where N is the number of subjects at a given age.

LINEAR MEASUREMENTS

The analysis of the growth-related changes was carried out in a selected number of measurements (27): head, 4; face, 9; orbits, 4; nose, 5; lips and mouth, 3; ears, 2. Tables 6–1 to 6–12 record the mean value and the developmental level at age 1 year, together with the mean value at age 18, the TGI between ages 1 and 18, the accelerated growth in each region, the age of maturation, and the difference between the value at maturation and at age 18 years, in each measurement. The results of the relative annual increment analysis were presented only in those measurements in which the growth was rapid and above average. Measurements with small differences between the sexes were analyzed jointly. The number of subjects examined in the individual regions varied between 1537 and 1594, divided almost equally between the sexes.

Head

Developmental Level at One Year of Age

The main cranial measurements reached high developmental levels, with 75.8% in craniofacial height (v-gn) of males being the lowest and 87.6% in head length (g-op) of females being the highest (Table 6–1).

Total Growth Increments (TGI)-Relative Total Increments (RTI)

The mean RTI was the highest (28.2%) in both sexes in the least developed craniofacial height (v-gn) (78.1%), and it was the lowest (14.9%) in the most developed head length (g-op) (87.1%). The head height (v-n) and the head width (eu-eu) exhibited identical moderate RTI values (19.3%) in both sexes. The smallest mean TGI value was observed in the head height (v-n) (18.5 mm), and the largest mean TGI value was seen in the craniofacial height (v-gn) (49.6 mm) (Table 6–1).

TABLE 6–1. *Growth patterns in the linear measurements of the head*[a]

Measurement	Sex	Normal value at 1 year of age, mean (mm)	Developmental level at 1 year of age, mean (%)		Normal value at 18 years, mean (mm)	Total growth (TGI) between 1 and 18 years, mean (mm)		Maturation age (years)	Difference between values at maturation and 18 years, mean (mm)
Width of the head (eu-eu)	M	125.5	83.1	83.8	151.1	25.6	23.9	15	2.4
	F	122.0	84.5		144.1	22.1		14	1.7
Height of the head (v-n)	M	97.5	82.8	84.0	117.6	20.1	18.5	13	4.3
	F	95.8	85.1		112.6	16.8		13	2.8
Length of the head (g-op)	M	166.7	86.5	87.1	192.7	26.0	24.5	14	3.5
	F	162.0	87.6		184.9	22.9		10	2.2
Craniofacial height (v-gn)	M	177.5	75.8	78.1	234.3	56.8	49.6	15	2.7
	F	173.8	80.4		216.2	42.4		11	3.2

[a] Abbreviations: M, male; F, female; TGI, total growth increment.

Growth Spurt

Early rapid growth (EGI) was seen between 1 and 4 years of age in both sexes in head height (v-n) and head length (g-op). Late accelerated growth (LGI) was registered in females only, in the head (v-n) and in the craniofacial heights (v-gn), between 11 and 12 years of age. The head width (eu-eu) in both sexes and the craniofacial height (v-gn) in males showed continuous below-average and moderate growth (Table 6–2).

Maturation

In males, the head measurements matured between age 13 (head height, v-n) and 15 years (head width, eu-eu; craniofacial height, v-gn). In females, the maturation occurred at earlier ages, between 10 years (head length, g-op) and age 14 years (head width, eu-eu). Differences between the mean measurement values at maturation and 18 years of age were in range of 1.7–4.3 mm (Table 6–1).

Face

Developmental Level at One Year of Age

In both sexes, the maxillary arc (t-sn-t) was the most developed measurement of the face (mean 81.9%), whereas the height of the mandible (sto-gn) was the least developed (mean 66.6%). Among the measurements generally, the surface arcs (t-sn-t and t-gn-t) reached the highest level of development (mean 79.3%), followed by width measurements (zy-zy and go-go, with a mean of 76.2%) and depth measurements in the maxillary (t-sn left) and mandibular regions (t-gn left) (mean 75.4%); the lowest level of development (mean 67.2%) occurred in the facial profile heights (n-gn; n-sto; sto-gn) (Table 6–3). The width of the mandible is better developed than the width of the face, but the height of the upper face is slightly more advanced in development than the height of the mandible.

Total Growth Increments (TGI) and Relative Total Increments (RTI)

The most developed maxillary arc (t-sn-t) at age 1 year (mean 81.9% in both sexes) revealed the smallest RTI value (mean 22.2 percent in both sexes), whereas the least developed mandible height (sto-gn) (mean 66.6% in both sexes) revealed the largest one (mean 50.5% in both sexes). The two main facial arcs (t-sn-t and t-gn-t) needed about *one-fourth* (26.3%) of their mean size at 1 year to reach the growth level at 18 years. The width (zy-zy and go-go) and depth measurements (sn-t and gn-t) of the face increased about *one-third* (31.9% and 32.9%, respectively). The highest RTI was seen in the heights of the face (n-gn; n-sto; sto-gn), which increased by almost *one-half* (49%) of their size at 1 year of age.

Growth Spurt

In general, the accelerated growth took place between ages 1 and 4 years, producing increments in the range 16.3% (zy-zy) to 21.1% (go-go) of the appropriate TGI. The depth and arc measurements of the face with high developmental levels by age 1 year revealed mostly below-average and moderate annual increments. In contrast, most of the profile height measurements with the lowest developmental levels at age 1 year showed rapid growth periods which produced great increments

TABLE 6–2. *Accelerated growth in the head*[a]

			Periods of accelerated growth						
			Early			Late			
Measurement	Sex	TGI (mm)	Age (years)	EGI, mean (mm)	EGI (in % of TGI), mean	Age (years)	LGI, mean (mm)	LGI (in % of TGI), mean	Comments
Head width (eu-eu)	M	25.6							*In both sexes:* Below average and moderate growth
	F	22.1							
Head height (v-n)	M	20.1	2–3	6.8	33.8				
	F	16.8	3–4	4.1	22.4	11–12	4.8	28.8	
Head length (g-op)	M	26.0	2–3	7.0	26.9				
	F	22.9	1–2	6.6	28.8				
Craniofacial height (v-gn)	M	56.8							*In males:* Below-average and moderate growth
	F	42.4				11–12	4.5	10.6	

[a] Abbreviations: M, male; F, female; TGI, total growth increment between 1 and 18 years; EGI, early growth increment; LGI, late growth increment.

TABLE 6–3. *Growth patterns in the linear measurements of the face*[a]

Measurement	Sex	Normal value at 1 year of age, mean (mm)	Developmental level at 1 year of age, mean (%)		Normal value at 18 years, mean (mm)	Total growth (TGI) between 1 and 18 years, mean (mm)		Maturation age, (years)	Difference between values at maturation and 18 years, mean (mm)
Width of the face (zy-zy)	M	96.7	70.5	72.1	137.1	40.4	37.4	15	3.6
	F	95.6	73.6		129.9	34.3		13	3.1
Width of the mandible (go-go)	M	76.2	78.5	80.2	97.1	20.9	18.7	13	3.0
	F	74.6	81.9		91.1	16.5		12	4.0
Height of the face (n-gn)	M	80.6	66.4	67.8	121.3	40.7	37.7	15	0.4
	F	77.2	69.1		111.8	34.6		13	2.7
Height of the upper face (n-sto)	M	49.0	66.2	67.3	74.0	25.0	23.3	14	4.0
	F	46.5	68.3		68.1	21.6		12	1.8
Height of the mandible (sto-gn)	M	31.9	63.7	66.6	50.1	18.2	16.0	15	2.3
	F	31.4	69.5		45.2	13.8		12	1.1
Depth in maxillary region, left (t-sn lt)	M	95.2	75.8	76.6	125.6	30.4	28.6	14	4.2
	F	91.4	77.3		118.3	26.9		13	2.3
Depth in mandibular region, left (t-gn lt)	M	100.1	72.3	74.2	138.5	38.4	34.7	15	2.5
	F	98.1	76.0		129.1	31.0		13	3.7
Maxillary arc (t-sn-t)	M	226.4	81.1	81.9	279.3	52.9	49.6	14	2.3
	F	221.3	82.7		267.5	46.2		12	0.0
Mandibular arc (t-gn-t)	M	226.3	74.4	76.7	304.1	77.8	68.8	15	0.0
	F	225.0	79.0		284.7	59.7		14	0.0

[a] Abbreviations: M, male; F, female; TGI, total growth increment.

amounting to 32.4% (n-sto) to 55.8% (sto-gn) of their TGI. The width of mandible in males was the only measurement in which, in spite of the high developmental level by age 1 year, rapid growth was observed. It extended to three 1-year periods (ages 3–4, 7–8, and 12–13 years) and produced the second greatest annual increment (13.7 mm), increasing the width of mandible by 65.6% of its TGI (24). Periods of early accelerated growth (EGI) were found mostly between 1 and 4 years of age; late growth (LGI) was noted only in width of the mandible, in both sexes (Table 6–4).

Maturation

In males, the age of maturation of the facial measurements was determined between 13 years (go-go) and 15 years of age (zy-zy, n-gn, sto-gn, gn-t and t-gn-t). In females, the facial measurements matured between 12 years (go-go, n-sto, sto-gn, t-sn-t) and 14 years of age (t-gn-t) (Table 6–3).

Comparison of the mean measurement values at ages of maturation and at 18 years of age revealed the following mean differences in both sexes: in width measurements of the face (zy-zy, go-go), 3.4 mm; in depth measurements of the face (t-sn, t-gn), 3.2 mm; in maxillary arc (t-sn-t), 2.3 mm; in vertical profile measurements (n-gn, n-sto, sto-gn), 2.1 mm. In the mandibular arc (t-gn-t), no difference was found (Table 6–3).

Orbits

Developmental Level at One Year of Age

Among the main orbital measurements, the most developed was the eye fissure height of males (ps-pi) (90.4%), whereas the least developed was the eye fissure length (ex-en) of females (82.7%) (Table 6–5).

Total Growth Increments (TGI) and Relative Total Increments (RTI)

Because of high developmental levels of all orbital measurements (Table 6–5) at 1 year of age, the mean TGI values are modest, between 1.2 mm (ps-pi) and 12.5 mm (ex-ex), in both sexes. To reach the findings at 18 years, the height of the eye fissure (ps-pi) had to increase by 12.5% (1.2 mm), the biocular width (ex-ex) by 16.5% (12.5 mm), the intercanthal width (en-en) by 19.2% (5.2 mm), and the length of the eye fissure by 20.6% (5.3 mm) of their values at 1 year of age (RTI values).

Growth Spurt

During accelerated growth between 3 and 4 years of age, the EGI of the intercanthal width (en-en) took more than half of the TGI in males and almost half of the TGI in females. Early rapid growth in males between 2 and

TABLE 6–4. *Accelerated growth in the face*[a]

			Periods of accelerated growth						
			Early			Late			
Measurement	Sex	TGI (mm)	Age (years)	EGI, mean (mm)	EGI (in % of TGI), mean	Age (years)	LGI, mean (mm)	LGI (in % of TGI), mean	Comments
Width of the face (zy-zy)	M	40.4	3–4	8.8	21.8				
	F	34.3	3–4	5.6	16.3				
Width of the mandible (go-go)	M	20.9	3–4	4.4	21.1	7–8, 12–13	9.3	44.5	
	F	16.5				6–7	4.4	26.7	
Height of the face (n-gn)	M	40.7	1–2, 3–4	14.8	36.4				
	F	34.6	1–2, 3–5	16.2	46.8				
Height of the upper face (n-sto)	M	25.0	1–2, 3–4	8.1	32.4				
	F	21.6	1–4	9.6	44.4				
Height of the mandible (sto-gn)	M	18.2	1–2, 3–4	9.7	53.3				
	F	13.8	1–2, 3–4	7.7	55.8				
Maxillary depth, left (t-sn lt)	M	30.4	3–4	6.0	19.7				*In females:* Below average growth
	F	26.9							
Mandibular depth, left (t-gn lt)	M	38.4							*In both sexes:* Below average and moderate growth
	F	31.0							
Maxillary arc (t-sn-t)	M	52.9							*In both sexes:* Below average and moderate growth
	F	46.2							
Mandibular arc (t-gn-t)	M	77.8							*In both sexes:* Below average and moderate growth
	F	59.7							

[a] Abbreviations: M, male; F, female; TGI, total growth increment between 1 and 18 years; EGI, early growth increment; LGI, late growth increment.

TABLE 6–5. *Growth patterns in the linear measurements of the orbits*[a]

Measurements	Sex	Normal value at 1 year of age, mean (mm)	Developmental level at 1 year of age, mean (%)		Normal value at 18 years, mean (mm)	Total growth (TGI) between 1 and 18 years, mean (mm)		Maturation age (years)	Difference between values at maturation and 18 years, mean (mm)
Intercanthal width (en-en)	M	27.3	83.0	84.1	32.9	5.6	5.2	11	0.3
	F	26.9	85.1		31.6	4.7		8	0.5
Biocular width (ex-ex)	M	76.0	85.0	85.9	89.4	13.4	12.5	15	0.0
	F	75.3	86.8		86.8	11.5		13	0.4
Eye fissure length (ex-en lt)	M	25.9	83.0	82.9	31.2	5.3	5.3	15	0.2
	F	25.4	82.7		30.7	5.3		13	0.9
Eye fissure height (ps-pi lt)	M	9.4	90.4	88.9	10.4	1.0	1.2	11	0.8
	F	9.7	87.4		11.1	1.4		14	0.6

[a] Abbreviations: M, male; F, female; TGI, total growth increment.

TABLE 6–6. *Accelerated growth in the orbits*[a]

			Periods of accelerated growth						
			Early			Late			
Measurement	Sex	TGI (mm)	Age (years)	EGI, mean (mm)	EGI (in % of TGI), mean	Age (years)	LGI, mean (mm)	LGI (in % of TGI), mean	Comments
Intercanthal width (en-en)	M	5.6	3–4	3.1	55.4				
	F	4.7	3–4	2.0	42.6				
Biocular width (ex-ex)	M	13.4							*In both sexes:* Below-average and moderate growth
	F	11.5							
Eye fissure length (ex-en lt)	M	5.3							*In both sexes:* Below average and moderate growth
	F	5.3							
Eye fissure height (ps-pi lt)	M	1.0	2–3	0.4	40.0				
	F	1.4				8–9 13–14	0.8	57.1	

[a] Abbreviations: M, male; F, female; TGI, total growth increment between 1 and 18 years; EGI, early growth increment; LGI, late growth increment.

3 years also increased the eye fissure height (ps-pi). In females, the great increase in size of the eye fissure heights was achieved by the late accelerated growth period (LGI) between 8 and 14 years of age. The biocular width (ex-ex) and the eye fissure length (ex-en lt) developed gradually with below-average and moderate increments (Table 6–6).

Maturation

In males, the intercanthal width (en-en) and the eye fissure height (ps-pi) matured earlier (11 years) than the biocular width (ex-ex) or the eye fissure length (ex-en) (15 years). In females, the intercanthal width (en-en) reached its full maturation by 8 years of age, the biocular width (ex-ex) and the eye fissure length (ex-en) became mature at 13 years, and the eye fissure height (ps-pi) reached its full potential at 14 years of age (17). At maturation ages the mean values of the measurements in both sexes were smaller than those at 18 years of age, by a mean of 0.5 mm (Table 6–5).

Nose

Developmental Level at One Year of Age

The highest mean developmental level was reached in both sexes by the nose width (al-al) (79.5%), whereas the lowest was reached by the nasal tip protrusion (sn-prn) (51%) (Table 6–7).

TABLE 6–7. *Growth patterns in the linear measurements of the nose*[a]

Measurement	Sex	Normal value at 1 year of age, mean (mm)	Developmental level at 1 year of age, mean (%)		Normal value at 18 years, mean (mm)	Total growth (TGI) between 1 and 18 years, mean (mm)		Maturation age (years)	Difference between values at maturation and 18 years, mean (mm)
Width of the nose (al-al)	M	26.5	76.4	79.5	34.7	8.2	6.9	14	1.6
	F	25.9	82.5		31.4	5.5		12	0.5
Height of the nose (n-sn)	M	30.9	58.3	59.0	53.0	22.1	20.9	15	1.1
	F	29.2	59.7		48.9	19.7		12	1.7
Nasal bridge length (n-prn)	M	27.6	56.3	56.7	49.0	21.4	20.5	15	1.2
	F	25.9	57.0		45.4	19.5		13	1.0
Nasal tip protrusion (sn-prn)	M	10.1	49.0	51.0	20.6	10.5	9.8	16	0.5
	F	10.2	52.9		19.3	9.1		14	0.7
Nasal ala length, left (ac-prn lt)	M	19.7	57.1	59.9	34.5	14.8	13.2	15	0.0
	F	19.3	62.7		30.8	11.5		13	0.1

[a] Abbreviations: M, male; F, female; TGI, total growth increment.

Total Growth Increments (TGI) and Relative Total Increments (RTI)

Width of the nose (al-al) which showed the highest mean developmental level of 79.5% at 1 year in both sexes needed the smallest mean TGI of 6.9 mm, expressing 26.3% of the mean absolute values of RTI at 1 year of age in both sexes. The nasal tip protrusion (sn-prn) with the smallest developmental level by 1 year of age increased in the following years by a mean of 9.8 mm, representing the highest RTI values of 96.1%. The remaining nasal measurements (n-prn, n-sn, and ac-prn) possessing mean TGI values of 20.5 mm, 20.9 mm, and 13.2 mm (Table 6–7), expressed 76.5%, 69.4%, and 67.7% of the appropriate RTI values at 1 year of age.

Growth Spurt

In the nasal tip protrusion (sn-prn) the smallest developmental level at 1 year of age among the nasal measurements (Table 6–7) was well compensated by a large EGI obtained during the extended early growth period between 1 and 6 years in both sexes, and in males by an additional late rapid growth period between 11 and 16 years. The best-developed width of the nose (al-al) had only a short accelerated growth period between 3 and 4 years. Certain similarities were found in the developmental levels of the nose height (n-sn) and nasal bridge length (n-prn), and their early rapid growth periods were slightly longer in the bridge length between 1 and 6 years, with EGI producing around one-third of their TGI. In males, the nose height gained a little during the short late accelerated growth period between 11 and 12 years. The early rapid growth period between 1 and 4 years of age in both sexes brought only modest increments (EGI) for the length of the ala (ac-prn lt) (Table 6–8).

Maturation

In males, the width of the nose (al-al) reached its mature size at 14 years of age, the nose height (n-sn), bridge length (n-prn), and ala length (ac-prn) reached their full size at 15 years, and the nasal tip protrusion (sn-prn) reached full maturity at 16 years. In females, the nose width (al-al) and nose height (n-sn) matured at 12 years of age, the nasal bridge length (n-prn) and ala length (ac-prn) reached their full size at 13 years, and the nasal tip protrusion reached full maturity at 14 years. The mean differences between the findings at maturation time and at 18 years of age in both sexes were 1.1 mm in al-al, 1.4 mm in n-sn, 1.1 mm in n-prn, 0.6 mm in sn-prn, and 0.1 mm in ac-prn (Table 6–7).

Lips and Mouth

Developmental Level at One Year of Age

In both sexes the upper lip height (sn-sto) was the most developed measurement (mean 81.6%), followed by the height of the lower lip (sto-sl) (mean 73.4%) and the width of the mouth (ch-ch) (mean 66.1%) (Table 6–9).

TABLE 6–8. *Accelerated growth in the nose*[a]

			Periods of accelerated growth					
			Early			Late		
Measurement	Sex	TGI (mm)	Age (years)	EGI, mean (mm)	EGI (in % of TGI), mean	Age (years)	LGI, mean (mm)	LGI (in % of TGI), mean
Nose width (al-al)	M	8.2	3–4	2.3	28.0			
	F	5.5	3–4	1.9	34.5			
Nose height (n-sn)	M	22.1	1–2 3–4	7.0	31.7	11–12	2.5	11.3
	F	19.7	1–4	8.6	43.7			
Nasal bridge length (n-prn)	M	21.4	1–4 5–6	7.5	35.0			
	F	19.5	1–2 5–6	6.0	30.8			
Nasal tip protrusion (sn-prn)	M	10.5	1–4 5–6	4.7	44.8	11–12 15–16	2.5	23.8
	F	9.1	1–3 4–6	4.7	51.6			
Ala length (ac-prn-lt)	M	14.8	1–2 3–4	2.7	18.2			
	F	11.5	1–3	2.7	23.5			

[a] Abbreviations: M, male; F, female; TGI, total growth increment between 1 and 18 years; EGI, early growth increment; LGI, late growth increment.

TABLE 6–9. *Growth patterns in the linear measurements of the lips and mouth*[a]

Measurement	Sex	Normal value at 1 year of age, mean (mm)	Developmental level at 1 year of age, mean (%)		Normal value at 18 years, mean (mm)	Total growth (TGI) between 1 and 18 years, mean (mm)		Maturation age (years)	Difference between values at maturation and 18 years, mean (mm)
Height of the upper lip (sn-sto)	M	17.3	79.4	81.6	21.8	4.5	3.9	11	1.0
	F	16.4	83.7		19.6	3.2		5	0.7
Width of the mouth (ch-ch)	M	34.8	65.3	66.1	53.3	18.5	17.5	14	1.2
	F	33.3	66.9		49.8	16.5		14	0.5
Height of the lower lip (sto-sl)	M	13.2	70.2	73.4	18.8	5.6	4.8	13	1.3
	F	12.8	76.6		16.7	3.9		9	0.9

[a] Abbreviations: M, male; F, female; TGI, total growth increment.

Total Growth Increments (TGI) and Relative Total Increments (RTI)

In both sexes the well-developed upper lip height (sn-sto) had to increase only by 3.9 mm (TGI)—that is, 23.1% of its height at 1 year (RTI)—to reach the height at 18 years. To reach the same level, the lower lip height (sto-sl) needed a mean of 4.8 mm (TGI) [i.e., 36.5% of its size at 1 year of age (RTI)] and the width of the mouth (ch-ch) needed a mean of 17.5 mm (TGI) [i.e., 51.3% of its size at 1 year of age (RTI)], in both sexes.

Growth Spurt

The large EGI in relation to the TGI value in the height of the upper lip in females during the early rapid growth (1–3 years of age) was partly made up in males by early (1–2 years of age) and late (15–16 years of age) rapid growth periods. Similarly as in the upper lip of males, the height of the lower lip showed a great increase by EGI (between 3 and 6 years) and LGI (between 10 and 11 years). The lower lip height obtained almost one-third of its TGI in females during the short early rapid growth period (3–4 years of age). The width of the mouth (ch-ch) revealed rapid growth only in males between 3 and 4 years of age, producing a small portion of the TGI. In females, the growth was continuous and moderate (Table 6–10).

Maturation

The upper lip height (sn-sto) revealed the earliest maturation age, in females at 5 years of age and in males at 11 years. In females the lower lip height (sto-sl) matured at 9 years of age, whereas in males it matured at 13 years. The width of the mouth (ch-ch) reached its mature size at 14 years, in both sexes. At ages of maturation the mean measurements of the upper and lower lip heights and mouth width in both sexes differed only by a mean of 0.9 mm from the values at 18 years of age (Table 6–9).

Ears

Developmental Level at One Year of Age

The highly developed width of the ear (pra-pa) in both sexes (93.5%) was in contrast with the much lower developmental level (76.4%) of the ear length (sa-sba) (Table 6–11) (25).

Total Growth Increments (TGI) and Relative Total Increments (RTI)

The well-developed width of the ear (pra-pa) at 1 year needed only a mean of 2.3 mm (TGI) in both sexes to reach the findings at 18 years. This increment represents 7.1% of the mean ear width in both sexes at 1 year of age (RTI). To attain the level found in 18-year-olds, the length of the ear at 1 year of age needed a six times larger increment (a mean of 14.3 mm, TGI) than did the ear width. In terms of RTI, the length of the ear had to increase by 31% of its mean length at 1 year of age.

Growth Spurt

Early rapid growth in ear length (sa-sba) was noted only in males in a short period between 2 and 3 years of age, producing a mild increase (EGI). The length of the ear in females and the width of the ear (pra-pa) in both sexes exhibited below-average and moderate annual growth increments, interspersed with short intervals of no growth (25) (Table 6–12).

Maturation

The advanced developmental level of the width of the ear (pra-pa) led to early maturation in both sexes (in males at 7 years of age, in females at 6 years of age). The length of the ear (sa-sba) was fully developed at 13 years in males and at 12 years in females. The differences be-

TABLE 6–10. *Accelerated growth in lips and mouth*[a]

			Periods of accelerated growth						
			Early			Late			
Measurement	Sex	TGI (mm)	Age (years)	EGI, mean (mm)	EGI (in % of TGI), mean	Age (years)	LGI, mean (mm)	LGI (in % of TGI), mean	Comments
Upper lip height (sn-sto)	M	4.5	1–2	1.4	31.1	15–16	1.3	28.9	
	F	3.2	1–3	2.4	75.0				
Mouth width (ch-ch)	M	18.5	3–4	2.2	11.9				*In females:* Continuous moderate increments
	F	16.5							
Lower lip height (sto-sl)	M	5.6	3–4 5–6	2.7	48.2	10–11	1.1	19.6	
	F	3.9	3–4	1.2	30.8				

[a] Abbreviations: M, male; F, female; TGI, total growth increment between 1 and 18 years; EGI, early growth increment; LGI, late growth increment.

TABLE 6–11. *Growth patterns in the linear measurements of the ear*[a]

Measurement	Sex	Normal value at 1 year of age, mean (mm)	Developmental level at 1 year of age, mean (%)	Normal value at 18 years, mean (mm)	Total growth (TGI) between 1 and 18 years, mean (mm)	Maturation age, mean (mm)	Difference between values at maturation and 18 years, mean (mm)
Width of the ear, left (pra-pa lt)	M	32.7	92.4 } 93.5	35.4	2.7 } 2.3	7	0.5
	F	31.7	94.6	33.5	1.8	6	0.5
Length of the ear, left (sa-sba lt)	M	46.9	75.2 } 76.4	62.4	15.5 } 14.3	13	1.4
	F	45.4	77.6	58.5	13.1	12	1.1

[a] Abbreviations: M, male; F, female; TGI, total growth increment.

TABLE 6–12. *Accelerated growth in the ear*[a]

			Periods of accelerated growth						
			Early			Late			
Measurements	Sex	TGI (mm)	Age (years)	EGI, mean (mm)	EGI (in % of TGI), mean	Age (years)	LGI, mean (mm)	LGI (in % of TGI), mean	Comments
Ear width (pra-pa lt)	M	2.7							*In both sexes:* Below average and moderate growth
	F	1.8							
Ear length (sa-sba-lt)	M	15.5	2–3	2.6	16.8				*In females:* Below-average and moderate growth
	F	13.1							

[a] Abbreviations: M, male; F, female; TGI, total growth increment between 1 and 18 years; EGI, early growth increment; LGI, late growth increment.

tween the measurements at maturation time and 18 years were in the range of a mean of 0.5–1.3 mm (Table 6–11) (25).

ANGULAR MEASUREMENTS

Fourteen angular measurements (10 inclinations and 4 angles) were analyzed: 1 in the head, 3 in the face, 1 in the orbits, 5 in the nose, 2 in the lips and mouth, and 2 in the ears. Four of the 14 measurements showed a decreasing trend in values between 1 and 18 years of age (Table 6–13). The remaining 10 angular measurements (7 inclinations and 3 angles) increased with age (Table 6–14).

The tables record the findings in both sexes separately at 1 year and 18 years of age and report the mean differences between these age groups. The age of maturation was calculated, and the value of the inclination (angle) for that age was reported.

Angular Measurements Decreasing with Age

Head

Inclination of the Anterior Surface of the Forehead from the Vertical (Table 6–13)

On average, the squamous portion of the frontal bone tilted forward slightly in the youngest age groups (see Appendix A, Table I–S–13).

By 1 year, the mean inclination in both sexes was 5.3 degrees. By 10 years in males and 11 years in females the forehead inclination was almost vertical. In the following years the backward tilt gradually increased, reaching a mean of −10.1 degrees in 18-year-old males and −6.6 degrees in females. The forehead inclination reached maturity in both sexes at 16 years of age.

Orbits

Inclination of the Left Eye Fissure from the Horizontal (Table 6–13)

At 1 year of age the mean inclination was 5.2 degrees in males and 5.4 degrees in females. Inclination between 1 and 18 years of age decreased in both sexes by a mean of 2.2 degrees.

A drop in inclination values was observed at 6 years in males and 8 years in females (see Appendix A, Table III–P–20). In subsequent years the increments were minimal in males. In females, after 13 years of age, inclination stabilized at a slightly higher level. At 18 years of age the mean inclination of the left eye fissure axis was 4.1 degrees in females and 2.1 degrees in males. The eye fissure inclinations were regarded as mature at 7 years of age in females and at 11 years of age in males.

Nose

Nasal Tip Angle (Table 6–13)

At 1 year of age, the obtuse angle of the nasal tip (a mean 82.1 degrees in both sexes) exhibited a moderate reduction in size in the following years, ending up with a mean of 72.3 degrees at 18 years, in both sexes. The nasal tip angle reached its mature size in females at 15 years and in males at 16 years of age (see Appendix A, Table IV–S–15).

Upper Lip

Inclination of the Upper Lip from the Vertical (Table 6–13)

In both sexes and for all age groups studied, the skin surface of the upper lip was positively inclined ("protrud-

TABLE 6–13. *Inclinations and angles decreasing with age in the craniofacial complex*[a]

Region	Inclination (I) or angle (A)	Sex	Finding at 1 year, mean (degrees)	Finding at 18 years, mean (degrees)	Difference between 1 and 18 years, mean (degrees)	Maturation: Age (years)	Maturation: Finding, mean (degrees)
Head	Forehead (I)	M	4.9	−10.1	15.0	16	−7.0
		F	5.7	−6.6	12.3	16	−3.4
Orbit	Eye fissure, left (I)	M	5.2	2.1	3.1	11	2.7
	(en-ex line)	F	5.4	4.1	1.3	7	4.7
Nose	Nasal tip (A)	M	82.6	74.6	8.0	16	77.5
		F	81.5	69.9	11.6	15	77.8
Lip	Upper lip (I)	M	18.7	4.9	13.8	16	5.0
	(sn-ls line)	F	15.1	5.3	9.8	14	6.4

[a] Abbreviations: M, male; F, female.

TABLE 6–14. *Inclinations and angles increasing with age in the craniofacial complex*[a]

Region	Inclination (I) or angle (A)	Sex	Finding at 1 year, mean (degrees)	Finding at 18 years, mean (degrees)	Difference between 1 and 18 years, mean (degrees)	Maturation	
						Age (years)	Finding, mean (degrees)
Face	Upper face (I)	M	−5.2	0.6[b]	4.4	14	0.5
	(g-sn line)	F	−5.3	0.3[b]	5.0	13	0.2
	Lower face (I)	M	−16.8	−14.5	2.3	15	−15.8
	(sn-pg line)	F	−17.3	−14.1	3.2	14	−15.2
	Mandible (I) (li-pg line)	M	−24.8	−24.4	0.4	15	−25.9
		F	−24.4	−18.9	5.5	14	−21.6
Nose	Nasal bridge (I)	M	29.9	31.6	1.7	Cannot be determined	
		F	29.7	30.0	0.3	statistically	
	Nasofrontal (A)	M	125.8	130.5	4.7	6	129.2
		F	123.0	134.0	11.0	8	131.1
	Nasolabal (A)	M	94.9	98.9	4.0	13	98.6
		F	95.7	99.1	3.4	10	100.7
	Columella (I)	M	63.6	80.4[b]	16.8	16	73.5
		F	63.8	82.9	19.1	15	74.3
Lip	Lower lip (I) (li-sl line)	M	−74.4	−57.1[b]	17.3	15	−54.5
		F	−66.6	−58.7[b]	7.9	14	−58.7
Ear	Ear axis, left (I)	M	19.1	20.0	0.9	Cannot be determined	
		F	18.0	18.4	0.7	statistically	
	Ear protrusion, left (A)	M	16.3	23.0	6.7	6	22.8
		F	15.4	22.3	6.9	8	19.9

[a] Abbreviations: M, male; F, female.
[b] Value at 17 years.

ing") from the vertical line, drawn through the subnasale. At 1 year of age, the protrusion of the upper lip was moderately larger in males, indicated by greater inclination (mean 18.7 degrees) than in females (mean 15.1 degrees).

The drop in inclination between 1 and 18 years of age was marked in both sexes, more in males than in females. The upper lip inclination matured at 14 years in females and at 16 years in males (see Appendix A, Table V–S–12).

Angular Measurements Increasing with Age

Face

Inclination of the Upper Face Profile from the Vertical

At 1 year of age, the mean upper face inclination was almost identical in both sexes (−5.3 degrees). At 12 years of age, in both sexes, the moderately receding upper face profile became slightly protruding: a mean of 2.3 degrees in males and a mean of 1.0 degrees in females (see Appendix A, Table II–S–16).

In the following years the changes were minimal in both sexes. The mean values of the upper face inclination at 17 years of age were close to vertical in both sexes. The maturation in inclination occurred at 13 years of age in females and at 14 years of age in males (Table 6–14).

Lower Face Inclination from the Vertical

At 1 year of age the lower face profile line greatly receded; this is indicated by negative inclination values (−16.8 degrees in males and −17.3 degrees in females), almost identical in both sexes. By 18 years, the negative tilt of the lower face became less marked (a mean of −14.3 degrees in both sexes). The maturation time was reached at 14 years in females and at 15 years in males (Table 6–14) (see Appendix A, Table II–S–18).

Inclination of the Mandible from the Vertical

By 1 year of age, the mandible revealed almost identical mean negative inclination in both sexes (−24.6 degrees). With age, the recession of the mandible became milder, much more in females, indicated by the mean −18.9 degrees in 18-year-old females, in contrast with a mean of −24.4 degrees in 18-year-old males. Maturation of the mandible inclination occurred at 14 years of age in females and at 15 years of age in males (Table 6–14).

Nose

Nasal Bridge Inclination from the Vertical

By 1 year of age, the inclination of the nasal bridge was almost identical (mean 29.8 degrees) in both sexes.

Between 1 and 18 years, the changes in the bridge inclinations were minimal (Table 6–14). Because changes in degrees of inclination in the age groups were very small (see Appendix A, Table IV–S–9) the age of maturation could not be determined statistically.

Nasofrontal Angle

The slightly acute nasofrontal angle in early childhood in both sexes (a mean of 125.8 degrees in 1-year-old males, a mean of 123.0 degrees in 1-year-old females) became moderately flatter with age, more in females than in males (see Appendix A, Table IV–S–13), reaching 134.0 degrees and 130.5 degrees, respectively, at 18 years of age. The nasofrontal angle matured in males at 6 years of age and in females at 8 years of age (Table 6–14).

Nasolabial Angle

The mean values of the nasolabial angle were almost identical in both sexes at 1 year of age (95.3 degrees). The angle displayed a moderate increase between 1 and 6 years of age (see Appendix A, Table IV–S–17), showing a mean 99-degree angle at 18 years in both sexes. The angle reached its mature size at 10 years of age in females and at 13 years of age in males (Table 6–14).

Columella Inclination from the Vertical (Table 6–14)

The small inclination of the columella at 1 year of age in both sexes (a mean of 63.7 degrees) became moderately larger with age, reaching at 17 years of age a mean of 82.9 degrees in females and 80.4 degrees in males (see Appendix A, Table IV–S–10). The inclination of the columella matured at 15 years in females and at 16 years in males (Table 6–14).

Lip

Lower Lip Inclination from the Vertical

By 1 year of age, the mean negative inclination was larger in males (−74.4 degrees) than in females (−66.6 degrees). In later years the negative inclination of the lower lip moderately decreased, reaching almost the same level at 17 years of age in both sexes (mean −57.9 degrees) (see Appendix A, Table V–S–13). Maturation of the lower lip inclination was observed at 14 years in females and at 15 years in males (Table 6–14).

Ear

Inclination of the Left Ear's Axis from the Vertical

The degree of inclination of the medial longitudinal axis of the ear was continuously smaller in females: At 1 year of age it was smaller by a mean of 1.1 degree, and by 18 years it showed a 1.6-degree difference in comparison with males (see Appendix A, Table VI–P–18).

The mean inclination value in both sexes at 1 year of age was 18.5 degrees, and at 18 years it was 19.2 degrees. The age of maturation of the ear axis inclination cannot be determined statistically due to minimal growth changes in the age groups (Table 6–14).

Protrusion of the Left Ear from the Head

At 1 year of age the protrusion was slightly larger in males than in females, and the mean protrusion angle in both sexes was 15.9 degrees. The mean protrusion angle became larger (by 6.8 degrees) in both sexes between 1 and 18 years, reaching 22.7 degrees at 18 years of age. The age-related changes were small, gradually increasing the protrusion (see Appendix A, Table VI–P–19). The angle of the ear protrusion matured at 6 years of age in males and at 8 years of age in females (Table 6–14).

DISCUSSION AND CONCLUSIONS

The linear and angular measurements chosen for the study represent only a small part of the complete set (41 of 132). They define the morphological framework of the regions. The study of the postnatal development was confined to subjects ranging between 1 and 18 years of age. Data from birth to 1 year were omitted due to small sample size and the uncertain quality of measurements obtained from uncooperative infants.

That part of the study relating to linear measurements showed substantial differences in the levels of development in the individual facial regions reached by age 1 year. In the head and orbits the measurements reached, on average, 78–88.9% of their adult size. With most of the nasal and labio-oral measurements, however (a mean 51–73.4% in both sexes), only the nose width and the upper lip height exhibited high developmental levels (mean 79.5% and 81.6%, respectively).

Although TGI records useful information about the entire growth increment (between 1 and 18 years), RTI values reveal the real rate of growth by expressing TGI as a percentage of the measurement value at 1 year of age. Thus, a poorly developed mandible height has to increase by a mean of 50.5% in both sexes, a mouth width by 51.5% and the small nasal tip protrusion by 96.1% of their initial sizes, at 1 year of age, to achieve their respective adult sizes. The well-developed measurements, such as head length, mandibular arc, eye fissure height, or width of ear, displayed a much lower RTI (7–22%) in the same period.

It was observed that in measurements which displayed small early developmental levels, the periods of early accelerated growth were longer and were often extended by late rapid growth. Thus, in the poorly developed (51% at age 1 year) nasal tip protrusion, rapid growth increments during the first 6 years accounted for a mean of 60% of the TGI values in both sexes (Table 6–8). In contrast, most of the measurements with high early developmental levels [e.g., head width (with a mean of 83.8% in both sexes); depth and arc measurements of the face (with a mean of 74.2–81.9% in both sexes); and in the orbits the intercanthal width (with a mean of 84.2% in both sexes) and biocular width (with a mean of 85.9% in both sexes)] showed only continuous growth with below-average and moderate increases (Tables 6–2, 6–4, and 6–6).

Those craniofacial regions having features vital to basic functions were observed to possess early developmental levels (at age 1 year) markedly higher than others. These included: head width, length, and height (brain); intercanthal and biocular width and eye fissure height in the orbits (visus); nose width (air-flow exchange); upper lip height (breast-feeding) in the orolabial region.

Suggestions for avoiding major surgical trauma during periods of rapid growth were expressed by some investigators (26–30). It is our feeling that, especially in the nasolabial region, extensive early rapid growth should be taken into account when planning a surgical revision (31).

Postnatal changes in angular measurements were small to moderate, slightly greater in decreasing than in increasing measurements. Among the decreasing angular measurements in both sexes, age-related decline in the forehead inclination was greater (mean 13.7 degrees) than in the upper lip (mean 11.9 degrees) and nasal tip angle (mean 9.8 degrees). Among the increasing angular measurements the greatest change occurred in the columella inclination (mean 17.8 degrees), followed by the lower lip inclination (12.6 degrees) and nasofrontal angle (mean 7.9 degrees). In the remaining angular measurements, mean increments were between 0.4 degrees and 6.9 degrees (Table 6–14).

Detailed monitoring of changes in linear and angular measurements during postnatal development showed that in addition to individual active growth, indirect forces can influence the size or position of measurements. Such factors include the effects caused by alterations in neighboring measurements. Thus it was observed that an increase in the length of the head was associated with a reduction in forehead inclination. The increase in length of the nasal bridge contributed to an increase in the columella inclination and a decrease in the nasolabial angle. The increase in the biocular width was accompanied by an increase in eye fissure length and a decrease in the inclination of the longitudinal axis (17). The increase of the maxillary depth led to a change in the upper face inclination from negative values to positive. Further analysis is needed to discover more comprehensive details about the interactive qualities of the relationships between the measurements of a growing face.

The study of linear measurements revealed certain links between the extent of development reached by age 1 year and that found at the age of maturity. Measurements showing a high developmental level by 1 year of age reached maturity generally at an earlier age. Thus the ear width (pra-pa) with the highest developmental level of 93.5% matured in females at age 6 and in males at age 7. Linear measurements with a developmental level between 80% and 88.9% (e.g., go-go, sn-sto, eu-eu, v-n, en-en, g-op, ps-pi) reached mature sizes between ages 5 and 15, whereas those showing developmental levels between 72.1% and 79.5% (e.g., zy-zy, sto-sl, sa-sba, t-sn, v-gn, al-al) matured between ages 11 and 15. The remaining linear measurements exhibited developmental levels between 66.1% and 67.8% (ch-ch, sto-gn, n-sto, and n-gn) or in the range 56.5–59.9% (n-prn, n-sn, and ac-prn), reaching their mature sizes between the ages of 12 and 15 years (Tables 6–1 to 6–12).

Generally, angular measurements reached maturity at a later age than did linear measurements. Full maturity in angular measurements was reached between 6 and 16 years in both sexes, usually occurring 1–4 years earlier in females. Late maturation was observed (in females only) in the inclination of the forehead and columella and in the nasal tip angle. The earliest age of maturation was recorded in the nasofrontal angle and the ear protrusion angle in males at 6 years and in females at 8 years.

Because of the small variations in the reading (degrees) of angular measurements during postnatal development, the precise determination of the age of maturity is often difficult when compared with linear measurements which have more well-established growth trends. Thus, the determination of maturity in the inclination of the nasal bridge and the inclination of the longitudinal axis of the auricle was impossible due to the small differences registered between mean values during postnatal development (Tables 6–13 and 6–14).

Population studies in children with congenital anomalies of the head and face, even if fractional, may disclose (a) the presence or absence of early rapid growth periods or (b) the extension of growth periods beyond those found in a healthy population. As a result of the extended growth period, the time taken for measurements to reach maturity could change to a later age than would be accepted as normal. Our follow-up examinations of young adults after surgical treatment occasionally revealed unexpected late facial growth for their age. Systematic searching for these changes would help to reinforce our beliefs concerning the influence of "natural forces" on the development of surgically treated faces.

REFERENCES

1. Hajniš K. Nose of patients with cheilo-, gnatho-, palatoschisis unilateralis immediately after lip suture. *Acta Fac Rerum Nat Univ Comenianae Anthropol* 1976;22:87–92.
2. Farkas LG. *Anthropometry of the head and face in medicine.* New York: Elsevier, 1981.
3. Farkas LG, Posnick JC. Growth and development of regional units in the head and face based on anthropometric measurements. *Cleft Palate Craniofac J* 1992;29:301–302.
4. Subtelny JD. A longitudinal study of soft-tissue facial structures and their profile characteristics, defined in relation to underlying skeletal structures. *Am J Orthod* 1959;45:481–507.
5. Goldstein MS. Changes in dimensions and form of the face and head with age. *Am J Phys Anthropol* 1936;22:37–89.
6. Brodie AG. Behavior of normal and abnormal facial growth patterns. *Am J Orthod* 1941;27:633–647.
7. Björk A. The face in profile. An anthropological x-ray investigation on Swedish children and conscripts. *Svensk Tandläk-T* 1947; 49(Suppl).
8. Tanner JM. The assessment of growth and development in children. *Arch Dis Child* 1952;27:10–33.
9. Savara BS, Singh IJ. Norms of size and annual increments of seven anatomical measures of maxillae in boys from three to sixteen years of age. *Angle Orthod* 1968;38:104–120.
10. Twiesselmann Fr. *Dévelopment biométrique de l'enfant à l'adulte.* Bruxelles: Presses Universitaires de Bruxelles, 1969.
11. Hajniš K. Das Wachtstum von Nase, Oberlippe und Mundspalte bei Deutschen Kindern und Jungendlichen und seine Bedeutung für die Operationen der Spaltmissbildungen des cavum oris. *HOMO* 1970;21:142–149.
12. Figalová P, Šmahel Z. Irregularities of facial growth within region exposed to the risk of cleft. *Acta Univ Carol (Biol) Praha* 1975;1973:199–207.
13. Krogman WM, Meier J, Canter H, Ross P, Mazaheri M, Mehta S. Craniofacial serial dimensions related to age, sex, and cleft-type from six months of age to two years. *Growth* 1975;39:195–208.
14. Pruzansky S. Time: the fourth dimension in syndrome analysis applied to craniofacial malformations. *Birth Defects* 1977;13(3C): 3–28.
15. Farkas LG. Otoplastic architecture. In: Davis J, ed. *Aesthetic and reconstructive otoplasty.* New York: Springer-Verlag, 1987;13–52.
16. Poswillo D. The aetiology and pathogenesis of craniofacial deformity. *Development* 1988;103(Suppl):207–212.
17. Hreczko T, Farkas LG, Katic M. Clinical significance of age-related changes of the palpebral fissures between age 2 and 18 years in healthy Caucasians. *Acta Chir Plast* 1990;32:194–204.
18. Farkas LG, Ngim RC, Lee ST. The fourth dimension of the face: a preliminary report of growth potential in the face of the Chinese population of Singapore. *Ann Acad Med Singapore* 1988;17:319–345.
19. Farkas LG. Age- and sex-related changes in facial proportions. In: Farkas LG, Munro IR, eds. *Anthropometric facial proportions in medicine.* Springfield, IL.: Charles C Thomas, 1987;29–56.
20. Hellman M. Changes in the human face brought about by development. *Int J Orthod* 1927;13:475–516.
21. Minot CS. On the weight of guinea pigs. *J Physiol* 1891;12:97–153.
22. Singh R. A cross sectional study of growth in five somatometric traits of Punjabi boys aged eleven to eighteen years. *Am J Phys Anthropol* 1970;32:129–138.
23. Blalock HM Jr. *Social statistics.* New York: McGraw-Hill, 1960;158–162.
24. Farkas LG, Posnick JC, Hreczko TM. Growth patterns of the face: a morphometric study. *Cleft Palate Craniofac J* 1992;29:308–315.
25. Farkas LG, Posnick JC, Hreczko TM. Anthropometric growth study of the ear. *Cleft Palate Craniofac J* 1992;29:324–329.
26. Brodie A. On the growth pattern of the human head, from the third month to the eighth year of life. *Am J Anat* 1941;68:209–262.
27. Burian F, Farkas LG, Hajniš K. The use of anthropology in the observation of facial clefts [in Czech]. *Anthropologie* 1964;1:41.
28. Farkas LG, Dobisiková M, Hajniš K. Quelque considerations anthropometriques concernant le temp de la reconstruction du pavilon de l'oreille lors d'une plasie congénitale. *Rev Laryngol Otol Rhinol (Bord)* 1966;87:873–879.
29. Hajniš K, Hajnišová M. Determination of the time for corrective operations of the face according to growth dynamics [in Czech]. *Rozhledy Chir* 1966;45:533–544.
30. Figalová P, Šmahel Z. *The growth of the skull and face in children from 3 months to 6 years of age* [in Czechoslovakian]. Praha, Czechoslovakia: Burian Laboratory of Plastic Surgery, Czech Academy of Science, 1972.
31. Farkas LG, Posnick JC, Hreczko TM, Pron GE. Growth patterns of the nasolabial region: a morphometric study. *Cleft Palate Craniofac J* 1992;29:318–324.

Chapter 7

Asymmetry of the Head and Face

Leslie G. Farkas

Asymmetries between the right and left half of the face, and between the position and size of the paired features of the face, are common findings in the craniofacial complex of a healthy population (1). Some investigators have examined the face visually (2), whereas others have studied photographs (3–5). Studies based on quantitative evaluation of the differences between the paired measurements obtained directly from the surface of the head and face in healthy subjects using anthropometric methods are few (1,6–8).

ANTHROPOSCOPY

Symmetry of the face is judged by viewing the *quality* of outer contours (framework) of the face (9) from the front (Fig. 7–1). Flatter facial contours give the impression of a smaller facial half. Positioning of the facial profile line in the vertical facilitates viewing any uneven levels (*vertical asymmetry*) of eyebrows, endocanthions, eye fissures, labial insertion points of alae (subalaria), and commissures of the labial fissure (cheilions).

Visual assessment of asymmetries of the facial depth (which is grossly the space between the midline of the face and each of the ears) can be made difficult because precise points of ear insertion (obs-obi line) are hard to see.

Horizontal asymmetry between the facial midline and the para-axially located paired features in the face (orbits, ala lengths, mouth halves) are barely discernible if the distances show small differences.

ANTHROPOMETRY

Determination of asymmetry is established where differences occur between paired linear (projective and tangential) and angular measurements of the head and

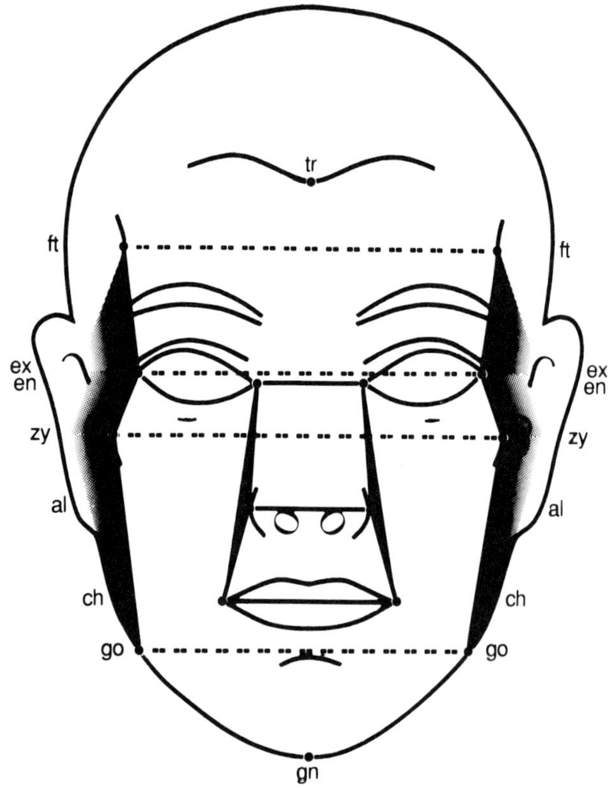

FIG. 7–1. Schematic drawing of the face showing the "sensitive" area—frontal view. The central area is outlined by two para-axial vertical lines connecting the frontotemporale points of the forehead with the gonions of the mandible. (From ref. 9, with permission.)

face. In the three-dimensional relief of the face, the linear asymmetries can be found in horizontal, vertical, and sagittal directions. When searching for inclination asymmetries, related to the vertical, the head is maintained in Frankfurt horizontal (FH) during the examinations. When examining the asymmetry in the horizontal, the face midaxis must be in the vertical.

Factors Leading to Asymmetry

Causes of asymmetry are very complex. In the face they can be produced by defects of the underlying skeleton which may lead to dislocations of the bony landmarks along the facial framework (ear canal, tragion, gonion) and/or midaxis of the face (nasion, subnasale, gnathion), or by disfigurements in the soft-tissue structures affecting the location of the soft landmarks (e.g., endocanthion, exocanthion, subalare, cheilion). Differences in the position of points determining the directions of lines (e.g., the midaxis of eye fissures, nostrils) lead to asymmetries in inclinations. Asymmetries in angles are caused by differences in the mutual position of two surfaces (e.g., ear protrusion from the head). Errors of measurement or failure to identify correctly the points from which to measure can also cause apparent asymmetry.

Quantitative Assessment of Asymmetry

Paired measurements were regarded as asymmetrical if the findings obtained from both sides differed by one or more millimeters or degrees (8). In a healthy population the maximum normal asymmetry between the paired measurements is given by the mean difference plus two standard deviations (mean + 2 SD). The asymmetry is supernormal if the difference between the paired measurements is larger than the maximum normal difference of the norms, at the same age, sex, and ethnic origin of the patient. The degree of asymmetry is expressed as a percentage of the maximum normal difference. The extent of asymmetry may be classified as mild, moderate, or severe (10).

Analysis of the Asymmetries of the Head and Face in a Healthy Caucasian Population

Of the 56 paired measurements analyzed for asymmetries between 1 and 18 years (see Appendix A–1), 30 were selected in the following regions as indicated: head, 1; face, 5; orbits, 6; nose, 4; lips and mouth, 5; and ears, 9. Tables 7–1 to 7–6 record the frequency of asymmetries in the entire study group with the mean values of the maximum normal differences between the asymmetrical measurements, separately in both sexes. The differences in the findings between the sexes were analyzed using the method of the *standard error of difference* (SED) (11).

Head

Table 7–1 reports the asymmetries between the right and left auricular heights of the head measured between the vertex of the head and the porion of the ear canals. In almost one-fifth of the samples the head heights were asymmetrical. The frequencies of asymmetries and the maximum normal asymmetry values were close in each of the sexes.

In clinical praxis, some measurements introduced by our laboratory may be helpful, even if norms for those measurements are not available at present. Asymmetry in *lateral heights of the forehead* is measured between the hairline (tr′) and the ipsolateral superciliare point (sci) of the eyebrow. Asymmetry can occur if the hairline is located lower on one side or if the eyebrow levels are uneven (Fig. 7–2).

In cases of unilateral protrusion or flatness of the forehead, three head-length measurements are recommended in order to obtain information about the extent of asymmetry.

TABLE 7–1. *Asymmetry—head*[a]

Paired measurement	Sex	*N*	Asymmetry *n*	Asymmetry Percent	Asymmetry Mean + 2 SD[b] (mm)	Difference (SED)
Auricular height of the head (v-po)[c]	M	652	129	19.8	6.5	NS
	F	657	113	17.2	7.0	

[a] Abbreviations: M, male; F, female; *N*, total number in sample; *n*, number bearing asymmetries; NS, non-significant difference in frequency.
[b] Mean + 2 SD = maximum normal asymmetry.
[c] In the age range 6–18 years only.

First, the classical head-length measurement (g-op) between the glabella and the most posterior point of the head is taken. To establish the extent of any unilateral forehead protrusion or flatness, two further oblique diameter measurements, using the spreading caliper, are required. The first should span the line from the point of greatest brow protrusion to the appropriate point of the occiput on the contralateral side of the head. The second is taken from the other brow point (where it is most flattened), again to its contralateral occiputal point (see Chapter 2, Fig. 2–2). The difference between these two oblique measurements should indicate the approximate extent of the asymmetry.

Face

Facial Depth Measurements

The frequency of asymmetries in the projective depth measurements was slightly greater in males than in females, with a mean 4.2-mm maximum difference between the paired measurements. The mean asymmetries in both sexes gradually increased from 43.8% in the upper third (t-n) of the face, to 47.9% in the middle third (t-sn) and 53.5% in the lower third (t-gn) (Fig. 7–3). The frequency of asymmetries was significantly greater in the mandibular (lower third) region than in the upper third of the face, in both sexes (*males:* SED 2.5, difference 9.8; *females:* SED 2.5, difference 9.7) (Table 7–2).

Maxillary and Mandibular Half-Arcs

The mean maximum difference between the paired tangential measurements was 5.4 mm in both sexes, slightly higher in males than in females. Asymmetries were significantly more frequent in the mandibular region, in both sexes (*males:* SED 2.4, difference 6.4; *females:* SED 2.4, difference 13.7). In the maxillary region, the males revealed significantly more asymmetries than did the females (SED 2.5, difference 8.1). In the mandibular region the asymmetries were slightly higher in males.

Comparison of the projective and tangential asymmetries showed a significantly higher percentage of asymmetries as tangential, both in maxillary and mandibular regions and in both sexes (*maxillary region, males:* SED 2.5, difference 10.4; *maxillary region, females:* SED 2.5, difference 5.3; *mandibular region, males:* SED 2.4, difference 11.8; *mandibular region, females:* SED 2.4, difference 12.8) (Table 7–2).

Orbits

Asymmetrical Eye Fissure Measurements

The frequency of the eye fissure height and length asymmetries were below 2% in both sexes, slightly less frequent in females. Asymmetrical eye fissure inclinations reached a mean of 2.3% in both sexes. The maxi-

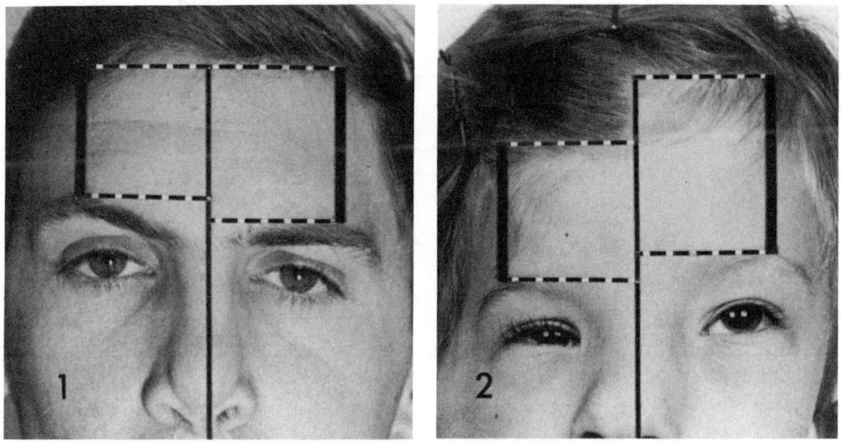

FIG. 7–2. Asymmetry in lateral heights of the forehead: (**1**) asymmetry caused by the uneven level of the eyebrows; (**2**) asymmetry caused by dislocation of the hairline downwards and by the uneven level of the eyebrow on the same side.

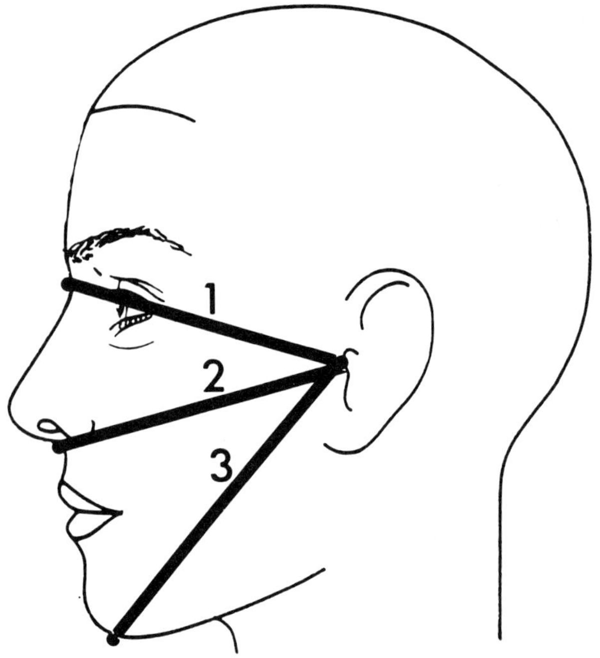

FIG. 7–3. Schematic drawing of the face showing depth measurements: **(1)** upper third, **(2)** middle third, **(3)** lower third.

mum mean difference between the asymmetrical paired measurements was in the range of 0.3–1.4 mm in linear measurements and between 0.7 and 1.5 degrees in inclinations (Table 7–3).

Projective Measurements Locating the Orbits

Of the three paired measurements defining the position of the orbits (Fig. 7–4) in the face, the orbitogonial distances revealed the highest percentage (mean 32.2%) of asymmetries in both sexes, with the highest maximum difference (mean 5.4 mm). Distances locating the orbits laterally (ex-t rt,lt) showed in about one-quarter (mean 26.2%) of the population asymmetries in both sexes, with a maximum mean difference of 4 mm. Least asymmetrical was the position of the orbits in relation to the midaxis of the face (endocanthion–sellion distance), with a mean of 17.2% and 1.7-mm maximum difference in both sexes.

Statistical analysis revealed a significantly higher percentage of asymmetries in the ex-go than in the ex-t distances, in both sexes (SED 1.7, difference 6). Similarly, the asymmetries were significantly higher in the ex-t than in the en-se distances (SED 1.4, difference 9) (Table 7–3).

Nose

The mean maximum normal differences in the paired asymmetrical nasal measurements ranged from 0.7 mm to 1.9 mm. The lowest frequencies were observed in asymmetries of the nasal root depth (mean 2.7%), followed by ala length measurements (mean 3.2%). Significantly greater was the frequency of asymmetries between the lengths of columellae (mean 8.9%) (SED 0.9, difference 5.7). The frequency of asymmetrical nostril floor width measurements was even higher (mean 12.8%), significantly greater than in columella asymmetries (SED 1.1, difference 3.9) (Table 7–4).

Upper Lip and Mouth

Labio-oral Asymmetries

In both sexes the differences between the frequencies of asymmetries were mild to moderate. The mean maximum normal differences revealed small variations between the sexes, with the smallest value in the lateral

TABLE 7–2. *Asymmetry—face*[a]

Paired measurement	Sex	*N*	Asymmetry *n*	Asymmetry Percent	Asymmetry Mean + 2 SD[b] (mm)	Difference (SED)
Tragion–nasion depth (t-n)	M	792	353	44.6	4.1	NS
	F	800	343	42.9	4.0	
Tragion–subnasale depth (t-sn)	M	792	391	49.4	4.1	NS
	F	800	371	46.4	4.3	
Tragion–gnathion depth (t-gn)	M	792	431	54.4	4.2	NS
	F	798	420	52.6	4.3	
Maxillary half-arc (t-sn surf)	M	791	473	59.8	5.7	SIG
	F	793	410	51.7	4.8	
Mandibular half-arc (t-gn surf)	M	791	524	66.2	5.8	NS
	F	793	519	65.4	5.1	

[a] Abbreviations: M, male; F, female; *N*, total number in sample; *n*, number bearing asymmetries; NS, nonsignificant difference in frequency; SIG, significant difference.
[b] Mean + 2 SD = maximum normal asymmetry.

TABLE 7-3. *Asymmetry—orbits*[a]

Paired measurement	Sex	*N*	Asymmetry *n*	Asymmetry Percent	Asymmetry Mean + 2 SD[b] (mm)	Difference (SED)
Eye fissure length (ex-en)	M	793	14	1.6	1.4	NS
	F	801	10	1.2	0.8	
Eye fissure height (ps-pi)	M	793	9	1.1	0.4	NS
	F	791	5	0.6	0.3	
Eye fissure inclination from the horizontal (in degrees)	M	788	17	2.2	0.7	NS
	F	800	19	2.4	1.5	
Orbitotragial distance (ex-t)	M	792	220	27.7	3.9	NS
	F	796	197	24.7	4.1	
Orbitogonial distance (ex-go)[c]	M	654	205	31.3	5.5	NS
	F	658	218	33.1	5.2	
Endocanthion-sellion distance (en-se)	M	791	142	18.0	1.9	NS
	F	793	130	16.4	1.5	

[a] Abbreviations: M, male; F, female; *N*, total number in sample; *n*, number bearing asymmetries; NS, nonsignificant difference in frequency.
[b] Mean + 2 SD = maximum normal asymmetry.
[c] In the age range 6–18 years only.

heights of the upper lip (1.5 mm), followed by labial fissure inclination (2.1 degrees), and with the highest differences between halves of the labial fissure (4.4 mm). Least frequent was the oblique labial fissure (2.2% of 1310 subjects). The percentage of asymmetries in both sexes increased to a mean of 7.7% in lateral heights of the upper lip (Fig. 7–5). This percentage was exceeded by the frequency of asymmetries between the halves of the labial fissure (mean 15.9% in both sexes) (Table 7–5).

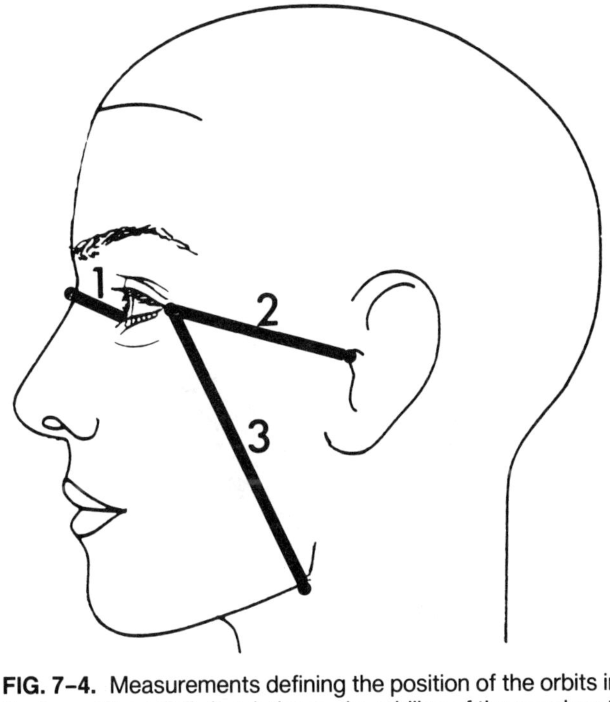

FIG. 7–4. Measurements defining the position of the orbits in the face: (**1**) medially in relation to the midline of the nasal root; (**2**) laterally to the tragion point; (**3**) distally from the gonion point of the mandible.

Mouth Location by Measurement

The projective distance between the commissure of the mouth and the tragion landmark (ch-t) (Fig. 7–6) showed asymmetries in about half of the population sample, in both sexes (mean 51.4%). The mean maximum normal difference was slightly larger in males (5.5 mm) than in females (4.8 mm). The tangential measurement (ch-t surf) between the cheilion and tragion points showed asymmetries in significantly higher percentage in both sexes (mean 58%) (SED 2.2, difference 6.6). The mean maximum difference in both sexes was moderately larger in tangential (6.4 mm) than in projective asymmetries (5.2 mm) (Table 7–5).

Ears

Asymmetries Between the Measurements of the Ears

Frequency in asymmetry as well as in the maximum normal differences in the sizes of the ears (sa-sba, pra-pa) were minimal between the sexes. The combined percentage of the ear size asymmetries in both sexes revealed that one-fifth (mean 20.8%) of the population sample had length and/or width asymmetries of the ear. The ear protrusion asymmetry showing a small difference in degree was significantly more frequent in males than in females (SED 2.4, difference 5.2). The ear protrusion asymmetries in both sexes (mean 31.7%) were significantly more frequently found than those of size (SED 1.6, difference 10.9). The highest percentage in asymmetries was reached in inclinations of the ears, observed in almost half (47.8%) of the combined male and female populations. The frequency of the inclination asymmetries significantly exceeded the percentage of protrusion

TABLE 7–4. *Asymmetry—nose*[a]

Paired measurement	Sex	*N*	Asymmetry *n*	Asymmetry Percent	Asymmetry Mean + 2 SD[b] (mm)	Difference (SED)
Depth of the nasal root (en-se sag)	M	701	19	2.7	1.1	NS
	F	668	18	2.7	0.7	
Length of the ala (ac-prn)	M	793	24	3.0	1.1	NS
	F	796	27	3.4	0.9	
Width of the nostril floor (sbal-sn)	M	793	112	14.1	1.9	NS
	F	795	91	11.4	1.6	
Length of the columella (sn-c′)	M	793	80	10.1	1.4	NS
	F	798	61	7.6	1.2	

[a] Abbreviations: M, male; F, female; *N*, total number in sample; *n*, number bearing asymmetries; NS, non-significant difference in frequency.
[b] Mean + 2 SD = the maximum normal asymmetry.

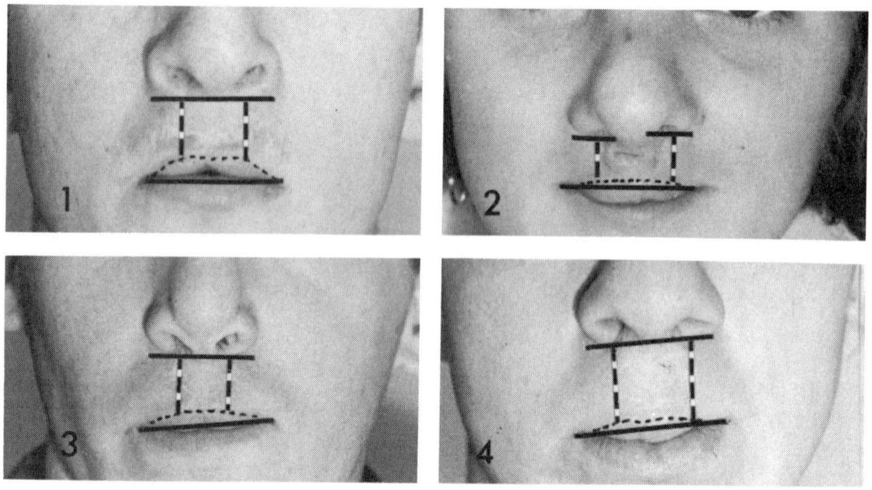

FIG. 7–5. Variations in sizes of the lateral heights of the upper lip. The lateral height is determined by the projective distances between the labial insertions of the ala (sbal rt,lt) and the points on the vermilion line (ls′ rt,lt) located directly beneath the alar bases: (**1**) symmetrical lateral heights; (**2**) asymmetry in lateral heights is caused by the higher location of the left alar base; (**3**) asymmetry in lateral heights is caused by the oblique labial fissure which led to an uneven level of landmarks on the vermilion line; (**4**) symmetry in lateral lip heights is due to identical level dislocation both in the alar bases and in the vermilion points.

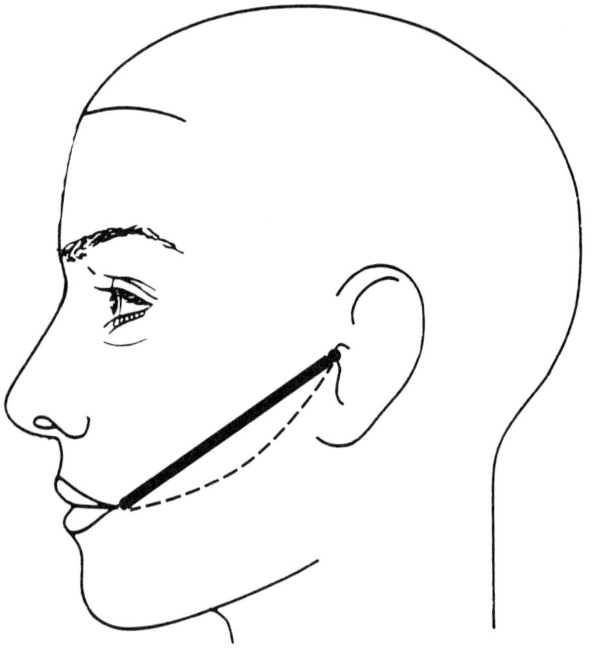

FIG. 7–6. Locating the commissure of the labial fissure in relation to the tragion landmark of the ear by a projective distance and a semi-arc measurement (*dotted line*).

TABLE 7–5. *Asymmetry—upper lip and mouth*[a]

Paired measurement	Sex	*N*	Asymmetry *n*	Asymmetry Percent	Asymmetry Mean + 2 SD[b] (mm)	Difference (SED)
Lateral upper lip height (sbal-ls′)	M	793	80	7.6	1.5	NS
	F	796	62	7.8	1.5	
Halves of the labial fissure length (ch-sto)[c]	M	654	99	15.1	4.0	NS
	F	656	109	16.6	4.8	
Labiotragial distance (ch-t)[c]	M	500	263	52.6	5.5	NS
	F	473	237	50.1	4.8	
Tangential labiotragial arc (ch-t surf)[c]	M	500	305	61.0	7.1	NS
	F	473	260	55.0	5.6	
Labial fissure inclination (ch-ch line) from the horizontal[c]	M	654	11	1.7	1.7	NS
	F	656	18	2.7	2.4	

[a] Abbreviations: M, male; F, female; *N*, total number in sample; *n*, number bearing asymmetries; NS, nonsignificant difference in frequency.
[b] Mean + 2 SD = maximum normal asymmetry.
[c] In the age range 6–18 years only.

asymmetries (SED 1.8, difference 16.1). The mean maximum normal difference in angular measurements of the ears was larger (6.5 degrees) than in linear measurements (3 mm), both calculated in joint male and female populations (Table 7–6).

Position Asymmetries

Anteriorly, the location of the ears is determined by measuring the linear distances between the upper and lower insertions (obs and obi) of the ears and the subnasale and the gnathion landmarks in the midaxis of the face (Fig. 7–7).

Posteriorly, the ear canal location (porion) is determined sagittally in relation to the most posterior point (opisthocranion) of the skull in the FH position of the head (Fig. 7–8).

Anterior Location. Projective distances taken between the upper ear insertion points and the gnathion (obs-gn) revealed significantly higher percentages of asymmetries (61.2% in both sexes) than those of the lower insertions of the ear and the gnathion (49.9% in both sexes) (SED 1.8, difference 11.3). The difference be-

TABLE 7–6. *Asymmetry—ears*[a]

Paired measurement	Sex	*N*	Asymmetry *n*	Asymmetry Percent	Asymmetry Mean + 2 SD[b] (mm)	Difference (SED)
Length of the auricle (sa-sba)	M	793	183	23.1	3.0	NS
	F	796	173	21.7	3.3	
Width of the auricle (pra-pa)	M	793	151	19.0	2.7	NS
	F	796	155	19.5	3.0	
Upper gnathion–aural distance (gn-obs)	M	793	494	62.3	4.8	NS
	F	797	479	60.1	4.9	
Lower gnathion–aural distance (gn-obi)	M	792	405	51.1	4.7	NS
	F	796	387	48.6	4.6	
Lower subnasale–aural distance (sn-obi)	M	792	402	50.8	4.8	NS
	F	796	381	47.9	4.4	
Lower gnathion–aural surface distance (gn-obi surf)	M	791	488	61.7	5.6	NS
	F	793	464	58.5	6.0	
Occipito–aural distance (op-po)[c]	M	566	48	8.5	6.3	SIG
	F	527	29	5.5	5.4	
Ear inclination from the vertical (in degrees)	M	730	341	46.7	5.7	NS
	F	731	353	48.3	5.9	
Ear protrusion from the head (in degrees)	M	732	251	34.3	6.8	SIG
	F	741	216	29.1	7.7	

[a] Abbreviations: M, male; F, female; *N*, total number in sample; *n*, number bearing asymmetries; NS, nonsignificant difference in frequency; SIG, significant difference.
[b] Mean + 2 SD = maximum normal asymmetry.
[c] In the range 6–18 years only.

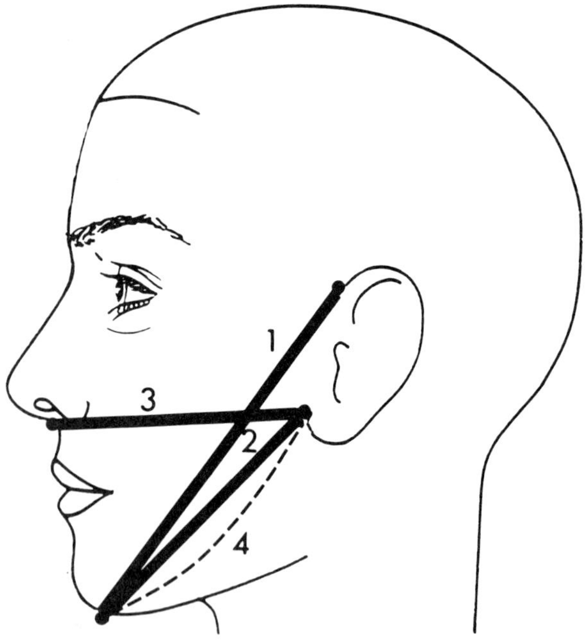

FIG. 7–7. Schematic drawing showing the major measurements locating the ear in relation to the facial midline. (**1**) The upper insertion of the ear is measured from the gnathion. (**2**) The position of the lower insertion of the ear is determined projectively from the gnathion and (**3**) from the subnasale and (**4**) tangentially from the gnathion.

tween the maximum asymmetry values was minimal between the sexes.

Projective distances measured between the ears' lower insertions (obi) and the subnasale point (sn-obi) in the face showed almost the same frequency of asymmetries (mean 49.4% in both sexes) as the obi-gn asymmetries (49.9%). Asymmetry was more frequent between the subnasale and the lower ear insertion in distance measured tangentially (mean 60.1% in both sexes). The tangential (gn-obi surf rt,lt) asymmetry showed a moderately larger maximum difference (mean 5.8 mm in both sexes) than did the asymmetry between the projective distances (sn-obi rt,lt) (mean 4.6 mm) (Table 7–6).

Posterior Location. An asymmetry in the sagittal position of the ear canal on the head (op-po) was found significantly more frequently in males (8.5%) than in females (5.5%) (SED 1.5, difference 3.0). The maximum normal difference between the asymmetrical measurements reached a mean 5.9 mm in both sexes, the largest among the asymmetrical linear measurements in the ear (Table 7–6).

DISCUSSION AND CONCLUSIONS

The determination of asymmetry on the head and face should be based on quantitive analysis—that is, comparison of data obtained from paired measurements. Prior to examination, the quality of the mutual position of the landmarks (which determine the size of the linear distance), or inclination of the line outlined by them, should be carefully established. There is a great difference between anthroposcopic and anthropometric determination of symmetry. At visual examination, as compared to anthropometry, the difference between the paired measurements should reach a critical point if asymmetry is to be noticed.

Asymmetries found in the healthy population do not present visually striking defects. The mean differences between the paired measurements of the soft orbits, nose, and upper lip and mouth ranging between 0.3 mm and 1.5 mm, and only between halves of the mouth reaching almost 5 mm, do not present an aesthetic problem. The mean frequency of these asymmetries, located in the "sensitive" area of the face (9), calculated from the asymmetries in both sexes, was the lowest in the soft orbits (1.6%), followed by the nose (6.9%) and the upper lip and mouth (11.7%) (Table 7–7). In contrast, the paired measurements which determine the position of the orbits, mouth, and ears on the face, as well as the distance defining the depth of the face, revealed much higher percentages of asymmetries: The frequency of asymmetries was smallest in the orbits (25.2%), followed by the depth

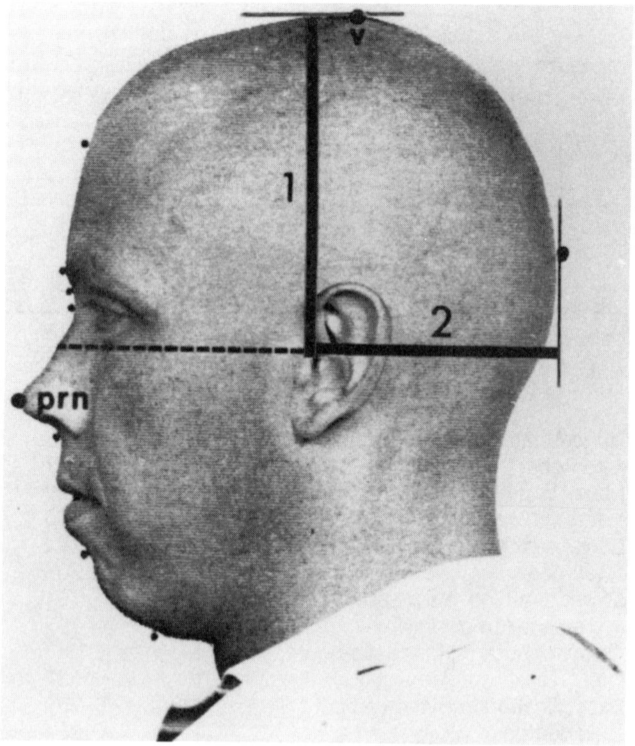

FIG. 7–8. Locating the ear canal on the head: (**1**) proximally from the vertex (v); (**2**) posteriorly from the opisthocranion point (op). Both measurements are taken in FH position of the head. The projective distance between the vertex and the ear canal (porion) defines also the auricular height of the head (v-po).

measurements of the face (53.2%), mouth (54.7%), and ears (55.1%) (Table 7–8). Among the regions, the orbits showed the smallest degree and frequency of asymmetries; least asymmetrical was the height of the eye fissure (ps-pi), and among the measurements locating the orbits in the face was the distance between the endocanthion and the sellion in the midline of the nasal root. The projective depth measurements located in the lower third of the face (mandibular region) revealed more asymmetries than those in either the middle third or the upper third, while the maximum difference between the asymmetries did not show great differences. Asymmetries were more frequent in the tangential than in the projective measurements of the face, especially in the mandibular region. The maximum differences were slightly larger in tangential than in projective measurements (Table 7–2).

In 27 of 30 paired measurements, frequency of asymmetry did not show significant differences between the sexes, and no major variations were observed in the values of the maximum normal differences (mean + 2 SD).

Asymmetries in the positions of facial features may or may not indicate dislocation. In the orbits, the mandible angle (ex-go) is more often responsible for asymmetry than the tragion (ex-t). It appears that the "infinite variety of anatomical variations of the human pinna" (12) is the most frequent cause of asymmetry, in which one of the "locating" measurements is the tragion (t), the upper (obs) or lower (obi) insertion point of the ear. Our findings confirm this observation, revealing the highest percentage of asymmetries (55.1%) in those measurements which determine the position of the auricles in relation to the facial midline. Our study showed that the dislocation of the ear canal related vertically to the vertex is more frequent (18.5%) than the dislocation related sagittally to the opisthocranion (7%) (Table 7–8). The asymmetrical length of the ears (22.4% in both sexes) and the high percentage (47.5%) of asymmetries in inclination of the ears greatly increase the list of anatomical variations.

The degree and frequency of asymmetry in a healthy population can help when evaluating the extent of asymmetry in facial syndromes. The analysis showed that greater attention must be paid when the causes of asymmetry need to be detected. Anthropometry provides the objective means to establish which factors contribute (and to what degree) to the formation of asymmetries in the face.

TABLE 7–7. *Mean frequencies showing maximum differences by individual region of the face and head*

Region	Paired measurement	Asymmetry: Percent[a]	Asymmetry: Mean + 2 SD[b] (mm, deg)
Orbits	ex-en, ps-pi, eye fissure inclination	1.6	0.9[c]
Nose	en-se sag, ac-prn, sbal-sn, sn-c′	6.9	1.2
Lip-mouth	sbal-ls, ch-sto	11.7	5.9
Ears	sa-sba, pra-pa	20.8	3.0

[a] The data reported were calculated from all measurements and both sexes. See Tables 7–1 to 7–6 for size of samples.
[b] Mean + 2 SD = maximum normal symmetry.
[c] In the orbits the mean maximum difference between the asymmetrical measurements is calculated from both the linear and angular measurements.

TABLE 7–8. *Mean frequencies of asymmetries and their maximum differences showing facial depth measurements and feature location measurements*

Region	Paired measurement	Asymmetry: Percent[a]	Asymmetry: Mean + 2 SD[b] (mm)
Head	v-po	18.5	6.8
Face	t-n, t-sn, t-gn, t-sn surf, t-gn surf	53.2	4.7
Orbits	ex-t, ex-go, en-se	25.2	3.7
Mouth	ch-t, ch-t surf	54.7	5.9
Ears[c]	a: gn-obs, gn-obi, sn-obi, gn-obi surf	55.1	5.0
	b: op-po	7.0	5.9

[a] The data reported were calculated from all measurements and both sexes. The size of the population samples is reported in Tables 7–1 to 7–6.
[b] Mean + 2 SD = maximum normal symmetry.
[c] Ears: a, anterior ear location; b, posterior ear canal location.

REFERENCES

1. Farkas LG, Cheung G. Facial asymmetry in healthy North American Caucasians. An anthropometrical study. *Angle Orthod* 1981;51:70–77.
2. Sutton PRN. Lateral facial asymmetry-methods of assessment. *Angle Orthod* 1968;38:82–92.
3. Busse H. Über normale Asymmetrien des Gesichts und im Körperbau des Menschen. *Z Morphol Anthropol* 1936;35:412–445.
4. Peck H, Peck S. A concept of facial esthetics. *Angle Orthod* 1970;40:284–318.
5. Roggendorf E. Die Symmetrieanalyse des Gesichts. *Anat Anz* 1972;132:178–188.
6. Hajniš K, Hajnišová M. Timing of corrective operations of the face according to the dynamics of growth [in Czech]. *Rozhledy Chir* 1966;45:533–544.
7. Figalová P. Asymmetry of the face. *Anthropologie* 1969;7:31–34.
8. Farkas LG. *Anthropometry of the head and face in medicine.* New York: Elsevier, 1981.
9. Farkas LG, Kolar JC. Anthropometrics and art in the aesthetics of women's faces. *Clin Plast Surg* 1987;14:599–616.
10. Farkas LG. The proportion index. In: Farkas LG, Munro IR, eds. *Anthropometric facial proportions in medicine.* Springfield, IL: Charles C Thomas. 1987;5–8.
11. Hughes DR. Finger dermatoglyphics from Nuristan, Afghanistan. *Man* 1967;2:119–125.
12. Webster GV. The tail of the helix as a key to otoplasty. *Plast Reconstr Surg* 1969;44:445–461.

Chapter 8

Anthropometry in Cleft Lip and Palate Research

Janusz Bardach

When designing clinical studies which assess the aesthetic results of various treatment techniques in patients with cleft lip and palate, it is always difficult to select the most appropriate evaluation method. In the majority of cases, the aesthetic appearance of cleft patients results from combined surgical and orthodontic treatment. Both specialties, in different time periods and in varying degrees, participate in the treatment of this congenital deformity. In all patients with complete unilateral and bilateral cleft lip, alveolus, and palate, surgical and orthodontic treatment has the greatest effect on appearance. It is also important to understand that in cleft patients, the nasal deformity must be considered as one of the most serious impairments with regard to aesthetics.

The vast amount of literature on clinical research in cleft lip and palate indicates a great interest in evaluating the immediate and late results of treatment procedures. Numerous publications and presentations have created a false impression that there is enough information about the validity of surgical procedures used in reconstructive surgery of cleft lip, nose, and palate, that we can easily evaluate aesthetic results of various surgical procedures, and that we can investigate all factors which may influence the outcome of treatment.

The measurement of aesthetic results is one of the most difficult problems in clinical research related to clefts. First of all, it is a problem with aesthetics because currently there is no uniform definition of what constitutes the optimal aesthetic outcome for cleft patients. Balance and harmony of the facial features are usually the goals of the plastic surgeon when performing reconstructive and corrective procedures, although some highly attractive faces happen to be unbalanced and inharmonious. Many clinical researchers refer to accepted standards as indicators for measuring the aesthetic attractiveness of the face. Those standards may be difficult to apply because features found within the limits of balance and harmony do not guarantee attractiveness.

Therefore, the accepted standards of facial aesthetics must be viewed within the limits of racial, regional, and cultural concepts of beauty.

Tobiasen and Hiebert (1) emphasized that "there are no standard techniques to assess cleft impairment for aesthetic acceptability. Therefore, it is not possible to evaluate objectively either the need for or the benefits of treatment." According to these authors, the personal judgment of a group of people can be used as the basis for developing methods to quantify the relationship between the aesthetic appearance of people born with clefts and the social and psychological consequences. The same authors (2) investigated the combined effects of the severity of cleft impairment and facial attractiveness on social perception. The authors concluded the following: "The social perception of cleft impairment is a complex process that includes the severity of the oral–nasal impairment and possibly overall facial attractiveness. The findings also suggest that a surgical intervention that significantly reduces the severity of an oral–nasal impairment should improve social desirability." The authors are correct in indicating the numerous limitations to their research. Unfortunately, they do not underscore two main limitations: The principal one is the standard of attractiveness, and the second one is their approach to measurements. The authors do not acknowledge that their standard of attractiveness lacks wide acceptance, because there is no established standard that can be used as the criterion for valid research. Clifford (3) underscores the fact that psychologists working with cleft patients have neglected research related to the influence of the facial appearance on psychological status: "Only recently, for example, have psychologists begun to address issues of physical attractiveness that have been the concern of social psychologists for the past fifteen years."

Reviewing the contemporary literature in the area of plastic surgery, orthodontics, and psychology, we did not find an accepted method for reliable measurement of facial attractiveness. We also did not find a standardized approach which would allow the ranking of aesthetic appearance following surgical and orthodontic treatment of cleft patients. It is necessary to emphasize that the aesthetic outcome of treatment of a cleft patient cannot be directly correlated with treatment methods or surgical proficiency, because it is highly dependent on the initial deformity and the degree of hypoplasia in the soft tissue and underlying maxillary complex. Those difficulties in assessing facial appearance experienced by clinical investigators imply a need to use any technique available which will be helpful in documenting the symmetry, balance, and harmony of the face.

The most common technique for evaluation of facial appearance is visual assessment based on the rating scale. When used by a single researcher, this technique is highly subjective and unreliable. However, the same technique used by a large group of people of different ages and genders may currently prove to be more reliable than other available techniques. Among other techniques, X-ray cephalometry is applicable for the assessment of skeletal deformities, but it does not provide valuable information about the soft tissue. The value of indirect cephalometry, which provides data from photographs, is strictly limited to certain facial areas but does not incorporate the entire head and face. Because assessment of facial aesthetics is very difficult and techniques are limited, direct physical anthropometry presents a viable option in the area of clinical research of patients with craniofacial deformities, and especially of patients with cleft lip and palate.

Physical anthropometry or surface measurement of the head and face has become an important supplement to other quantitative techniques as an assessment of patients with congenital or acquired craniofacial deformity. This technique allows for objective documentation of facial features and gathering of quantitative data regarding deviation from the established norms. In cleft lip and palate patients, physical anthropometry also allows the comparison of specific facial measurements before and after surgical intervention and/or orthodontic treatment, thus providing information about the validity of treatment procedures. Physical anthropometry fills an important gap in the activity of the plastic surgeon, because it is concerned with morphological changes in the facial features which influence the aesthetic results of surgery. This technique is helpful in determining the need for surgical correction in patients with clefts after comparative assessment is performed according to the norms established for specific groups in relation to age, gender, and ethnicity.

The number of anthropometric measurements and proportions vary depending on the area measured and the purpose of study. Whether a particular measurement or set of measurements is considered a deviation from the norm must be interpreted according to the established range of variations within a specific population. Using this approach allows one to determine which cleft lip and palate patients fit within the range of variations accepted for the normal population. However, it is necessary to understand that findings based on anthropometric measurements do not indicate a complete optimal outcome. The decision of whether there is a need for corrective surgery must be based on visual inspection and anthropometric measurements.

The use and popularization of anthropometric measurements in clinical evaluation of cleft patients must be attributed to Farkas, whose primary studies have been carried out for almost 30 years. One of the first publications in his extensive scientific activities in this area concerns the use of anthropology in the observation of facial clefts (4). In this paper, for the first time, the authors suggested the use of direct anthropometry to evaluate anatomical changes of facial features in patients with cleft lip and palate, particularly as a primary means of assessing the initial cleft defect as compared to the final result of treatment. Suggestions presented in this paper became a milestone in developing clinical anthropometry for craniofacial deformities. At that time, measurements according to selected points on the face and skull appeared to be the only evidence for precise conclusions of the surgical outcome. The authors concluded that it is necessary to establish a set of normative measurements in different age groups according to gender and ethnicity. The norms will facilitate orientation when deciding how much the face of a cleft child differs from the face of a normal individual of the respective group. Those findings could serve in the planning of surgical procedures at a certain age, and even more for postoperative control and evaluation of the results. The main significance of creating the set of normative measurements is determining the periods of accelerated growth in which surgical treatment may be postponed.

Direct anthopometric technique can serve as an exact evaluation of the results and subsequent decisions based on the findings of the type and degree of asymmetry or disharmony. Taking into account that the measurements are performed on both sides of the face and the fact that there is usually slight asymmetry between both sides, the measurements may help to establish to what degree the secondary cleft lip–nasal deformity influences facial asymmetry.

Since the initial stage of introducing anthropometric measurements in clinical observations and studies in patients with cleft lip and palate, Farkas collected unique data from people with normal facial features (age, gender, and race-related) which serve as the comparative yardstick for studying pre- and postoperative results of cleft patients (5). As a result of his efforts and devotion, Farkas established a unique facial measurement laboratory at the Hospital for Sick Children in Toronto. Close cooperation of Dr. Farkas with the outstanding plastic surgeons in this institution (Drs. Lindsay, Munro, and Posnick) resulted in most interesting collaborative studies in which anthropometry was raised to a valuable technique in the evaluation of children with craniofacial deformities, and particularly with facial clefts. In the last 25 years, Farkas accumulated important and valuable data regarding normal children as well as concerning the variety of pre- and postoperative cleft deformities. His studies in this area must be considered as classical and deserve the special attention of all clinicians interested in studying the aesthetic results and attractiveness of the cleft patients.

Over 30 years ago, Farkas had an innovative idea to create a uniform coding system for patients with cleft lip and palate. His goals were to facilitate comparative studies from various centers as well as within the same center when treatment is performed by different specialists. To apply this concept, he and his colleagues created the first coding system for cleft patients (6). However, at that time, there were insufficient data concerning head and face measurements to serve as a basic standard for normal children. In 1969, Hajniš and Farkas (7) introduced a new anthropometric coding system for registration of pre- and postoperative measurements. In clinical practice, Farkas repeatedly used anthropometric examination prior to and after each surgical procedure until facial growth and development was complete. By doing so, he could compare the facial features of a child with a cleft in different stages of surgical and orthodontic treatment. It was helpful to determine to what degree the severity of the initial deformity and the hypoplastic changes in soft and hard tissue affect the final outcome of treatment.

It is interesting that currently the concept of having a uniform coding chart for cleft patients is widely acknowledged and considered a necessity, but no agreement has been reached as to which proposal is most practical, and therefore no uniform system is implemented. The lack of comparable information, which can only be obtained by using uniform charts, hampers clinical research and, especially, any comparative studies from the various centers. Many attempts have been made in the last few years to establish a system for collecting data in an identical manner, to enable valid and controlled clinical research. Most unfortunately, each cleft and craniofacial center uses its own classification and coding systems, which in no way allows the information obtained from those sources to be practically useful for comparative clinical studies. Farkas deserves great credit for his innovative initiatives and persistence in introducing direct anthropometry into the treatment of patients with cleft lip and palate.

The growing interest in studying facial growth and aesthetic appearance in cleft patients as well as in patients with craniofacial deformities led Farkas and co-workers (8,9) to study the growth and development of regional units in the head and face based on anthropometric measurements. Information extracted from those studies enhances our knowledge about the normal growth of the craniofacial skeleton and soft tissue and about the relationship between particular facial regions. Those measurements were obtained through cross-sectional studies of growing Caucasian children and proved to be useful in quantitative evaluation of growth and development in specific regions of the head and face. In the studies, only 21 measurements out of 149 applied for complete facial assessment were used to provide basic information about each region. Those measurements are strongly affected by age-related morphologic changes in the craniofacial complex; according to the authors' experience, the selected measurements proved to be the most clinically useful because they allow pre- and postoperative assessment in the growing patient.

Of special interest for plastic surgeons and orthodontists participating in the treatment of cleft patients is a morphometric study of growth patterns of the nasolabial region by Farkas et al. (10). In this study, age-related growth changes for the upper lip and nose were analyzed on the basis of six measurements taken between 1 and 18 years of age in 1593 North American Caucasians. The findings may serve in planning the timing of surgical treatment. The study revealed that by 1 year of age, the length of the cutaneous portion of the upper lip and the width of the nose showed the highest levels of development compared with their adult size: 80.2% and 79.5%, respectively. By 5 years of age, the developmental level of the nasolabial region, except nasal tip protrusion, was found to be at the maturation level. The study also revealed that the growth of the nose between 5 and 18 years of age was significantly greater than that of the upper lip. Lip growth was greater between 1 and 5 years of age. This study also provided information that the cutaneous upper lip reaches its adult size in 3-year-old females and 6-year-old males, while the nasal width and height were fully developed in 12-year-old females and in 14- and 15-year-old males. This information about the age-related morphologic variations within the nose and upper lip is definitely helpful in establishing the time and surgical technique in lip and nose reconstruction and in secondary corrective procedures. It may also be helpful in predicting the possible changes which may occur within the lip and nose following primary repair.

A substantial effort was made by Farkas and Lindsay in studying the late results of treatment outcome in adult patients with cleft lip and palate. In a series of studies, they investigated the morphology of the adult face following primary repair of unilateral cleft lip and palate in childhood (11), after repair of bilateral cleft lip and palate (12), and following repair of isolated cleft palate (13). In addition, the morphology of the orbital region in adults following cleft lip and palate repair in childhood was assessed (14). The control data for all four studies were obtained from 100 young, healthy Canadians (50 men and 50 women).

This series of studies was initiated to assess the growth and development of the face following surgical treatment of bilateral cleft lip and palate in infancy. Twenty-nine patients were operated at age 2 weeks to 3 months using LeMesurier one-stage technique (12). The palate was repaired by a modified Dorrence pushback technique at about 2 years of age. Of the 29 patients, 28 had orthodontic treatment. Secondary lip repair was carried out in eight patients, and secondary correction of the nose was performed in seven patients. The findings revealed that the patients with symmetrical bilateral clefts had a number of face and nose asymmetries which the authors attributed to either unilateral embryological deficits or inadequate surgical procedures. A direct relationship was found between the shape of the nostrils and the length of the columella: The shorter the columella, the more horizontal the orientation of the nostrils.

The morphology of the adult face following the repair of unilateral cleft lip and palate in childhood was studied in 74 patients who had complete (57 cases) and incomplete (17 cases) forms of unilateral clefts (11). The LeMesurier lip repair was performed in 70 patients at 2 weeks to 3 months after birth. A modified Dorrence pushback operation was performed at about 2 years of age. Of the 74 patients, 73 had orthodontic treatment. Fifteen patients had secondary lip revision, and 32 patients had correction of the nasal deformity. The findings indicated that the general features of the face in adults with repaired unilateral cleft lip and palate are similar to the features of adults with bilateral cleft lip and palate—that is, a longer facial profile, a horizontally narrower face, and a narrower labial fistula. These findings suggest that the prenatal and postnatal developments of the middle third of the face are similar in all patients with cleft lip and/or palate of whatever form.

Another study in this series included 42 patients with isolated cleft palates (13). The palates were repaired by a modified Dorrence pushback operation at the age of approximately 2 years. Nineteen patients had orthodontic treatment. Nine patients had a secondary cleft palate repair: Eight patients had closure of the palatal fistula, and one patient had a pharyngeal flap. The high narrow face found in patients with isolated cleft palate is very similar to features in adult faces following repair of unilateral and bilateral clefts. The authors assume that this may be a combined result of original embryonal damage and of surgical repair. This study showed a high number of facial anomalies associated with isolated cleft palate, a finding in accordance with the experience of some other investigators (15,16). Quite surprising was the relatively high frequency of nasal deformities in patients with isolated cleft palate, which may be considered as a microform of cleft lip and palate. An interesting finding was that in children with cleft palate only, the mandible is usually smaller during early childhood than in the control group; however, at a later age, growth of the mandible accelerates to the normal size.

The fact that cleft lip and palate affects not only the lip, maxilla, palate, and nose, but also the entire facial region and especially the orbital region, stimulated Farkas and Lindsay to investigate the morphology of the orbital region in adults following the repair of cleft lip and palate in childhood (14). The control data for this study were obtained from 100 normal Caucasian Canadians. The authors concluded that the incidence of hypertelorism was not influenced by the extent of the cleft. However, they hypothesized that the complete cleft lip and palate defect, especially bilateral, may contribute to the creation of a wider interorbital space. The above-mentioned series of studies investigating the morphology of the adult face following the surgical repair of cleft lip and palate in childhood, and orthodontic treatment in the majority of these patients, reveals the effectiveness of treatment procedures as well as morphological changes from the initial defect to the completion of the growth and development of the facial structures.

Farkas et al. (17) performed a unique morphometric study of the nose which served as the basis for many of his studies related to cleft lip–nasal deformities. In the study entitled "Geography of the Nose: A Morphometric Study," the noses of 34 attractive North American Caucasian women were analyzed quantitatively, based on 19 nasal measurements and 15 craniofacial measurements taken directly from the face of the women under study. The relationship between the nasal measurements was studied in 16 proportion indices, and the relationship between the nasal and other craniofacial measurements was studied in 13 interareal indices. These findings were compared with those in 21 women with below-average faces. Two types of facial harmony disruption were identified: disharmony and disproportion. The study revealed that in the group of 21 women, the percentage of disharmony and disproportions was significantly higher than that in the other group. The greatest disproportion in the attractive-face group was a moderately short columella. Disproportions were associated with a combination of normal and abnormal measurements. The study may serve as a classic example of the clinical value of (a) anthropometric measurements in the assessment of attractiveness and (b) the measurement of disharmony and disproportions found in individuals. The study by Farkas et al. indicates to what degree direct anthropometry may be used in a clinical setting for planning a corrective rhinoplasty procedure.

In several studies, Farkas investigated the nostril asymmetry in patients with unilateral and bilateral clefts. He began by studying the microform of clefts (18) to define variation in nostril size and shape and to establish to what degree nostril asymmetry may be considered a microform of the cleft lip. It was also his intention to establish the incidence of various types of nasal asymmetries in the North American normal population, and thus to contribute to anthropogenetic research of the healthy population. The same topic was presented by Farkas et al. (19) regarding a study of the asymmetry of nostril size secondary to disfigurements of the surrounding structures.

The nasal deformities present in unilateral and bilateral clefts vary in terms of extent, severity, and involvement of nasal structures. The most severe nasal deformities are usually associated with unilateral clefts. Those deformities are not only severe, but also asymmetrical, which makes correction more difficult than in cases when the deformity is symmetrical. To assess nasal deformities associated with clefts, Farkas et al. (19) performed several studies using direct anthropometry. They concentrated on the assessment of cleft lip–nasal deformity in unilateral and bilateral clefts, and a separate study was performed to assess deformities of the columella in cleft patients. With its central position in the face, the nose strongly influences the total facial appearance. In cleft lip and palate patients, secondary nasal deformities are usually more disturbing to patients than secondary lip deformities. Their observations suggest that the cleft lip–nasal deformity in adults is caused by displacement of cartilaginous and soft tissue elements of the ala, nasal tip, and columella, rather than by a tissue deficiency. The only true initial deficiency was found in the columella in both unilateral and bilateral clefts. The high incidence of nasal tip deformity and of nasal bridge deviation indicates the persistence of this deformity despite primary cleft lip surgery. With unilateral cleft lip and nose, the most common deformity of the nostril is a round shape in contrast to the elongated shape of the normal nostril. The authors suggest that the abnormal positions of the particular nasal structures must be corrected at the time of primary cleft lip repair to avoid severe secondary nasal deformities. This idea, expressed in 1972, was implemented several years later by Salyer (20), Anderl (21), and McComb (22), who initiated aggressive correction of the nasal deformity at the time of the primary cleft lip repair.

A study of the columella by Farkas and Lindsay (23) was based on the examination of 145 adults of both sexes with unilateral (74 patients) or bilateral (29 patients) cleft lip and palate, or with cleft palate only (42 patients). The study revealed that in unilateral cleft patients, the length of the columella on the cleft side is significantly shorter than on the noncleft side, and that it is significantly shorter than in normal controls. The columella in patients with bilateral clefts may not be shortened and has a considerable growth potential. The extent of the cleft did not significantly affect either the length of the columella or the incidence of asymmetry in a particular type of cleft. The results of this study suggest that a true primary defect of the columella on the cleft side in unilateral cleft patients occurs in less than 25% of the cleft population. The authors suggest that in the repair of unilateral cleft lip, precise placement of the base of the columella on the cleft side at the same level as the normal side is most important to ensure its normal growth and development. With respect to bilateral cleft lip–nasal deformity, the authors found that prior to surgery the columella may not be shortened and also has considerable growth potential. Because the columella is purely a nasal structure, it will develop and grow adequately if it remains in its initial position. Those findings did not concur with our findings at the University of Iowa, in which the following characteristics were noted: The columella was usually short in complete bilateral clefts, and after bilateral cleft lip repair the columella remained shorter than that in the normal population (24).

Farkas et al. (25) continued their active anthropometric research and reported new material, findings, and conclusions; these were not only important from the standpoint of anthropometry, but also essential to clinical practice. Their study entitled "Anthropometric and Anthroposcopic Findings of the Nasal and Facial Region in Cleft Patients Before and After Primary Lip and Palate Repair" is most interesting in terms of findings as well as the discussion and conclusions. As the authors indicate, "This study may overestimate the true incidence of cleft lip–nose stigmata generally found after primary lip repair, since many of our surgically-repaired patients were seeking surgical revision of cleft soft tissue deformities. The use of anthropometric measurements or quantitative analysis of the nose and its relationship to the rest of the face before cleft soft tissue revision, including a comparison of the patients' measurements to age-matched control values, is a first step in the planning of surgical reconstruction." This comparative study is unique in terms of design, because the authors aimed to quantitatively analyze the status before and after surgery, thus assessing the changes induced by surgical procedures. Careful analysis of the papers published and presented at meetings related to the evaluation of results of treatment procedures in patients with cleft lip and palate indicates great difficulty in establishing standards for evaluation of facial attractiveness in cleft patients. All techniques applied for this purpose have serious limitations and are rather subjective, because there is no objective methodology which allows for valid assessment that can be repeated and reproduced. There is great interest in developing a standard technique; however, at the present time, the only technique which can be considered objective and scientifically conclusive is direct anthropometry. This area of clinical investigation is helpful in terms of planning surgical procedures and assessing pre- and postoperative results.

Direct anthropometry of the head and face developed over the last 30 years is a result of the tireless efforts and immense contributions of Farkas and his colleagues. The specialists participating in the surgical and orthodontic treatment of clefts may benefit from becoming well-acquainted with publications by Farkas et al., and may apply the Farkas technique as one of the methods in the evaluation of cleft patients.

REFERENCES

1. Tobiasen JM, Hiebert JM. Reliability of esthetic ratings of cleft impairment. *Cleft Palate J* 1988;25:313–317.
2. Tobiasen JM, Hiebert JM. Combined effects of severity of cleft impairment and facial attractiveness on social perception: an experimental study. *Cleft Palate Craniofac J* 1993;30:82–86.
3. Clifford M. The state of what art? [Editorial] *Cleft Palate J* 1988;25:174–175.
4. Burian F, Farkas LG, Hajniš K. The use of anthropology in the observation of facial clefts [in Czechoslovakian]. *Anthropologie* 1964;1:41–44.
5. Farkas LG. *Anthropometry of the head and face in medicine.* New York: Elsevier, 1981.
6. Farkas LG, Hajniš K, Kliment L. Codes of surgical charts for patients with cleft lip, alveolus and palate [in Russian]. *Acta Chir Plast* 1967;9:100–110.
7. Hajniš K, Farkas LG. Anthropological record for congenital developmental defects of the face (especially clefts). *Acta Chir Plast* 1969;11:261–267.
8. Farkas LG, Posnick JC. Growth and development of regional units in the head and face based on anthropometric measurements. *Cleft Palate Craniofac J* 1992;29:301–302.
9. Farkas LG, Posnick JC, Hreczko TA. Growth patterns of the face: a morphometric study. *Cleft Palate Craniofac J* 1992;29:308–315.
10. Farkas LG, Posnick JC, Hreczko TA, Pron GE. Growth patterns of the nasolabial region: a morphometric study. *Cleft Palate Craniofac J* 1992c;29:318–324.
11. Farkas LG, Lindsay WK. Morphology of the adult face following repair of unilateral cleft lip and palate in childhood. *Plast Reconstr Surg* 1973;52:652–655.
12. Farkas LG, Lindsay WK. Morphology of the adult face following repair of bilateral cleft lip and palate in childhood. *Plast Reconstr Surg* 1971a;47:25–32.
13. Farkas LG, Lindsay WK. Morphology of the adult face after repair of isolated cleft palate in childhood. *Cleft Palate J* 1972;9:132–142.
14. Farkas LG, Lindsay WK. Morphology of the orbital region in adults following the cleft lip/palate repair in childhood. *Am J Phys Anthropol* 1972;37:65–74.
15. Coupe TB, Subtelny JD. Cleft palate-deficiency or displacement of tissue. *Plast Reconstr Surg* 1960;26:600–612.
16. Jaworska M, Reszke S. Genetic analysis of the primary cleft palate and secondary cleft palate on the basis of 459 cases [in Polish]. *Pol Pediatr* 1963;9:819.
17. Farkas LG, Kolar JC, Munro IR. Geography of the nose: a morphometric study. *Aesthetic Plast Surg* 1986;10:191–223.
18. Farkas LG, Cheung GCK. Nostril asymmetry: microform of cleft lip palate? An anthropometrical study of healthy North American Caucasians. *Cleft Palate J* 1979;16:351–357.
19. Farkas LG, Deutsch CK, Hreczko TA. Asymmetry in nostril size secondary to disfigurements of surrounding structures. *Dtsch Z Mund Kiefer Gesichts Chir* 1983;7:258–262.
20. Salyer KE. Unilateral cleft lip and cleft lip nasal reconstruction. In: Bardach J, Morris HL, eds. *Multidisciplinary management of cleft lip and palate.* Philadelphia: WB Saunders, 1990;173–183.
21. Anderl H. Primary unilateral cleft lip and nose reconstruction. In: Bardach J, Morris HL, eds. *Multidisciplinary management of cleft lip and palate.* Philadelphia: WB Saunders, 1990;184–196.
22. McComb H. Anatomy of the unilateral and bilateral cleft lip nose. In: Bardach J, Morris HL, eds. *Multidisciplinary management of cleft lip and palate.* Philadelphia: WB Saunders, 1990;144–149.
23. Farkas LG, Lindsay WK. The columella in cleft lip and palate anomaly. In Hueston JT, ed. *Transactions of the fifth international congress on plastic reconstructive surgery.* Melbourne: Butterworths, 1971;373–381.
24. Bardach J, Salyer KE. *Surgical techniques in cleft lip and palate,* 2nd ed. Chicago: Mosby–Year Book, 1991.
25. Farkas LG, Hajniš K, Posnick J. Anthropometric and anthroposcopic findings of the nasal and facial region in cleft patients before and after primary lip and palate repair. *Cleft Palate Craniofac J* 1993;30:1–12.

Chapter 9

Craniofacial Anthropometry in Clinical Genetics

Richard E. Ward

Genetic disease remains one of the major threats to children, accounting for a large portion of neonatal mortality and pediatric hospitalizations in developed nations. There are now over 6500 recognized inherited birth defects (1), and many have some effect on craniofacial development (2). The face, like the brain and heart, is a complex structure arising from the confluence of a delicately orchestrated embryology. Disruption of this development can leave scars that are disfiguring and revealing, because the defect often provides the "fingerprints" by which an underlying cause (pathophysiology) is defined. It is the study of this link between the physical defect and disruption of normal development that has fueled the emergence of clinical dysmorphology and medical genetics.

While in recent years much of the visible drama in these fields has focused on the identification of the molecular (genetic) agents of these birth defects, a quiet revolution has been brewing in the wings. This revolution began with the work of Hajniš (3) and Farkas (4), who provided important new comparative data that allowed for the quantitative description of the human face. Together, Hajniš and Farkas collected normative data on more than 100 measures of the head and face, a significant advance over the handful of craniofacial measurements theretofore available. More importantly, the new measurements were explicitly developed for clinical application, whereas those previously in use had been derived from failed anthropological attempts to define racial typologies. Thus, with the means at hand to describe quantitatively the surface anatomy of the face, craniofacial anthropometry has joined the ranks of morphometric tools such as roengentometric cephalometry and metacarpal phalangeal pattern analysis as important adjuncts to the clinically trained eye. In the remainder of this chapter we will explore some of the ways this technique has been employed in clinical genetics and discuss some of the potential applications that have yet to be realized.

SYNDROME DESCRIPTION AND DELINEATION

In clinical genetics, a central problem is pattern recognition. Genetic syndromes, by definition, are identified by a set of key criteria or features that serve to distinguish one syndrome from another. Accurate description of these key features is critical in defining previously unrecognized syndromes, developing a useful diagnostic key, placing syndromes in a meaningful taxonomy, and ultimately, as noted above, in defining a syndrome's pathophysiology.

There are two approaches to pattern recognition, one subjective and the other metric. The subjective approach relies on the clinician's ability to describe the appearance of the craniofacial complex. For example, in the 1982 edition of Smith (5), a standard reference work in clinical genetics, the craniofacial characteristics of Down syndrome are described as follows: "Brachycephaly with relatively flat occiput and a tendency toward midline parietal hair whorl. Mild microcephaly with up-slanting palpebral fissures. Thin cranium with late closure of fontanels. Hypoplasia to aplasia of frontal sinuses, short hard palate. Small nose with low nasal bridge and a tendency to have intercanthal folds. . . . [Ears] Small; overfolding of angulated upper helix; sometimes prominent; sometimes absent earlobes." This descriptive approach emphasizes the gestalt and relies in large measure on the clinician's cumulative experience, judgment, and understanding of the developmental process. Terms such as "small," "short," or "hypoplasia" are subjective assessments based upon the clinician's internalized criteria and practiced eye. A good clinical description should convey enough information so that other clinicians can readily identify similar cases.

In contrast, the metric approach seeks to identify quantitative patterns based on linear, angular, circumferential, or proportional measurements derived from cephalometric or tomographic x-rays (cephalometry), photographs (photogrammetry), or directly from the head and face (anthropometry). Thus, Farkas et al.'s (6) anthropometric description of Down syndrome children includes the following "key" features: "A shallow upper-third of the face depth (between the nasal root and the ear's tragi) and a short right auricle were the most striking subnormal findings (each 71.5%), with subnormal palpebral fissures in third place (68.8%)." Here subnormal has a specific statistical meaning, referring to measurements that fall at least two standard deviations below the mean, in obvious contrast to the subjective description. Such measurements usually convey little of the facial gestalt and focus primarily on size differences rather than those of shape. On the other hand, they rely much less on clinical experience and subjective judgment. Metric approaches have the added advantage of being inherently comparative because individual measurements must be converted to standard scores (*z*-scores) before they can be interpreted. Thus, by subtracting each individual score from the age- and sex-matched normative or control mean and dividing that sum by the age- and sex-matched standard deviation, we have an immediate comparison with normal values and a quotient (*z*-score) which is unitless and therefore comparable to *z*-scores derived from other individuals, other variables, or other syndromes.

When describing patients who have no known diagnosis or in characterizing potentially new conditions (provisionally unique syndromes), the metric approach can provide concrete and easily transferable measurements of features thought to be definitive of the condition (7). These data can even be presented graphically so that similar conditions can be contrasted and compared across the key measurements. This approach, called *pattern profile analysis* (8), will be discussed in greater detail below.

Ideally the delineation of genetic syndromes should rely on both approaches; however, there are very few such complementary descriptions (9–11). More commonly, anthropometric descriptions are presented separately from those of the dysmorphologist, or only a few cursory measurements are presented with the clinical description. In any case, it is readily apparent from the scant references cited above that only a fraction of more than 6000 recognized genetic syndromes have been described anthropometrically.

PATTERN RECOGNITION AND SYNDROME DIAGNOSIS

The obvious goal of syndrome description is to make accurate diagnosis possible. Thus, once a pattern of traits has been described for a condition, this pattern must be recognizable in other individuals with the same condition. This is essentially an exercise in taxonomy or classification and is analogous to the process faced by biologists when trying to place an individual organism (unknown case) into its proper genus and species (disease category or syndrome). Like biological taxonomy, disease taxonomy or nosology is complicated by the fact that individual members of a natural grouping vary tremendously in the expression of diagnostic traits. This variability in expression is caused by the complexities of allelism, pleiotropy, and gene–environment interaction. Key features may also vary with developmental stage, ethnicity, and sex. Variability is the greatest hindrance to accurate diagnosis.

Because the subjective approach to syndrome identification relies by definition on archetypical traits, it is particularly prone to the problems introduced by individual variation in the expression of these traits. In contrast, the metric approach offers an opportunity to construct a diagnostic key from a "population" of affected individuals and to thereby incorporate naturally occurring variability. For example, in a study of X-linked hypohidrotic ectodermal dysplasia (HED), 20 craniofacial variables were analyzed in 16 affected individuals and 31 of their first-degree relatives (12,13). The latter group included 15 obligate gene carriers and 16 normal family members. Means and standard deviations were calculated from the *z*-scored data. When these data are presented in graphic form, as a pattern profile, a clear distinction can be seen between the affected and unaffected means, with those of the obligate carriers falling between these two profiles (Fig. 9–1).

Univariate analysis was used to identify the variable means that differed significantly between the two groups. When combined with multivariate (stepwise) discriminant function analysis, a small subset of four variables was isolated that accurately classified 97% of the sample. It is worth noting that none of the four variables selected were those usually cited as diagnostic in the subjective descriptions of the syndrome. These findings illustrate the complementary nature of the metric and descriptive approaches: Each contributes in a different way to syndrome identification. Using a combination of the two approaches would be most useful in conditions with poorly defined phenotypes and wide variation in the expression of the component features.

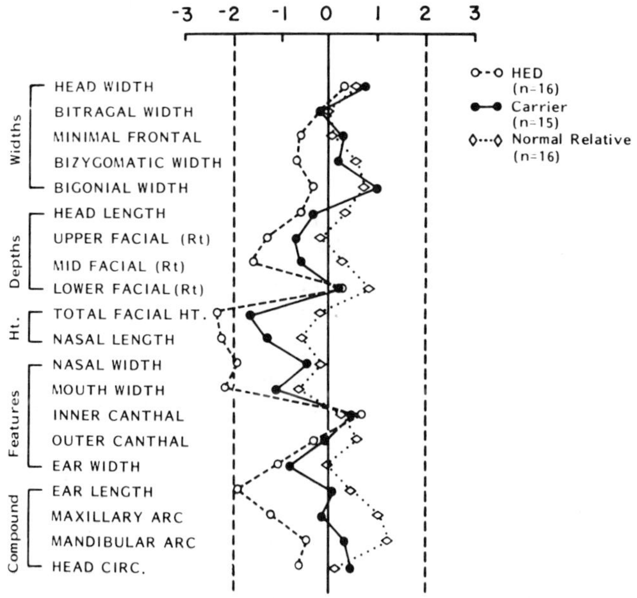

FIG. 9–1. Craniofacial pattern profile comparing mean *z*-scores of affected, carrier, and normal groups from HED families. Zero baseline represents mean values from published standards. (From ref. 13, with permission.)

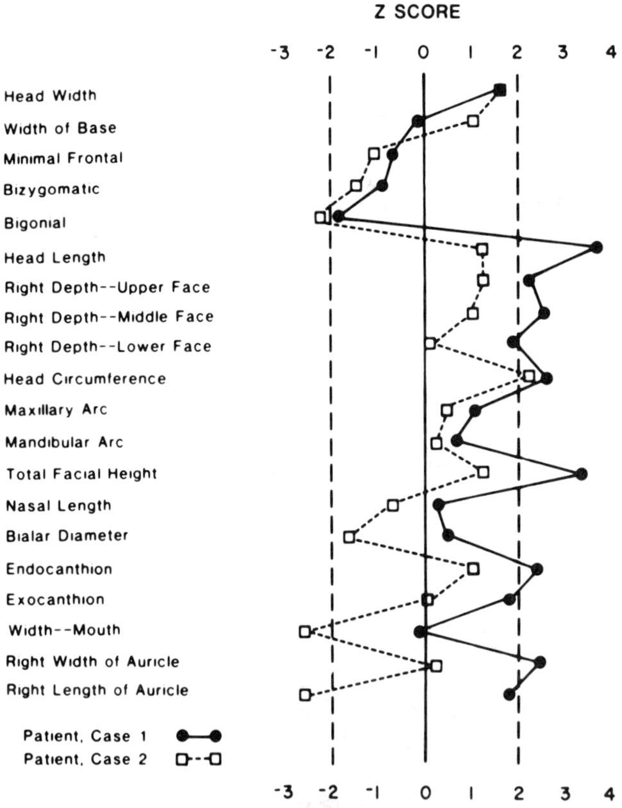

FIG. 9–2. Craniofacial pattern profile of two patients diagnosed as having the same condition (Soto's syndrome) on the basis of pattern similarity. (Modified from ref. 11, with permission.)

Craniofacial anthropometry can also be used as a diagnostic aid for isolated cases of a particular condition. Figure 9–2 depicts craniofacial pattern profiles for two unrelated children with a similar set of clinical traits including early overgrowth, developmental delay, and nystagmus. The diagnosis of Soto's syndrome was eventually established in the male after much ancillary investigation. This diagnosis was not made on the basis of the facial measurements but on the combination of clinical signs and a positive diagnostic score from a metacarpal phalangeal pattern profile (MCPP) analysis (11). The second child was seen at a later date; based on the similarity of the facial pattern and clinical symptoms with those in the first case, she was determined to have Soto's syndrome as well (the MCPP analysis in this case was inconclusive).

CLARIFICATION OF INHERITANCE PATTERN AND IDENTIFICATION OF MINIMALLY AFFECTED GENE CARRIERS

As the defining characteristics of a syndrome are clarified, its pattern of inheritance must be established before accurate genetic counseling can be provided. This entails

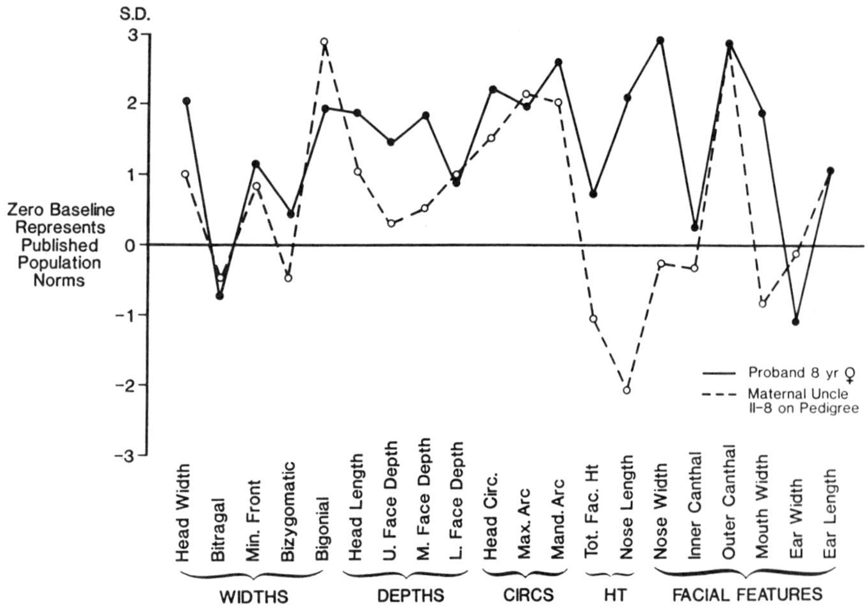

FIG. 9–3. Craniofacial pattern profile of an individual with Beckwith–Wiedemann syndrome (BWS) superimposed on that of her maternal uncle with ear pits, abdominal wall defects, and giganticism (r_2 = .50). On the basis of the pattern similarity and subsequent medical findings, it is believed that the uncle has a mild, previously undiagnosed expression of the condition.

identifying other people in the family who may be carrying the gene responsible for the syndrome but who may show a "subclinical" pattern of expression. In the study of HED described above, (Fig. 9–1) female relatives who were probable gene carriers of this X-linked condition were found to express a "pattern" of metric results that fell between those of affected and normal individuals. This indicates that it should be possible to define a useful metric diagnostic key for detecting minimally affected gene carriers of HED as Saksena and Bixler (14) have already done using a similar approach with cephalometric variables.

In a previously unpublished study of another genetic condition, the Beckwith–Wiedemann syndrome (BWS), craniofacial anthropometry was also used to identify undiagnosed family members. BWS is characterized clinically by neonatal somatic overgrowth, visceromegaly, metabolic disorders, and an unusual (though poorly defined) facial phenotype (15). Figure 9–3 presents the pattern profile of an affected child and contrasts this with that of her maternal uncle. Using the pattern similarity index (r_z) as described by Garn et al. (8), it was found that these two had a much higher degree of pattern similarity (r_2 = .50) than would be expected by their degree of genetic relationship (.25). One other maternal uncle showed a similar degree of pattern similarity. When the medical histories of these two individuals were checked, it was found that they both had other features consistent with BWS (large size at birth, ear pits, and umbilical hernias). It is probable that these two individuals have mild cases of the syndrome. More importantly, this finding lends support to the still controversial assumption that this is a dominantly inherited condition (1).

MOLECULAR GENETICS, GENETIC LINKAGE ANALYSIS, AND CRANIOFACIAL ANTHROPOMETRY

Given the dramatic successes in identifying specific genetic defects made possible by revolutionary approaches in molecular genetics, it is perhaps not surprising that morphometric approaches appear to some clinicians as outdated and no longer necessary. However, the new technologies have a number of serious limitations that have served to make careful morphometric analysis all the more important. For example, the molecular techniques are most effective when the pattern of inheritance is well known and when all gene carriers in a family can be accurately identified. This may be relatively easy in dominantly inherited conditions (such as Huntington's chorea) where penetrance is high, but it is much more difficult for that vast majority of syndromes where variable expression, reduced penetrance, and allelism confound the picture. For such conditions, careful facial measurements might be useful in defining a lower threshold of penetrance, thereby improving the possibility of success for subsequent linkage studies.

It is seldom appreciated that until a particular disease-causing gene has been precisely located and chemically described, it is usually not possible to develop a simple test for the presence of that gene in an individual. Instead, several first-degree relatives in two or more generations must also be analyzed so that the specific linkage between a molecular marker and genetic defect (as it occurs in that family) can be identified. The less precise the localization of the gene, the greater the chance of error and the greater the need to analyze additional family members. In those cases where additional family members are unavailable or unwilling to cooperate, a molecular diagnosis may not be possible. However, if care has been taken previously to establish a morphometric profile from affected individuals with well-established patterns of linkage, then the facial measurements may be the best available mode of diagnosis.

It is also worth noting that the facial dimensions described in an anthropometric analysis are, to a large extent, under genetic control. Specific forms of morphometric variation (such as macrocephaly or micrognathia) that are sometimes components of genetic syndromes might themselves be studied by molecular linkage analysis and eventually ascribed to specific genetic defects. Such information would provide valuable insight into the mechanisms of normal development and could lead one day to more effective treatment and prevention of such defects.

LIMITATIONS AND UNEXPLORED POTENTIALS OF CRANIOFACIAL ANTHROPOMETRICS IN CLINICAL GENETICS

As the examples cited above indicate, craniofacial anthropometry provides a simple means of describing important aspects of facial morphology when used in a properly designed study (13). It can objectify clinical observation, reveal patterns of variation common to specific syndromes, and provide a means of comparing unknown cases to various known disorders. However, even these applications barely scratch the technique's potential. Craniofacial anthropometry, like other forms of whole-body measurement, is probably best suited for large-scale population studies. Thus, just as it has been possible to develop national standards of height, weight, and head circumference that are used in screening for abnormal growth and development, craniofacial assessment could become a generalized screening device for developmental disruption in newborns, much as Garn et al. (16) have done with the "dysmorphic index" derived from cephalometric X-rays. There are a variety of environmental agents known to affect craniofacial development (fetal alcohol syndrome, fetal alcohol effects, fetal hydantoin effect, fetal warfarin effects). Routine application of even a few craniofacial measures such as facial depth, inner and outer canthal distances, minimal frontal width, and mandibular and maxillary arcs could provide an early indication of problems in neural development (reduced inner canthal or minimal frontal distances) or skeletal development (midfacial growth) or could detect potential airway problems (mandibular arc). Additional studies might attempt to correlate specific etiological agents with specific types of craniofacial problems. Thus, chromosomal defects may be associated with a different pattern of disruption than the one associated with a single gene or with environmental agents.

In spite of the many ways in which craniofacial anthropometry has proven useful in clinical genetics and its untapped potential, it remains underutilized. While the clinical literature often includes one or two standard anthropometric measurements such as head circumference, inner canthal distances, or ear lengths as routine components of a case description, more systematic applications are rare. The most obvious explanation for this is that few clinical geneticists have taken the time or have had the opportunity to acquire the skills necessary to utilize the technique effectively. In addition, the basic instruments are difficult to acquire and may seem excessively expensive (although at a cost of a little over $1000.00 for a complete set, it would seem a bargain against the costs of the more exotic imaging techniques now routinely used in clinical settings). In addition, effective anthropometric studies require careful planning with attention toward documenting measurement error and collecting local control data (17). Finally, unlike stature, weight, and head circumference, which have long been part of National Center for Health Statistics studies, the craniofacial standards developed by Farkas have not had wide circulation. Thus, it would seem that without concerted effort to expand the circle of its practitioners, craniofacial anthropometry will remain a useful but little used technique in the expanding realm of clinical genetics.

REFERENCES

1. McKusick VA. *Mendelian inheritance in man,* 10th ed. Baltimore: The Johns Hopkins University Press, 1991.
2. Salinas CF. An approach to an objective evaluation of the craniofacies. *Birth Defects* 1980;16(5):47–74.
3. Hajniš K. Ohrmuschel- und Handwachstum (Verwendung bei den Operationen der angeborenen Missbildungen und Unfallsfolgen). *Acta Univ Carol* [*Biol*] *Praha* 1972;2–4.
4. Farkas LG. *Anthropometry of the head and face in medicine.* New York: Elsevier, 1981.
5. Smith DW. *Recognizable patterns of human malformation: genetic, embryologic and clinical aspects,* 3rd ed. Philadelphia: WB Saunders, 1982.
6. Farkas LG, Munro IR, Kolar JC. Abnormal measurements and disproportions in the face of Down's syndrome patients: preliminary report of an anthropometric study. *Plast Reconstr Surg* 1985;75:159–167.
7. Goldstein DJ, Ward RE, Nichols WC, Palmer CG. Familial 8;15 (p23.3, q22.3) translocation: report of two cases with 15 (q22.3-qter) duplication. *J Med Genet* 1987;24:684–687.
8. Garn SM, Smith BH, LaVelle M. Applications of pattern profile analysis to malformations of the head and face. *Radiology* 1984;150:683–690.
9. Ward RE, Goldstein DJ: Case report 130: hypotonia, small genitalia, hypoplastic pinnae and seizures. *Dysmorph Clin Genet* 1987;1: 24–28.
10. Bavinck JNB, Weaver DC, Ellis FD, and Ward RE. Brief clinical report: a syndrome of microcephaly, eye anomalies, short stature and mental deficiency. *Am J Med Genet* 1987;26:825–831.

11. Goldstein DJ, Ward RE, Moore E, Fremion AS, Wapner RS. Overgrowth, congenital hypotonia, nystagmus, stabisimus, and mental retardation: variant of dominantly inherited Soto's sequence? *Am J Med Genet* 1988;29:783–792.
12. Ward RE, Bixler D. Anthropometric analysis of the face in hypohidrotic ectodermal dysplasia: a family study. *Am J Phys Anthropol* 1987;74:453–458.
13. Ward RE. Facial morphology as determined by anthropometry: keeping it simple. *J Craniofac Dev Biol* 1989;9:45–60.
14. Saksena SS, Bixler D. Facial morphometrics in the identification of carriers of X-linked hypohidrotic ectodermal dysplasia. *Am J Med Genet* 1990;35:105–114.
15. Cohen MM. Comment on the macroglossia-omphalocele syndrome. *Birth Defects* 1969;5(2):197.
16. Garn SM, LaVelle M, Smith BH. Quantification of dysmorphogenesis: pattern variability index, σ_z. *Am J Radiol* 1985;144:365–369.
17. Ward RE, Jamison PL. Measurement precision and reliability in craniofacial anthropometry: implications and suggestions for clinical applications. *J Craniofac Genet Dev Biol* 1991;11:156–164.

Chapter 10

The Application of Anthropometric Surface Measurements in Craniomaxillofacial Surgery

Jeffrey C. Posnick and Leslie G. Farkas

Anthropometry is the objective analysis that replaces subjective visual judgment (anthroposcopy) with quantitative measurement of the soft tissues of the face which comprise the outside layer through which we judge skeletal morphology. Anthropometric surface measurements have been obtained through cross-sectional study in a large group of growing children and young adults (1). They have proven useful in the quantitative evaluation of postnatal development in the head and face (2,3). Growth trends and relationships between aspects of the head and face based on anthropometric findings are predictable. The data gathered from a specific patient can be compared to normal values controlled for age and gender (4–8). The analysis of the patient's measurements can be used for planning the timing and type of reconstructive surgery for those with congenital, developmental, traumatic, or neoplastic disorders (9–11).

Examination of the entire craniofacial region should be meticulous and systematic. The skeleton and soft tissue are assessed in a standard way to identify all normal and abnormal anatomy. Specific findings tend to occur in particular deformities, but each patient is unique. A quantitative analysis may be carried out by taking measurements from computerized tomographic (CT) scans, surface anthropometry, cephalometry, and dental models (12–20). Each form of analysis will play a different role in each patient. The use of symmetry, proportionality, and balance, combined with the reconstruction of specific aesthetic units, is critical to achieve an unobtrusive or attractive face in a child or adult with a craniomaxillofacial problem.

The purpose of a quantitative assessment—whether by CT scan analysis, anthropometric measurements, cephalometric analysis, or dental model analysis—is to help predict growth patterns, confirm or refute clinical impressions, aid in treatment planning, and provide a framework in the objective assessment of immediate and long-term results. The measurement of linear distances, angles, and proportions based on specific anatomic landmarks is a useful step in patient evaluation.

There are specific regions within the head and face for which I find anthropometric surface measurements clinically useful (4–8). They include the head, face, orbital region, nasolabial region, and the ears.

The measurement within each region of specific landmarks (linear, angular, and proportional) and the relationship of the parts to each other comprise an important aspect of the analysis. Abnormal development of one region of the face may affect another, and irregularities of growth and shape in either the cranial vault or face will disturb the harmonious relationship between them and the orbits. When reconstructive surgery seems indicated, it is important for the surgeon and the craniofacial team to know (a) the mean and standard deviations of key facial measurements at varied ages, (b) the rates of growth of each facial region, (c) the extent of growth completed at key age intervals, and (d) the times of maturation. This information can help determine both the extent and preferred timing of surgery within specific regions of the craniofacial skeleton.

REGIONAL UNITS IN THE HEAD AND FACE

Head

The relatively thin soft-tissue cover with few contour inequalities makes the skull ideal for utilizing surface measurements when reliable data about the basic dimensions of the head are required (5). Head circumference is the most frequently reported measurement in the medical literature. Head width and length have been used to calculate the "cephalic index" in which the ratio between the measurement defines the proportional quality of the head.

Five anthropometric surface measurements in particular can be taken directly from the head of the subject and effectively used in clinical practice: head width (eurion–eurion, eu-eu), forehead width (frontotemporal–frontotemporal, ft-ft), height (vertex–nasion, v-n), length (glabella–opisthocranion, g-op), and circumference (on g-op plane). The technique of taking the measurements has been described previously (5). Tables of mean values and standard deviations of all measurements controlled for age and gender are available (5). Normal growth trends and relationships between aspects of the head and those of the face based on anthropometric findings are predictable. The patient's data can be compared to normal values and then used for planning the timing and type of reconstructive surgery and in the immediate and long-term assessment of reconstructive results.

Face

We tend to identify the facial framework transversely by the width of the upper face (zygomatic region) and the lower face (mandibular region) (4), vertically by the facial heights, and horizontally (sagitally) by the facial depth in the maxillary and mandibular regions.

There are seven key anthropometric surface measurements that can be taken directly from the face to provide clinically useful information. The three vertical facial measurements are as follows: *facial height* (n-gn), measured between the nasion point at the root of the nose (n) and the chin point (gnathion, gn); *upper face height* (n-sto), taken between the nasion (n) landmark and the stomion point (sto) in the middle of the labial fissure between the gently closed lips; and the *mandibular height* (sto-gn), measured between the stomion and the gnathion landmarks. The two transverse facial measurements are (i) *face width* (bizygomatic diameter, zy-zy), defined by the distance between the right and left zygion (zy) landmarks positioned at the most lateral point of the zygomatic arches, and (ii) *mandible width* (go-go), measured between the gonion (go) points of the mandible. The two horizontal (sagittal) measurements define the facial depth when viewed in profile. The depth is determined in the *maxillary region* of the face (t-sn), between the tragion (t) point (located just above the tragus of the ear) and the subnasale landmark (sn) (which is at the midpoint of the base of the columella). The depth of the lower third of the face (*mandibular region*) (t-gn) is measured as the distance between the tragion landmark (t) and the chin point (gn).

Orbital Region

The orbits are situated between the cranial vault and the face (8). They are the dominant aesthetic elements within the craniofacial complex. They greatly influence visual judgment of a person's overall facial appearance.

Two anthropometric surface measurements within the orbital–adnexal region are particularly useful to the clinician: the *intercanthal width* (endocanthion–endocanthion, en-en) and the *biocular width* (exocanthion–exocanthion, ex-ex). Tabulation of the means and standard deviations of these measurements controlled for age and gender between 1 and 18 years of age are available (8).

Nasolabial Region

Knowledge of the rate of growth of the nose, upper lip, and their individual parts can help the surgeon appreciate variations from normal, understand altered proportions, and select an optimal time for reconstruction of a specific facial deformity in the nasolabial region (7).

Six anthropometric surface measurements have been selected as the most useful to provide a quantitative framework for understanding deformities of the nose and upper lip. The three measurements from the nose include: *nose height* (nasion–subnasale, n-sn), *nose width* (alare–alare, al-al), and *nasal tip protrusion* (subnasale–pronasale, sn-prn). The three measurements from the upper lip include: *cutaneous upper lip height* (subnasale–labiale superius, sn-ls), *vermilion upper lip height* (labiale superius–stomion, ls-sto), and *total upper lip height* (subnasale–stomion, sn-sto). Tables reporting the means and standard deviations of these measurements controlled for age and gender are available (7).

Ear

Knowledge of the postnatal growth of the ear, in both width and height, is important in the determination of the optimal time for reconstruction of the ear and in planning the morphology of the new ear (6,21–23).

Two anthropometric surface measurements are especially useful in defining ear morphology. The *ear width* (preaurale–postaurale, pra-pa) and *length* (superaurale–subaurale, sa-sba). The means and standard deviations of these two measurements between 1 and 18 years of age in both sexes are available in table form for quick reference (6).

Because of the discrepancy between the developmental levels of ear width and length, the ear is relatively wide in relation to its length in early childhood. Among the craniofacial syndromes, disproportionally wide ears are observed most often in patients with Apert and Crouzon syndromes, whereas disproportionately narrow ears are usually found in cleft lip and palate patients. The impression of small ears in Down syndrome patients results from the disharmony between reduced ear length and face height rather than from a small ear width.

Successful reconstruction of a microtic ear is dependent on accurate knowledge of development, maturation, and growth qualities of the normal ear. Selection of the optimal timing of ear surgery in a child's life is a major factor in its success.

CLINICAL EXAMPLES

Patient 1: Anterior Plagiocephaly (Unilateral Coronal Synostosis) (Fig. 10–1)

Procedure: Anterior cranial vault and three-quarter orbital osteotomies and reshaping

An 8-month-old infant had a craniofacial assymetry at birth (24). The left (ipsilateral) forehead was flat and retruded. The left supraorbital ridge and lateral orbital rim were displaced posteriorly. The root of the nose was constricted to the left side. The left infraorbital rim and anterior zygoma also displayed a degree of flatness. The right (contralateral) side of the forehead bulged forward with inferior and anterior displacement of the orbital roof and supraorbital ridge. The tip of the nose was shifted to the side away from the flat forehead.

A CT scan of the craniofacial skeleton revealed a left coronal suture fusion. The left sphenoid wing was elevated superiorly (harlequin shape), and the anterior cranial base was short in the anterior–posterior dimension on the left side.

Preoperative surface anthropometric measurements confirmed a decreased head length (130 mm, mean 149 mm), a diminished head circumference (403 mm, mean 434 mm), a slightly increased medial canthal distance (29 mm, mean 25 mm), and a decreased distance from lateral canthus to lateral canthus (69 mm, mean 74 mm).

At operation, a coronal (skin) incision was carried out with soft-tissue dissection and exposure for a bifrontal craniotomy and then retraction of the frontal and temporal lobes of the brain. The three-quarter orbital osteotomies were completed across the orbital roof, lateral orbital wall, and orbital floor into the inferior orbital fissure and down to the mid-infraorbital rim region. The symmetrically reshaped orbital osteotomy units were then re-inset between the ocular globes and the brain. The previously removed anterior cranial vault bone was sectioned to reconstruct a new forehead. Stabilization was with direct transosseous wires and bone microplates and screws as needed.

As a result of the cranial vault and three-quarter orbital osteotomies with reshaping, the head length increased (174 mm, mean 181 mm), the head circumference increased but did not normalize (489 mm, mean 518 mm), the medial canthal separation improved (32 mm, mean 30 mm), and the lateral canthal–lateral canthal distance normalized (76 mm, mean 76 mm).

Patient 2: Metopic Suture Synostosis (Trigonocephaly) (Fig. 10–2)

Procedure: Anterior cranial vault and three-quarter orbital osteotomies and reshaping

An 18-month-old boy presented with premature closure of the metopic suture, which had caused (a) a vertical midline ridging of the forehead, (b) flatness and recession to the supraorbital ridges that were more severe laterally, and (c) a degree of orbital hypotelorism (15). The bitemporal width was constricted, as was the anterior cranial vault width. Overall, these features gave a triangular shape to the anterior cranial base and vault.

Preoperative surface anthropometric measurements confirmed a decreased forehead width (81 mm, normal 83 mm), a decreased head length (160 mm, mean 166 mm), a diminished head circumference (475 mm, normal 491 mm), and a decreased lateral canthal–lateral canthal distance (69 mm, normal 76 mm).

At operation, the patient underwent a coronal (skin) incision, through which anterior cranial vault and three-

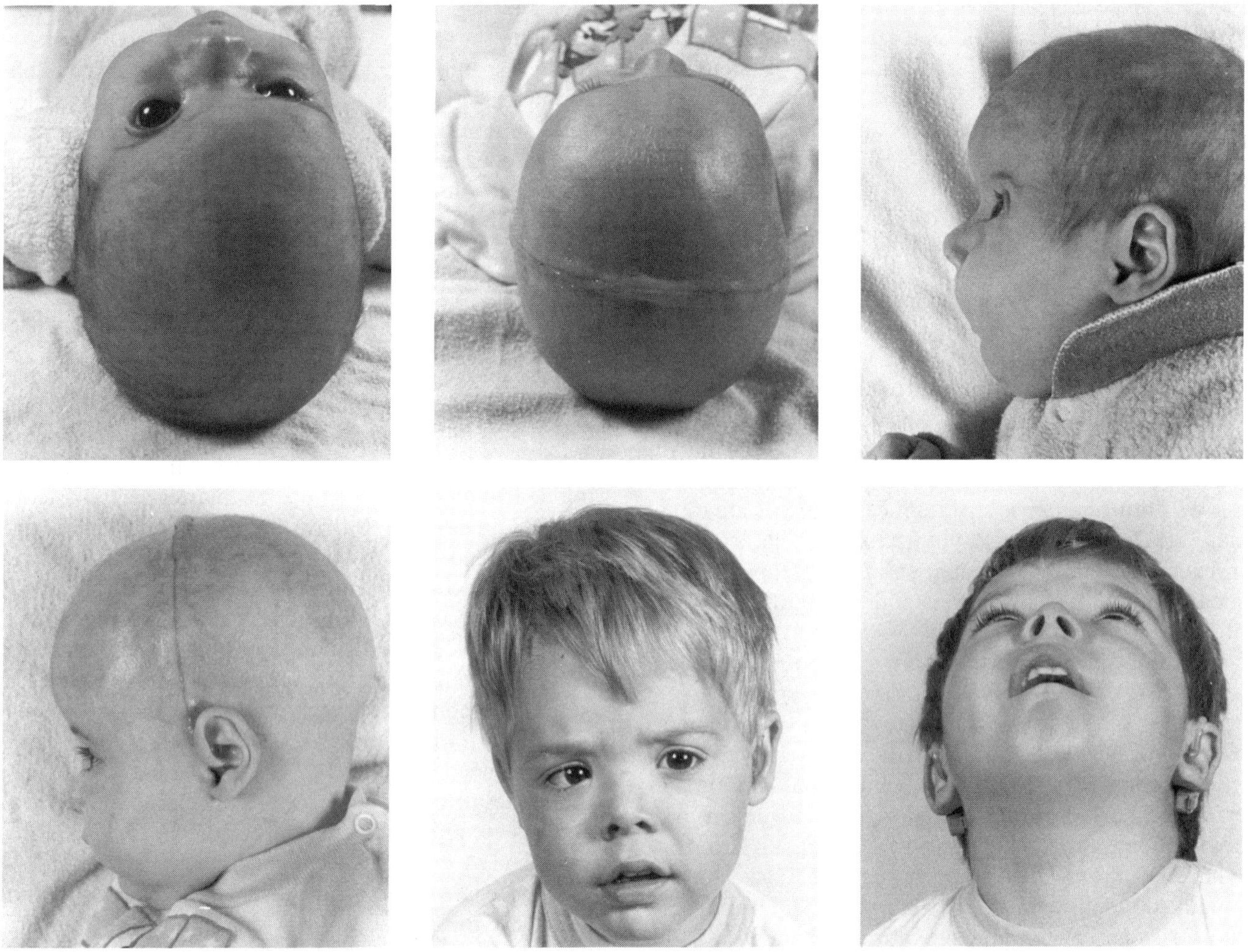

A,B C

D,E F

FIG. 10–1. An 8-month old infant with unrepaired unilateral coronal synostosis required suture release and anterior cranial vault and three-quarter orbital osteotomies with reshaping and advancement. **A:** Preoperative bird's-eye view. **B:** Bird's-eye view early after reconstruction. **C:** Preoperative profile view. **D:** Profile view early after reconstruction. **E:** Frontal view at 3 years of age. **F:** Worm's-eye view at 3 years of age. (From ref. 24, with permission.)

quarter orbital osteotomies with reshaping were carried out using an intracranial approach. The orbital units, which were badly misshapen, required three-dimensional reshaping. In addition, the orbits were separated in the midline to correct his hypotelorism. A bone miniplate was helpful in stabilizing the orbital units in the midline, and direct wires were used in many locations to stabilize weakened areas of the orbital rims. The total orbital unit was re-inset, and the anterior cranial vault was reconstructed.

As a result of the cranial vault and three-quarter orbital osteotomies with reshaping, the surface anthropometric measurements improved. The forehead width increased (90 mm, normal 87 mm), the head length increased (170 mm, normal 170 mm) the head circumference increased (510 mm, normal 518 mm), and the lateral canthal–lateral canthal distance normalized (75 mm, normal 77 mm).

Patient 3: Scaphocephaly (Sagittal Suture Synostosis) (Fig. 10–3)

Procedure: Total cranial vault and upper orbital osteotomies with reshaping

Shortly after birth, a patient was noted to have an elongated and constricted cranial vault (24). By the age of 9 years, the child was being teased at school and in the neighborhood to the extent that he was referred for a craniofacial evaluation and treatment.

The boy had a normal psychology profile but with anterior–posterior elongation of the skull and diminished bitemporal width. The midface seemed normal with an Angle class I occlusion. Ophthalmologic examination revealed a strabismus and amblyopia and upper eyelid ptosis, but funduscopic examination was normal.

A CT scan carried out in the axial and coronal planes

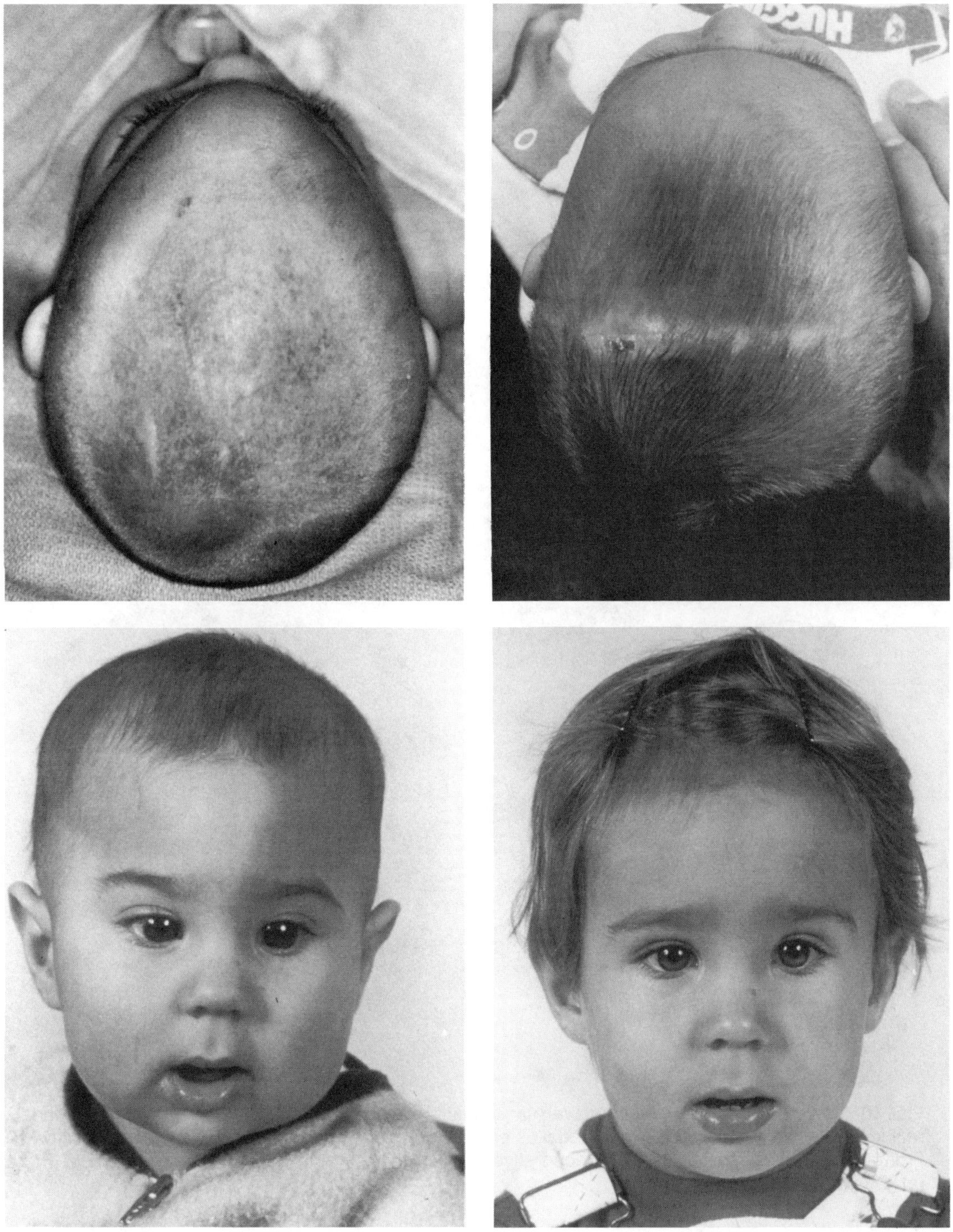

A B C D

FIG. 10–2. Anterior cranial vault and three-quarter orbital osteotomies and reshaping in an 18-month-old boy with metopic synostosis. **A:** Preoperative bird's-eye view of skull. **B:** Early postoperative bird's-eye view of skull. **C:** Preoperative frontal view. **D:** One-year postoperative frontal view. (From ref. 15, with permission.)

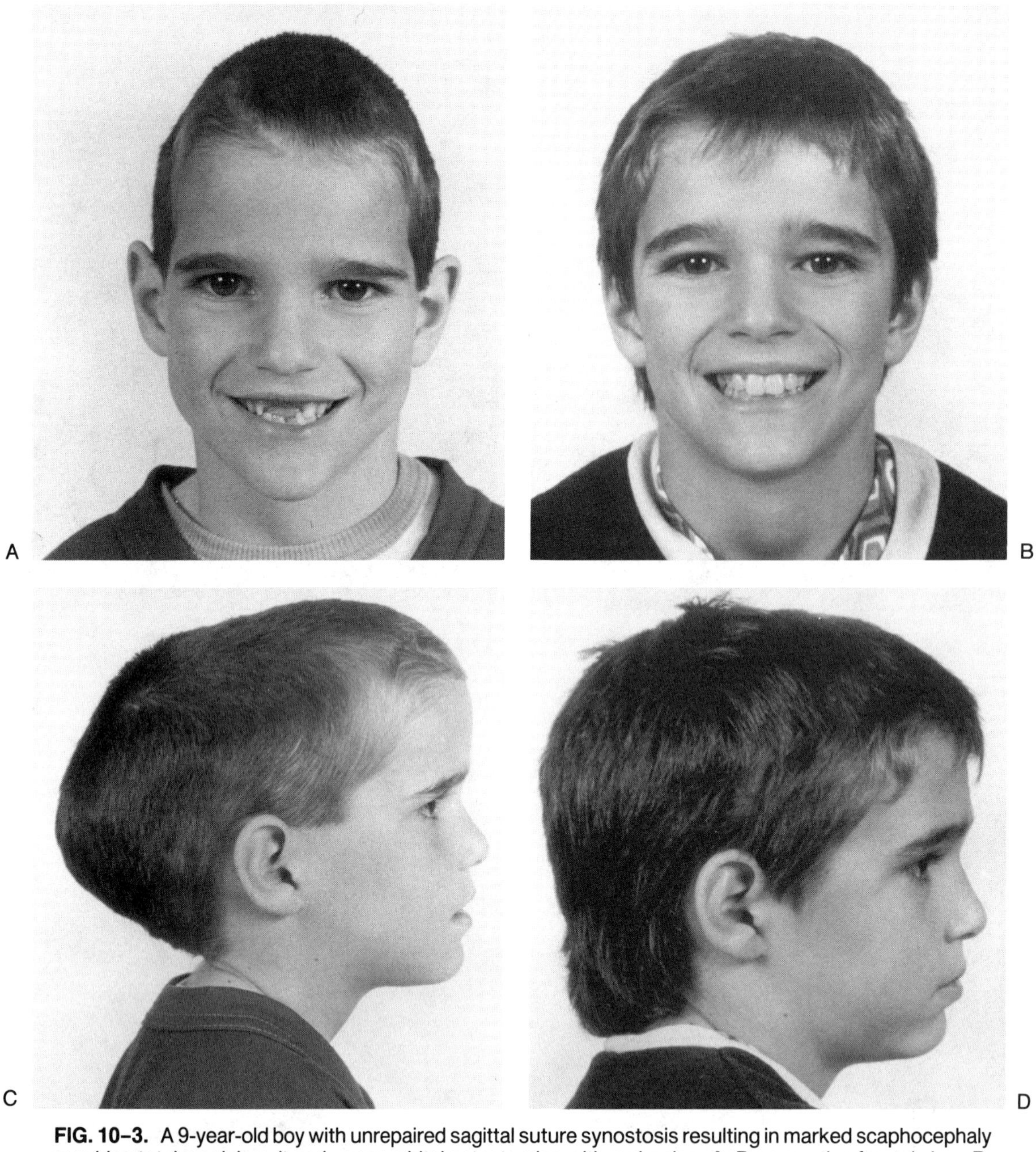

FIG. 10–3. A 9-year-old boy with unrepaired sagittal suture synostosis resulting in marked scaphocephaly requiring total cranial vault and upper orbital osteotomies with reshaping. **A:** Preoperative frontal view. **B:** Frontal view after reconstruction. **C:** Preoperative profile view. **D:** Profile view after reconstruction. (From ref. 24, with permission.)

showed ridging over the sagittal suture. The shape of the cranial vault was consistent with a neglected sagittal suture synostosis.

Preoperative surface anthropometric measurements confirmed a decreased head width (129 mm, normal 140 mm), a decreased forehead width (94 mm, normal 104 mm); an increased head length (215 mm, normal 183 mm), and an overall increased head circumference (563 mm, normal 519 mm).

Working through a coronal (skin) incision and multiple craniotomies the total cranial vault, upper orbits, and squamous temporal bone were removed and then reshaped to widen the bitemporal and biparietal width and shorten the anterior–posterior length of the skull. Stabilization was with multiple titanium miniplates and screws.

When surface anthropometric measurements were made 2 years after the cranial vault and upper orbital reshaping, the head width was improved (136 mm) as was the forehead width (105 mm); the head length was now closer to normal (190 mm) while the overall head circumference had increased to 570 mm.

Patient 4: Crouzon Syndrome (Fig. 10–4)

Procedure: Anterior cranial vault and monobloc osteotomies with reshaping and advancement

An 8-year-old boy who was born with Crouzon syndrome underwent bilateral "lateral Canthal advancements" at 6 months of age (25). The family history was positive: The father had severe oxycephaly, proptosis, and midface deficiency. Later the boy required placement of a ventriculoperitoneal shunt for management of hydrocephalus. Although he was lost to follow-up for a number of years, when he returned he had a history of recent onset of severe headaches and bedwetting; he was 1 year behind in school and had been subjected to teasing by his peers because of his "bulging eyes and flat face."

Physical examination revealed a flat and wide forehead. The supraorbital ridges as well as the infraorbital rims were recessed, resulting in shallow orbits and bulging eyes. The total midface was flat with an Angle class III malocclusion. Psychologic testing showed a degree of developmental delay. Funduscopic examination revealed mild optic atrophy, strabismus, and amblyopia.

A CT scan was completed in both the axial and coronal planes. The ventricles were mildly enlarged. The orbits were shallow and the zygomas were retruded.

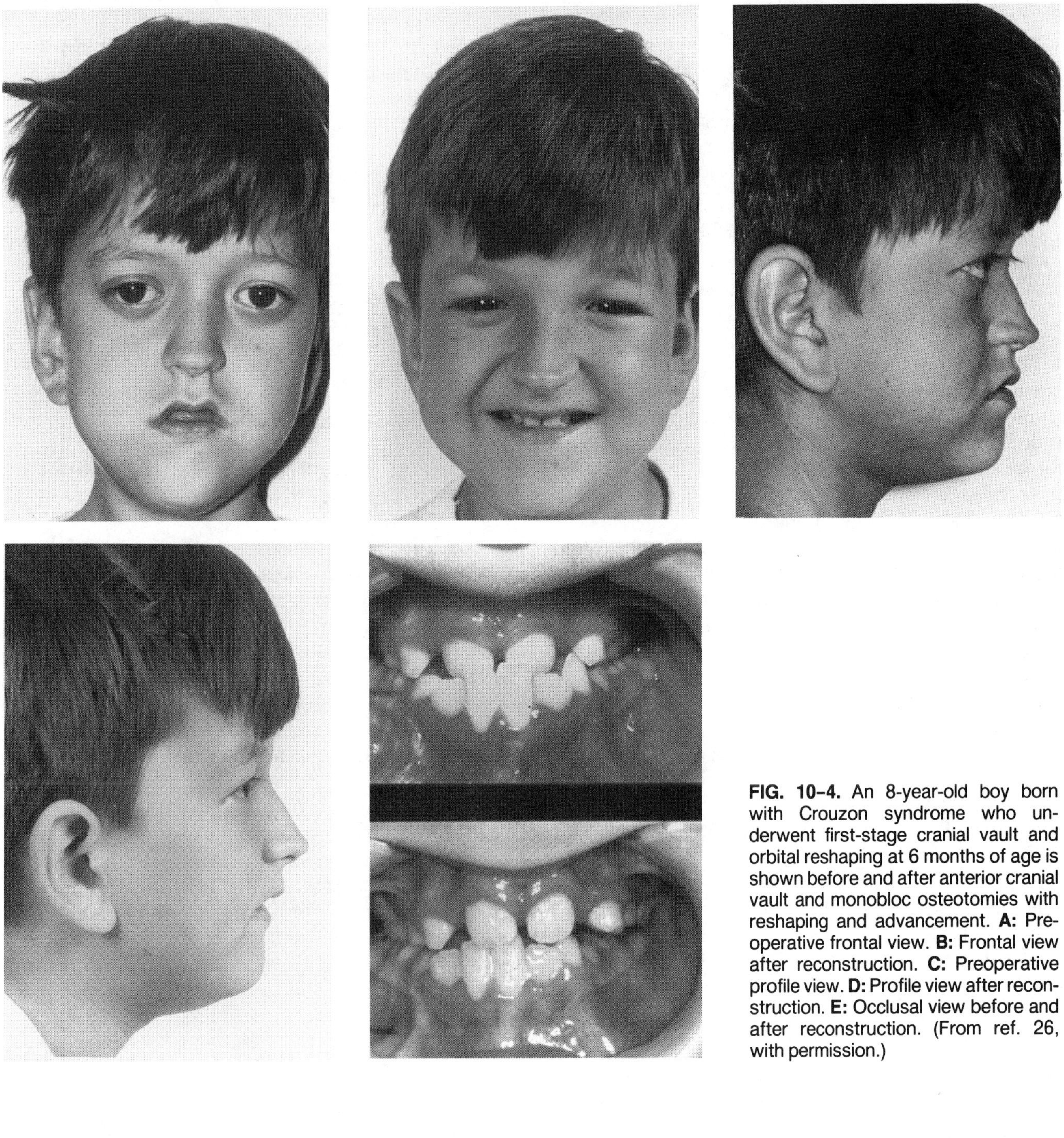

A,B C D,E

FIG. 10–4. An 8-year-old boy born with Crouzon syndrome who underwent first-stage cranial vault and orbital reshaping at 6 months of age is shown before and after anterior cranial vault and monobloc osteotomies with reshaping and advancement. **A:** Preoperative frontal view. **B:** Frontal view after reconstruction. **C:** Preoperative profile view. **D:** Profile view after reconstruction. **E:** Occlusal view before and after reconstruction. (From ref. 26, with permission.)

Preoperative surface anthropometric measurements confirmed a decreased head circumference (485 mm, normal 529 mm), a decreased head length (160 mm, normal 186 mm), and a decreased maxillary arc (230 mm, normal 250 mm).

Anterior cranial vault and monobloc osteotomies with reshaping and advancement were carried out through an intracranial approach (25,26). Bone grafts were from the cranium, and stabilization of the osteotomies and grafts was with titanium miniplates and screws. The patient's recovery was uneventful, and he assumed all age-appropriate activities.

As a result of the monobloc and cranial vault osteotomies with horizontal advancement, all three surface anthropometric measurements improved: Head circumference increased to 514 mm, head length increased to 175 mm, and the maxillary arc increased to 254 mm.

The patient will require a LeFort I osteotomy and genioplasty in combination with orthodontic treatment at the time of skeletal maturation as part of his staged reconstruction.

Patient 5: Apert Syndrome with Total Cranial Vault Dysplasia and Midface Hypoplasia (Fig. 10–5)

Procedure: Facial bipartition and cranial vault reshaping

This 5-year-old girl was born with Apert syndrome. She underwent suture release and forehead reshaping at 6 months of age (26). She presented with a retruded and wide anterior cranial base (residual brachycephaly). The forehead was flat at the supraorbital ridge level with bitemporal constrictions and bulging superiorly. The orbits were shallow with eye proptosis and a moderate degree of orbital hypertelorism. The midface lacked the normal convexity when viewed from above. The total midface was deficient, with a marked anterior overbite and a full Angle class III malocclusion. Nasal airflow was poor, and the patient habitually breathed through her mouth.

A craniofacial CT scan confirmed that the anterior cranial base was short, with bitemporal constrictions and recessed supraorbital ridge. The superior forehead bulged anteriorly. The orbits were shallow with moderate orbital hypertelorism. The midface was deficient both vertically and horizontally, and the nasal air passages were diminished.

Preoperative surface anthropometric measurements confirmed a head width that was greater than normal (144 mm, normal 136 mm), a forehead width that was greater then normal (118 mm, normal 94 mm), and a greatly diminished head length (163 mm, normal 175 mm). The orbits demonstrated hypertelorism of both the medial orbital walls (40 mm, normal 29 mm) and the lateral orbital walls (94 mm, normal 75 mm). The bizygomatic width was increased (120 mm, normal 107 mm), the maxillary depth was diminished (93 mm, normal 100 mm), and the upper face height was decreased (53 mm, normal 61 mm).

At operation, working through a coronal (skin) incision, the dysplastic cranial vault bone was removed through multiple craniotomies. With the frontal and temporal lobes of the brain retracted, exposure was gained to the cranial base for completion of facial bipartition osteotomies (26). With the midface and cranial vault reshaped, stabilization was with autogenous cranial bone and miniplate and screw fixation. She has maintained good function and facial aesthetics.

After undergoing facial bipartition and cranial vault osteotomies that included horizontal advancement and medial repositioning of the facial bones to diminish upper face hypertelorism, anthropometric measurements were again made. The head length improved (185 mm), and the separation between the lateral orbital walls moved closer to the normal range (89 mm). The upper face height improved to 61 mm, and the maxillary depth increased to 102 mm. These measurements confirmed the procedure's effectiveness to horizontally advance the forehead and midface while increasing the upper face vertical height and diminishing the upper face width.

Routine orthognathic surgery and orthodontic treatment are planned for the teenage years to complete her reconstruction.

Patient 6: Isolated Cleft Palate (Fig. 10–6)

Procedure: Maxillary LeFort I osteotomy, combined with vertical reduction and advancement genioplasty

A 19-year-old woman with an isolated cleft palate had undergone cleft palate repair at 18 months of age (27). No further revisions were required. On examination, she had a marked maxillary hypoplasia with an Angle class III maloccusion and a vertically long and retrognathic chin.

Preoperative surface anthropometric measurements confirmed a decrease in the maxillary depth (113 mm, normal 118 mm), an increased total facial height (127 mm, mean 111 mm), and a greatly increased mandibular length (60 mm, mean 45 mm).

Preoperative orthodontic therapy was carried out. She underwent a maxillary LeFort I osteotomy with advancement and posterior intrusion. The chin was vertically reduced and horizontally advanced. Stabilization was with titanium miniplates and screws.

As a result of the LeFort I osteotomy with horizontal advancement and posterior intrusion, mandibular autorotation, and vertical reduction with advancement geni-

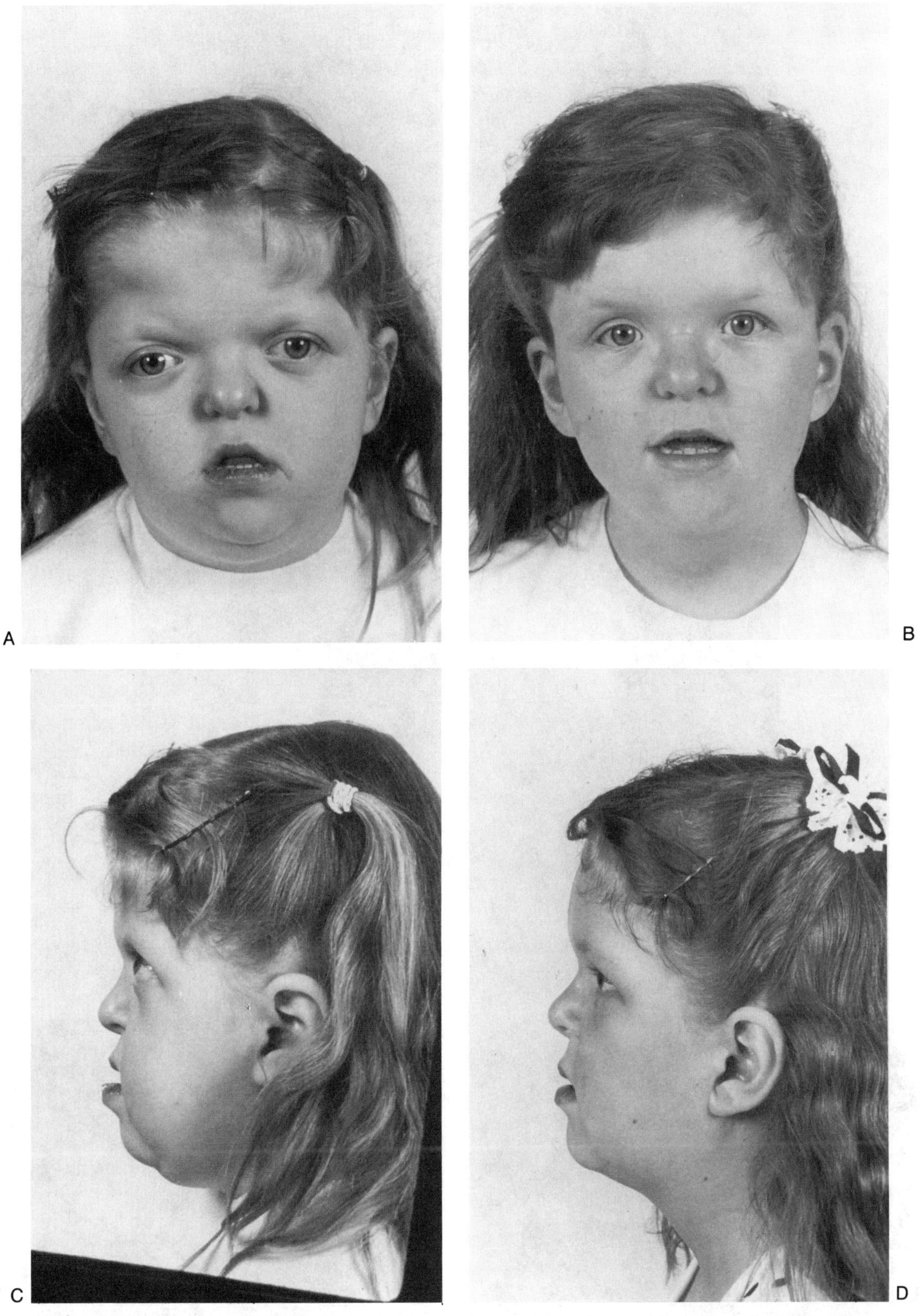

FIG. 10–5. A 5-year-old girl with Apert syndrome who presented for cranial vault and facial bipartition osteotomies with reshaping and anterior repositioning. **A:** Preoperative frontal view. **B:** Postoperative frontal view 2 years after reconstruction. **C:** Preoperative lateral view. **D:** Postoperative lateral view 2 years after reconstruction. (From ref. 26, with permission.)

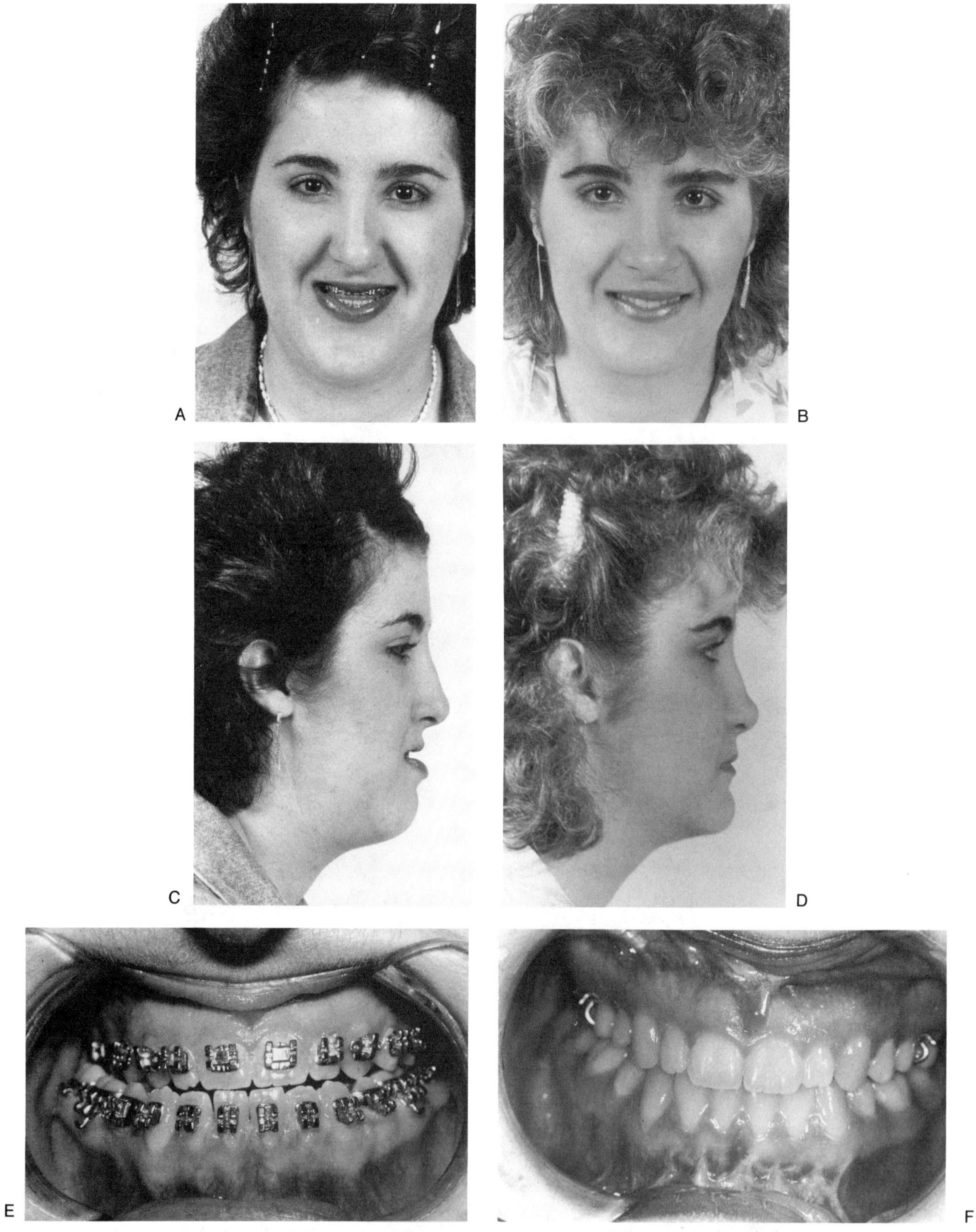

FIG. 10–6. A 19-year-old woman with isolated cleft palate who presented with maxillary hypoplasia and a vertically long and retrognathic chin. **A:** Preoperative frontal view with smile. **B:** Frontal view 1 year later. **C:** Preoperative lateral view. **D:** Lateral view 1 year later. **E:** Preoperative occlusal view. **F:** Occlusal view 1 year later. (From ref. 27, with permission.)

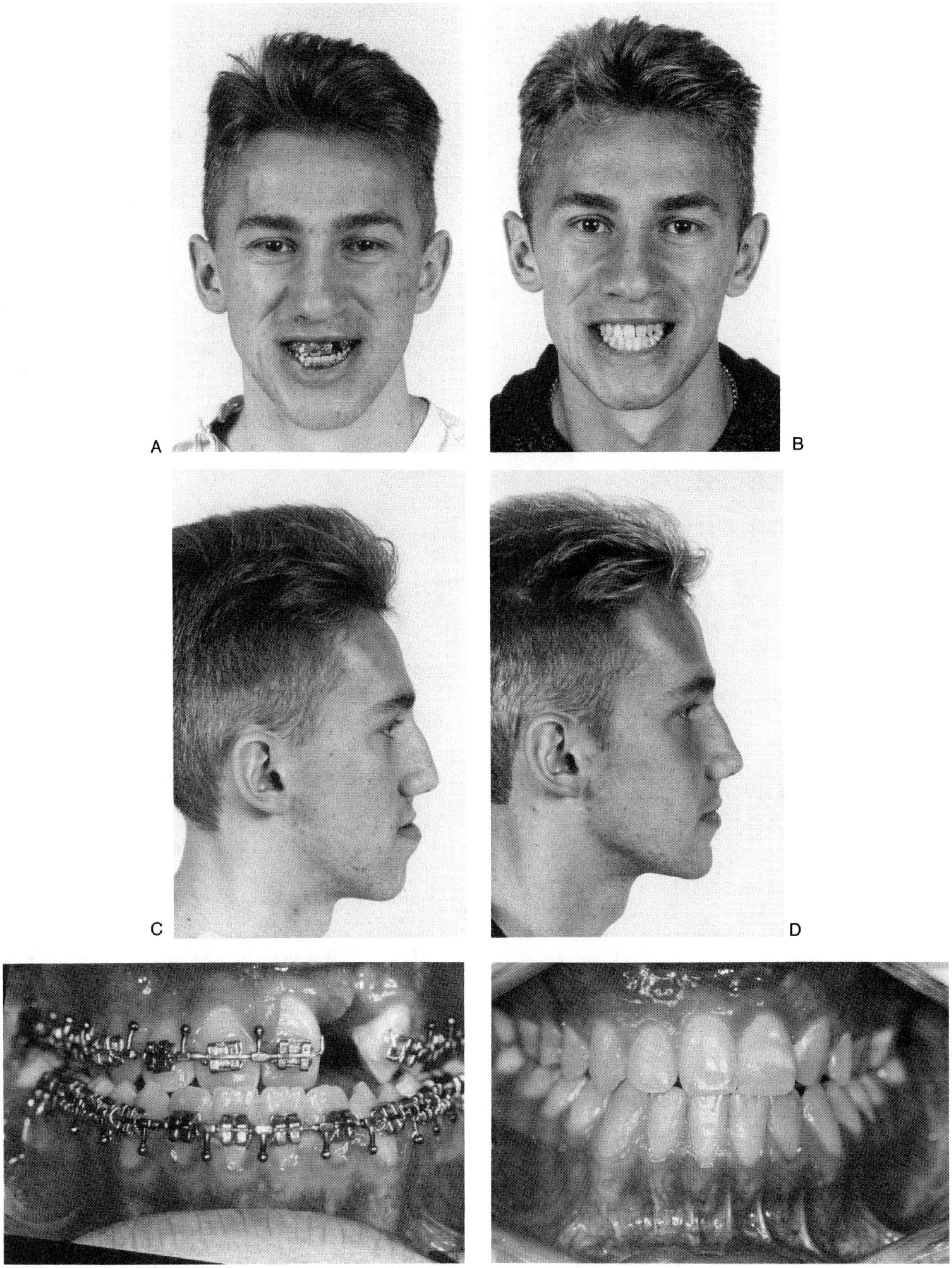

FIG. 10–7. A 19-year-old boy with UCLP undergoing a modified LeFort I osteotomy in two segments. Bilateral sagittal splits of the mandible and a vertical reduction and advancement genioplasty are shown before surgery and 1 year later. **A:** Preoperative frontal view. **B:** Postoperative frontal view. **C:** Preoperative lateral view. **D:** Postoperative lateral view. **E:** Preoperative occlusal view. **F:** Postoperative occlusal view. (From ref. 28, with permission.)

oplasty, the abnormal surface anthropometric measurements improved. The maxillary depth increased (117 mm, normal 118 mm), the overall facial height diminished (116 mm, normal 111 mm), and the mandibular height, which is a reflection of vertical chin height, diminished (49 mm, normal 45 mm).

The patient completed her orthodontic treatment and has maintained a positive overjet and overbite.

Patient 7: Jaw Deformity with Malocclusion in Association with Unilateral Cleft Lip and Palate (UCLP) (Fig. 10–7)

Procedure: Modified maxillary LeFort I osteotomy (2 segments), sagittal split osteotomies of the mandible, and vertical reduction and advancement genioplasty

A 19-year-old boy born with left complete cleft lip and palate had undergone surgery for lip repair in infancy, followed by cleft palate repair in early childhood (28). His residual clefting problems included (a) maxillary hypoplasia with an Angle class III malocclusion, (b) a negative overbite and overjet with crossbite of the lesser maxillary segment, (c) shift of the maxillary dental midline off the facial midline, (d) a perialveolar oronasal fistula, and (e) a cleft-dental gap resulting from a congenitally absent lateral incisor tooth. Orthodontic brackets had been placed at the age of 16 in preparation for orthognathic surgery, which was performed at the age of 19. Speech assessment, including nasoendoscopy, confirmed adequate velopharyngeal closure and sibilant distortions secondary to malocclusion.

A modified maxillary LeFort I osteotomy was carried out in two segments with differential repositioning to close the cleft-dental gap and perialveolar oronasal fistula. Through the bony- and soft-tissue periodontal support of the teeth adjacent to the cleft improved, a positive overbite and overjet was created and the lateral crossbite and class III malocclusion were corrected. Achievement of these improvements also required sagittal split osteotomies of the mandible, along with a vertical reduction and advancement genioplasty. The septum of the nose was straightened and the inferior turbinates were reduced. Stabilization was accomplished with iliac bone grafts, titanium miniplate and screw fixation, and the use of an interocclusal splint.

Two and one-half years after surgery, the patient's general appearance, smile, and profile are improved.

CONCLUSION

Craniomaxillofacial surgery remains both an art form and a science. Although there is no substitute for experience and sound clinical judgment, a quantitative assessment of the facial deformity will often assist the craniofacial surgeon. The purpose of the quantitative assessment—whether by anthropometric surface measurements, cephalometric analysis, CT scan analysis, or the study of articulated dental models—is to help predict growth patterns, confirm or refute clinical impressions, aid in treatment planning, and provide a framework in the objective assessment of intraoperative, immediate, and long-term results.

When used selectively, anthropometric surface measurements of the head and face are not only reliable and accurate but also inexpensive and without X-ray exposure and can easily be performed by the clinician without the need for additional consultations and appointments. Once the basic technique is mastered, facial anthropometry can be used successfully in everyday practice.

ACKNOWLEDGMENT

The anthropometric examination and the evaluation of the preoperative and postoperative findings in patients were carried out by Leslie G. Farkas, M.D., Director of the Craniofacial Measurement Laboratory, The Hospital for Sick Children, Toronto, Ontario, Canada.

REFERENCES

1. Farkas LG. *Anthropometry of the head and face in medicine.* New York: Elsevier, 1981.
2. Farkas LG, Munro IR. *Anthropometric facial proportions in medicine.* Springfield: Charles C Thomas, 1987.
3. Farkas LG, Posnick JC. Growth and development of regional units in the head and face based on anthropometric measurements. *Cleft Palate Craniofac J* 1992;29:301–302.
4. Farkas LG, Posnick JC, Hreczko T. Growth patterns of the face: a morphometric study. *Cleft Palate Craniofac J* 1992;29:308–314.
5. Farkas LG, Posnick JC, Hreczko T. Anthropometric growth study of the head. *Cleft Palate Craniofac J* 1992;29:303–307.
6. Farkas LG, Posnick JC, Hreczko T. Anthropometric growth study of the ear. *Cleft Palate Craniofac J* 1992;29:324–329.
7. Farkas LG, Posnick JC, Hreczko T, Pron G. Growth patterns of the nasolabial region: a morphometric study. *Cleft Palate Craniofac J* 1992;29:318–323.
8. Farkas LG, Posnick JC, Hreczko T, Pron G. Growth patterns in the orbital region: a morphometric study. *Cleft Palate Craniofac J* 1992;29:315–317.
9. Farkas LG. Ear morphology in Treacher Collin's, Apert's and Crouzon's syndromes. *Arch Otorhinolaryngol* 1978;220:153–157.
10. Farkas LG, Munro IR, Kolar JC. Abnormal measurements and disproportions in the face of Down's syndrome patients: preliminary report of an anthropometric study. *Plast Reconstr Surg* 1985;75:159–167.
11. Farkas LG. Anthropometry of the normal and defective ear. *Clin Plast Surg* 1990;17:213–221.
12. Waitzman AA, Posnick JC, Armstrong D, et al. Craniofacial skeletal measurements based on computed tomography, part 1: accuracy and reproducibility. *Cleft Palate Craniofac J* 1992;29:112–117.
13. Waitzman AA, Posnick JC, Armstrong D, et al. Craniofacial skeletal measurements based on computed tomography, part 2: normal values and growth trends. *Cleft Palate Craniofac J* 1992;29:118–128.

14. Carr M, Posnick J, Armstrong D, et al. Cranio-orbito-zygomatic measurements from standard CT scans in unoperated Crouzon and Apert infants: comparison with normal controls. *Cleft Palate Craniofac J* 1992;29:129–136.
15. Posnick JC, Bite U, Nakano P, et al. Indirect intracranial volume measurements using CT scans: clinical applications for craniosynostosis. *Plast Reconstr Surg* 1992;89:1;34–35.
16. Posnick JC, Goldstein JA, Waitzman A. Surgical correction of the Treacher–Collins malar deficiency with quantitative CT scan analysis of long-term results. *Plast Reconstr Surg* 1993;92:1;12–22.
17. Posnick JC, Lin KY, Chen P, et al. Sagittal synostosis: Quantitative assessment of presenting deformity and surgical results based on CT Scans. *Plast Reconstr Surg* 1993;92:6;1015–1024.
18. Posnick JC, Lin KY, Jhawar BJ, et al. Crouzon syndrome: quantitative assessment of presenting deformity and surgical results based on CT scans. *Plast Reconstr Surg* 1993;92:6;1027–1037.
19. Posnick JC, Lin KY, Chen P, et al: Metopic synostosis: quantitative assessment of presenting deformity and surgical results based on CT scans. *Plast Reconstr Surg* 1994;93:16.
20. Posnick JC, Lin KY, Jhawar BJ, et al. Apert syndrome: quantitative assessment of presenting deformity and surgical results after first-stage reconstruction by CT scan. *Plast Reconstr Surg* 1993;in press.
21. Farkas LG, Lindsay WK. Ear morphology in cleft lip and palate anomaly. *Arch Otorhinolaryngol* 1973;206:57–68.
22. Farkas LG. Growth of normal and reconstructed auricles. In: Tanzer RC, Edgerton MT, eds. *Symposium on reconstruction of the auricle.* St Louis: CV Mosby, 1974;24–31.
23. Farkas LG. Anthropometry of normal and anomalous ears. *Clin Plast Surg* 1978;5:401–412.
24. Posnick JC. Craniosynostosis: surgical management in infancy. In: Bell WH, ed. *Orthognathic and reconstructive surgery,* Vol 52. Philadelphia: WB Saunders, 1992;1839–1887.
25. Posnick JC. Craniofacial dysostosis: management of the midface deformity. In: Bell WH, ed. *Orthognathic and reconstructive surgery,* Vol 52. Philadelphia: WB Saunders, 1992;1888.
26. Posnick JC. Craniofacial dysostosis: staging of reconstruction and management of the midface deformity. *Neurosurg Clin North Am* 1991;2(3):683–702.
27. Posnick JC, Witzel MA, Dagys AP. Management of jaw deformities in the cleft patient. In: Bardach J, Morris HL, eds. *Multidisciplinary management of the cleft lip and palate,* 1st ed. Philadelphia: WB Saunders, 1990;65:530–543.
28. Posnick JC, Tompson B. Modification of the maxillary LeFort I osteotomy in cleft-orthognathic surgery: the unilateral cleft lip and palate deformity. *J Oral Maxillofac Surg* 1992;50:666–675.

Chapter 11

Anthropometric Measurements in Rhinoplasty

A Clinical Approach

Rollin K. Daniel and Leslie G. Farkas

Anthropometric measurements are becoming an accepted part of rhinoplasty planning (1,2). Guyuron (3,4) has applied these measurements in over 1700 rhinoplasties, using them to define the problem, establish the operative plan, and evaluate the results. Recently, Byrd and Hobar (5) have defined a critical relationship between ideal nasal bridge length and midfacial height which allows ideal norms to be determined for a specific individual. Thus, a new era is emerging which will allow the surgeon to more accurately plan and achieve an aesthetically attractive nose.

RHINOPLASTY EVOLUTION

For decades, aesthetic surgeons have examined the patient's nose and focused on what was wrong with it—that is, an obvious hump, a broad tip, or a wide base. The goal was to eliminate the negatives and reduce the overall size of the nose, with the final result limited by the patient's own tissues. During the last decade, two advances have occurred to change this approach. First, Sheen (6) introduced the concept of "balance," in which one part of the nose may be augmented which will decrease the amount of reduction necessary and therefore result in a more natural nose. Second, an enormous number of new techniques have been introduced which allow the surgeon to change the fundamental character of a nose (7). Thus, surgeons are no longer focusing on just the negatives, but rather trying to define what is ideal for both the overall nose and its specific subunits. An example is the use of radix grafts which will affect the root of the nose while having a powerful effect on specific angles (nasofrontal, nasofacial) and dorsal–base disproportions. Three applications for anthropometric measurements have evolved, including the following: (i) clinical surface measurements, (ii) photographic analysis for operative planning, and (iii) in-depth scientific studies.

TABLE 11-1. *Terminology of conventional landmarks used in aesthetic surgery translated into the terms used in medical anthropometry*[a]

In aesthetic surgery		In medical anthropometry	
Name	Symbol	Name	Symbol
Nasion	N	Nasion	n
Subnasale	SN	Subnasale	sn
Tragion	T	Tragion	t
Alare	AL	Alare	a
Alar base point (alar crest)	AC	Alar crest, facial insertion point of the alar base	ac
Nasal tip point	T	Nasal tip point = pronasale	prn
Columella breakpoint	C^1	Top point on each edge of the columella	c^1
Endocanthion	EN	Endocanthion	en
Cornea point	C	Pupil point	p
Menton (chin point)	M	Gnathion, menton	gn
Stomion	S	Stomion	sto
Glabella	G	Glabella	g
Philtrum crest point	PC	Labiale superius	ls
Intercanthal midpoint—that is, the intersection between the vertical facial midline plane and a line drawn between the medial canthi (en-en)	IN	Sellion—that is, the midpoint on the nasal root at the level of endocanthions	se

[a] A lowercase "i" after the symbol or the landmark indicates an ideally located anatomical point (e.g., Ti, Ni, C^1i).

TERMINOLOGY

The combination of classical anthropometric measurements and rhinoplasty surgery must reconcile two different backgrounds: (i) the traditional anthropometry adjusted for the diverse needs of medicine and (ii) the goals of the clinical aesthetic surgeon interested in the morphological changes in a greatly reduced anatomical area of the craniofacial complex. In comparison with the changes in the subtle surface expressions of the soft tissues, reconstructive surgery of the head and face with striking changes sometimes requires drastic interventions in both soft and hard tissues. The diversity of goals in these two directions led to the changes initiated by aesthetic surgeons in terminology of the medical anthropometry. More simplistic terms and abbreviations were introduced for some landmarks, and new methods were developed or old ones were modified to smooth and shorten the process of preoperative and postoperative analysis. The simpler terms, fewer in number and more logical in their application, promised their easier acceptance. An important factor was the need for "ideals." In interpretation of the findings regarding the face of a child with facial syndrome, correctly normative values, ranges, and standard deviations are used, all derived from large population samples. In contrast, the aesthetic surgeon must deal with a specific patient on whom surgery is scheduled the next day. The goal is to have an individualized operative plan, not a data table. For these reasons, a terminology conflict between the aesthetic surgeon and the conventional medical anthropometry appears inevitable. For mutual understanding by both parties, a "dictionary" of terminology is offered, pairing aesthetic surgical terms with their anthropometric definition (Tables 11-1 to 11-4).

CLINICAL SURFACE MEASUREMENTS

Why would a busy surgeon bother to take anthropometric surface measurements? In my practice, the answer is that it provides essential information for making an accurate diagnosis, allows an objective start for refining the operative plan, and occasionally corrects a misdiagnosis (1,8). For example, a moderate-sized dorsal hump may be apparent to both patient and surgeon, but direct measurements demonstrate that tip and radix projection are significantly decreased, leading one to minimize the hump by augmenting both the tip and the radix without significant dorsal reduction. The technique which I have used in over 500 cases during the past 5 years consists of markings, measurements, and analysis.

TABLE 11-2. *Terminology used in aesthetic surgery: new landmarks*

Name	Symbol	Definition
Base bony width point	X	Landmark at the most laterally located point of the bony lateral nasal wall on each side marked on the frontal view of the face
Domes	D	The most lateral point of the nasal tip, marking the junction between tip and lobule (alar) on each side

TABLE 11–3. *Terminology of measurements used in aesthetic surgery translated into the terms used in medical anthropometry*[a]

In aesthetic surgery		In medical anthropometry	
Name	Symbol	Name	Symbol
Nasal bridge length	N-T	Nasal bridge length	n-prn
Tip projection	AC-T	Alar length	ac-prn
Interalar width	AL-AL	Nose width (morphological width of the nose) [Farkas et al. 1993 (18)]	al-al
Interalar base width	AC-AC	Anatomical width of the nose [Farkas et al. 1993 (18)]	ac-ac
Columella length	SN-C^1	Columella length measured on right and left crest	sn-c^1
Nasofacial angle	NFA	Inclination of the nasal bridge	
Nasofrontal angle	NFR	Nasofrontal angle	
Columella labial angle	CLA	Nasolabial angle	
Stomion to menton height	SME	Mandible height	sto-gn
Intercanthal width	EN-EN	Intercanthal width	en-en
Relative projection of the radix nasi	EN-N	Nasal root depth (close to EN-N) measured right and left	en-se sag
Columella inclination: the angle is measured from the vertical below the columella		Columella inclination: the angle is measured from the vertical above the columella	
Frankfort horizontal plane	FHP	Frankfort horizontal	FH

[a] A small "i" after the symbol indicates an ideal measurement (e.g. N-T*i*, AC-T*i*).

Markings

With the patient sitting upright and the head in straight forward gaze, the following dots are marked: (a) top of tragus, (b) infraorbital rim, (c) nasion (N), (d) base bony width (X), (e) tip (T), (f) right (RD) and left dome (LD), (g) columella breakpoint (C^1), (h) subnasale (SN), and (i) alar crease (AC) (Fig. 11–1). Then using a sliding caliper, measurements are taken and recorded in the patient's chart as shown in Table 11–5.

Clinical Interpretation

What is their clinical significance? By comparing intercanthal width (EN-EN) to base bony width (X-X) to interalar width (AL-AL), the author can decide on the indication for, and type of, lateral osteotomies. In general, the following guideline is recommended: (a) If X-X is greater than EN-EN by >2 mm, then low-to-low osteotomies with complete movement of the lateral wall are done; (b) if X-X ranges from +2 to −2 mm compared to EN-EN, then transverse greenstick fractures with tilt of the lateral walls is sufficient and can be accomplished with low-to-high lateral osteotomies; and (c) if X-X is significantly narrower than EN-EN > −2 mm, then no osteotomies are indicated. Equally, the type of alar base modification is evaluated by first determining the need for surgery based on the relationship of alar flare (AL-AL) to intercanthal width (EN-EN) and then comparing alar flare (AL-AL) to interalar base width (AC-AC). As a general guideline, alar base surgery is recommended when alar flare (AL-AL) is significantly greater (>2 mm) than intercanthal width. But what type of excision is indicated? The answer is often determined by comparing alar flare (AL-AL) (which is measured at the widest point on the morphologic nose) to interalar width (AC-AC) (which is the anatomical width of the nose as it reflects the underlying pyriform aperture). In general, a two-step plan is used. Step 1 consists of comparing EN-EN to AC-AC, which determines whether a decrease in width is indicated. Specifically, if AC-AC is greater than EN-EN, then width should be reduced and a portion of the nostril floor will be excised or used as a cinch to narrow the base width. Step 2, the relationship between AL-AL and AC-AC, determines whether direct excision of the alar side wall is done. In most cases, if the AC-AC–EN-EN relationship is acceptable but AL-AL is greater than EN-EN,

TABLE 11–4. *Measurements newly introduced by aesthetic surgeons*

Name	Symbol	Definition
Tip angle (Daniel)	TA	Measured on the lateral print oriented in FH. The angle is formed by the line connecting the AC point and the nasal tip point (T), and the vertical is drawn through the AC point. The angle is measured below the AC-T line.
Radix projection (Byrd)	C-N	Measured on the lateral print oriented in FH. The radix projection is given by the horizontal between the nasion and a point on the vertical tangent to the cornea.
Base bony width	X-X	The width between the most lateral points of the bony lateral walls of the nose.
Infralobular distance (Daniel)	C^1-T	The projective distance between the columella break point and the tip of the nose.
Midface height (Byrd)	MFH	Given by the vertical distance between the glabella and the alar crease (G-AC).

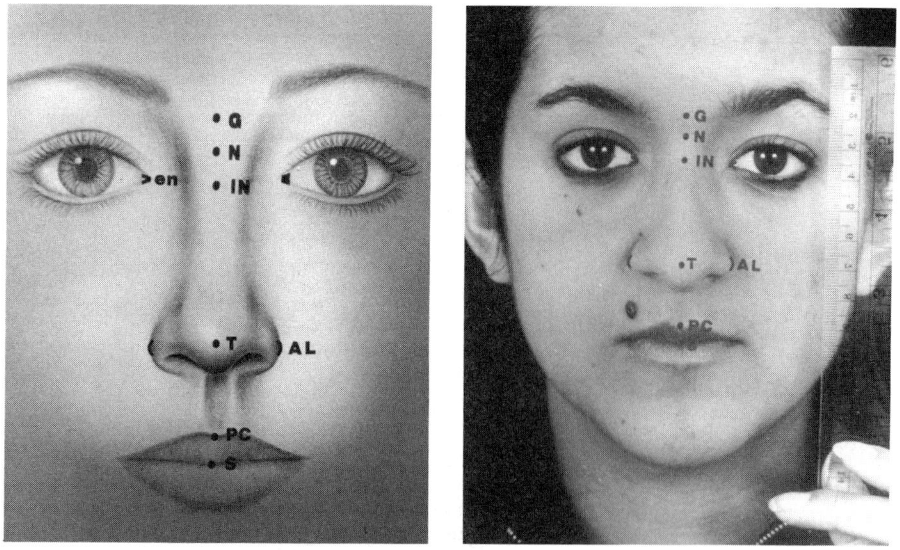

A B

FIG. 11–1. The markings on anterior view (**A, B**) consist of the following: glabella (G), nasion (N), endocanthion (EN), intercanthal midpoint (IN), tip (T), alar flare (AL), midpoint of the philtral columns (PC), and stomion (S).

then a direct excision of only the side wall is done (type I wedge, no nostril component). However, when AC-AC is greater than EN-EN, and AL-AL is greater than AC-AC, then both alar side wall and nostril floor are excised (type II or III wedge excision extending into the nostril). Thus, in less than 5 minutes, the surgeon can decide on osteotomies and alar base modification using a sliding caliper (Fig. 11–2).

On lateral view, the critical factors become angles and projections rather than comparative widths. Additional equipment includes (a) a projector with measuring limbs and (b) the modified sliding caliper with level. The critical angles are nasofrontal, nasofacial, tip angle, and columella labial, while projection is measured at the radix, keystone area, and tip (Fig. 11–3). In addition, midfacial height (MFH) and stomion to menton (SME) are recorded. The most important angles are (a) tip angle (TA) measured along a vertical facial plane through the alar crease to the tip (T) and (b) the columella labial angle with its subunit, columella inclination. The ability to compare tip angle to columella inclination is important in determining specific surgical techniques. For example, the nasal tip of a female patient may appear plunging and masculine. Yet the measured tip angle is 105 degrees, thus forcing the need for caudal resection of the columella. Relative projection of the radix (EN-N) and tip (AC-T) can be measured on an inclination with a caliper or straight from the facial surface using a special depth gauge. One can then relate this to the ideal using Byrd's criteria that NTi = .67 MFH = SME, which then

TABLE 11–5. *Case study: photographic operation planning*

Analysis	Presentation	Objective	Desired change	Result 2 years postoperatively	Actual change
Points					
N	↑→	Eyelash	↓6 ←4 mm	Eyelash	
T	↓	Tip angle	↑4 mm	Tip angle	
C[1]	↑→	Nostril	↑3 ←2 mm	Nostril	
SN	→	Upper lip	←4 mm	Upper lip	
Angles					
NFR	155 deg	138 deg	−17 deg	140 deg	−15 deg
NFA	25 deg	36 deg	+11 deg	31 deg	+6 deg
TA	90 deg	100 deg	+10 deg	105 deg	+15 deg
COL LAB	92 deg	100 deg	+8 deg	109 deg	+17 deg
Distances					
C-N	15 mm	11 mm	−4 mm	13	−2 mm
N-T	62 mm	53 mm	−9 mm	48	
AC-T	32 mm	30 mm	−2 mm	30	
P/C/I	11/8/15 mm	10/10/11 mm	−1/−2/−4 mm	10/11/12	

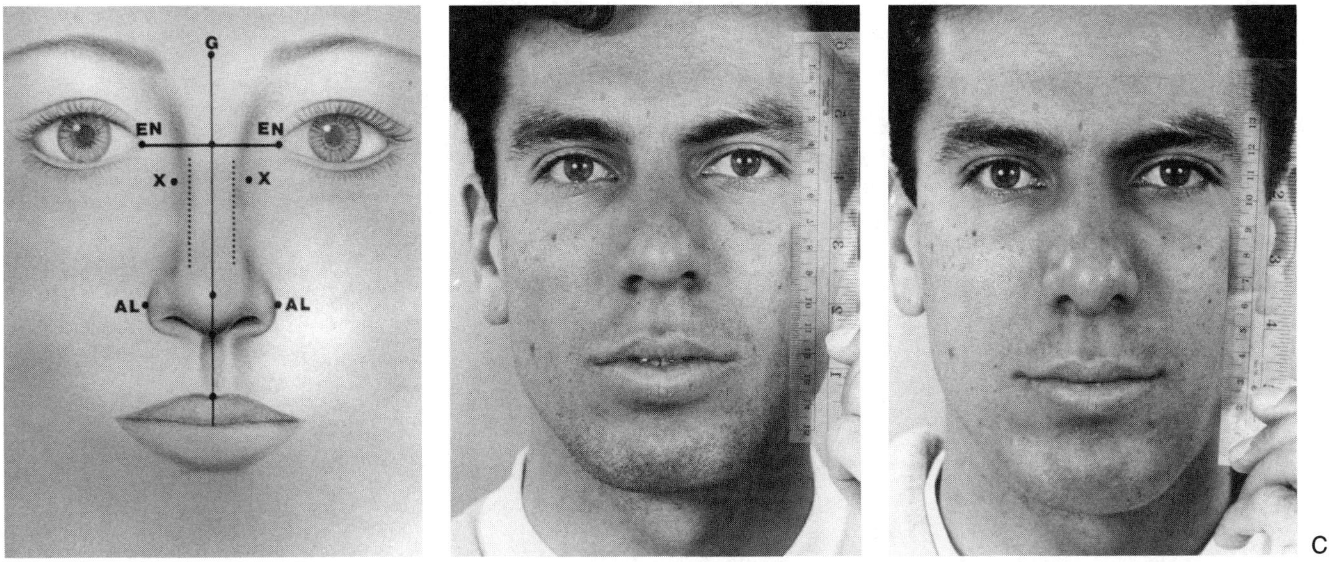

A,B C

FIG. 11–2. Role of anterior measurements in determining the need for osteotomies and alar wedges. The standard comparisons (**A**) are between intercanthal width (EN-EN), base width of the bony vault (X-X), and alar flare (AL-AL). The preop (**B**) and postop (**C**) comparisons are as follows: intercanthal width (32 mm, 32 mm), base bony width (36 mm, 32 mm) following low-to-low osteotomies, and alar flare (34 mm, 31 mm) following type I wedges. *Note:* AL-AL (34 mm), AC-AC (26 mm), preoperatively.

allows C-Ni = .28 NTi for radix projection and AC-Ti = .68 NTi for tip projection. Thus one can determine the need for reduction, augmentation, or remodeling at each point (Fig. 11–4). Surface measurements are excellent for making a diagnosis in both views. Anterior measurements also allow definitive operative planning as regards osteotomies and basilar modifications, and life-size lateral photographs provide more opportunities for operative planning than do surface measurements.

Evaluation

For the first 5 years, all measurements were recorded in the tabular form. At 1 year postoperatively, repeat measurements were made with the paper folded to obscure the original data. Certain measurements, including EN-EN, MFH, and SME, served as internal checks for accuracy. Although a large number of cases have not been analyzed, certain subgroups are available to answer specific questions. For example, many surgeons feel that an "open roof" in the dorsum must exist for lateral osteotomies to be effective in narrowing the nose. Specifically, they recommend medial osteotomies even when no dorsal work is deemed necessary. This question can be easily answered by comparing preoperative and postoperative X-X values in patients having lateral osteotomies without dorsal surgery, a maneuver that occurs in certain ethnic and secondary rhinoplasties. As shown in Fig. 11–5, the patient measured 33 mm at X-X preoperatively and 29 mm at X-X at 18 months following low-to-low lateral osteotomies plus dorsal augmentation (i.e., no dorsal reduction and no open roof). Additional cases have proven that lateral osteotomies are effective and permanent without the need for an open dorsum. Another useful assessment is to determine the effectiveness and permanency of alar wedge excisions. Utilizing sequential measurements preoperatively, at 1 month, at 3 months, at 6 months, and annually thereafter, one can get a general impression of their efficacy at about 50%; that is, for every 4 mm excised, the nose is narrowed 2 mm at 1 year. Given the enormous number of variations in presenting problems and in surgical techniques, large series are required. Direct surface measurements are more accurate than those taken from photographs (9).

PHOTOGRAPHIC ANALYSIS

The use of photographic analysis for rhinoplasty planning is not new. Beginning in the 1920s, plaster casts gave way to surface profileometers and then to photographs (10,11). By the 1970s, sketches were made on an acetate placed over the photograph to indicate the desired changes (12,13). Currently, analysis using accurate life-size photographs have achieved critical importance for two reasons (1,6,14). First, the number of surgical techniques has grown exponentially, allowing each of the cardinal nasal reference points (N, T, C^1, SN) to be changed and also allowing a wide range of totally different operative plans to be possible. Photographic analysis allows one to outline and compare the various operative plans. Second, the use of inexpensive scanners to load the photographs, along with the use of graphic programs

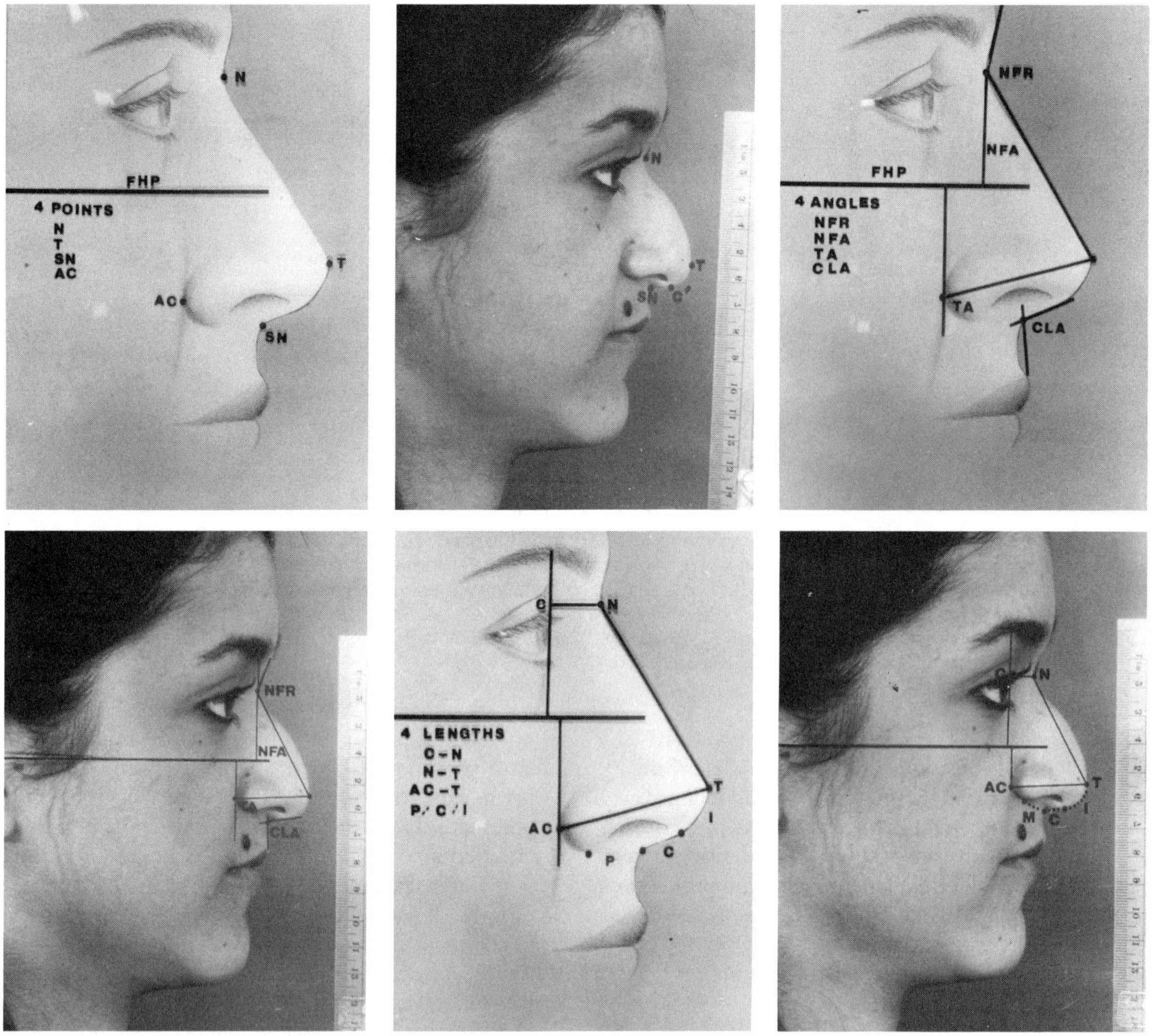

A,B C D,E F

FIG. 11-3. Lateral view markings. **A, B:** The four landmarks are the nasion (N), tip (T), subnasale (SN), and alar crease (AC). **C, D:** The four angles are the nasofrontal (NFR), nasofacial (NFA), tip (TA), and columella labial (CLA). **E, F:** The important distances are radix projection (C-N), dorsal bridge length (N-T), and tip projection (AC-T).

for drawing, allows multiple computer images for both analysis and record-keeping. These images are valuable in providing the surgeon and patient with various alternatives prior to surgery. Ultimately, photographic analysis consists of the following five steps: (i) Define the nasal parameters, (ii) superimpose the ideal, (iii) analyze the alternative, (iv) establish the operative goals, and (v) assess the result.

This sequence will be illustrated using a lateral photograph of a patient seeking a secondary rhinoplasty. The technique utilized is a personal one which integrates the data of Farkas and co-workers (15,16) with many of the measurements of Guyuron (3,4) and Byrd and Hobar (5). It basically consists of four landmarks (N, T, C^1, SN), four angles (NFR, NFA, TA, CLA), and four distances (C-N, N-T, AC-T, P/C/I), all applied in a five-step sequence.

Define the Nasal Parameters

The four reference points (N, T, C^1, SN) plus the alar crease (AC) are marked as they exist on the patient. Then the reference lines are drawn, including the Frankfort horizontal plane (FHP) and vertical axes through the cornea, alar crease, and nasion. A table is drawn on the

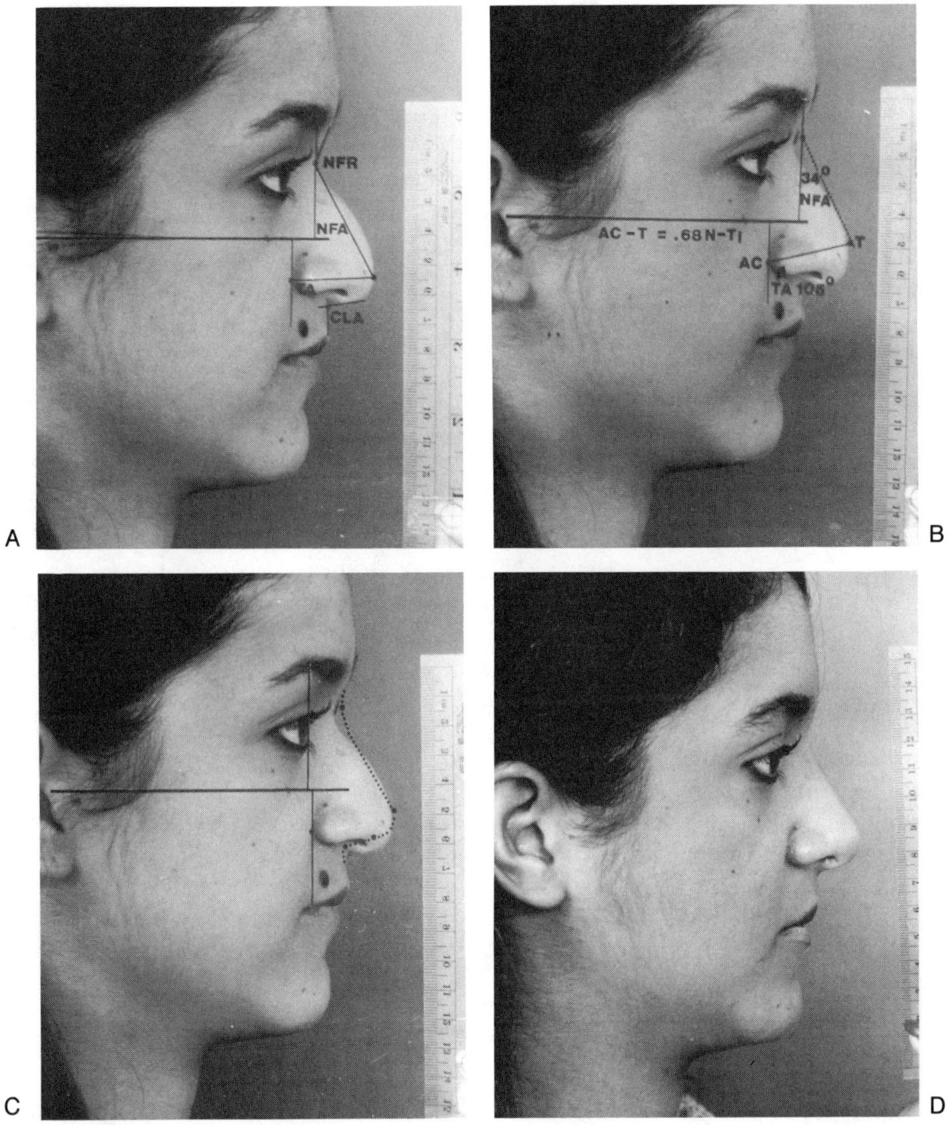

FIG. 11–4. Operative planning using lateral measurements. **A:** The angles are measured on the patient (NFR 127 degrees, NFA 28 degrees, TA 92 degrees, CLA 93 degrees). **B:** The ideal angles are superimposed using the ideal nasion (Ni) as a reference point. **C:** The ideal objective is drawn over the patient's existing profile. **D:** The result at 1 year.

bottom of the acetate to allow recording of the data (Table 11–6). Then the angles are drawn (including NFR, NFA, TA, and CLA) with extension of the columella limb to the vertical facial plane, allowing determination of columella inclination. Next, the distances are recorded—including C-N, N-T, AC-T, and P/C/I, with the latter being premaxilla (P = ACJ-SN), columella (C = SN-C^1), and infralobule (I = C^1-T).

Superimpose the Ideal

The first column records the actual characteristics of the patient's nose, the second column lists the ideals which are relatively standard, the third column establishes a realistic operative plan, and the fourth column is left blank for follow-up evaluation. The ideal nasion (Ni) has a level between eyelash and tarsal crease on the superior eyelid, while its height is governed by ideal radix projection (C-Ni = .28N-Ti). The position of ideal tip is governed by a tip angle of 105 degrees for females and 100 degrees for males, with projection influenced by Ac-Ti = .67 N-Ti. Alternatively, one could set the ideal nasofacial angle of 34 degrees for females and 36 degrees for males from the ideal nasion (Ni) and then the ideal tip angle. Where the two lines intersect is the ideal tip. Cross-checking can be done using the ideal lengths of N-Ti and Ac-Ti. The ideal columella breakpoint C^1i occurs

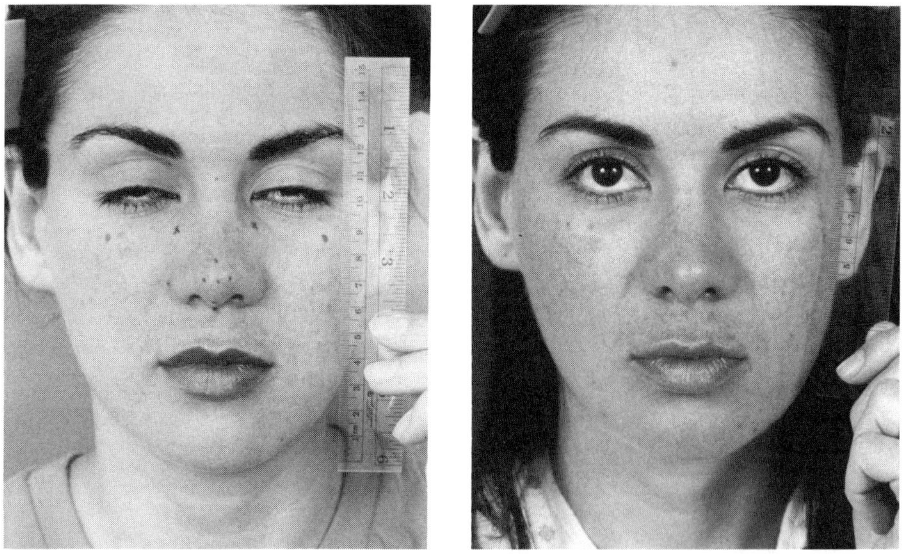

A B

FIG. 11–5. The value of anthropometric measurements in assessing surgical cause and effect. **A:** Preoperative patient with a base bony width (X-X) of 33 mm. **B:** At 18 months postoperatively, the base bony width (X-X) is 29 mm. Definite narrowing of the nose using low-to-low osteotomies, but with *no* open roof dorsally.

at the top of the nostrils along the ideal columella inclination of 105 degrees for females and 95–100 degrees for males. The ideal subnasale (SNi) has been defined by Guyuron as being located in the following steps: (i) Divide Ni-SO into thirds; (ii) at junction of middle and lower one-third, go down 2 mm and then back 2 mm from a vertical plane through the lip border; and (iii) this intersection represents the ideal SN. The ideal angles have been defined as follows: (a) nasofrontal (NFR: 134 degrees for females, 134 degrees for males), (b) nasofacial (NFA: 34 degrees for females, 36 degrees for males), (c) tip angle (TA: 105 degrees for females, 100 degrees for males), (d) columella labial (CLA: 102 degrees for females, 98.5 degrees for males). The ideal distances are as follows: (a) C-Ni = .28NTi, (b) N-Ti = .67MFH = SME, (c) AC-Ti = .67N-Ti, and (d) P = C = I on a curved line. Once the patient's existing points, angles, and distances are recorded, then the ideal is superimposed and the difference can be startling. The difference between actual and ideal allows one to define the problem.

Analyzing the Alternatives

This step is essentially testing the various alternatives in lessening the differences between actual and ideal. For

TABLE 11–6. *Case study: photographic operation planning*

Analysis	Presentation	Objective	Desired change	Result 2 years postoperatively	Actual change
Points					
N	↑→	Eyelash	↓6 ← 4 mm	Eyelash	
T	↓	Tip angle	↑4 mm	Tip angle	
C^1	↑→	Nostril	↑3 ← 2 mm	Nostril	
SN	→	Upper lip	← 4 mm	Upper lip	
Angles					
NFR	155 deg	138 deg	−17 deg	140 deg	−15 deg
NFA	25 deg	36 deg	+11 deg	31 deg	+6 deg
TA	90 deg	100 deg	+10 deg	105 deg	+15 deg
COL LAB	92 deg	100 deg	+8 deg	109 deg	+17 deg
Distances					
C-N	15 mm	11 mm	−4 mm	13	−2 mm
N-T	62 mm	53 mm	−9 mm	48	−5 mm
AC-T	32 mm	30 mm	−2 mm	30	0
P/C/I	11/8/15 mm	10/10/11 mm	−1/−2/−4 mm	10/11/12	0/+1/+1

most aesthetic surgeons, the greatest attention is focused on radix projection and tip angle/projection. During the last decade, surgeons have realized both (a) the importance of the radix as a counterpoise to the base and (b) the devastation wrought by overreduction of the dorsum. To avoid these problems, a dorsum-to-tip sequence is currently favored rather than the traditional tip-to-dorsum. Also, by setting the ideal nasion (Ni), one can then draw the ideal nasofacial angle (NFAi), which is considered the most important nasal angle. Frequently, one may look at several different nasion levels and heights while leaving tip position constant. Millimeter changes in the nasion have enormous multiplier effect on the nasofacial angle. Actual tip angle is almost a fixed aesthetic imperative, whereas tip position is governed by dorsal length (N-Ti) and tip projection (AC-Ti) and can be influenced by the nasofacial angle. Various combinations of nasion and tip alternatives are analyzed.

Establish the Operative Goals

Because tip inclination and projection are often the most difficult to achieve technically, a realistic appraisal must be made (17). The author sets the ideal Ni, NFAi, and TAi and then looks at the resulting projection AC-Ti. Frequently, it is limitations in tip projection that become the determining factor, within the constraints imposed by the skin envelope. Essentially, one then analyzes the changes required at each point, angle, and distance. Are these changes realistic, and will the patient's tissues allow it? Obviously, the surgeon's previous experience governs this decision. Eventually, specific goals are defined. Guyuron (3) has gone one step further by combining life-size photographs and facial xeroradiography. The result is a subdivision of the nose into seven zones, each with a percentage response rate between deep structure excision and surface expressions. For each 1-mm change desired in the radix area, 4 mm of bone must be excised (25% response rate); over the dorsum, however, 2 mm is excised (50% response rate) for each 1 mm desired. Thus, Guyuron has definitively linked photographic analysis with surgical technique, thereby quantitating the operative goal.

Assessing the Result

With the preoperative deformity defined so clearly, one can evaluate both the results achieved and the efficacy of the surgical procedure. Both Byrd and Guyuron have indeed taken photographic analysis to this ultimate conclusion. In a series of 100 cases, Guyuron has analyzed his results with repeat photographs and xeroradiograms and determined the various response rates in each zone of the nose.

CLINICAL CASE STUDY

The value of anthropometric measurements will be demonstrated in the clinical case study of a 27-year-old male executive who presented for secondary rhinoplasty (Fig. 11–6). Two years previously, he had a septoplasty and lowering of the dorsal hump but remained dissatisfied with both his respiration and appearance. Life-size photographs and anthropometric surface measurements were done, followed by surgery and sequential follow-up over a 2-year period (see Table 11–6). The patient requested a smaller, more aesthetic nose to complement his 5-ft 9-in. (172.5 cm) stature.

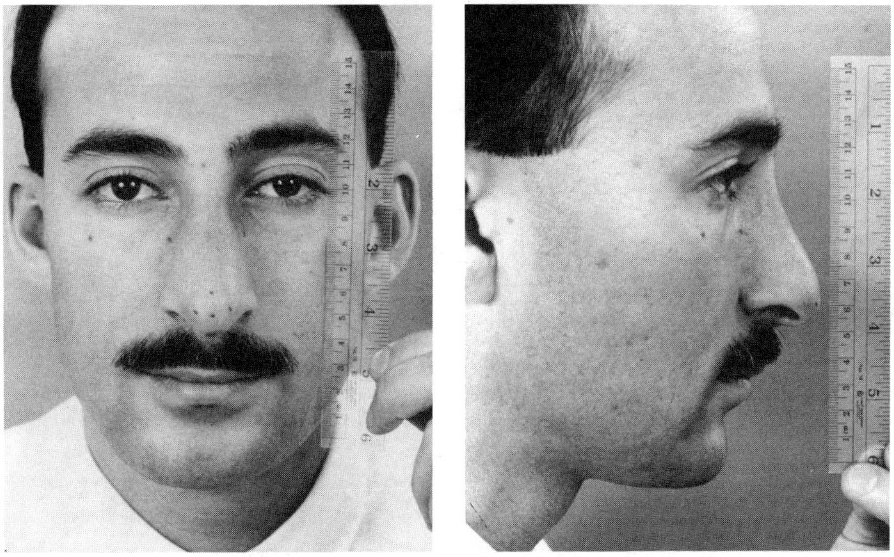

A B

FIG. 11–6. A 27-year-old male 2 years post rhinoplasty who considers the nose too long (**A, B**).

Anthropometric Surface Measurements

On anterior view, the width relationship (EN-EN 31, X-X 31, AL-AL 35, AC-AC 33) indicated the need for low-to-high osteotomies (X-X same as EN-EN) and alar wedge resection (AL-AL greater than EN-EN) (Table 11–7). Because AC-AC was also greater than EN-EN, it was elected to do 4-mm type II wedge resections to narrow the alar base width and nostril size. The lateral measurements indicated that the nose was far too long for both the nasal bridge length and the nasal length. The etiology was both a nasion at the level of the eyebrow and a dependent tip.

The role of anthropometric surface measurements is to make definitive decisions with regard to osteotomies and nasal base modifications. Life-size photographs are more valuable for defining profile changes and various surgical alternatives.

TABLE 11–7. *Case study: anthropometric surface measurements*

	Preoperatively	Two years postoperatively
EN-EN	31	31
MF-MF	20	19.5
X-X	31	30
AL-AL	35	36
AC-AC	33	34
N-T	59	47
N-C^1	65	51
N-SN	65	52
C-N	16	13
AC-T	30	31

Photographic Planning

Utilization of life-size photographs requires a four-step sequence: (i) Define the presenting problems, (ii) superimpose the ideal, (iii) evaluate the alternatives, and (iv) reconcile the objectives with the patient's tissues to determine the final operative plan (Fig. 11–7). The major problem with most techniques is a failure to define the deformity at the four critical landmarks (N, T, C^1, SN). In this case, all four landmarks will require significant change, which means a complicated operative plan. When the angles are analyzed, it is obvious that an NFA of 25 degrees and a TA of 90 degrees indicate a very dependent tip. Equally, the distance N-T = 62 mm is extraordinary long. Yet, it is the interrelationship of landmarks, angles, and distances which define the major problems: (a) Nasion (N) is too high relative to the lash line; this means that nasal bridge length (N-T) will be excessive and too anterior, which means that the nasofacial angle (NFA) will be low; (b) the tip (T) is dependent, which means a small tip angle (TA) and contributes to excessive nasal bridge length (N-T). Thus, the critical first step will be to alter the location of the nasion (down 6 mm, back 4 mm) and to raise the tip angle (+10 degrees) while decreasing tip projection (−2 mm). Next, the columella labial angle and SN location will be assessed. Because radix reduction can be difficult, one set of alternatives is to leave nasion constant and evaluate what would happen if only tip changes are made. The effect is a definite improvement but without major change in the all-important nasofacial angle (NFA), and the nose would still appear long.

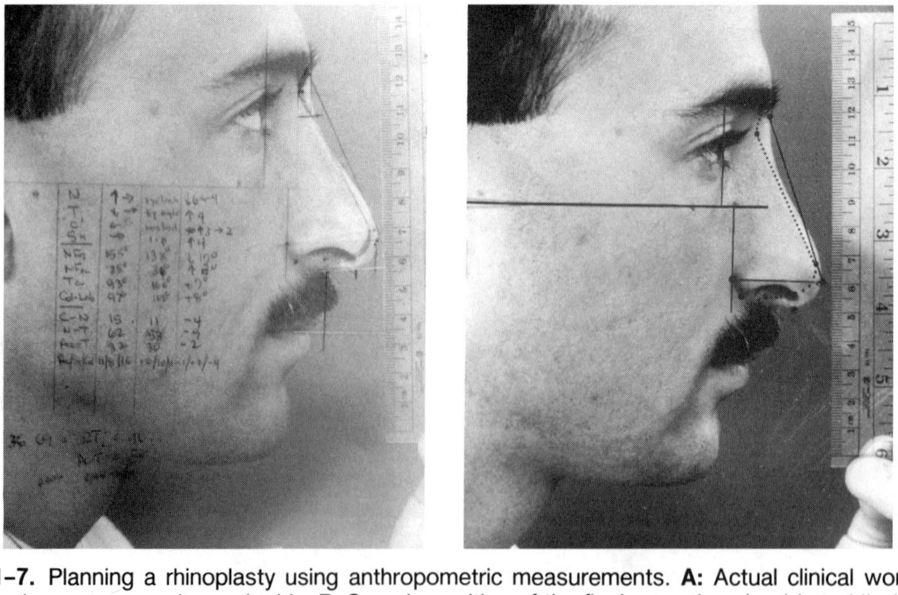

A B

FIG. 11–7. Planning a rhinoplasty using anthropometric measurements. **A:** Actual clinical worksheet showing the acetate overlay and table. **B:** Superimposition of the final operative plan (*dotted line*) on the nose.

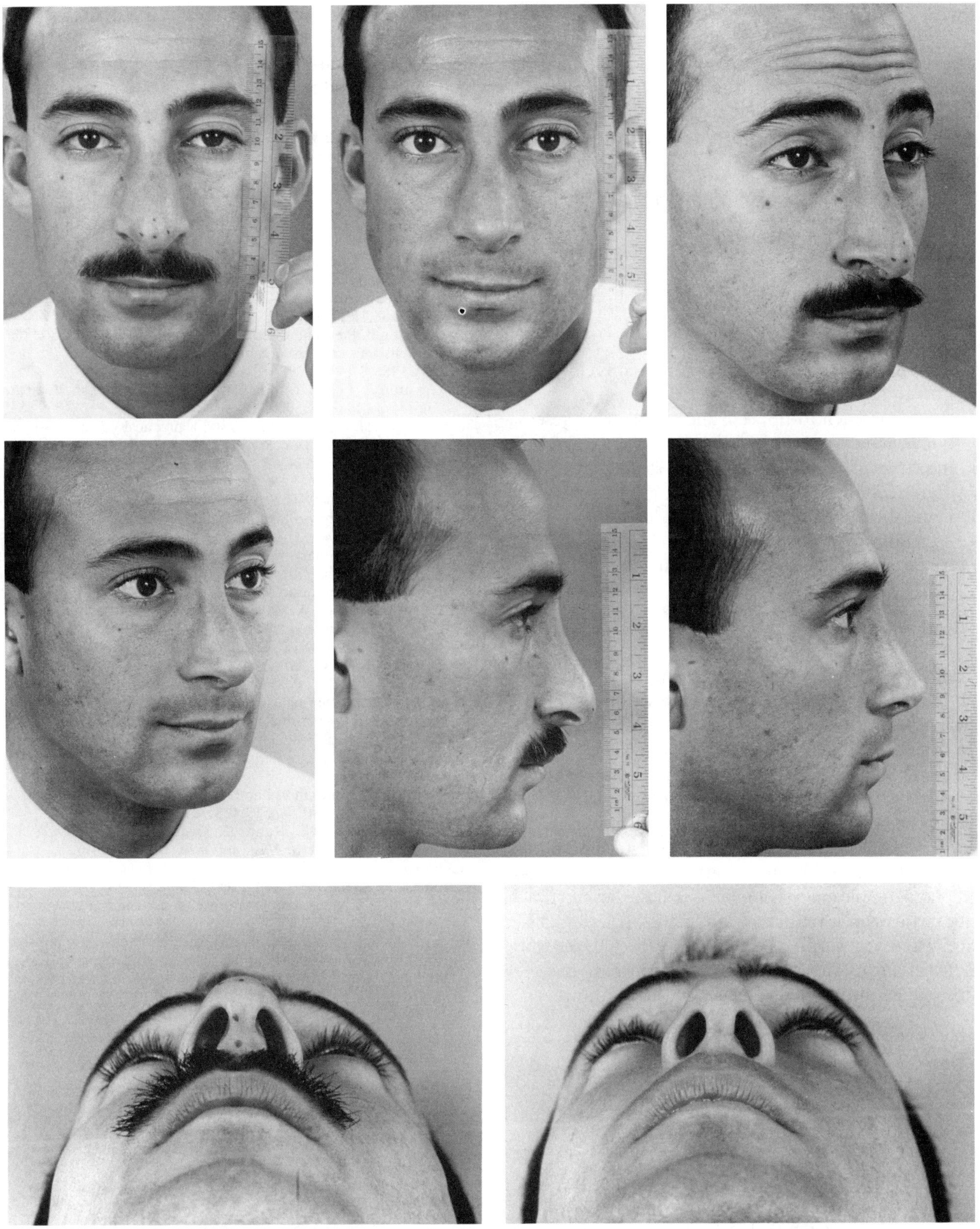

A,B C

D,E F

G H

FIG. 11–8. Comparison of preoperative and postoperative results at 2 years. **A:** Preoperative frontal view of the face. **B:** Postoperative frontal view of the face. **C:** Preoperative right oblique view of the face. **D:** Postoperative right oblique view of the face. **E:** Preoperative right lateral view of the face. **F:** Postoperative right lateral view of the face. **G:** Preoperative base view of the nose. **H:** Postoperative base view of the nose.

Surgical Technique

Because the operative plan requires establishment of the ideal profile, it was elected to use a closed/open approach setting the dorsal profile first, then the caudal septum, and finally the nasal tip. The actual surgical sequence was:

1. Intercartilaginous incision with complete unilateral transfixion incision
2. Hump reduction: minimum bony rasp, 3- to 4-mm cartilage excision
3. Maximum radix reduction, but only 3–4 mm achieved
4. Caudal septum/ANS resection, 3–4 mm
5. Septoplasty with graft harvest, all incisions closed
6. Low-to-high lateral osteotomies
7. Open approach to tip; cephalic lateral crura resection, leaving 6-mm strip
8. Insertion of a shaped columella strut with a 40-degree columella lobular angle
9. Domal creation sutures with 5-0 nylon
10. Tip closure
11. Alar wedge resection: type II, 4 mm

Evaluation

At 2 years postoperatively, the patient is quite pleased with the result and there has been no revisional surgery (Fig. 11–8). What do the repeat anthropometric surface measurements show? A slight narrowing of the bony base (X-X) did occur following the lateral osteotomies. Surprisingly, there has been no change of interalar width or alar crease width despite aggressive 4-mm type II wedges. Yet, the change in nasal bridge length (−12 mm) and nasal length (−13 mm) is rather phenomenal. Repeat photographic analysis confirms that the desired dramatic changes occurred at each landmark, angle, and distance. However, surprises did occur, especially in the nasofacial angle. The 31-degree NFA occurred because radix reduction did change level to eyelash; however, depth was reduced only 2 mm, not the desired 4 mm, resulting in a smaller NFA. Tip position was virtually ideal, and slight upward rotation compensates for failure to deepen the radix. Obviously, accurate measurements allow surgeons to objectively evaluate their results and move toward analysis of surgical cause and effect.

REFERENCES

1. Daniel RK. Rhinoplasty planning. In: Daniel, RK, ed., *Aesthetic Plastic Surgery: Rhinoplasty*. Boston: Little, Brown and Co., 1993.
2. Daniel RK, Farkas LG. Rhinoplasty: image and reality. *Clin Plast Surg* 1988;15:1–10.
3. Guyuron B. Precision rhinoplasty, part I: the role of life-size photographs and soft-tissue cephalometric analysis. *Plast Reconstr Surg* 1988;81:489–499.
4. Guyuron B. Precision rhinoplasty, part II: prediction. *Plast Reconstr Surg* 1988;81:500–505.
5. Byrd HS, Hobar PC. Rhinoplasty: a practical guide for surgical planning. *Plast Reconstr Surg* 1993;91:642–654.
6. Sheen, JH and Sheen, AP. *Aesthetic Rhinoplasty*. St. Louis: CV Mosby, 1987.
7. Regnault P, Daniel RK. Septorhinoplasty. In: Daniel RK, ed. *Aesthetic plastic surgery*. Boston: Little, Brown and Co, 1984;101–171.
8. Daniel RK. Primary rhinoplasty: emphasis on approach. In: Vistnes L, ed. *Procedures in plastic and reconstructive surgery: how they do it*. Boston: Little, Brown and Co, 1991;120.
9. Farkas LG, Bryson W, Klotz J. Is photogrammetry of the face reliable? *Plast Reconstr Surg* 1980;66:346–355.
10. Brown JB, McDowell F, eds. *Plastic surgery of the nose*. Springfield, IL: Charles C Thomas, 1965.
11. Fomon S, Bell JW, eds. *Rhinoplasty—new concepts: evaluation and application*. Springfield, IL: Charles C Thomas, 1970.
12. Anderson JR. A personal technique of rhinoplasty. *Otolaryngol Clin North Am* 1975;8:559–562.
13. Bernstein L. Esthetics and rhinoplasty. *Otolaryngol Clin North Am* 1975;8:705–715.
14. Daniel RK, Hodgson J, Lambros VS. Rhinoplasty: the light reflexes. *Plast Reconstr Surg* 1990;85:859–866.
15. Farkas LG, Kolar JC, Munro IR. Geography of the nose: a morphometric study. *Aesth Plast Surg* 1986;10:191–223.
16. Farkas LG, Munro IR, eds. *Anthropometric facial proportions in medicine*. Springfield, IL: Charles C Thomas, 1987.
17. Daniel RK. The nasal tip: anatomy and aesthetics. *Plast Reconstr Surg* 1992;89:216–224.
18. Farkas LG, Hajniš K, Posnick JC. Anthropometric and anthroposcopic findings of the nasal and facial region in cleft patients before and after primary lip and palate repair. *Cleft Palate Craniofac J* 1993;30:1–12.

Chapter 12

Quantitative Methods of Dysmorphology Diagnosis

Curtis K. Deutsch and Leslie G. Farkas

This chapter outlines schemata for diagnosing dysmorphic features, using the quantitative methods described in Chapter 2. These anthropometric measures described in this book permit objective and reliable assessment of head and face anomalies.

This overview is designed to serve as a guide to the clinician. Medical geneticists, teratologists, and embryologists adduce patterns of these abnormalities to indicate general classes of altered morphogenesis or specific malformation syndromes. The quantitative methods described here provide a means of standardized assessment of anomalies of the head and face. These methods also find application in plastic and reconstructive surgery, permitting diagnosis of abnormalities, surgical planning for their correction, and assessment tools for postoperative follow-up.

In the past, clinicians have relied on visual inspection (*anthroposcopy*) rather than direct measurement (*anthropometry*) to diagnose dysmorphic features (1–4). But it is often difficult to attain agreement on anthroposcopic diagnosis among clinicians, even when experts standardize their nomenclature and methods (5). Protocols have been proposed for the diagnosis of craniofacial anomalies, but there have been many difficulties rendering diagnoses objective and reliable [e.g., see critiques by Krouse and Kaufman (6) and by Pomeroy et al. (7)]. Another limitation to anthroposcopic methods is that they yield qualitative diagnoses, usually dichotomous. Yet most craniofacial anomalies constitute *extremes of graded variation.* In order to fully describe these abnormalities, it is necessary to describe them on a continuous scale. Dr. Farkas' anthropometric methods (8) permit morphological description along a continuum, conditionalized on sex, age, and ethnicity (see also Appendix A).

SYSTEMATICS IN DYSMORPHOLOGY

This book's quantitative methods provide an objective, reliable method of the standardizing diagnosis of dysmorphic signs. Our studies indicate good to excellent intra- and interrater reliability (Chapter 18, *this volume;* D'Agostino et al., *in preparation;* Deutsch et al., *in preparation*).

By using a standard nomenclature and metric, it is possible to compare measurements across subjects. This is essential to conducting research studies of recognizable or unknown medical conditions. Without standardized diagnosis, it is impossible to determine criteria for pooling data, and thus there are no means of rank-ordering the prevalence of stigmata. The knowledge of the relative frequency of these abnormalities would be of considerable use to training and practicing dysmorphologists.

A systematic approach to dysmorphology diagnosis would also allow clinicians to discriminate among recognizable patterns of malformation (9). An anthropometric approach seems promising in this regard. It has identified craniofacial abnormalities in several conditions:

Acrocephalosyndactyly (10)
Apert syndrome (10,11)
Carpenter syndrome (10)
Crouzon syndrome (12)
Down syndrome (13)
Johanson–Blizzard syndrome (14)
Lateral facial dysplasia (15,16)
Noonan syndrome (17)
Pfeiffer syndrome (10)
Saethre–Chotzen syndrome (10)
Treacher Collins syndrome (18)

A standardized approach is needed for the diagnosis of unknown syndromes. There have been several attempts to develop syndrome identification systems, such as POSSUM (Pictures of Standard Syndromes and Undiagnosed Malformations) (19) and BDIS (Birth Defects Information Services) (20). Yet these databases are limited by difficulties in standardizing nomenclature and describing anomalies (21). The approach described in this chapter may help to overcome these problems.

NOMENCLATURE FOR MAJOR AND MINOR CRANIOFACIAL ANOMALIES

Our quantitative approach provides an objective basis for studying dysmorphology along the continuum of severity. The protocol gives the practitioner the option of studying dysmorphic features on a continuous scale, or using cutoff points to describe "major" or "minor" degrees of dysmorphology. Minor anomalies have several synonyms in the literature, including *forme fruste, Kleinform,* minor anomaly, minor defect, minute expression, micromanifestation, microsymptom, and informative morphogenetic variant (22–27).

This nomenclature can be confusing. For instance, the term "microform" does not necessarily imply that an anomaly is a minor form of a major dysmorphic sign. Rather, it can designate unusual morphological features contiguous to a clearly identified anomaly (23).

The distinction between *major* and *minor* anomalies is an arbitrary one, because they generally lie on a continuum. The term "minor physical anomaly" more generally denotes stigmata that are, to use Smith's frequently cited definition, "unusual morphologic features that are of no serious medical or cosmetic consequence to the patient" (3), but this is difficult to operationally define.

Quantitative cutoff points have also been proposed to differentiate between major and minor grades. For instance, Marden et al. (1) felt that an anomaly should be considered "minor" only if its population prevalence is less than 4%. Farkas (28) has defined minor manifestations of major abnormalities by introducing the categories "borderline-small" (between -1 SD and -2 SD) and "borderline-large" (between $+1$ SD and $+2$ SD) to describe scalar measurements. For facial proportions, he proposed the term "disharmony" for minor disproportion (between -1 and -2 SD, or between $+1$ and $+2$ SD); he reserved the term "disproportion" for extreme scores ($> +2$ SD or < -2 SD) (28–30).

Some anomalies represent qualitatively abnormal differentiation (e.g., preauricular fistulae, commissural lip pits, pigmented medial crests of the upper lip and nose). Qualitative measures have revealed a spectrum of microforms of cleft lip and/or palate in the general population (23,31,32). Holmes et al. (5) have established a list of qualitative measures that can be reliably diagnosed by anthroposcopy, but this approach requires methodologic rigor and appears to be applicable to a minority of anomalies. In this chapter, we will discuss only quantitative assessment.

DIAGNOSTIC UTILITY OF MINOR ANOMALIES

Both minor and major anomalies can provide important diagnostic information. Minor anomalies may signify a *major* malformation complex or a syndrome comprised of only minor anomalies (e.g., see refs. 26 and 33–36). Even syndromes involving major malformations are often diagnosed by minor anomalies in aggregate rather than by major abnormalities; this is the case for Down syndrome (37). Minor anomalies also may indicate classes of altered morphogenesis (26,38).

It turns out that minor anomalies are prevalent in many illnesses of unknown etiology (39). Several research groups have found dysmorphic features to be overrepresented in neuropsychiatric disorders (see ref.

TABLE 12-1. *Head (neurocranium)*

Dysmorphic sign	Anthropometric measurement or index	Measurement abbreviation
Small head circumference (microcephaly in major form)	<N *z* circumference (through glabella and opisthocranion)	<N *z* [circ (g-op)]
Large head circumference (macrocephaly in major form)	>N *z* circumference (through glabella and opisthocranion)	>N *z* [circ (g-op)]
Brachycephaly (short-wide head)	>N *z* cephalic index (C-1)	>N *z* (eu-eu/g-op)
Dolichocephaly (long narrow head)	<N *z* cephalic index (C-1)	<N *z* (eu-eu/g-op)
High cranium (relative to face height) (turricephaly in major form)	>N *z* head–face height index (C-22)	>N *z* (v-n/n-gn)
Short cranium (relative to face height)	<N *z* head–face height index (C-22)	<N *z* (v-n/n-gn)
High forehead (relative to head height)	>N *z* forehead–head height index (C-11)	>N *z* (tr-n/v-n)
Low forehead (relative to head height)	<N *z* forehead–head height index (C-11)	<N *z* (tr-n/v-n)
Receding forehead (posteriorly-tilted)	<N *z* inclination of the anterior surface of the forehead	<N *z* (∠ tr-g)
Prominent forehead	>N *z* inclination of the anterior surface of the forehead	>N *z* (∠ tr-g)
Bitemporal narrowing	<N *z* forehead–face width index (C-14)	<N *z* (ft-ft/zy-zy)
Bitemporal widening	>N *z* forehead–face width index (C-14)	>N *z* (ft-ft/zy-zy)

38 for review). How might this relationship between dysmorphology and brain disorder arise? The brain and face derive from common embryological origins and are molded by common morphogenetic forces. It seems plausible that genetic deviations or environmental insults during embryogenesis could manifest themselves both in the brain and in delicate craniofacial formations. Indeed, diseases in which embryological development has gone awry are often associated with both distinctive facial dysmorphology and mental retardation or other behavioral pathology.

A standardized protocol facilitates analysis of these minor anomalies. One can focus on analyses of *embryologically determined combinations* of individual anomalies rather than single anomalies, which are of limited theoretical interest. Combinations can be chosen to enable hypothesis-testing about the nature of dysmorphology in a neuropsychiatric syndrome. Thus, specific classes of dysmorphology may delineate brain maldevelopment.

DIAGNOSTIC METHODS

Because the distinction between "minor" and "major" is difficult to render objective and to provide a logical rationale, perhaps quantitative studies might best analyze anomalies on a continuous scale. We propose methods for this approach below. We also describe how to apply cutoff points to operationally define anomalies, should the investigator or clinician desire (see section entitled "Assignment of Extreme Scores").

The list of anomalies assessed by this protocol was gleaned from classical clinical genetics and teratology references (3–5,26,40–42). Using the anthropometric landmarks and measurement procedures described in Chapter 2, we operationally define craniofacial anomalies of the head (here, indicating the neurocranium; Table 12–1), face (Table 12–2), orbits (Table 12–3), nose (Table 12–4), lips and mouth (Table 12–5), and ears (Table 12–6).

These definitions are based on measurements of cra-

TABLE 12-2. *Face*

Dysmorphic sign	Anthropometric measurement or index	Measurement abbreviation
Long-narrow face	>N *z* facial index (F-1)	>N *z* (n-gn/zy-zy)
Short-wide face	<N *z* facial index (F-1)	<N *z* (n-gn/zy-zy)
Maxillary hypoplasia: narrow midface	<N *z* lower face arcs index (F-29)	<N *z* (t-sn-t/t-gn-t)
Mandibular hypoplasia: narrow, relative to face width	<N *z* mandible–face width index (F-2)	<N *z* (go-go/zy-zy)
Mandibular hypoplasia: short, relative to face height	<N *z* mandible–face height index (F-11)	<N *z* (sto-gn/n-gn)
Mandibular hypoplasia: retrusive, relative to maxillary depth	>N *z* middle-lower third face depth index (F-21)	>N *z* (t-sn/gn-t)
Small chin height (relative to mandible height)	<N *z* chin–mandible height index (AF-3)	<N *z* (sl-gn/sto-gn)

TABLE 12-3. *Orbits*

Dysmorphic sign	Anthropometric measurement or index	Measurement abbreviation
Hypertelorism	>N *z* intercanthal index (O-1)	>N *z* (en-en/ex-ex)
Hypotelorism	<N *z* intercanthal index (O-1)	<N *z* (en-en/ex-ex)
Small palpebral fissure[a] height relative to palpebral fissure width	<N *z* eye fissure index (O-4)	<N *z* (ps-pi/ex-en)
Long upper eyelid (vertically, relative to lower eyelid)	<N *z* eyelid height index (AO-4)	<N *z* (pi-or/ps-os)
Highly inclined eye fissure (mongoloid)	>N *z* eye fissure inclination from horizontal	>N *z* (∠ en-ex)
Negatively inclined eye fissure (antimongoloid)	<N *z* eye fissure inclination from horizontal Angle of inclination is negative	<N *z* (∠ en-ex) ∠en-ex negative
Protrusive supraorbital ridge (relative to suborbital ridge)	>N *z* upper-lower orbital rim inclination	>N *z* (∠ midpoints of upper and lower orbital rims) (os′/or′)
Depressed supraorbital ridge (relative to suborbital ridge)	<N *z* upper-lower orbital rim inclination	<N *z* (∠ midpoints of upper and lower orbital rims) (os′/or′)
Flat supraorbital contour	>N *z* unilateral supraorbital contour index (AF-6)	>N *z* (t-g/t-g surface)
Highly curved supraorbital contour	<N *z* unilateral supraorbital contour index (AF-6)	<N *z* (t-g/t-g surface)

[a] Ptosis is indicated by the presence of a small height of the palpebral fissure in a relatively long fissure.

TABLE 12-4. *Nose*

Dysmorphic sign	Anthropometric measurement or index	Measurement abbreviation
Narrow-long nose	<N *z* nasal index (N-1)	<N *z* (al-al/n-sn)
Wide-short nose	>N *z* nasal index (N-1)	>N *z* (al-al/n-sn)
Long nose (relative to face height)	>N *z* nose–face height index (N-26)	>N *z* (n-sn/n-gn)
Short nose (relative to face height)	<N *z* nose–face height index (N-26)	<N *z* (n-sn/n-gn)
Wide nose (relative to face width)	>N *z* nose–face width index (N-24)	>N *z* (al-al/zy-zy)
Narrow nose (relative to face width)	<N *z* nose–face width index (N-24)	<N *z* (al-al/zy-zy)
Deep nasal root (sagittal, relative to nasal root width)	>N *z* nasal root depth–width index (N-6)	>N *z* (en-se sag/mf-mf)
Shallow nasal root (sagittal, relative to nasal root width)	<N *z* nasal root depth–width index (N-6)	<N *z* (en-se sag/mf-mf)
Wide nasal root (horizontal, relative to intercanthal width)	>N *z* nasal root–intercanthal width index (N-30)	>N *z* (mf-mf/en-en)
Narrow nasal root (horizontal, relative to intercanthal width)	<N *z* nasal root–intercanthal width index (N-30)	<N *z* (mf-mf/en-en)
Anteverted nasal tip	>N *z* nasolabial angle	>N *z* (∠ [(sn-c) − (sn-ls)])
	Upper lip inclination is not negative (i.e., vertical or protruding)	∠ sn-ls > 0
Protrusive nasal tip (relative to nose width)	>N *z* nasal tip protrusion–width index (N-7)	>N *z* (sn-prn/al-al)
Small nasal tip protrusion (relative to nose width)	<N *z* nasal tip protrusion–width index (N-7)	<N *z* (sn-prn/al-al)
Long columella (relative to nasal protrusion)	>N *z* columella length–nasal tip protrusion index (N-18)	>N *z* (c′-sn/sn-prn)
Short columella (relative to nasal protrusion)	<N *z* columella length–nasal tip protrusion index (N-18)	<N *z* (c′-sn/sn-prn)
Long nasal alae (relative to nose width)	<N *z* nose width–ala length index (N-9)	<N *z* [al-al/(ac-prn, r + 1)]
Short nasal alae (relative to nose width)	>N *z* nose width–ala length index (N-9)	>N *z* [al-al/(ac-prn, r + 1)]
Protrusive nasal bridge	>N *z* inclination of nasal bridge	>N *z* ∠ nasal bridge[a]
Flat nasal bridge	<N *z* inclination of nasal bridge	<N *z* ∠ nasal bridge

[a] The straight side of the angle meter rests on the nasal bridge, along a line which approximates but does not strictly follow the line between the nasion (n) and the pronasale (prn), measured in FH. The inclination is measured from the vertical (see Chapter 2).

TABLE 12–5. *Lips and mouth*

Dysmorphic sign	Anthropometric measurement or index	Measurement abbreviation
Long upper lip (both vermilion and cutaneous, relative to upper face height)[a]	>N *z* upper lip–upper face height index (L-11)	>N *z* (sn-sto/n-sto)
Short upper lip (both vermilion and cutaneous, relative to upper face height)	<N *z* upper lip–upper face height index (L-11)	<N *z* (sn-sto/n-sto)
Long cutaneous upper lip (relative to upper lip height)	>N *z* cutaneous–total upper lip height index (L-5)	>N *z* (sn-ls/sn-sto)
Short cutaneous upper lip (relative to upper lip height)	<N *z* cutaneous–total upper lip height index (L-5)	<N *z* (sn-ls/sn-sto)
High upper vermilion (relative to upper lip height)	>N *z* vermilion–total upper lip height index (L-6)	>N *z* (ls-sto/sn-sto)
Small upper vermilion height (relative to upper lip height)	<N *z* vermilion–total upper lip height index (L-6)	<N *z* (ls-sto/sn-sto)
High lower lip (both vermilion and cutaneous, relative to mandible height)	>N *z* lower lip–mandible height index (AL-10)	>N *z* (sto-sl/sto-gn)
Small lower lip height (both vermilion and cutaneous, relative to mandible height)	<N *z* lower lip–mandible height index (AL-10)	<N *z* (sto-sl/sto-gn)
High lower cutaneous lip (relative to total lower lip height)	>N *z* cutaneous–total lower lip height index (AL-7)	>N *z* (li-sl/sto-sl)
Small lower cutaneous lip height (relative to total lower lip height)	<N *z* cutaneous–total lower lip height index (AL-7)	<N *z* (li-sl/sto-sl)
High lower vermilion (relative to total lower lip height)	>N *z* vermilion–total lower lip height index (AL-5)	>N *z* (sto-li/sto-sl)
Small lower vermilion height (relative to total lower lip height)	<N *z* vermilion–total lower lip height index (AL-5)	<N *z* (sto-li/sto-sl)
Protrusive upper lip (premaxillary prominence)	>N *z* inclination of the upper lip	>N *z* (∠sn-ls)
Retrusive upper lip	<N *z* inclination of the upper lip	<N *z* (∠sn-ls)
Wide mouth (horizontal, relative to face width)	>N *z* mouth–face width index (L-10)	>N *z* (ch-ch/zy-zy)
Narrow mouth (horizontal, relative to face width)	>N *z* mouth–face width index (L-10)	>N *z* (ch-ch/zy-zy)
Highly curved mouth contour	<N *z* mouth width contour index (L-2)	<N *z* [(ch-ch)/(ch-sto-ch)]
Flat mouth contour	>N *z* mouth width contour index (L-2)	>N *z* [(ch-ch)/(ch-sto-ch)]

[a] In contrast to popular usage, the term "lip" refers not only to the vermilion portion but also to the cutaneous (i.e., skin) area adjoining the vermilion. The philtrum is equivalent to the cutaneous upper lip (skin portion).

TABLE 12–6. *Ears*

Dysmorphic sign	Anthropometric measurement or index	Measurement abbreviation
Wide/short ears	>N *z* ear index (E-1)	>N *z* (pra-pa/sa-sba)
Narrow/long ears	<N *z* ear index (E-1)	<N *z* (pra-pa/sa-sba)
Long ears (vertical) (relative to facial height)	>N *z* ear–face height index (E-5)	>N *z* (sa-sba/n-gn)
Short ears (vertical) (relative to facial height)	<N *z* ear–face height index (E-5)	<N *z* (sa-sba/n-gn)
High-set ears	<N *z* auricular head–craniofacial height index (C-8)	<N *z* (v-po/v-gn)
Low-set ears	>N *z* auricular head–craniofacial height index (C-8)	>N *z* (v-po/v-gn)
Increased ear protrusion	>N *z* ear protrusion (between posterior aspect of pinna and mastoid plane)	>N *z* ear protrusion angle
Ear inclination high (backward tilt)	>N *z* inclination of the ear's medial longitudinal axis from the vertical	>N *z* (∠ terminal points of the midaxis) (see Chapter 2)

niofacial linear distances, arcs, angles of inclination, and angles of deviation. Most measurements are expressed as combinations of measurements, in the form of proportions. For all operational definitions, standard (z-) scores are computed, based on the normative data for single measurements in this book's appendixes and for proportions in Farkas and Munro's *Anthropometric Facial Proportions in Medicine* (43). The z-score, conditionalized on age and sex, is expressed as

$$z = \frac{\text{subject score} - \text{normative group mean}}{\text{normative group standard deviation}}$$

All tables give the Farkas (28) reference numbers for proportion indices [e.g., "N-1" for nasal index, for which norms are published in Table 47 (page 212)]. The first entry in these tables, **dysmorphic sign,** identifies the craniofacial anomaly. The second entry, **anthropometric measurement or index,** is the operational definition for the dysmorphic sign, in which "<N z" signifies "lower-than-normal z-score" and ">N z" signifies "higher-than-normal z-score." The third entry, **measurement abbreviation,** gives the proportion, angle, or circumference in terms of landmarks. For instance, the abbreviation ">N z (eu-eu/g-op)" signifies "greater-than-normal z-score for the cephalic index (C-1), formed by the proportion of the eurion-to-eurion measurement over the glabella-to-opisthocranion measurement." The ratios of individual measurements are multiplied by 100 (43). The symbol "<" indicates an angle. In some cases, two conditions must both be met in an operational definition (e.g., anteverted nasal tip, Table 12–4).

The use of craniofacial proportions to define dysmorphology is described by Deutsch (27). In some cases, proportion measurements are used to describe shape. For instance, the cephalic index can be used to operationally define a disproportionately short-wide (brachycephalic) or long-narrow (dolichocephalic) head (see Table 12–1). In other cases, expressing an individual measurement as a proportion of another measurement adjusts for size. For example, a mouth can be defined as wide if the cheilion-to-cheilion (or ch-ch) measurement is wide. Because the dimensions of facial features are correlated with face size, wider faces tend to have wider mouths. By expressing this measurement as a function of facial width (zygion-to-zygion, or zy-zy), an internal control is made for size.

Using facial proportions to define dysmorphology "has inherent appeal, but some limitation" (44, page 56). Jorgenson (44) argues that interpretation is complicated when the dimension against which one compares a measurement is itself abnormal. We discuss this case in the Statistical Appendix (Appendix D).

ASSIGNMENT OF EXTREME SCORES

What constitutes a high or low score? This decision is arbitrary. In order to extract the maximum information from this score, it should be used on a continuous scale (see above). However, it is possible to obtain qualitative diagnoses based on one or more cutoff points. Conventionally, the normal range of a measurement is between a z of -2 (two standard deviations below the mean) and a z of $+2$ (two standard deviations above the mean); measurements below a z of -2 and above a z of $+2$ are considered dysmorphic (*subnormal* and *supernormal,* respectively).

The assignment of cutoff points for major and minor anomalies has historically been based on the expected frequency in the general population. One basis of choice would be the often-cited specification that a "minor" anomaly be present in not more than 4% of the general population (see ref. 3). A 4% cutoff point (one-tailed) for a normal distribution would correspond to a $z \geq 1.75$ or ≤ -1.75 (see Appendix D). Scores more extreme than $z > 2$ or < -2 might then be termed "major" anomalies.

The investigator or clinician may prefer to adopt alternative cutoff points for dichotomizing variables. Appendix D contains a table of cumulative standard normal probabilities, giving z-scores and their corresponding areas under the normal curve.

Worked Example. An anthropotrist aims to operationally define highly inclined palpebral fissures in a 7-year-old female. He or she measures the angle of inclination between the endocanthion (en) and exocanthion (ex)—here, for the left orbit. The fissure is slanted at 9 degrees above the horizontal. This would constitute a z-score of $(9 - 4.7)/2.4 = 1.79$. Using a cutoff of 1.75 (but not 2.00), this child would meet the criterion for a minor physical anomaly.

Combination Scores. Combinations of individual anomalies can be computed for individual dichotomous or continuous variables. For dichotomous measures, in which an anomaly is considered present or absent, a weighted or unweighted sum can be computed. For continuously distributed scores, the absolute values of z-scores or their squared values may be summed for each combination.

APPLICATION OF THREE-DIMENSIONAL IMAGING TO THE ASSESSMENT OF DYSMORPHOLOGY

The development of high-resolution surface scanners has opened the possibility of capturing and displaying anthropometric data in three dimensions. The measurements described in this book can be obtained directly from a structured-light scan of the head and face (see Chapter 17). This exciting development soon may allow the dysmorphologist and surgeon to obtain quantitative assessments in a matter of seconds.

We have found the agreement between direct anthropometry and digital imaging to be excellent. Thus, the normative data in this volume and the schemata for dysmorphology diagnosis find application to these new computer-imaging methods (see Chapter 18, and Deutsch et al., *in preparation*).

Surface scanning also lends itself to craniofacial morphometrics to describe the geometry of multiple coordinates (45). Using these methods, one can assess shape changes in dysmorphology by studying the intercoordinated distortions of landmark configurations.

CONCLUSION

The protocol outlined here is a first step towards quantitative diagnosis of major and minor anomalies. These methods are designed to provide objective, reliable means of diagnosing craniofacial dysmorphology, as an adjunct to clinical judgment. As Buyse (40) points out, "our continuing search for scientific methods to supplement the art of clinical dysmorphology . . . has become all the more pressing as [it] continues to move from an academic discipline practiced in relative isolation to a key actor in basic science research and advanced treatment procedures."

REFERENCES

1. Marden PM, Smith DW, McDonald MJ. Congenital anomalies in the newborn infant, including minor variations. *J Pediatr* 1964;64:357–371.
2. Feiglová-Shaumann B, Peagler FD, Gorlin RJ. Minor craniofacial anomalies among negro populations. *Oral Surg Oral Med Oral Pathol* 1970;29:566–575.
3. Jones KL. *Smith's recognizable patterns of human malformation,* 4th ed., New York: WB Saunders, 1988.
4. Aase, JM. *Diagnostic dysmorphology.* New York: Plenum Press, 1990.
5. Holmes LB, Kleiner BC, Leppig KA, Cann CI, Munoz A, Polk BF. Predictive value of minor anomalies. II. Use in cohort studies to identify teratogens. *Teratology* 1987;36:291–297.
6. Krouse JP, Kaufman JM. Minor physical anomalies in exceptional children: a review and critique of research. *J Abnorm Child Psychol* 1982;10:247–264.
7. Pomeroy JC, Sprafkin J, Gadow KD. Minor physical anomalies as a biological marker for behavior disorders. *J Am Acad Child Adolesc Psychiatry* 1988;27:466–467.
8. Farkas LG. *Anthropometry of the head and face in medicine, 2nd ed.* New York: Raven Press, 1994.
9. Hall JG, Froster-Iskenius UG, Allanson JE. *Handbook of normal physical measurements.* New York: Oxford University Press, 1989.
10. Young SC, Kolar JC, Farkas LG, Munro IR. Acrocephalosyndactyly: comparison of morphometric measurements in Pfeiffer, Saethre-Chotzen, Carpenter, and Apert syndromes. *Dtsch Z Mund Kiefer Gesichts Chir* 1986;10:436–443.
11. Farkas LG, Kolar JC, Munro IR. Craniofacial disproportions in Apert's syndrome: an anthropometric study. *Cleft Palate J* 1985;22:253–265.
12. Kolar JC, Munro IR, Farkas LG. Patterns of dysmorphology in Crouzon syndrome: an anthropometric study. *Cleft Palate J* 1988;25:235–244.
13. Farkas LG, Posnick JC, Hreczko T. Anthropometry of the head and face in ninety-five Down syndrome patients. In: Epstein CJ, ed. *The morphogenesis of Down syndrome,* New York: Wiley-Liss, 1991, 53–97.
14. Deutsch CK, Hreczko T, Holmes LB. Objective diagnosis of dysmorphology: A case study of Johanson-Blizzard Syndrome (abstract). *Am J Hum Genetics* 1991;49(4):141.
15. Farkas LG, Ross RB, James J. Anthropometry of the face in lateral facial dysplasia: the bilateral form. *Cleft Palate J* 1977;14:41–51.
16. Farkas LG, James JS. Anthropometry of the face in lateral facial dysplasia: The unilateral form. *Cleft Palate J* 1977;14:193–199.
17. Duncan WJ, Fowler RS, Farkas LG et al. A comprehensive scoring system for evaluating Noonan syndrome. *Am J Med Genet* 1981;10:37–50.
18. Kolar JC, Farkas LG, Munro IR. Surface morphology in Treacher Collins syndrome: an anthropometric study. *Cleft Palate J* 1985;22:266–274.
19. Edwards CN, Buyse ML. *Subscriber's guide to the computerized Birth Defects Information System.* Dover, MA: Center for Birth Defects Information Services, 1985.
20. Pitt DB, Bankier A, Skoroplas T, Rogers JG, Danks DM. The role of photography in syndrome identification. *J Clin Dysmorphol* 1984;2:2–4.
21. Bankier A, McGill JJ, Danks JJ, McGill JA, Danks DM. Dysmorphology: Problems in nomenclature. *Dysmorphol Clin Genet* 1988;2:24–50.
22. Fukuhara T, Saito S. Genetic consideration on the dysplasia of the nasopalatal segments as "formes frustes" radiologically found in patients of cleft children: a preliminary report. *Jpn J Hum Genetics* 1962;7:234–237.
23. Tolarová M, Havlová Z, Ružičková J. The distribution of characters considered to be microforms of cleft lip and/or palate in a population of normal 18–21 year old subjects. *Acta Chir Plast* 1967;9:1–14.
24. Pashayan H, Fraser FC. Nostril asymmetry not a microform of cleft lip. *Cleft Palate J* 1971;8:185–188.
25. Waldrop MF, Halverson CF. Minor physical anomalies and hyperactive behavior in young children. In: Hellmuth J, ed. *The exceptional infant.* New York: Brunner & Mazel, 1971.
26. Pinsky L. Informative morphogenetic variants: minor congenital anomalies revisited. In: Kalter H, ed. *Issues and reviews in teratology.* New York: Plenum Press, 1985.
27. Deutsch CK. Disproportion in psychiatric syndromes. In: Farkas LG, Munro IR, eds. *Anthropometric facial proportions in medicine.* Springfield, IL: Charles C Thomas, 1987.
28. Farkas LG, Munro IR, eds. *Anthropometric facial proportions in medicine.* Springfield, IL: Charles C Thomas, 1987.
29. Farkas LG. Basic anthropometric measurements and proportions in various regions of the craniofacial complex. In: Brodsky D, Ritter-Schmidt D, Holt L, eds. *Craniofacial anomalies: an interdisciplinary approach.* St Louis: Mosby-Year Book, 1992;41–57.
30. Farkas LG, Hajniš K, Posnick JC. Anthropometric and anthroposcopic findings of the nasal and facial region in cleft patients before and after primary lip and palate repair. *Cleft Palate Craniofac J* 1993;30:1–12.
31. Tolarová M, Havlova Z, Ruzickova J. The distribution of characters considered to be microforms of cleft lip and/or palate in a population of normal 3–6 year old subjects. *Acta Chir Plast* 1967;9:184–194.
32. Farkas LG, Cheung GCK. Nostril asymmetry: microform of cleft lip and palate—an anthropometric study of healthy North American Caucasians. *Cleft Palate J* 1979;16:351–357.
33. Fogh-Anderson P. *Inheritance of harelip and cleft palate.* Copenhagen: Busck, 1942.
34. Burian F. *Surgery of the cleft lip and palate* [in Czech]. Prague: SZdn, 1954.
35. van der Woude A. Fistula labii inferioris congenita and its association with cleft lip and palate. *Am J Hum Genet* 1954;6:244–256.
36. Fraser FC, Sproule JR, Halal F. Frequency of the orachio-ororenal syndrome in children with profound hearing loss. *Am J Med Genet* 1980;7:341–349.
37. Preus M, Ayme S. Formal analysis of dysmorphism: objective methods of syndrome identification. *Clin Genet* 1983;23:1–16.
38. Deutsch CK, Matthysse S, Swanson JM, Farkas LG. Genetic latent structure analysis of dysmorphology in attention deficit disorder. *J of the Am Acad Child Adolesc Psychiatry* 1990;29:189.
39. Deutsch CK, Kinsbourne M. Genetics and biochemistry in attention deficit disorder. In: Lewis M, Miller SM, eds. *Handbook of developmental psychopathology.* New York: Plenum Press, 1990.

40. Buyse ML. Measurement and beyond. *Dysmorphol Clin Genet* 1988;2:23–24.

41. Mehes K. *Minor malformations in the neonate.* Budapest: Akademia Kiado, 1983.

42. Smith DW, Bostian K. Congenital anomalies associated with idiopathic mental retardation. *J Pediatr* 1964;65:189.

43. Farkas LG. The proportion index. In: Farkas LG, Munro IR, eds. *Anthropometric facial proportions in medicine.* Springfield, IL: Charles C Thomas, 1987;5–8.

44. Jorgenson RJ. The real problems in dysmorphology. *Dysmorphol Clin Genet* 1988;2:54.

45. Bookstein FL. *Morphometric tools for landmark data: geometry and biology.* New York: Cambridge University Press, 1991.

CHAPTER 13

Anthropometry of the Attractive North American Caucasian Face

Leslie G. Farkas

Everyday opinion of what constitutes an attractive face can be summed up as "one (which is) pleasing to the eye" (1). We have used a combination of objective and subjective methods in our efforts to define the morphological characteristics of the "attractive" and "unattractive" face, minimizing oversubjective judgments wherever possible by recording mean values from teams of observers rather than from individuals. The adoption of these methods has provided valuable data, not only for aesthetic surgery but also for surgical correction of congenitally or traumatically damaged faces. Artists, portraitists, and facial illustrators may also benefit from our study.

CATEGORIES OF THE FACIAL QUALITIES

Categorizing facial quality has been assisted by adopting a system based on Lickert's 7-point method (2), which uses a scale of numbers from 1 to 7 (low representing "poor," high denoting "best") in judging facial quality from photographs. Judgments are taken from groups of viewers rather than individuals so as to diminish any excess of subjective opinion as far as possible. Our modified method (3) adopted the numbers 1 and 2 to denote *below-average* facial quality, the *average* being numbered 3 and 4, while the *above-average* (attractive) face was recorded by numbers 5 and 6.

The goal of our study was to collect anthropometric data of male and female faces judged to be of above- or below-average quality, and six teams were chosen to assist in the task. They were drawn from varied groups which comprised fashion photographers, plastic surgeons, psychologists, medical illustrators, hospital clerks, and students of fine art and anthropology. It was felt that a fair range of aesthetic evaluation from teams having such a wide variety of experience and interests would result.

Colored photographic slides of standard size were projected on screen, in random sequence, offering frontal and lateral views of selected male and female subjects. Each team member was asked to "number" the subject, and then mean values were adduced to represent the team's collective judgment of that face. The goal was to identify the significant characteristics of those individuals having above- or below-average facial quality.

SUBJECTS

Three hundred fifteen healthy young adults were selected at random to participate in our study of a North American Caucasian group. One hundred fifteen were male and 200 were female, with a proportion of fashion models forming part of each gender group (women 50, men 70). Analysis of the ethnic origin of the selectees disclosed the following: Anglo-Saxon male (M) 46% (53 of 115), female (F) 45.5% (91 of 200); Slavic M 8.7% (10 of 115), F 13.5% (27 of 200); Germanic M 14.8% (17 of 115), F 13% (26 of 200); Latin M 20.9% (24 of 115), F 12.5% (25 of 200). Those not categorized in the preceding groups were placed in a miscellaneous category comprising participants of Hungarian, Finnish, Estonian, and Greek origin: men 9.6% (11 of 115) and women 15.5% (31 of 200).

Applying the modified Lickert method, 18.3% (21 of 115) of male faces were seen as above-average (attractive, A subgroup) and 16.5% (19 of 115) were classified as below-average (B subgroup). In females, 17% (34 of 200) were judged above-average (attractive, A subgroup) and 10.5% (21 of 200) were rated below-average (B subgroup).

ANTHROPOMETRY

Measurements. In males and females, 135 and 130 measurements, respectively, were taken from each subject by one investigator (LGF), in the A and B subgroups. The linear measurements were expressed in millimeters, the angular ones in degrees.

Canons (C). The relationship between certain measurements was investigated with the help of nine neoclassical facial canons (4) in both sexes and in both A and B subgroups.

Craniofacial proportion indices. The quality of relationship between the measurements was analyzed with the help of 156 proportion indices in males and 155 proportion indices in females (3).

Asymmetries. The differences between 36 paired linear and angular measurements were determined in both sexes, in both facial quality subgroups: seven in the face, eight in the orbits, nine in the nose, four in the lips and mouth, and eight in the ears. A 1-mm or 1-degree difference would indicate asymmetry. The mean value for the frequency of asymmetry was calculated between the paired measurements of each region (except head with no paired measurement), and in both sexes and both facial quality subgroups it was calculated separately.

Control groups. The findings from the study group were compared to the North American Caucasian population norms (3,5) and to norms obtained from a special facial study (3,6).

Statistical analysis. For statistical analysis of data, the method of standard error of the difference (SED), based on differences between the percentages in the findings compared, and the Student's t test were used. The degree of significancy was regarded as *mild* ($p = 0.05$), *moderate* ($p = 0.01–0.04$), or *high* ($p \leq 0.001$).

RESULTS

In each sex, region, and facial subgroup, the identical main linear and angular measurements and all measurements significantly differing in the two facial subgroups are shown. Also, the main indentical facial proportions and all significantly differing proportion indices are reported in the two subgroups separately in each sex. The mean, standard deviation (SD), and the optimal range (mean ± 1 SD) are reported in those linear and angular measurements and proportion indices of the A subgroups of males and females which may greatly influence our visual judgment.

Males

Of the 135 craniofacial measurements, 33 (24.4%) were identical, 26 (19.3%) were significantly different, and 76 (56.3%) were nonsignificantly different in A and B subgroups (Table 13–1). Of the 156 facial proportion indices, 36 (23.1%) were identical, 37 (23.7%) were significantly different, and 83 (53.2%) were nonsignificantly different (Table 13–2). Nine neoclassical canons of the face were examined in each member of the above-average subgroup (21 times 9 = 189 findings) and in each member of the below-average subgroup (19 times 9 = 171 findings). In the above-average group, seven of the nine canons were confirmed in 16 of 189 (8.5%) cases. In the below-average subgroup, five of the nine canons were reported in 10 of 171 (5.8%) cases (Table 13–3).

Head

Linear measurements. Identical measurements: the head width (eu-eu) and the skull-base width (t-t). *Significantly different* were 4 of the 13 measurements (30.8%), all vertical measurements (Table 13–4). The only highly significant difference was seen in the larger height of the calvarium (v-tr) in the attractive faces.

Angular measurements. The only measurement, the forehead inclination, was nonsignificantly more tilted in the A subgroup.

TABLE 13–1. *Identical and significantly differing craniofacial measurements in males with above- and below-average faces*

			Findings in the A and B subgroups[a]					
			Identical		Differing			
					SIG		NS	
Region	Measurement	*N*	*n*	Percent	*n*	Percent	*n*	Percent
Head	Linear	13	2	15.4	4	30.8	7	53.8
	Angular	1	0	0.0	0	0.0	1	100.0
Face	Linear	29	7	24.1	4	13.8	18	62.1
	Angular	6	0	0.0	4	66.7	2	33.3
Orbits	Linear	14	9	64.3	0	0.0	5	35.7
	Angular	4	2	50.0	0	0.0	2	50.0
Nose	Linear	22	5	22.7	6	27.3	11	50.0
	Angular	9	0	0.0	2	22.2	7	77.8
Lips and mouth	Linear	18	6	33.3	2	11.1	10	55.6
	Angular	3	1	33.3	1	33.3	1	33.3
Ears	Linear	12	1	8.3	2	16.7	9	75.0
	Angular	4	0	0.0	1	25.0	3	75.0
Subtotal	Linear	108	30	27.8	18	16.7	60	55.5
	Angular	27	3	11.1	8	29.6	16	59.3
Total		135	33	24.4	26	19.3	76	56.3

[a] A, above-average (attractive) facial quality; B, below-average facial quality; SIG, significantly; NS, not significantly.

Canons. The two-section facial profile canon was seen in 19% (4 of 21) of the A subgroup and in none of the B subgroup. The three-section and four-section facial profile canons were not observed in the two facial subgroups.

Cranial proportions. In both facial subgroups there were two *identical* proportions: (i) the relationship between the skull-base width and the maxillary arc (t-t/t-sn-t) and (ii) the ratio between the sum of the right and left facial depth and the slightly curved line going from the right ear canal to the mouth along the labial fissure and ending at the left ear canal (t-sn rt) + (t-sn lt)/(t-ch rt) + (ch-ch) + (ch-t lt). About one-quarter (7 of 27) of the indices in the two subgroups showed *significant differences* from one another (Table 13–4).

Highly significant differences were present in the following four proportions: In the attractive faces, the calvarium was relatively higher than the head or forehead heights (v-tr/v-n and v-tr/tr-g); the forehead height was smaller in relation to the head and face heights (tr-n/v-n and tr-n/n-gn).

Measurements which may influence the visual impression. Seven linear and one angular measurements were noted (Table 13–5).

Cranial proportions which may influence the visual impression. Five indices were found: four between vertical measurements and one between horizontal measurements (Table 13–5).

Face

Linear measurements. In the two facial subgroups, almost one-quarter (24.1%, 7 of 29) of the measurements

TABLE 13–2. *Identical and significantly differing craniofacial proportion indices in males with above- and below-average faces*

		Findings in the A and B subgroups[a]					
				Differing			
		Identical		SIG		NS	
Region	Number of proportion indices, *N*	*n*	Percent	*n*	Percent	*n*	Percent
Head	27	3	11.1	7	25.9	17	63.0
Face	44	16	36.4	9	20.4	19	43.2
Orbits	13	6	46.2	0	0.0	7	53.8
Nose	43	7	16.3	12	27.9	24	55.8
Lips and mouth	25	3	12.0	6	24.0	16	64.0
Ears	4	1	25.0	3	75.0	0	0.0
Total	156	36	23.1	37	23.7	83	53.2

[a] A, above-average (attractive) facial quality; B, below-average facial quality; SIG, significantly; NS, not significantly.

TABLE 13–3. *Neoclassical facial proportion canons in males with above- and below-average faces*

	Facial quality			
	Above-average (*N* = 21)		Below-average (*N* = 19)	
Canon	*n*	Percent	*n*	Percent
Vertical canon				
Two-section facial profile canon (v-en = en-gn)	4	19.0	0	0.0
Three-section facial profile canon (tr-n = n-sn = sn-gn)	0	0.0	0	0.0
Four-section facial profile canon (v-tr = tr-g = g-sn = sn-gn)	0	0.0	0	0.0
Nasoaural canon (n-sn = sa-sba)	1	4.8	0	0.0
Horizontal canon				
Orbitonasal proportion canon (en-en = al-al)	5	23.8	1	5.3
Orbital proportion canon (en-en = ex-en)	1	4.8	4	21.0
Naso-oral proportion canon (ch-ch = 1½al-al)	1	4.8	1	5.3
Nasofacial proportion canon (al-al = ¼zy-zy)	3	14.3	3	15.8
Angular canon				
Nasoaural inclination canon (nasal bridge inclination = ear axis inclination)	1	4.8	1	5.3

were *identical* (Table 13–1). The main identical findings were the face width (zy-zy), mandible width (go-go), chin height (sl-gn), and the halves of the maxillary arc (sn-t, rt and lt). Four measurements were found *significantly differing* (13.8%, 4 of 29) in the two facial subgroups (Table 13–6). The differences were not highly significant between the two subgroups.

Angular measurements. Identical measurements were not found in the two facial subgroups (Table 13–1). In four of the six measurements, the difference was *significant* (66.7%), but only in two of them with high degree: The lower face and the general facial profile line in the attractive faces were less receding (Table 13–6).

Canon. The nasofacial proportion canon was confirmed in three subjects of each facial subgroup (Table 13–3).

Facial proportions. The proportion indices were *identical* in both facial subgroups in 36.4% (16 of 44). These indexes were found in the *vertical* direction, between the height of the face (n-gn) and the craniofacial height (v-gn) and also between the height of the chin (sl-gn) and the height of the face (n-gn); *horizontally,* they were found between the width of the mandible (go-go) and the width of the face (zy-zy) and also between the supraorbital and maxillary arcs (t-g-t/t-sn-t), and the ratio between the mandible width and the sum of the mandibular depth measurements (gn-t rt and lt) was determined. Other *identical* relationships were also found between the upper face height (n-sto) and the maxillary arc (t-sn-t) and between the mandibular height (sto-gn) and the mandibular arc (t-gn-t).

One-fifth (20.4%, 9 of 44) of the proportion indices showed *significantly different* values in the two facial subgroups (Table 13–7).

Three proportion indices showed highly significant differences: In the attractive faces, the height of the lower face in relation to the face height was larger (sn-gn/n-gn), the height of the upper profile was smaller in relation to the lower profile height (tr-prn/prn-gn), and the maxillary depth was shallower in relation to the mandibular depth (sn-t/gn-t).

Measurements which may influence the visual impression. Eleven linear and five angular measurements are shown in Tables 13–8 and 13–9.

Facial proportions which may influence the visual impression. Ten proportion indices indicate the relationships between the main vertical, horizontal, and arc measurements, as well as between depth measurements, of the face (Table 13–10).

Orbits

Linear measurements. A high proportion of measurements (64.3%, 9 of 14) (Table 13–1) in both facial subgroups were found to be *identical,* the most important being (a) the intercanthal width (en-en), (b) the length (ex-en) and height (ps-pi) of the eye fissures, and (c) the distance (en-se) of the orbits from the middle line of nasal root. *Significantly differing* measurements were not present.

Angular measurements. Two of the four measurements were *identical* in both facial subgroups: the inclination of the longitudinal axis of the eye fissures. *Significantly differing* angular measurements were not found.

Canons. The orbital proportion canon indicating identical intercanthal width (en-en) and eye fissure length (ex-en) measurements was seen in about one-fifth of the B subgroup (4 of 19) and in only one male in the A subgroup (4.8% of 21).

The orbitonasal canon, which assumes that the intercanthal width (en-en) is equal to the width of the soft nose (al-al), was present more often in attractive male faces (23.8%, 5 of 21) than in below-average faces (B subgroup) (5.3%, 1 of 19) (Table 13–3).

Orbital proportions. Identical proportion indices were recorded in 46.2% (6 of 13) in the two facial subgroups (Table 13–2). Among the most important proportions were (a) the relationship of the intercanthal width (en-en) to the biocular width (ex-ex) and the width

TABLE 13–4. *Significantly differing cranial measurements and proportion indices in males with above- and below-average faces*

Measurement	Facial subgroup (A-B)[a]	Mean (mm)	SD	Difference (*p* value)
Auricular height of the head lt (v-po)	A	135.8	6.8	0.04
	B	131.5	6.1	
Height of the calvarium (v-tr)	A	49.4	6.5	0.006
	B	41.4	10.6	
Height of the forehead I (tr-g)	A	56.2	10.6	0.01
	B	62.2	7.0	
Height of the forehead II (tr-n)	A	66.2	7.4	0.01
	B	72.2	7.1	
Proportion index	**Facial subgroup (A-B)[a]**	**Mean**	**SD**	**Difference (*p* value)**
v-tr × 100/v-n	A	43.2	5.0	0.007
	B	37.4	7.9	
v-tr × 100/tr-g	A	89.8	18.8	0.005
	B	68.8	25.0	
tr-n × 100/v-n	A	58.1	6.7	0.002
	B	65.9	8.0	
tr-n × 100/n-gn	A	52.9	5.5	0.004
	B	58.6	6.2	
tr-n × 100/n-sto	A	87.1	9.7	0.04
	B	94.3	12.2	
v-gn × 100/body height	A	125.7	3.9	0.02
	B	129.1	4.9	
t-t × 100/gn-t rt & lt	A	48.8	1.4	0.03
	B	50.0	1.9	

[a] A, above-average (attractive) facial quality (A subgroup, *N* = 21); B, below-average facial quality (B subgroup, *N* = 19).

of the forehead (ft-ft) and (b) the relationship of the biocular width (ex-ex) to the skull-base width (t-t).

Indices *significantly differing* in the two facial subgroups were not recorded.

Measurements which may influence the visual impression. Six measurements were selected. The four linear measurements are those of the soft-tissue orbits. The two angular measurements are the inclinations of the eye fissures (Table 13–11).

Orbital proportions which may influence the visual impression. The intercanthal index is essential, together with the orbital index and the index showing the relationship of the intercanthal width (en-en) to the width of the soft nose (al-al) (Table 13–11).

Nose

Linear measurements. Identical measurements in the two facial subgroups were present in 22.7% (5 of 22) of the whole (Table 13–1). The most important are the nasal tip protrusion (sn-prn) and the lengths of the columella (sn-c′ rt and lt). Measurements *significantly differing,* localized in the root and the soft nose, were identified in 27.3% (6 of 22). The nose of the attractive men had highly significantly deeper root, narrower columella, and thinner alae (Table 13–12).

Angular measurements. Identical measurements were not seen. Two measurements (22.2% of 9) were *significantly differing;* in the A subgroup, the bridge inclination was smaller and the nasal tip angle larger than in the B subgroup (Table 13–12).

Canons. Naso-oral, nasofacial, and nasoaural proportion canons were observed only in small percentages (Table 13–3).

Nasal proportions. Values *identical* in both facial subgroups were found in 16.3% (7 of 43) of the subjects. Among the main indices, the nasal index (al-al/n-sn), the nose width–ala length index (al-al/ac-prn rt & lt), the ala length–nose height index (ac-prn/n-sn), and the facial alar base width–face width index (ac-ac/zy-zy) were the same.

Significantly differing proportion index values were recorded in 27.9% (12 of 43 subjects) in the two facial subgroups (Table 13–13). In comparison to the B subgroup, the nose of the attractive men exhibited highly significant differences in the following seven proportion qualities: The nasal root was deeper in relation to its width (en-se sag/mf-mf), to the intercanthal width (en-se

TABLE 13–5. *Cranial measurements and proportion index values of attractive men and women which may influence the visual impression*

Measurement	Sex[a]	Mean (mm)	SD
Craniofacial height (v-gn)	M	231.1	5.7
	F	213.9	7.5
Head height (v-n)	M	114.2	7.4
	F	106.7	5.0
Calvarium height (v-tr)	M	49.4	6.5
	F	46.3	7.5
Forehead height (tr-n)	M	66.2	7.5
	F	61.9	5.4
Head width (eu-eu)	M	151.3	5.3
	F	143.0	4.3
Forehead width (ft-ft)	M	117.5	5.2
	F	111.7	4.5
Head length (g-op)	M	199.4	6.9
	F	188.5	6.0
Forehead inclination[b]	M	−11.2	4.8
	F	−6.0	5.1
Proportion index	**Sex[a]**	**Mean**	**SD**
v-n × 100/v-gn	M	49.4	2.6
	F	49.9	1.8
tr-n × 100/v-n	M	58.1	6.7
	F	58.2	6.6
v-n × 100/n-gn	M	91.4	6.2
	F	95.9	5.8
tr-n × 100/n-gn	M	52.9	5.5
	F	55.6	4.7
ft-ft × 100/zy-zy	M	84.5	3.0
	F	86.8	2.9

[a] M, attractive male subgroup ($N = 21$); F, attractive female subgroup ($N = 34$).
[b] In degrees.

TABLE 13–6. *Significantly differing facial measurements in males with above- and below-average faces*

Measurement	Facial subgroup (A-B)[a]	Mean (mm)	SD	Difference (p value)
Height of the lower face (sn-gn)	A	73.7	3.2	0.01
	B	70.1	5.2	
Height of the mandible (sto-gn)	A	51.3	3.0	0.03
	B	49.2	3.1	
The upper profile height (tr-prn)	A	114.2	7.3	0.02
	B	120.7	8.8	
Mandibular depth rt (gn-t)	A	150.1	4.8	0.04
	B	146.8	5.1	
Inclination of the lower face (sn-pg line)[b]	A	−8	4.3	0.004
	B	−13	5.9	
Inclination of the mandible (li-pg line)[b]	A	−12.4	4.8	0.02
	B	−17.6	8.2	
Inclination of the general profile (g-pg line)[b]	A	−1.8	2.9	0.004
	B	−5.1	3.9	
Inclination of the chin[b]	A	18.6	8.5	0.02
	B	10.5	12.8	

[a] A, above-average (attractive) facial quality (A subgroup, $N = 21$); B, below-average facial quality (B subgroup, $N = 19$).
[b] In degrees.

sag/en-en), and to the slope length of the nasal root (en-se sag/en-se); the nose was higher in relation to the craniofacial height, face height, and upper face height (n-sn/v-gn, n-sn/n-gn, and n-sn/n-sto); the nasal root was wider in relation to the nose width (mf-mf/al-al).

Measurements which may influence the visual impression. Seventeen measurements are recommended to be taken: 12 linear and 5 angular (Tables 13–14 and 13–15). The linear measurements inform us about the main measurements of the nasal framework (n-sn, n-prn, al-al) and its root (en-se sag and mf-mf), as well as about the main components of the soft nose (sn-prn, sn-c′, sn′-sn′, sbal-sbal, ac-ac, ac-prn, sbal-sn). Among the angular measurements listed are (a) the inclinations of the nasal bridge and the columella and (b) the angles of the nasal tip (only in males) and the nasofrontal and nasolabial angles.

Nasal proportions which may influence the visual impression. Among the 15 proportion indices are the nasal index (al-al/n-sn), the nasal root index (en-se sag/mf-mf), and the soft nose index (sn-prn/al-al). In addition, the main nasal measurements are related to the selected measurements of the orbits (en-en), face (vertical profile measurements and zy-zy), and mouth (ch-ch) (Table 13–16).

Lips and Mouth

Linear measurements. In 33.3% (6 of 18) of the subjects (Table 13–1) the measurements were *identical* in

TABLE 13–7. *Significantly differing facial proportion index values in males with above- and below-average faces*

Proportion index	Facial subgroup (A-B)[a]	Mean	SD	Difference (p value)
n-sto × 100/n-gn	A	60.8	1.9	0.04
	B	62.4	2.6	
sn-gn × 100/n-gn	A	58.9	2.5	0.008
	B	56.7	2.4	
sto-gn × 100/n-gn	A	41.0	1.7	0.03
	B	39.8	1.5	
tr-prn × 100/prn-gn	A	124.4	8.3	0.001
	B	135.8	11.9	
t-sn-t × 100/t-gn-t	A	88.6	1.6	0.05
	B	92.4	8.6	
n-t lt × 100/sn-t lt	A	97.2	2.4	0.02
	B	95.3	2.5	
sn-t lt × 100/gn-t lt	A	87.5	2.2	0.007
	B	89.5	2.3	
ex-t × 100/n-t lt	A	66.6	2.2	0.05
	B	68.0	2.0	
ch-t × 100/ch-t surf lt	A	91.7	1.3	0.05
	B	90.7	2.0	

[a] A, above-average (attractive) facial quality (A subgroup, $N = 21$); B, below-average facial quality (B subgroup, $N = 19$).

TABLE 13–8. *Facial linear measurements in attractive men and women which may influence the visual impression*

Linear measurements	Sex[a]	Mean (mm)	SD
Face height (n-gn)	M	125.1	4.6
	F	111.4	4.6
Upper face height (n-sto)	M	76.1	3.1
	F	69.1	2.5
Lower face height (sn-gn)	M	73.7	3.3
	F	64.9	3.9
Mandible height (sto-gn)	M	51.3	3.0
	F	43.6	3.1
Chin height (sl-gn)	M	32.6	2.9
	F	28.0	2.5
Face width (zy-zy)	M	139.1	5.3
	F	128.8	4.3
Mandible width (go-go)	M	106.4	4.6
	F	94.5	4.6
Maxillary arc (t-sn-t)	M	301.8	8.2
	F	280.5	9.1
Mandibular arc (t-gn-t)	M	340.6	11.4
	F	306.6	11.3
Maxillary depth of the face (t-sn lt)	M	131.1	3.9
	F	120.1	4.3
Mandibular depth of the face (t-gn lt)	M	149.9	5.0
	F	135.5	5.2

[a] M, attractive male subgroup (*N* = 21); F, attractive female subgroup (*N* = 34).

TABLE 13–10. *Facial proportion indices in attractive men and women which may influence the visual impression*

Proportion index	Sex[a]	Mean	SD
n-sto × 100/n-gn	M	60.8	1.9
	F	62.1	2.1
sn-gn × 100/n-gn	M	58.9	2.5
	F	58.2	2.0
sto-gn × 100/n-gn	M	41.0	1.7
	F	39.1	1.7
sto-gn × 100/sn-gn	M	59.8	2.6
	F	67.2	2.7
sl-gn × 100/sto-gn	M	63.5	4.2
	F	64.3	4.2
go-go × 100/zy-zy	M	76.6	3.8
	F	73.4	3.4
n-sto × 100/zy-zy	M	54.7	2.0
	F	53.7	2.1
sto-gn × 100/go-go	M	48.4	4.0
	F	46.2	3.7
t-sn-t × 100/t-gn-t	M	88.6	1.6
	F	91.5	1.5
t-sn × 100/t-gn lt	M	87.5	2.2
	F	88.7	1.7

[a] M, attractive male subgroup (*N* = 21); F, attractive female subgroup (*N* = 34).

the two subgroups. The main measurements included the upper lip height (sn-sto) and the heights of the vermilion (sto-li), skin portion (li-sl), and total lower lip (sto-sl). *Significant difference* was revealed in two measurements (11.1% of 18); in the A subgroup, the height of the upper lip vermilion (ls-sto) was moderately larger and the upper vermilion line (ch-ls-ch) length was mildly smaller than in the B subgroup (Table 13–17).

Angular measurements. Of the three angular measurements, one was *identical* (the horizontal direction of the labial fissure) and one was moderately *significantly differing* (the less receding lower lip inclination in the A subgroup); the upper lip inclination in the A subgroup indicated a more protruding upper lip (Table 13–17).

Canon. Only in one member of each facial subgroup (4.8% and 5.3%) was the width of the mouth (ch-ch)

TABLE 13–9. *Facial angular measurements in attractive men and women which may influence the visual judgment*

Angular measurements	Sex[a]	Mean (degrees)	SD
Upper face inclination	M	0.9	3.7
	F	2.7	3.0
Lower face inclination	M	−8.0	4.2
	F	−12.4	3.4
Mandible inclination	M	−12.4	4.8
	F	−19.0	5.0
Chin inclination	M	18.6	8.5
	F	13.2	6.6
General profile line inclination	M	−1.8	2.9
	F	−3.0	2.7

[a] M, attractive male subgroup (*N* = 21); F, attractive female subgroup (*N* = 34).

TABLE 13–11. *Orbital measurements and proportion indices in attractive men and women which may influence the visual impression*

Measurements	Sex[a]	(mm)	SD
Intercanthal width (en-en)	M	33.0	1.9
	F	31.9	1.7
Biocular width (ex-ex)	M	90.9	2.6
	F	87.9	2.6
Eye fissure length lt (ex-en)	M	31.1	1.2
	F	31.0	1.0
Eye fissure height lt (ps-pi)	M	10.9	0.9
	F	11.8	1.1
Eye fissure inclination lt[b]	M	4.6	1.3
	F	6.3	1.4
Orbital rim inclination lt[b]	M	18.2	3.9
	F	12.5	2.8
Proportion index	**Sex[a]**	**Mean**	**SD**
en-en × 100/ex-ex	M	36.3	1.7
	F	36.3	1.5
ex-en × 100/en-en	M	94.9	7.6
	F	97.3	5.6
en-en × 100/al-al	M	96.3	6.9
	F	102.9	7.4

[a] M, attractive male subgroup (*N* = 21); F, attractive female subgroup (*N* = 34).
[b] In degrees.

TABLE 13–12. *Significantly differing nasal measurements in males with above- and below-average faces*

Measurement	Facial subgroup[a]	Mean (mm)	SD	Difference (p value)
Nasal root width (mf-mf)	A	19.9	1.4	0.03
	B	18.8	1.5	
Nasal root depth (en-se sag rt)	A	16.9	2.3	0.0003
	B	14.4	1.5	
Nasal root depth (en-se sag lt)	A	17.0	2.3	0.0001
	B	14.4	1.5	
Columella width (sn′-sn′)	A	6.6	0.6	0.0006
	B	7.3	0.6	
Ala thickness (al′-al′ rt)	A	5.6	0.6	0.0001
	B	6.5	0.6	
Ala thickness (al′-al′ lt)	A	5.6	0.6	0.0001
	B	6.5	0.6	
Nasal bridge inclination[b]	A	28.3	3.1	0.01
	B	31.6	4.9	
Nasal tip angle[b]	A	74.5	7.2	0.01
	B	68.2	8.2	

[a] A, above-average (attractive) facial quality (A subgroup, $N = 21$); B, below-average facial quality (B subgroup, $N = 19$).
[b] In degrees.

equal to 1.5 times the width of the nose (naso-oral C) (Table 13–3).

Labio-oral proportions. Identical proportion index values were found in 12% (3 of 25 subjects). In the two subgroups, the relationships between the height of the upper lip (sn-sto) and the upper face (n-sto), as well as between the heights of the lower lip (sto-sl) and the chin (sl-gn), were the same. The index values were *significantly different* in 24% (6 of 25). Among them, highly significant difference was revealed only in the quality of relationship between the height of the vermilion and that of the entire upper lip (ls-sto/sn-sto): The vermilion was relatively higher in the A subgroup than in the B subgroup (Table 13–17).

TABLE 13–13. *Significantly differing nasal proportion indices in males with above- and below-average faces*

Proportion index	Facial subgroup[a]	Mean	SD	Difference (p value)
en-se sag lt × 100/mf-mf	A	85.9	12.5	0.007
	B	76.4	7.3	
en-se sag lt × 100/en-en	A	52.0	8.5	0.0004
	B	43.3	4.9	
en-se sag lt × 100/sn-prn	A	86.6	11.6	0.0002
	B	72.9	8.5	
en-se sag lt × 100/en-se lt	A	67.9	8.8	0.001
	B	56.0	5.0	
n-sn × 100/v-gn	A	23.5	1.3	0.04
	B	24.4	1.4	
n-sn × 100/n-gn	A	43.3	2.0	0.009
	B	45.0	1.9	
n-sn × 100/n-sto	A	73.7	5.9	0.002
	B	79.5	4.6	
mf-mf × 100/en-en	A	60.4	5.2	0.03
	B	56.7	5.1	
mf-mf × 100/al-al	A	58.0	4.0	0.001
	B	53.4	4.3	
mf-mf × 100/ac-ac	A	60.2	3.3	0.04
	B	56.4	4.3	
sbal-sn rt & lt × 100/al-al	A	77.0	9.6	0.03
	B	70.0	9.4	
sn′-sn′ × 100/al-al	A	19.2	2.0	0.03
	B	20.6	1.9	

[a] A, above-average (attractive) facial quality (A subgroup, $N = 21$); B, below-average facial quality (B subgroup, $N = 19$).

TABLE 13–14. *Nasal linear measurements in attractive men and women which may influence the visual judgment*

Linear measurements	Sex[a]	Mean (mm)	SD
Nose height (n-sn)	M	54.2	3.2
	F	50.8	2.6
Nasal bridge length (n-prn)	M	49.3	3.9
	F	44.6	3.0
Nasal root width (mf-mf)	M	19.9	1.4
	F	18.9	1.8
Nasal root depth lt (en-se sag)	M	17.0	2.3
	F	14.7	1.6
Nose width (al-al)	M	34.3	1.9
	F	31.1	1.8
Nasal tip protrusion (sn-prn)	M	19.8	1.6
	F	19.3	1.3
Columella length (sn-c′ lt)	M	11.2	1.4
	F	11.2	1.8
Columella width (sn′-sn′)	M	6.6	0.6
	F	6.6	0.6
Ala length lt (ac-prn)	M	34.9	1.4
	F	31.0	1.8
Nostril floor width lt (sbal-sn)	M	13.1	1.6
	F	10.9	1.5
Facial alar base width (ac-ac)	M	32.9	1.2
	F	—	—
Labial alar base width (sbal-sbal)	M	22.1	2.0
	F	—	—

[a] M, attractive male subgroup (N = 21); F, attractive female subgroup (N = 34).

Measurements which may influence the visual impression. Eight measurements were selected: five linear and three angular. The linear measurements refer to (a) the main vertical measurements of the upper and lower lip and (b) the mouth width. The angular measurements include the inclinations of the upper and lower lip and the labiomental angle (Table 13–18).

Labio-oral proportions which may influence the visual impression. The main relationships between the basic measurements of the upper lip, lower lip, and mouth and the measurements of the face and nose are presented in seven proportion indices (Table 13–19).

TABLE 13–15. *Nasal angular measurements in attractive men and women which may influence the visual judgment*

Angular measurement	Sex[a]	Mean (degrees)	SD
Nasal bridge inclination	M	28.3	3.1
	F	30.8	3.9
Columella inclination	M	76.8	6.4
	F	76.1	8.5
Nasal tip angle	M	74.5	7.2
	F	—	—
Nasofrontal angle	M	133.0	8.9
	F	133.9	6.5
Nasolabial angle	M	98.5	10.5
	F	102.1	8.2

[a] M, attractive male subgroup (N = 21); F, attractive female subgroup (N = 34).

TABLE 13–16. *Nasal proportion indices in attractive men and women which may influence the visual judgment*

Proportion index	Sex[a]	Mean	SD
al-al × 100/n-sn	M	63.4	4.5
	F	61.4	5.1
n-prn × 100/n-sn	M	91.0	3.3
	F	87.8	3.1
en-se sag lt × 100/mf-mf	M	85.9	12.5
	F	78.5	11.1
sn-prn × 100/al-al	M	58.0	6.2
	F	62.2	5.5
mf-mf × 100/al-al	M	58.0	4.0
	F	60.8	6.0
en-se sag lt × 100/sn-prn	M	86.6	11.6
	F	76.6	10.0
en-se lt × 100/ac-prn	M	71.7	4.2
	F	72.9	3.9
sn-c′ lt × 100/sn-prn	M	56.7	7.6
	F	58.2	7.8
sn′-sn′ × 100/sn-c′ lt	M	59.8	10.3
	F	60.5	12.8
sbal-sn rt & lt × 100/al-al	M	77.0	9.6
	F	70.6	10.1
al-al × 100/ac-prn rt & lt	M	49.3	3.0
	F	50.3	3.5
al-al × 100/ch-ch	M	63.9	5.3
	F	61.5	5.8
al-al × 100/zy-zy	M	24.7	1.6
	F	24.2	1.4
n-sn × 100/n-sto	M	73.7	5.9
	F	73.5	3.2
n-sn × 100/n-gn	M	43.3	2.0
	F	45.6	2.1

[a] M, attractive male subgroup (N = 21); F, attractive female subgroup (N = 34).

Ears

Linear measurements. The only *identical* measurement was the projective distance between the lower insertion of the ear and the subnasale point at the base of the columella (obi-sn). The findings were moderately *significantly different* in 16.7% (2 of 12 subjects) (Table 13–20).

Angular measurements. Three of the four measurements were not significantly differing (Table 13–1). The protrusion of the right ear from the head was highly *significantly larger* in the A subgroup (Table 13–20).

Canon. The nasoaural inclination canon was confirmed in one subject of each facial subgroup (4.8% and 5.3%) (Table 13–3).

Auricular proportions. In the two facial subgroups the index value was *identical* in one of the four proportions: the relationship of the ear width (pra-pa) to the ear length

TABLE 13–17. *Significantly differing labio-oral measurements and proportion indices in males with above- and below-average faces*

Measurement[a]	Facial subgroup[b]	Mean (mm)	SD	Difference (p value)
Upper lip vermilion height (ls-sto)	A	8.4	1.1	0.02
	B	7.5	1.3	
Upper lip vermilion arc (ch-ls-ch)	A	77.9	4.6	0.05
	B	81.0	5.1	
Lower lip inclination[c]	A	−36.8	22.4	0.04
	B	−49.2	12.2	
Proportion index	**Facial subgroup[b]**	**Mean**	**SD**	**Difference (p value)**
ls-sto × 100/sn-ls	A	54.2	10.0	0.01
	B	46.6	8.1	
ls-sto × 100/sn-sto	A	38.0	4.7	0.009
	B	33.7	5.3	
ls-sto × 100/sto-li	A	97.1	12.9	0.05
	B	87.2	17.9	
sn-ls × 100/sbal-ls′ lt	A	90.2	11.3	0.02
	B	97.7	9.8	
cph-cph × 100/ch-ch	A	19.9	2.0	0.02
	B	17.8	3.1	
(t-ch rt) + (ch-ch) + (ch-t lt) × 100/t-gn-t	A	89.3	1.6	0.04
	B	93.1	8.4	

[a] Linear measurements are in millimeters.
[b] A, above-average (attractive) facial quality (A subgroup, $N = 21$); B, below-average facial quality (B subgroup, $N = 19$).
[c] In degrees.

(sa-sba). The remaining three proportion index values in the A subgroup were significantly different from those in the B subgroup, but a highly significant difference was recorded only in a smaller ear (sa-sba) in relation to the lower face height (sn-gn) in the attractive men (Table 13–20).

TABLE 13–18. *Labio-oral measurements in attractive men and women which may influence the visual judgment*

Measurements	Sex[a]	Mean (mm)	SD
Upper lip height (sn-sto)	M	22.2	1.8
	F	20.0	1.6
Upper lip's vermilion height (ls-sto)	M	8.4	1.1
	F	8.9	0.9
Lower lip's vermilion height (sto-li)	M	8.8	1.2
	F	9.7	1.1
Lower lip height (sto-sl)	M	19.9	1.5
	F	18.9	10.7
Mouth width (ch-ch)	M	53.9	3.6
	F	50.9	3.5
Upper lip inclination[b]	M	2.7	6.3
	F	1.8	6.3
Lower lip inclination[b]	M	−36.8	22.4
	F	−51.2	10.3
Labiomental angle[b]	M	123.5	14.8
	F	—	—

[a] M, attractive male subgroup ($N = 21$); F, attractive female subgroup ($N = 34$).
[b] In degrees.

Measurements which may influence the visual impression. Four measurements were selected for each auricle: width, length, inclination, and protrusion (Table 13–21).

Ear proportions which may influence the visual impression. Three proportions were selected: the relationship between the width and length of the ear (auricular index) and the indices showing the relative length of the ear (sa-sba) related to the heights of the upper face (n-sto) and face (n-gn) (Table 13–21).

TABLE 13–19. *Labio-oral proportion indices in attractive men and women which may influence the visual judgment*

Proportion index	Sex	Mean	SD
sn-sto × 100/ch-ch	M	41.5	4.8
	F	39.4	3.7
sn-sto × 100/n-sn	M	41.2	4.3
	F	39.5	3.4
sto-sl × 100/sn-sto	M	90.2	10.9
	F	86.2	8.5
sto-sl × 100/sto-gn	M	38.9	3.7
	F	43.1	21.1
ls-sto × 100/sn-sto	M	38.0	4.7
	F	44.9	5.4
sto-li × 100/sto-sl	M	44.5	8.0
	F	55.4	9.9
ch-ch × 100/zy-zy	M	38.7	2.6
	F	39.5	2.5

[a] M, attractive male subgroup ($N = 21$); F, attractive female subgroup ($N = 34$).

TABLE 13–20. *Significantly differing ear measurements and proportion indices in males with above- and below-average faces*

Measurement	Facial subgroup[a]	Mean (mm)	SD	Difference (p value)
Lower gnathion–aural distance (gn-obi rt)	A	122.9	5.5	0.04
	B	119.3	5.2	
Lower gnathion–aural distance (gn-obi lt)	A	123.3	5.1	0.03
	B	119.7	4.7	
Ear protrusion from the head rt[b]	A	21.8	3.3	0.005
	B	17.6	3.9	
Proportion index	**Facial subgroup[a]**	**Mean**	**SD**	**Difference (p value)**
sa-sba × 100/v-gn	A	27.3	1.7	0.05
	B	28.2	1.0	
sa-sba × 100/n-gn	A	50.4	2.9	0.05
	B	52.1	2.5	
sa-sba lt × 100/sn-gn	A	85.7	6.5	0.004
	B	92.1	6.6	

[a] A, above-average (attractive) facial quality (A subgroup, $N = 21$); B, below-average facial quality (B subgroup, $N = 19$).
[b] In degrees.

Females

Of the 130 measurements, 40 (30.8%) were *identical,* 23 (17.7%) were *significantly differing,* and 67 (51.5%) were *nonsignificantly differing* in the two subgroups (Table 13–22). An analysis of differences in quality between 155 proportion indices of the head and face showed that the index values were *identical* in 20 (12.9%), *significantly differing* in 15 (9.7%), and *nonsignificantly differing* in 120 (77.4%), of the two facial subgroups (Table 13–23). Nine neoclassical canons of the face were examined in each member of the above-average subgroup (34 times 9 = 306 findings) and the below-average subgroup (21 times 9 = 189 findings). In the above-average subgroup, three of the nine canons were confirmed in 4.2% (13 of 306). In the below-average subgroup, four of the nine canons were reported in 2.1% (4 of 189) (Table 13–24).

TABLE 13–21. *Ear measurements and proportion indices in attractive men and women which may influence the visual judgment*

Measurement	Sex[a]	Mean (mm)	SD
Ear width lt (pra-pa)	M	35.9	1.8
	F	33.4	2.1
Ear length lt (sa-sba)	M	63.0	3.5
	F	59.3	2.5
Ear inclination lt[b]	M	18.4	3.6
	F	17.9	4.4
Ear protrusion lt[b]	M	21.8	3.3
	F	18.6	4.1
Proportion index	**Sex[a]**	**Mean**	**SD**
pra-pa × 100/sa-sba lt	M	57.0	2.8
	F	56.4	3.7
sa-sba × 100/n-sto lt	M	82.9	4.9
	F	85.3	3.3
sa-sba × 100/n-gn	M	50.4	2.9
	F	53.3	2.8

[a] M, attractive male subgroup ($N = 21$); F, attractive female subgroup ($N = 34$).
[b] In degrees.

Head

Linear measurements. The left auricular height of the head (v-po) and the forehead width (ft-ft) were found to be *identical* in 15.4% (2 of 13 cases) in the two facial subgroups (Table 13–22). In the case of women deemed attractive, 30.8% (4 of 13) of the measurements showed mild-to-moderate *significant differences* (Table 13–25).

Angular measurement. The mean inclination indicated a slightly more tilted forehead in the A subgroup (−6 degrees) than the B subgroup (−4 degrees).

Canons. Of the three canons involving head measurements (v-en, tr-n and tr-g), none could be applied in the two female subgroups. These comprise the two-section facial profile canon, the three-section facial profile canon, and the four-section facial profile canon (Table 13–24).

Cranial proportions. Identical cranial proportion indices were revealed in 6.9% (2 of 29): in forehead–head height index (tr-g/v-n) and in the relationship between the height of the calvarium (v-tr) and the combined height of the forehead and face (tr-gn). In the subgroup of attractive females, 13.8% (4 of 29) of the proportions *differed significantly* from the subgroup of below-average facial qualities. The craniofacial height (v-gn) and the

TABLE 13–22. *Identical and significantly differing craniofacial measurements in females with above- and below-average faces*

			Findings in the A and B subgroups[a]					
					Differing			
			Identical		SIG[b]		NS[c]	
Region	Measurement	*N*	*n*	Percent	*n*	Percent	*n*	Percent
Head	Linear	13	2	15.4	4	30.8	7	53.8
	Angular	1	0	0.0	0	0.0	1	100.0
Face	Linear	29	9	31.0	1	3.4	19	65.5
	Angular	6	0	0.0	2	33.3	4	66.7
Orbits	Linear	14	1	7.1	6	42.9	7	50.0
	Angular	4	2	50.0	2	50.0	0	0.0
Nose	Linear	18	12	66.6	3	16.7	3	16.7
	Angular	8	1	12.5	1	12.5	6	75.0
Lips and mouth	Linear	18	6	33.3	4	22.2	8	44.4
	Angular	3	1	33.3	0	0.0	2	66.7
Ears	Linear	12	6	50.0	0	0.0	6	50.0
	Angular	4	0	0.0	0	0.0	4	100.0
Subtotal	Linear	104	36	34.6	18	17.3	50	48.1
	Angular	26	4	15.4	5	19.2	17	65.4
Total		130	40	30.8	23	17.7	67	51.5

[a] A, above-average (attractive) facial quality; B, below-average facial quality.
[b] SIG, significantly.
[c] NS, not significantly.

head height (v-n) were highly significantly smaller in relation to the body height in the attractive facial subgroup (Table 13–25).

Measurements which may influence the visual impression. These are the same as in the subgroups of the facial quality in males (Table 13–5).

Cranial proportions which may influence the visual impression. These are identical to those listed in the facial subgroups of males (Table 13–5).

Face

Linear measurements. The measurements were *identical* in the two subgroups in 31% (9 of 29) (Table 13–22). Among the vertical measurements were the heights of the face and of the upper and lower face (n-gn, n-sto, and sn-gn), while others were found to include the following: the depth of the upper third (n-t), middle third (sn-t), and lower third (gn-t) of the face on the right side. The only *significantly differing* measurement was the width of the face (zy-zy), which was smaller in the A subgroup (Table 13–26).

Angular measurements. None of the six measurements were *identical* in the two facial subgroups. Two of the six measurements (33.3%) were mildly *significantly different* (Table 13–26).

Canons. None of the three facial profile canons was present in the two facial subgroups for females (Table 13–24).

TABLE 13–23. *Identical and significantly differing craniofacial proportion indices in females with above- and below-average faces*

		Findings in the A and B subgroups[a]					
				Differing			
		Identical		SIG[b]		NS[c]	
Region	Number of proportion indices, *N*	*n*	Percent	*n*	Percent	*n*	Percent
Head	29	2	6.9	4	13.8	23	79.3
Face	45	10	22.2	1	2.2	34	75.6
Orbits	16	2	12.5	2	12.5	12	75.0
Nose	34	5	14.7	2	5.9	27	79.4
Lips and mouth	25	0	0.0	6	24.0	19	76.0
Ears	6	1	16.7	0	0.0	5	83.3
Total	155	20	12.9	15	9.7	120	77.4

[a] A, above-average (attractive) facial quality; B, below-average facial quality.
[b] SIG, significantly.
[c] NS, not significantly.

TABLE 13–24. *Neoclassical facial proportion canons in females with above- and below-average faces*

	Facial quality			
	Above-average ($N = 34$)		Below-average ($N = 21$)	
Canon	*n*	Percent	*n*	Percent
Vertical canon				
Two-section facial profile canon (v-en = en-gn)	0	0.0	0	0.0
Three-section facial profile canon (tr-n = n-sn = sn-gn)	0	0.0	0	0.0
Four-section facial profile canon (v-tr = tr-g = g-sn = sn-gn)	0	0.0	0	0.0
Nasoaural canon (n-sn = sa-sba)	0	0.0	1	4.8
Horizontal canon				
Orbitonasal proportion canon (en-en = al-al)	4	12.0	1	4.8
Orbital proportion canon (en-en = ex-en)	7	20.6	1	4.8
Naso-oral proportion canon (ch-ch = 1½al-al)	0	0.0	1	4.8
Nasofacial proportion canon (al-al = ¼zy-zy)	0	0.0	0	0.0
Angular canon				
Nasoaural inclination canon (nasal bridge inclination = ear axis inclination)	2	5.9	0	0.0

Facial proportions. Surprising was the high percentage of the proportion qualities (97.8%), which indicated either *identical* index values (22.2%) or *nonsignificantly differing* (75.6%) proportion indices in the two facial subgroups (Table 13–23). Among the identical indices were those which define the general framework of the upper face (n-sto/zy-zy) and mandible (sto-gn/go-go and sto-gn/t-gn-t). No difference was found in relation to the upper face height, the face height (n-sto/n-gn), or the height of the face and the maxillary and mandibular arcs (n-gn/t-sn-t and n-gn/t-gn-t), nor was a difference found in ratio between the width of the face and head (zy-zy/eu-eu).

The only *significantly differing* (1 of 45, 2.2%) index showed a moderately smaller distance between the ear canal and the orbit (ex-t) relative to the depth in the upper third of the face (t-n) (Table 13–26).

Measurements which may influence the visual impression. These are identical with those listed in the facial subgroups of males (Table 13–8 and 9).

Facial proportions which may influence the visual impression. These were reported in the facial qualities in men (Table 13–10).

Orbits

Linear measurements. One measurement, the intercanthal width (en-en), was *identical* in both subgroups (7.1%, 1 of 14) (Table 13–22). *Significant differences* were found in 42.9% (6 of 14) of the measurements; the eye fissures of the attractive women were mildly greater in length (ex-en) and highly significantly larger in height (ps-pi); also, the eye fissures were moderately more laterally positioned from the facial midline (en-se) than in women of the B subgroup.

Angular measurements. Of the four inclinations, two were *identical* (the inclination between the upper and lower orbital rims on each side) and two were highly *significantly different:* In attractive women the inclination of the longitudinal axis of the eye fissures was larger (Table 13–27).

Canons. The orbital proportion canon was present in about one-fifth of the attractive women (20.6%, 7 of 34) but only in one member (4.8% of 21) of the B subgroup. Equal intercanthal width (en-en) and soft nose width (al-al) (orbitonasal proportion canon) were reported in the attractive subgroup in 12% (4 of 34) and 4.8% (1 of 21) in the B subgroup (Table 13–24).

Orbital proportions. Two proportion indices out of 16 (12.5%) were highly *significantly different* and the same percentage *identical* in the two facial subgroups of women. The total width of the orbits determined between the outer canthi (ex-ex) was greater in the attractive subgroup in relation to the face width (zy-zy). In the attractive women subgroup, the eyes appeared larger because of the greater eye fissure heights (ps-pi) in relation to their length (ex-en) (Table 13–27). The intercanthal index (en-en/ex-ex) indicated mesoteloric orbits in both groups. A similar relationship was noted between the eye fissure length (ex-en) and the intercanthal width (en-en).

Measurements which may influence the visual impression. These are the same as in the facial subgroups in men (Table 13–11).

Orbital proportions which may influence the visual impression. These are identical to the proportion indices in the two male subgroups (Table 13–11).

Nose

Linear measurements. Two-thirds (12 of 18) of the measurements were *identical* in the two facial subgroups (Table 13–22). These comprised the nose width (al-al) and height (n-sn), the nasal root width (mf-mf), the nasal bridge length (n-prn) and the main soft nose measurements, the columella length (sn-c′), nasal floor width (sbal-sn), and ala length (ac-prn), bilaterally. Three measurements (16.7% of 18) *differed* mildly to moderately *significantly* in the A subgroup and B subgroup: the right and left nasal root (en-se sag) was deeper and the columella (sn′-sn′) was thinner in attractive women.

Angular measurements. In the two facial subgroups, the nasolabial angle was *identical* and the bridge inclination was moderately *significantly* larger in the group of

TABLE 13–25. *Significantly differing cranial measurements and proportion indices in females with above- and below-average faces*

Measurement	Facial subgroup[a]	Mean (mm)	SD	Difference (p value)
Height of the head (v-n)	A	106.7	5.0	0.02
	B	110.3	6.6	
Length of the head (g-op)	A	188.5	6.0	0.05
	B	185.4	5.5	
Circumference of the head	A	551.8	16.2	0.05
	B	543.4	10.6	
Head and upper face height (v-sn)	A	156.5	6.3	0.01
	B	162.0	7.2	
Proportion index	**Facial subgroup[a]**	**Mean**	**SD**	**Difference (p value)**
eu-eu × 100/g-op	A	75.9	3.5	0.01
	B	78.7	4.4	
v-gn × 100/body height	A	12.5	0.5	0.001
	B	13.4	0.7	
v-n × 100/body height	A	6.2	0.4	0.001
	B	6.9	0.5	
t-t × 100/gn-t rt & lt	A	50.9	1.4	0.05
	B	51.9	2.1	

[a] A, above-average (attractive) facial quality (A subgroup, $N = 34$); B, below-average facial quality (B subgroup, $N = 21$).

attractive women (Table 13–28); six angular measurements out of eight (75%) were nonsignificantly differing.

Canons. Nasal measurements were present in six canons: (a) the nose height (n-sn) in three- and four-section profile canons and in the nasoaural inclination canon and (b) the nose width (al-al) in the orbitonasal proportion, the naso-oral proportion, and the nasofacial proportion canons. In the A subgroup only the orbitonasal proportion canon was seen in a small number of attractive women (12%, 4 of 34). The nasoaural inclination canon, in which the inclination of the nasal bridge was equal to the inclination of the medial longitudinal axis of the auricle, was recorded in 2 of 34 (5.9%) attractive women. The latter two canons in the B subgroup of females were observed in one and none, respectively, of the 21 subjects (Table 13–24).

Nasal proportions. In 14.7% (5 of 34) the indices were *identical* in both A and B subgroups. In the two facial subgroups, identical relationships were found between the nose height (n-sn) and the combined head and face height (v-gn), face height (n-gn), and maxillary arc (t-sn-t). Nose width (al-al) and face width (zy-zy) were also found to be identical. The ala contours determined with the ac-prn/ac-prn surf ratio were also the same in the two subgroups (Table 13–23).

Significantly differing nasal proportion indices were

TABLE 13–26. *Significantly differing facial measurements and proportion indices in females with above- and below-average faces*

Measurement	Facial subgroup[a]	Mean (mm)	SD	Difference (p value)
Width of the face (zy-zy)	A	128.8	4.3	0.05
	B	131.0	3.8	
Inclination of the upper face[b] (g-sn line)	A	2.7	3.0	0.05
	B	0.8	3.9	
Inclination of the general profile line (g-pg line)[b]	A	−3.0	2.7	0.05
	B	−4.6	3.4	
Proportion index	**Facial subgroup[a]**	**Mean**	**SD**	**Difference (p value)**
ex-t × 100/t-n lt	A	66.4	2.1	0.01
	B	68.2	2.4	

[a] A, above-average (attractive) facial quality (A subgroup, $N = 34$); B, below-average facial quality (B subgroup, $N = 21$).
[b] In degrees.

TABLE 13–27. *Significantly differing orbital measurements and proportion indices in females with above- and below-average faces*

Measurement	Facial subgroup[a]	Mean (mm)	SD	Difference (p value)
Eye fissure length rt (ex-en)	A	31.0	1.0	0.05
	B	30.1	1.4	
Eye fissure length lt (ex-en)	A	31.0	1.0	0.01
	B	30.1	1.4	
Eye fissure height rt (ps-pi)	A	11.8	1.1	0.001
	B	10.3	1.1	
Eye fissure height lt (ps-pi)	A	11.8	1.1	0.001
	B	10.3	1.1	
Endocanthion–facial midline distance rt (en-se)	A	22.4	1.2	0.02
	B	21.4	2.0	
Endocanthion–facial midline distance lt (en-se)	A	22.4	1.2	0.02
	B	21.4	1.8	
Eye fissure inclination rt[b]	A	6.3	1.4	0.001
	B	4.1	1.4	
Eye fissure inclination lt[b]	A	6.3	1.4	0.001
	B	4.1	1.4	
Proportion index	**Facial subgroup[a]**	**Mean**	**SD**	**Difference (p value)**
ex-ex × 100/zy-zy	A	68.3	2.1	0.001
	B	66.1	2.6	
ps-pi × 100/ex-en lt	A	38.1	3.5	0.001
	B	34.1	3.3	

[a] A, above-average (attractive) facial quality (A subgroup, $N = 34$); B, below-average facial quality (B subgroup, $N = 21$).
[b] In degrees.

found in 2 of 34 (5.9%) proportions in the two subgroups. In the faces of the attractive women the depth of the nasal root (en-se sag) was moderately significantly greater in relation to the nasal tip protrusion. The nasal root–slope length (en-se) was highly significantly greater in relation to the ala length (ac-prn) (Table 13–23).

Measurements which may influence the visual impression. These are as listed in the facial qualities of the males (Tables 13–14 and 15).

Nasal proportions which may influence the visual impression. These are the same as in men (Table 13–16).

TABLE 13–28. *Significantly differing nasal measurements and proportion indices in females with above- and below-average faces*

Measurement	Facial subgroup[a]	Mean (mm)	SD	Difference (p value)
Nasal root depth rt (en-se sag)	A	14.7	1.6	0.05
	B	13.7	2.0	
Nasal root depth lt (en-se sag)	A	14.7	1.6	0.05
	B	13.7	2.0	
Width of the columella (sn′-sn′)	A	6.6	0.6	0.02
	B	7.0	0.6	
Nasal bridge inclination[b]	A	30.8	3.9	0.01
	B	27.9	4.1	
Proportion index	**Facial subgroup[a]**	**Mean**	**SD**	**Difference (p value)**
en-se sag × 100/sn-prn	A	76.6	10.0	0.02
	B	70.1	11.0	
en-se × 100/ac-prn lt	A	72.9	3.9	0.001
	B	67.8	6.5	

[a] A, above-average (attractive) facial quality (A subgroup, $N = 34$); B, below-average facial quality (B subgroup, $N = 21$).
[b] In degrees.

TABLE 13–29. *Significantly differing orolabial measurements and proportion indices in females with above- and below-average faces*

Measurement	Facial subgroup[a]	Mean (mm)	SD	Difference (p value)
Upper vermilion arc (ch-ls-ch)	A	71.9	4.0	0.001
	B	67.0	6.8	
Labial fissue half rt (ch-sto)	A	30.1	2.5	0.05
	B	28.5	2.9	
Labial fissure half lt (ch-sto)	A	29.9	2.4	0.02
	B	28.2	2.6	
Height of the cutaneous lower lip (li-sl)	A	9.6	1.8	0.001
	B	12.1	2.0	
Proportion index	**Facial subgroup[a]**	**Mean**	**SD**	**Difference (p value)**
ch-ch × 100/zy-zy	A	39.5	2.5	0.01
	B	37.4	3.7	
cph-cph × 100/ch-ch	A	17.9	3.1	0.001
	B	20.5	2.7	
li-sl × 100/sn-ls	A	73.4	13.1	0.001
	B	90.5	20.3	
li-sl × 100/sto-sl	A	54.5	11.0	0.001
	B	70.6	11.8	
ch-li-ch × 100/ch-ls-ch	A	89.1	4.7	0.001
	B	94.5	5.9	
sto-li × 100/li-sl	A	1.1	0.3	0.01
	B	0.8	0.3	

[a] A, above-average (attractive) facial quality (A subgroup, $N = 34$); B, below-average facial quality (B subgroup, $N = 21$).

Lips and Mouth

Linear measurements. One-third (6 of 18) of the measurements were *identical* in the two facial subgroups (Table 13–22): the height of the upper lip (sn-sto) as well as the heights of the upper lip skin portion (sn-ls), upper lip vermilion (ls-sto), and lower lip vermilion (sto-li). Over one-fifth (22.2%, 4 of 18) of the measurements *differed significantly:* In the attractive women the upper vermilion line (ch-ls-ch) was highly significantly longer, and the height of the skin portion of the lower lip (li-sl) was highly significantly smaller in the A subgroup than in the B subgroup (Table 13–29).

Angular measurements. One of the three (33.3%) measurements, the mouth inclination, was *identical* (horizontal) in both facial subgroups. *Significantly differing* angular measurements were not seen.

Canons. The naso-oral proportion canon was not observed in any of the 34 attractive women and in only 1 of 21 (4.8%) subjects in the B subgroup (Table 13–24).

Labio-oral proportions. Identical indices were not recorded. Almost one-quarter (24%, 6 of 25) of the proportion indices *differed significantly.* Highly significant difference was found in four of the six proportion indices. In the attractive women, the philtrum compared with the mouth was narrower (cph-cph/ch-ch), and the height of the skin portion of the lower lip was smaller in relation to the lower lip height and the height of the skin portion of the upper lip (li-sl/sto-sl and li-sl/sn-ls); a fuller and more protruding upper lip vermilion was indicated by a relatively larger upper vermilion arc (ch-li-ch/ch-ls-ch). The moderately significantly larger ch-ch/zy-zy index in the A subgroup designated a relatively larger mouth in relation to the face in the attractive women (Table 13–29).

Measurements which may influence the visual impression. These are consistent with those reported as the facial qualities of men (Table 13–18).

Labio-oral proportions which may influence the visual impression. These are listed in facial qualities of males (Table 13–19).

Ears

Linear measurements. Half (6 of 12) of the measurements were *identical* in the two subgroups (Table 13–22): the width of both ears (pra-pa), the projective distances on each side between the lower insertion of the ear (obi) and the subnasale point at the base of the columella (sn), and the tangential measurements between the same landmarks (sn-obi surf rt & lt). *Significantly differing* measurements were not found.

Angular measurements. None of the four measurements was identical or significantly differing in the two facial subgroups.

Canons. The nasoaural canon was not present in the A subgroup, but the nasoaural inclination canon was seen in 2 of 34 (5.9%) subjects of the subgroup. In the B subgroup, equal height of the nose and length of the ear (nasoaural canon) was identified in one female (4.8% of 21). The nasoaural inclination canon was not observed (Table 13–24).

Auricular proportions. None of the six proportion indices showed *significant differences* between the two facial qualities in females. One proportion index, the average width (pra-pa)–length (sa-sba) relationship, was the *same* in both subgroups of the face.

Measurements which may influence the visual impression. These are identical to those in the facial qualities of men (Table 13–21).

Auricular proportions which may influence the visual impression. These are the same as those reported for the male subgroups (Table 13–21).

ASYMMETRIES

The frequency of a selected group of asymmetrical paired measurements together with the mean differences between them were compared with the appropriate findings for the general population (5). The aim of the study was to determine whether facial asymmetry, its frequency, and/or degree could be one of the decisive factors in defining facial quality. Special attention was paid to paired measurements in the "sensitive" area of the face located centrally—at the levels of the orbits, soft nose, and mouth (6). The study was completed by examining asymmetries of the face and ears.

Males

Degree of Asymmetries in Men of the General Population and of Those in the Above-Average Facial Group (A Subgroup)

Face

Asymmetries were analyzed between the linear depth measurements in the upper (n-t), middle (sn-t), and lower thirds (gn-t) of the face. The range of the mean differences between the right and left side measurements was similar: in the general population between 2.8 mm and 4.2 mm, and in the A subgroup from 1.9 mm to 4.3 mm.

Orbits

In the soft-tissue orbital measurements (the eye fissure length, height, and inclination), neither of the two groups showed asymmetry. The range of asymmetries was similar in the two groups in linear measurements which determine the position of the orbits, medially in relation to the midline of the nasal root (en-se), laterally to the ear canal (ex-t), and distally to the angle of the mandible (ex-go). The mean asymmetry in the general population group ranged from 1 mm to 3.6 mm, and in the A subgroup from 1 mm to 3 mm.

Nose

Asymmetries in the thickness of the nasal alae (al′-al′) and depth of the nasal root (en-se sag) were not found in the two groups. The mean asymmetries in projective linear measurements of the soft nose (the nostril floor widths, ala lengths, columella lengths) were close in the general population (range: 1–2.5 mm) and in the A subgroup [range: 1–4 mm (found in ala length)]. The mean deviation values of the nasal bridge and columella were slightly higher in the men with above-average face quality (6.7 degrees, both for bridge and columella) than in the general population (range: mean 4.6–4.7 degrees).

Lips and Mouth

The mean differences between the paired lateral upper lip heights (sbal-ls′), the halves of the labial fissure (ch-sto), and the measurements (ch-t, both projective and tangential) defining the position of the mouth in relation to the ear canal were similar in the two groups: In the general population they ranged between 1.8 mm and 3.6 mm, and in the A subgroup they ranged between 1 mm and 3.1 mm.

Ears

The mean difference between the relation of the width and length of the ears (pra-pa/sa-sba) was very close in the two groups, as was the position of the ears medially in relation to the subnasale point at the base of the columella (obi-sn) and distally to the chin point (gnathion) (obi-gn). In the general population the mean asymmetry ranged from 1.5 mm to 4 mm and in the A subgroup it ranged from 1 mm to 3.6 mm. The angular measurements (ear protrusion and ear inclination) revealed similar findings: asymmetries from mean 4.4 degrees to 4.7 degrees in the general population and from 2 degrees to 5 degrees in the A subgroup.

Asymmetries in the Above-Average (A Subgroup) and Below-Average (B Subgroup) Facial Qualities in Men

Comparison of the *extent* of asymmetries in the paired linear measurements in various regions of the face did not show statistical differences between the two male subgroups. Noteworthy differences were found in angular measurements of the nose. The mean nasal bridge deviation in men with below-average faces (4.6 degrees) was highly significantly larger ($p = 0.001$) than in men of the A subgroup (2 degrees). The columella deviation in the A subgroup (6.7 degrees) was moderately smaller than in the B subgroup (7.3 degrees).

Analysis of the *frequency* of asymmetrical paired measurements revealed significant differences between the two facial qualities in 11.1% (4 of 36).

In the mandibular region, significantly more asymmetries were seen in depth measurements (gn-t: 85.7%, 18 of 21) and tangential measurements in the men of the A subgroup (gn-t surf: 90.5%, 19 of 21) than in men with below-average face qualities (52.6%, 10 of 19 in both measurements) (SED 13.8, difference 33.1; SED 13.1, difference 37.9). In the orbits, there were significantly more asymmetries in the A subgroup (81%, 17 of 21) in the projective linear distances between the orbit and the mandible angle (ex-go) than in below-average faces (31.6%, 6 of 19) (SED 13.7, difference 49.4). In the nose, asymmetries in length of alae (ac-prn) and columella (sn-c') were moderately more frequent in the below-average faces, and only the nasal bridge deviation showed a significantly higher percentage (26.3%, 5 of 19) than in the A subgroup (4.8%, 1 of 21) (SED 10.2, difference 21.5). In the ears, asymmetries in protrusions and inclinations were observed only in men with below-average faces.

Surprisingly, symmetrical measurements between the lengths, heights, and inclinations of the eye fissures as well as in depth measurements of the nasal root (en-se sag) were a constant sign in men of both facial-quality subgroups and in the general population.

Females

Degrees of Asymmetries in Attractive Women (A Subgroup) Compared with Those Found in the General Population

Face

The extent of asymmetries between the right and left depth measurements in the upper, middle, and lower third of the face did not show statistically different values in attractive women and in the general population. In the general population the mean differences ranged from 3 mm to 3.8 mm, and in the attractive women they ranged from 2 mm to 4.5 mm.

Orbits

The mean range of asymmetries in the linear projective measurements which define the position of the orbits in the face (en-se, ex-t and ex-go) was smaller in the A subgroup (2–2.3 mm) than in the general population (1.7–3.8 mm). Mild asymmetry (mean: 1 mm) between the inclinations of the eye fissures was observed in 5.9% (2 of 34) of attractive women but in none of the general population. Both in the general population and in women of the A subgroup, the lengths of the eye fissures (ex-en) and their heights (ps-pi) were symmetrical.

Nose

The nasal root depth measurements (en-se sag), the lengths of the nasal alae (ac-prn), and the lengths of the columella (sn-c') were symmetrical both in the attractive women and in the general population. The mean asymmetry in the widths of the nostrils was almost identical in the A subgroup (1 mm) and in the general population (1.4 mm). The mean deviation of the columella in women of the general population (4.4 degrees) and in those of the A subgroup (4.5 degrees) was also found to be identical. Deviation of the nasal bridge in the general population was moderately greater (mean 3.4 degrees) than in attractive women (mean 2.8 degrees).

Lips and Mouth

Asymmetry in paired measurements involving the halves of the labial fissure (ch-sto), the lateral heights of the upper lip (sbal-ls'), and the measurements defining the position of the mouth in the face (ch-t and ch-t surf) was mild both in the general population (mean: 2 mm and 2.9 mm) and in the A subgroup (mean: 1 mm and 2.7 mm).

Ears

The mean values of asymmetries in the width (pra-pa) and length (sa-sba) of the ears were minimal and almost identical in the general population (1.7 mm and 1.9 mm) and in the attractive women (1.7 mm and 2.2 mm). The asymmetries increased in linear measurements determining the position of the ears in the face, horizontally (obi-sn, obi-sn surf) and vertically (obi-gn, obi-gn surf). For the attractive women, the extent of asymmetry between the ear (obi) and the subnasalae (sn) point was highly significantly smaller, both projectively (mean: 2.2 mm) and tangentially (mean: 2.4 mm), than in women of the general population (means: 3.4 mm and 3.7 mm) ($p = 0.001$). In the asymmetrical position of the ears in relation to the chin point (gnathion), the findings were moderately larger in the general population (means: 2.9 mm and 3.9 mm) than in the A subgroup (means: 2.4 mm and 3.1 mm).

The degree of asymmetry between the protrusion of the ears was mildly significantly larger ($p = 0.05$) in women of the general population (mean: 6 degrees) than in the attractively faced women (mean: 4.7 degrees). The mean difference in inclination of the ears in attractive women (5.5 degrees) proved to be moderately significantly greater ($p = 0.01$) than in women of the general population (4.1 degrees).

Asymmetries in the Above-Average (A Subgroup) and Below Average (B Subgroup) Facial Qualities in Women

The *extent* of facial asymmetry in women of both subgroups, determined by the mean value of differences between paired measurements, was mostly mild, being significant only in two of the 36 paired measurements. The deviation of the nasal bridge was highly significantly worse in the B subgroup (mean: 4.3 degrees) than in the attractive women (mean: 2.8 degrees) ($p = 0.001$). The degree of asymmetry in inclination of the ears was highly significantly larger in the A subgroup (mean: 5.5 degrees) than in the B subgroup (mean: 2 degrees) ($p = 0.001$).

Analysis of the *frequency* of asymmetries between the paired linear or angular measurements of the face did not show significant differences between above-average and below-average facial qualities. The face's paired linear depth measurements—that is, measurements determining the position of the orbits in the face, the position of the mouth in the face, the halves of the labial fissure, and the lateral upper lip height—were less asymmetrical in women of the A subgroup than in those of the B subgroup. The frequency of asymmetries noted as affecting the nose was mixed. Nasal floor width asymmetries were more frequently found, but those between the lengths of the columella were less numerous in the A subgroup than in the B subgroup. In the ears, the asymmetry in the width and length was seen more often in the A subgroup than in the B subgroup. The findings in asymmetrical positions of the ears relative to the subnasale point and chin landmark were mixed.

Analysis of the asymmetries in paired angular measurements revealed mixed results. In the orbits, asymmetry in inclination of the eye fissures (mean: 1 degree) was seen only in the attractive women (5.9%, 2 of 34). Asymmetrical inclinations between the upper and lower orbital rims were more frequent in women of the A subgroup (6.1%, 2 of 33) than in those of the B subgroup (4.8%, 1 of 21). The bridge of the nose was less frequently deviated in the A subgroup (11.8%, 4 of 34) than in the B subgroup (14.3%, 3 of 21). In contrast, the columella deviation was more frequent amongst the attractive women (11.8%, 4 of 34) than in the B subgroup (4.8%, 1 of 21). Asymmetry between the protrusion of the ears was seen in fewer instances in the A subgroup (29.4%, 10 of 34) than in the B subgroup (33.3%, 7 of 21). Asymmetry in inclination of the ears was a rare finding, occurring less often in the attractive women (5.9%, 2 of 34) than in the B subgroup (7.1%, 1 of 14).

Symmetry between the basic paired measurements of the soft-tissue orbits (eye fissure length and height) and nose (projective and tangential length measurements of alae, depth of the nasal root) was preserved in both the attractive and below-average categories of women.

DISCUSSION AND CONCLUSIONS

Detailed analysis of the physiognomy—by application of a large number of linear, tangential, and angular measurements and numerous craniofacial proportion indices, supplemented by data recording asymmetries in various regions of the face—did not help us to determine a simple and definitive difference, in either sex, between a healthy above-average (attractive) face and a below-average face. The complexity of facial morphology, be it in an attractive or unattractive but healthy face, is more sophisticated than a sum of quantitative data. Four hundred years ago, Albrecht Dürer (1471–1528) (7), who employed different canons and described in detail his ideas on proportions, admitted he could not define absolute beauty (8,9). Indeed, establishment of rigid measurements and proportions, as characteristics of the ideal face, is unnatural because it negates the variations essential to any creation of nature (6). Beauty may be a "personal affair" (10) but not the attractiveness of the face. Attractiveness can be expressed quantitatively (6), foreseen by the Italian orthodontist, Edmond Muzj, more than half a century ago (11).

The present study offers the results of a systematic search for differences between two differing facial qualities. The findings are not formulated to meet the demands of aesthetic surgeons looking for simple formulae for creation of an attractive face.

In spite of the detailed statistical analysis, not all morphologically weightly differences might have been discovered. More than half (54%, 143 of 265) of the measurements, from both sexes, did not show significant differences between the above-average and below-average facial qualities. The large number of identical measurements found in more than one-quarter (27.5%, 73 of 265) of the two facial quality subgroups, in both sexes, was surprising. Almost in two-thirds (64.6%, 201 of 311) of proportion indices, from both sexes, the differences between the two facial qualities were nonsignificant, and in less than one-fifth (18%, 56 of 311) the proportion indices were identical. These findings forced us to pay more attention to the various degrees of the significantly differing measurements and proportion indices in the study group.

Data significantly differing in the two facial qualities

were recorded in less than one-fifth of the total number of measurements (18.5%, 49 of 265) and proportions (16.7%, 52 of 311). Analysis showed that in more than two-thirds (69.4%, 34 of 49) of the significantly differing measurements and in more than half (55.8%, 29 of 52) of the index values, the degrees of significance were mild to moderate with questionable effect on visual appearances. Consequently, determination of facial regions with the most decisive effect on the visual appearance was based on the presence of the highest number of measurements and proportion indices, exhibiting highly significant differences between the facial qualities. The percentages of the highly significantly differing signs were very small: 5.7% of all measurements (15 of 265) and 7.4% of all proportion indices (23 of 311) calculated from both sexes.

In the attractive men the nasal region revealed the highest percentage of highly significantly differing values, both in measurements and in proportions. The nasal root was deep, the columella was narrower, and the alae were thinner. The nose was relatively short in relation to the craniofacial, facial, and upper facial heights. The nasal root was deep in relation to its width and to the nasal tip protrusion, and it was wider in relation to the nose width than in men in the B subgroup (Tables 13–12 and 13–13). In the labio-oral region of attractive men, no highly significantly differing measurements were found. In the orbits neither the measurements nor the proportion indices were revealed with highly significant differences, in comparison to the men in the B subgroup. In contrast, in the attractive women, it was in the orbital and labio-oral regions where the highest number of measurements and/or proportion indices were recorded, highly significantly differing from the women in the B subgroup (Tables 13–27 and 13–29). The higher eye fissures and the higher tilt made the eyes of attractive women more special. The larger eye fissure height in relation to its length gave the eyes a startling size. In addition, in the attractive women, the total width of the orbits (ex-ex), occupying most of the space between the zygomas (zy-zy), dominated in the upper face. In the lips of the women in the A subgroup the upper vermilion arc was larger than that in the B subgroup. The upper vermilion arc was larger in relation to the lower vermilion arc, which indicated a more protruding upper lip in the attractive women. In addition, the cutaneous portion of the upper lip in the A subgroup was larger than that of the lower lip.

Neoclassical canons, found in a small number in the two subgroups, did not reveal significant differences in the frequency of their occurrence either between the sexes or between the two facial qualities (Tables 13–3 and 13–24). Because "all canons are more or less idealizations, they do not purport to represent actual human beings"; instead, their value is artistic (12).

Analysis of asymmetries did not contribute greatly to the statistical determination of signs of the attractive face as expected by some authors (13). The degree of nasal bridge deviations in the below-average subgroup of men (4.6 degrees) and women (4.3 degrees), which was highly significantly greater ($p = 0.001$) than in attractive men (2 degrees) or women (2.8 degrees), was the only finding which may influence the visual judgment.

For all regions of the craniofacial complex, 62 measurements (42 linear and 20 angular) and 44 proportion indices were selected from our norms of attractive men and women which may influence the results of visual judgment (Tables 13–4, 13–8, 13–9, 13–11, 13–14). The data can serve only as a guide, not as rigid measures of attractiveness. The findings in the attractive subgroups might have been influenced by the inclusion of large groups of fashion models with highly significantly larger ($p = 0.001$) body height (in men 184 cm, in women 171.8 cm) than that of the below-average subgroup (in men 176.9 cm, in women 161.3 cm).

In spite of the limited number of signs significantly differing between the two facial qualities, we still believe in the value of the anthropometric approach in the quantitative determination of characteristics of the attractive face. It seems that the key to a more exact description of attractiveness is hidden in the large number of proportion indices, not differing significantly in the two facial qualities. In clinical practice, disharmonies, although belonging to the borderline values of the normal range statistically, may create visually the impression of moderate disproportions (14–16), which were categorized as abnormal relationship between the measurements. We assume that in the complex system of craniofacial proportions, the disharmonies in addition to disproportions will play a decisive role for the objective determination of attractiveness.

REFERENCES

1. Farkas LG, Munro IR, Kolar JC. Linear proportions in above- and below-average women's faces. In: Farkas LG, Munro IR, eds. *Anthropometric facial proportions in medicine.* Springfield, IL: Charles C Thomas, 1987;119–129.
2. Guilford JP. *Psychometric methods,* 2nd ed. New York: McGraw-Hill, 1954.
3. Farkas LG, Munro IR. *Anthropometric facial proportions in medicine.* Springfield, IL: Charles C Thomas, 1987.
4. Farkas LG, Hreczko TA, Kolar JC, Munro IR. Vertical and horizontal proportions of the face in young adult North American Caucasians: revision of neoclassical canons. *Plast Reconstr Surg* 1985;75:328–337.
5. Farkas LG. *Anthropometry of head and face in medicine.* New York: Elsevier, 1981.
6. Farkas LG, Kolar JC. Anthropometrics and art in the aesthetics of women's faces. *Clin Plast Surg* 1987;14:599–616.
7. Dürer A. *Vier Bücher von menschlichen Proportion.* Nürenburg: H Formschneider, 1528.
8. Dürer A. *Della simmetria dei corpi humani.* Venetia: Presso D Nicolini, 1591.

9. Wölfflin H. *Die Kunst Albrecht Dürers,* 5th ed. Münich: F Bruchmann, 1926.

10. Renooy L. In the eye of the beholder. *About Face* 1992;6:4.

11. Muzj E. The method for orthodontic diagnosis based upon the principles of morphologic and therapeutic relativity. *Angle Orthod* 1939;9:123–141.

12. Cobb WM. The artistic canons in the teaching of anatomy. *J Natl Med Assoc* 1944;36:3–14.

13. Powell N, Humphreys B, eds. *Proportions of the aesthetic face.* New York: Thieme–Stratton, 1984.

14. Farkas LG. The proportion index. In: Farkas LG, Munro IR, eds. *Anthropometric facial proportions in medicine.* Springfield, IL: Charles C Thomas, 1987;5–8.

15. Farkas LG, Posnick JC, Hreczko T. Anthropometry of the head and face in 95 Down syndrome patients. In: Epstein CHJ, ed. *The morphogenesis in Down syndrome.* New York: Wiley–Liss, 1991;53–97.

16. Farkas LG. Basic anthropometric measurements and proportions in various regions of the craniofacial complex. In: Brodsky L, Holt L, Ritter-Schmidt DH, eds. *Craniofacial anomalies. An interdisciplinary approach.* St. Louis: Mosby Year Book, 1992;41–57.

Chapter 14

The Use of Anthropometry in Forensic Identification

Leslie G. Farkas (Part I)

Bette Clark (Part II)

J. David Sills (Part III)

In the last decade the number of missing children has been gradually increasing. Police authorities and social services throughout the world are asked for assistance in finding them, sometimes years after their disappearance, which may have occurred in early childhood. Often the only description available is one obtained from an amateur snapshot, which can be several years old. In other circumstances, it may be necessary to establish identity by reconstructing appearance from the skull of a missing person, a complicated and difficult task. In 1986, the Forensic Identification Services of the Metropolitan Toronto Police introduced the Computer Assisted Recovery Enhancement System (C.A.R.E.S.), using a PC/82 specially constructed computer (1). This greatly improved the techniques of identification by introducing new methods of analysis and developing new approaches in image reconstruction both from skulls and from early childhood photographs.

Parts II and III of this chapter have been contributed by experienced specialists in forensic identification who have been successful in reconstructing 10 skulls leading to recognition and identification and the safe recovery of seven missing children following long-term absence.

In the three parts of this chapter, the use of anthropometry will be demonstrated as an aid in facial restoration leading to recognition. In the first part, the usefulness of the reliable measurements obtained from photogrammetry and completed with anthropometric population norms will be shown to provide objective parameters for facial reconstruction. In the second part, a method is presented for "aging" the face of a missing child to render outdated photographs in a more current form. The third part examines the complexity of methods used in reconstructing the soft-tissue contours on the skull, thereby assisting in the identification of a missing person.

Chapter 14
Part I

Anthropometric Guidelines in Photographic Reconstruction of a Missing Child's Face

Leslie G. Farkas

Detailed evaluation of the reliability of the surface measurements obtained from standard life-size photographs revealed the advantages and disadvantages of photogrammetry (2). The key elements in "aging" a photograph depend on those measurements which were proved to be reliable by our study and which possess clearly identifiable landmarks on the print.

Materials. Photograph (any) from the front and from the side; age and sex of the child/adult; national origin; race; photographs of parents, siblings, and relatives.

Methods. Do the following:

1. Measure the actual distance (in millimeters) between the orbits (endocanthion–endocanthion, en-en), from the frontal print.
2. Measure the actual height of the upper lip (sn-sto) between the midpoint at the base of the columella (subnasale, sn) (at the junction of the nose and the upper lip) and the midpoint (stomion, sto) between the closed lips, in millimeters.

PRODUCING THE LIFE-SIZE PHOTOGRAPH AT THE TIME OF DISAPPEARANCE

1. Population norms of the face: Find the en-en distance of the missing person at the age of disappearance. Do the same with the height of the upper lip. Use the mean and the standard deviation (SD) methods to produce the entire range of normal dimensions as appropriate.
2. Calculate the differences between the normal values of these two measurements and those taken from the pictures of the missing person (the actual measurements obtained from the original photographs).
3. The en-en and the sn-sto distances obtained from the snapshots, increased by the appropriate differences calculated from the norms, are the *numbers* which will help to enlarge the snapshot to a life-size picture of the missing person. The degree of enlargement of the snapshot will be given by the *new distance* of the en-en (for the frontal print) and the upper lip height (sn-sto) for the lateral print.
4. Measurements taken from life-size pictures:
   A. Those proven to be reliable (2) (see also Chapter 5):
      I. *Linear measurements, frontal view.* Upper face height (n-sto), eye fissure heights (ps-pi rt,lt), height of the orbit plus the eyebrow (or-sci rt,lt), height of the nose (n-sn), lateral heights of the upper lip (sbal-ls′ rt,lt), upper lip height (sn-sto), height of the lower vermilion (sto-li).
      II. *Linear measurements, lateral view.* Height of the eye fissure (ps-pi), height of the orbit plus eyebrow (or-sci), length of the columella (sn-c′), upper lip height (sn-sto), lower lip vermilion height (sto-li).
      III. *Angular measurements, lateral view.* Inclinations of the forehead, upper and lower face, nasal bridge, and upper lip. Angles: nasofrontal, nasolabial, mentocervical and nasal tip.
   B. Measurements which slightly vary without affecting appearance:
      I. *Linear measurements, frontal view.* Face width (zy-zy), biocular widths (ex-ex), nose

width (al-al), length of the nasal bridge (n-prn), and mouth width (ch-ch).

II. *Linear measurement, lateral view.* Face height (n-gn).

C. Measurements which are unsuitable:

I. *All head measurements.* The mandible width (go-go), the length of the body of the mandible (go-gn), and all depth measurements of the face (t-g, t-n, t-sn, t-gn).

ANALYSIS OF THE MEASUREMENTS TAKEN FROM ENLARGED PHOTOGRAPHS OF THE MISSING PERSON (DATED AT DISAPPEARANCE)

These measurements, both angular and linear, will relate to age at the time of disappearance and will be determined within the normal range, or relative to it (3). Proportion indices will be calculated to show whether the relationship between the individual measurements was harmonious, disharmonious, or disproportionate at age of the disappearance.

LIFE-SIZE PICTURE ADAPTED TO THE PRESENT TIME

This is done by an artist with help taken from anthropometric norms matching the current age of the missing person.

1. The general framework of the face is given by width and length. The upper face frame is determined by width of the face (zy-zy) and the height of the upper face (n-sto). The mandibular frame is defined by the width (go-go) and height of the mandible (sto-gn).
2. Features within the general framework of the face—namely, orbits (measurements: en-en, ex-ex, inclination of the eye fissures), nose (height and width of the nose, length of the nasal protrusion, length and inclination of the nasal bridge,), upper lip (height of the upper lip, inclination of the upper lip), and mouth (width of the mouth)—are reconstructed with the help of the appropriate measurements.
3. For recreation of the profile line, the following measurements are required: (a) the appropriate norms of the forehead, upper face, and lower face inclinations and (b) the nasofrontal, nasolabial, labiomental, and mentocervical angles.
4. The proportion indices observed on the enlarged photograph at the time of disappearance can help to reproduce similar proportion qualities.
5. Some characteristics seen on the relatives' photographs (i.e., size and shape of the nose, balding, glasses, etc.) may also be helpful.

These data alone would not be enough to redraw the new face. The creation of a "new" face appropriate to its current age group with the aid of the assembled pieces of information requires the special skills of the portrait artist.

NOTES ABOUT THE PROPORTIONS AND GROWTH

The proportion index values determined on the enlarged original photograph of the missing person can help to select the measurements for a similar relationship quality (4). Here the artist's aesthetic judgment will play a great role. If the face was harmonious at the earlier date, it is expected that it would be harmonious now. In case of disharmony, that quality could increase or decrease with age, but may not disappear completely. Thus, the disproportion must show up in the reconstructed face, together with any birth defects, scars, port-wine stains, or malformed bones.

A knowledge of the age-related changes in the head and face between birth and young adulthood can be of great help when the changes in the size of the head and facial frameworks are calculated. Data recently made available concerning age-related changes in these measurements (5–10) will help to establish the age groups where such changes are hardly discernible or are nonexistent (see also Chapter 6). Those other age periods where the changes in measurements may produce visible alterations in the quality of physiognomy can also be identified. The age of the missing person at the time of reconstruction of the face would indicate which changes are likely and where they can be expected.

CONCLUSIONS

1. The main contribution of anthropometry is in objective conversion of a snapshot picture into a life-size photograph.
2. The life-size photograph helps to obtain objective data from the face at the time of the disappearance of the person.
3. Data obtained will be increased according to the norms and for the age of the missing person adjusted to a current date.
4. Proportion indices will show the main proportion qualities in the face: the harmonies and disproportions, which would influence the selection of the new measurements for the face.
5. The age-related changes determined in the general population will assist in the selection of the new measurements.
6. The quantitative data, although important, would not be sufficient to produce a face with all details which are at least as important as the "frameworks." This difficult task requires the skills of the artist.

Chapter 14
Part II

Method Used in "Aging" the Photographic Records of Missing Children

Bette Clark

We have found that a series of tracings of the head and face is useful when attempting to enlarge or "age" the photographs of children. Each of these tracings should show the vital landmarks that will be proportionately altered with each progressive tracing. The alterations in each tracing should show approximately a 2- to 3-year difference in growth. Small age increments are used because areas of growth vary at different times (5–10).

MATERIAL AND METHODS

From the original photograph of a 5-year old Caucasian girl (Fig. 14–1), two measurements were taken as guiding references for enlargement to life size of the face: horizontally the intercanthal distance (en-en) and vertically the upper face height (n-sto). The measurements and proportions are carefully analyzed on the life-size picture of the subject. Measurements are taken from the regions in which they proved to be reliable (see Chapter 5). The next task would be the comparison of these quantitative findings with valid population norms for 5-year-old girls. It will show the "position" of the measurements within the normal range (e.g., close to the mean, close to the maximum normal or minimum normal, or out of the normal range). Based on these data the framework of the face and some of its regions for 8-year-old girls can be reconstructed using the norms for that age, with similar position of the measurements within the normal range at 5 years of age. For vertical heights of the face, a set of seven measurements was used (Fig. 14–2). The selection of the appropriate norms for planning the horizontal dimensions of the face (zy-zy), nose (al-al), and mouth (ch-ch) requires special evaluation and judgment by the artist. The intercanthal width serves as a guide (Fig. 14–3).

Using the appropriate North American population norms (3), a first "aging" process of the original photograph the face of an 8-year old girl was created [Fig. 14–4(1)]. Figure 14–4(2) displays the result of the final "aging," showing the face at 18 years of age. Remember to keep the child's face proportionate in width as well as in length. These measurements must be harmonious and continue to capture the character of the original photograph with consideration given to physical characteristics of look-alike family members.

When the tracing of the face meets the preconceived measurement (age) that is desired, then the final drawings may be done. Ancillary matters such as hairstyle, clothing, and/or eyeglasses are a personal choice and should reflect the trends of the age of the person in the drawing.

DISCUSSION AND CONCLUSIONS

Experience shows that most children's photographs are not ideal for use in the procedure of aging. Most photographs show the child smiling and/or in an awkward position. Practice with the studio-posed images of children, if available, will give the artist useful experience in dealing with the average snapshot photos that are most common.

Success is almost ensured when the artist remembers to:

1. Retain the proportions of the face in harmony with the real child.
2. Preserve the facial character of the child, without allowing the look-alike relatives to influence the final drawing to the extent that the drawing becomes one of them.

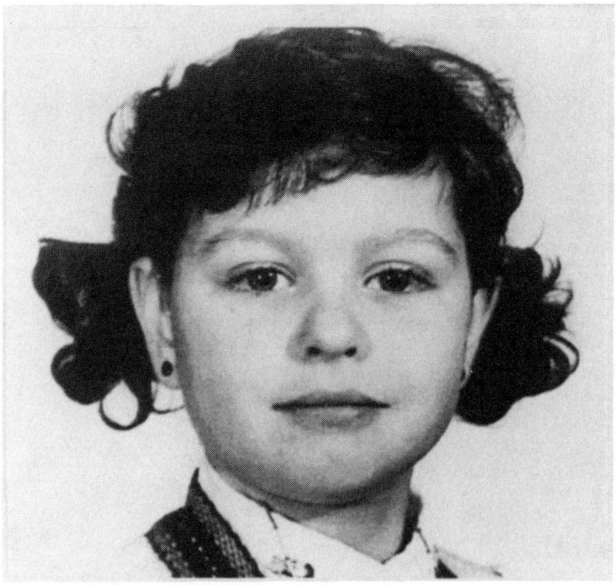

FIG. 14-1. Face of a 5-year-old Caucasian girl.

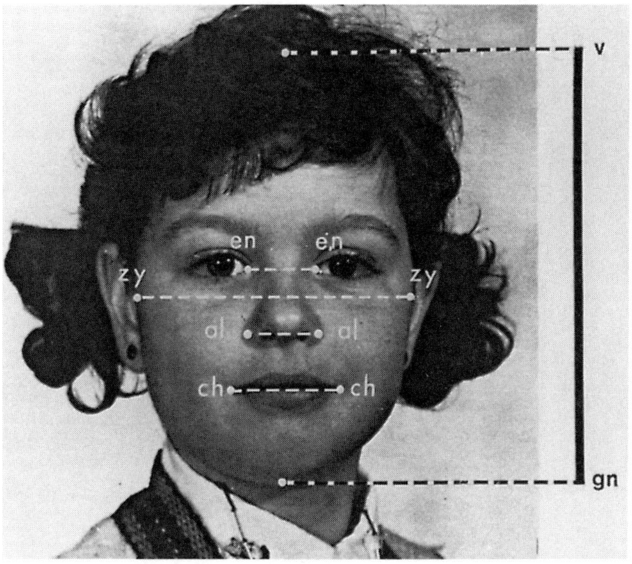

FIG. 14-3. Original photograph of the 5-year-old girl. This picture was used to determine the set of four horizontal frontal measurements in the face. The intercanthal width (en-en) is the key measurement, reliable in photogrammetry (mean en-en at 5 years: 29.4 mm). Measurements which must be evaluated and judged by the artist are as follows: the width of the face (zy-zy), the width of the soft nose (al-al), and the width of the mouth (ch-ch). The combined head and face height (v-gn) is one of the vertical measurements which must be judged by the artist.

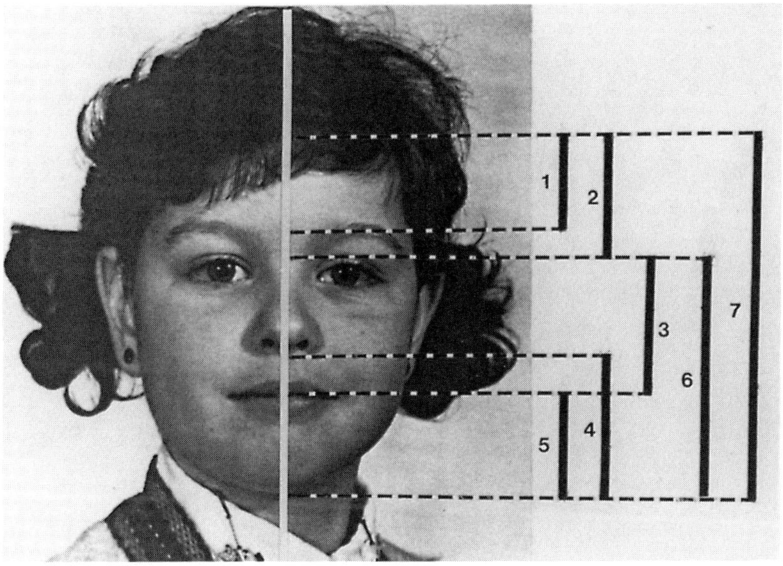

FIG. 14-2. The set of seven vertical profile measurements to recreate the life-size face at 5 years of age. **1:** Height of the forehead I (tr-g). **2:** Height of the forehead II (tr-n). **3:** Height of the upper face (n-sto). The key measurement, reliable in photogrammetry. **4:** Height of the lower face (sn-gn). **5:** Height of the mandible (sto-gn). **6:** Height of the face (n-gn). **7:** Height of the forehead and face (tr-gn).

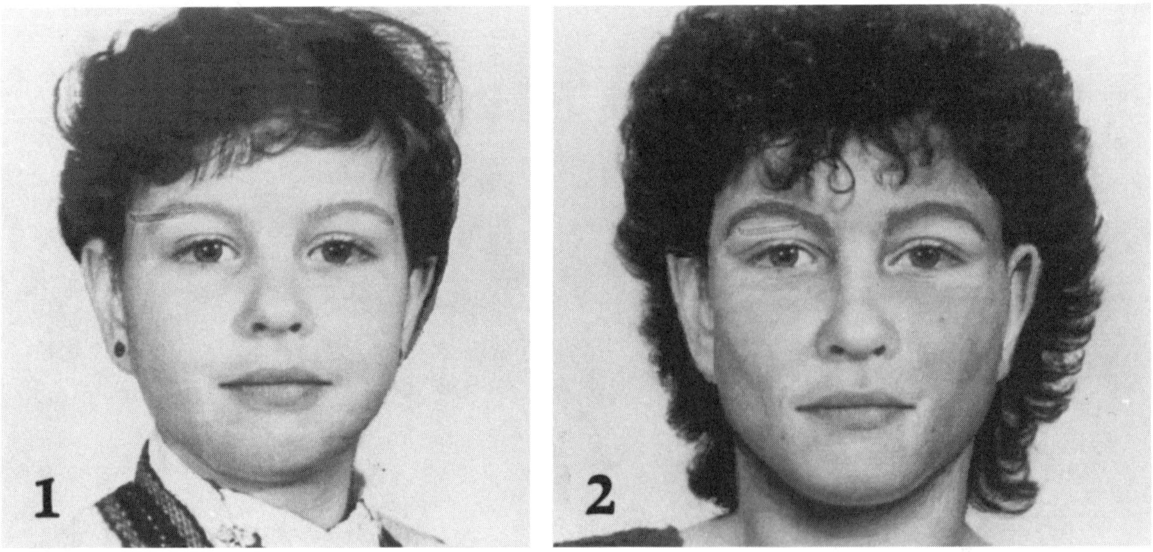

FIG. 14–4. 1: First "aging" using mean value measurements for 8-year-old girls. **2:** Final "aging" using the mean value measurements for 18-year-old girls. The mean of the combined forehead and face height (tr-gn) was 151.9 mm at 5 years of age, 159.3 mm at 8 years of age, and 172.5 mm at 18 years of age.

3. Avoid perpetuating any facial expression which distorts appearance. (If the mouth is open in the original photograph, it is best to close it. Teeth are an "anomaly" that could complicate the aging process.)
4. Maintain and preserve drawing skill (even taking a refresher course in anatomy if necessary) in order to produce a realistic face. Photographic reconstruction of the skull—in particular—will rely heavily on anatomical knowledge for successful results.

Chapter 14
Part III

Computer Photographic Skull Reconstruction (Methods Used in Facial Restoration)

J. David Sills

The photographic reconstruction of the skull of an unknown victim depends on a number of factors. Access to the pathologist's report is essential, while the findings of routine police investigations can be of considerable assistance (11).

The forensic artist's role is to weigh the importance of the collected data needed for performing his (her) task with the best possible result. His (her) artistic skills are enhanced by anatomical knowledge, awareness of the surface measurements of the head and face, and an understanding of the nature of the developmental changes affecting the face. The computer-created face is the result of the sum of all these factors.

MATERIALS

Photographs of the skull from the front, three-quarter right side, full right side, bottom looking up at the palate, and the mandible looking down (Fig. 14–5), all at right angles, with a ruler beside the skull but not obstructing the skull. Photographs should be life size and can be color or black and white. The pathologist's report will record age, sex, and racial origin, together with any supplementary information deemed helpful in facial reconstruction (e.g., hair color/length; body height/weight, etc). Scenes of crime photos and description of clothing and sizes if available will help to establish the height and weight of the victim. Additional help may be provided by mug-shot photographs of criminals from police files. Where the skull structure is found to be incomplete, initial procedure will require the bony framework to be corrected.

METHODS

1. Overlay paper is placed over the photographs of the skull for outlining the contours (Fig. 14–6). Landmarks are marked on the overlay to establish the size and shape of the features that will appear on the surface of the reconstructed face (e.g., eyes, mouth, ears, hairline) (3).
2. The following landmarks are marked on the overlay paper after the skull is outlined:
   - tr is the point on the hairline in the middle of the forehead.
   - g is the most prominent midline point between the eyebrows.
   - n is the point in the midline of the nasofrontal suture, on the nasal root.
   - sn is the midpoint of the columella base where the lower border of the nasal septum and the surface of the upper lip meet.
   - gn is the lowest landmark of the face, in the middle on the lower border of the mandible.
   - en is the point of the inner commissure of the eye fissure.
   - ex is the point of the outer commissure of the eye fissure.
   - zy is the most lateral point of each zygomatic arch.
   - sa is the highest point of the free margin of the auricle (top of the ear).
   - sba is the lowest point on the free margin of the ear lobe.
   - pra is the most anterior point of the ear, just in front of the helix attachment to the head.

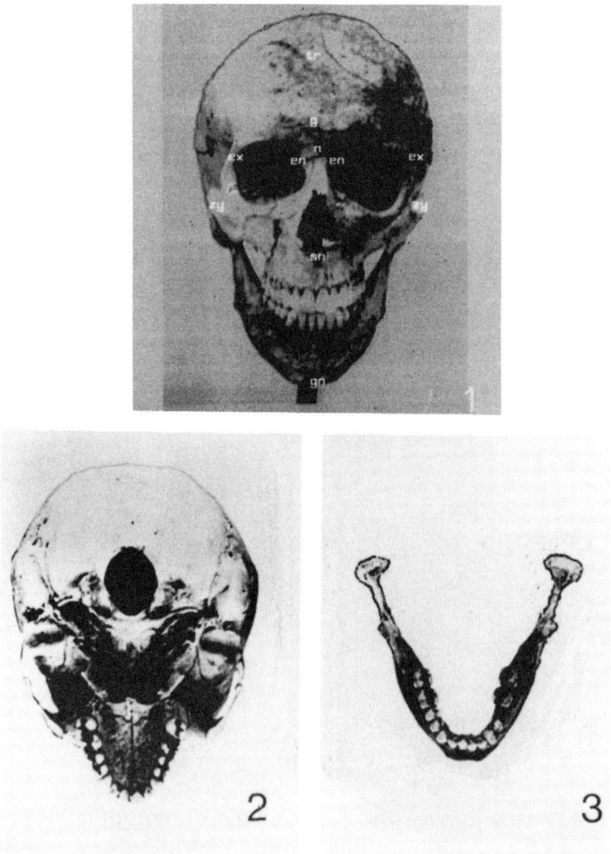

FIG. 14–5. Photographs of the skull from the front in Frankfurt horizontal (**1**), from beneath (**2**), and the mandible (**3**). For reconstruction of the face the photograph of the skull must be life-sized.

po is landmark of the ear canal opening.

FH is the Frankfurt horizontal line, connecting the bottom of the orbit (or) and the ear canal opening (po) or the tragion landmark of tragus (t) on the same side.

3. Once the landmarks have been established on the overlay paper (views: front; on the right lateral side; and on 45-degree semiprofile view of the skull), another sheet of overlay paper is placed over it; then the features are sketched in, leaving space for the soft tissues (e.g., muscles, fat) (Fig. 14–7). The size of the features such as the eyes are obtained from the data bank (3), using the average measurements for the sex and age determined by the pathologist. (Here the skull belonged to a young adult male.) An overlay is also done on the mandible and palate views to determine any aberration from the normal which may have a bearing on the facial reconstruction.
   - A. *The eyes.* The size of the eye can be determined by locating the endocanthion (en) 6–7 mm from the medial wall of the bony orbit. The position of the exocanthion (ex) is about 1 mm from the outer edge of orbit (Fig. 14–8). By using these measurements to locate the en and ex, space is left for the soft tissues protecting the eyeball in the bony orbit. The eye fissure length is defined by the distance between the new en and ex.
   - B. *The nose.* The size and shape of the nose is obtained by "triangulating" the area of the future nose by drawing a straight line from the bony bridge (nasal bones) guided by the inclination of its surface, in distal direction. A second line is drawn starting from a small bony protrusion indicating the subnasale (sn) point of the soft nose. This line defines the direction of the future columella of the nose. The columella inclination would be determined by the position of the surface of the upper jaw (alveolar process) in relation to the vertical. Where the lines drawn from the nasal bridge and from the subnasale point meet would be the nasal tip. The nasal tip angle ranges between 56 and 87 degrees in young adult Caucasian males. The width of the soft nose is found by allowing for an average amount of flesh on each side of the piriform aperture of the skull.
   - C. *The mouth.* Vertical lines are drawn from the middle of the right and left pupil, already located in the orbits, down to the area of the teeth establish the maximum width of the mouth. The ap-

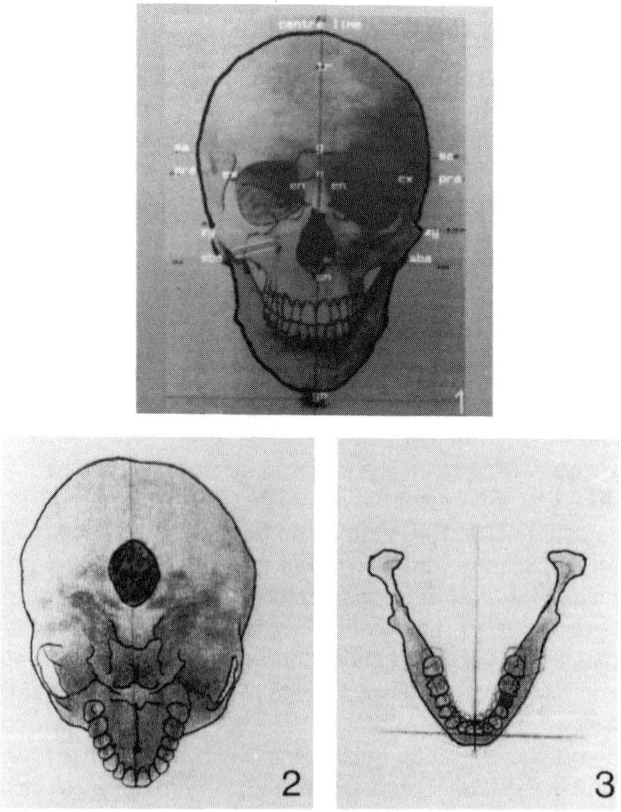

FIG. 14–6. The skulls with overlay. The new contours are outlined (**1–3**). Landmarks on the frontal view of the skull are also shown (**1**).

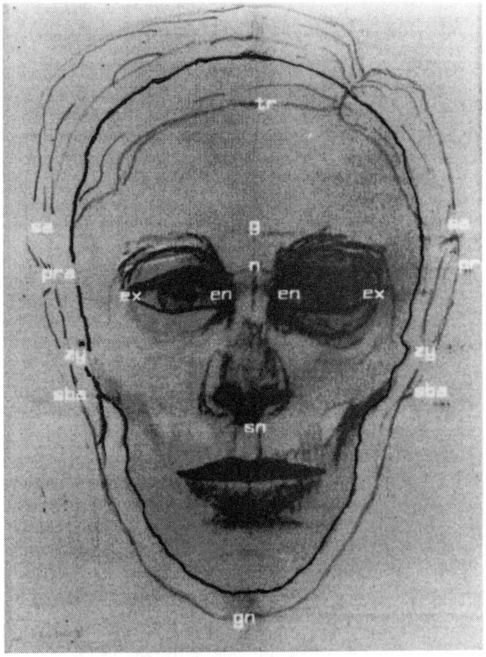

FIG. 14–7. Skull with soft-tissue features in place and landmarks.

propriate mouth size can be obtained by adjusting the width of the mouth to the center of the eyes. The population norms in young adults will help to find the required height for the upper lip.

D. *The ears.* The size of the ears can be determined by drawing a parallel line above the Frankfurt horizontal, along the line touching the upper edge of the eyebrow, and a second line, distally from the Frankfurt horizontal, through the base of the columella (the subnasale point) of the nose. The two horizontals indicate the top and bottom of the ear. For calculating the anterior border of the width of the ears a vertical line to the Frankfurt horizontal is drawn about 15 mm medially from the ear canal opening (po). The posterior border of the ear will be determined by the mean width measurement. By these four lines a rectangular box is formed in which the ear can be drawn.

E. *The hair.* To put hair on the skull, the position of the hairline is first determined, guided by an appropriate forehead height and the age of the victim. A suitable hair style from the period in which the deceased died would be used. In the absence of some evidence of hair left at the scene or attached to the skull to show the hair color, length, style, and so on, it becomes difficult to determine a style close to that which the deceased may have had. Facial hair on males also creates a problem. Again, unless there is some evidence of this at the scene or on the skull, it becomes guess work as to whether it should be added. A two-version alternative with and without facial hair may be considered.

4. Once the features have been established as described above, the photographs of the skull and the overlays are entered into the C.A.R.E.S. computer through a video camera.
5. A photographic facial image can now be created on the computer (Fig. 14–9). To create this image the photographs in the police files are searched in an effort to find the features and a head shape that is close to the drawn one and which can be adjusted without too much difficulty to fit over the skull. Once the files have been searched and the features found, each photograph is then entered into the C.A.R.E.S. computer. The photograph with the best overall head shape is then placed over the skull on the computer screen and adjusted to make sure it fits correctly by using overlay comparisons. When the base head is in place, the eyes, nose, mouth, and so on, are cut from the other photographs that had been entered into the computer and are placed individually onto the base head and adjusted. After each feature is placed, it is checked by using an overlay of the facial reconstruction over the skull to make sure each feature is in correct location. Should further adjustments be required of each feature, an overlay comparison is made after each adjustment. When the features are in position, any lines that may appear around the features that have been added are blended away on the screen by use of a stylus. The photographic image is then printed onto paper to be distributed to the requesting police agency. The same method can be used in the reconstruction of childrens' faces by using the appropriate norms

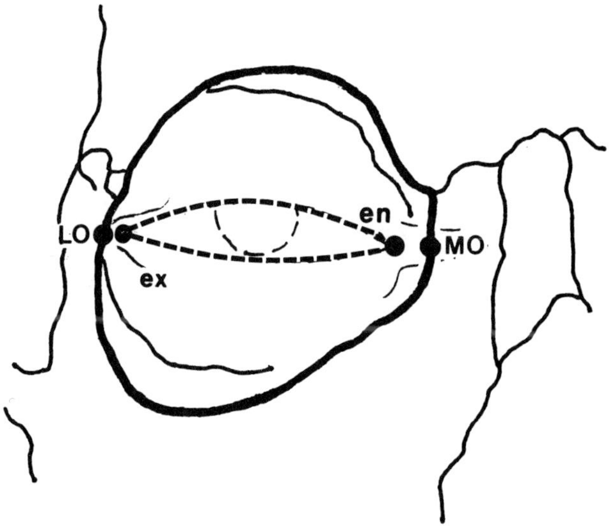

FIG. 14–8. Schematic drawing of the bony orbit. Tentative design of the eye fissure (*dotted lines*). MO, bony landmark of the medial orbital wall; en, endocanthion (medial commissure of the eye fissure); LO, bony landmark of the lateral orbital wall; ex, outer commissure of the eye fissure.

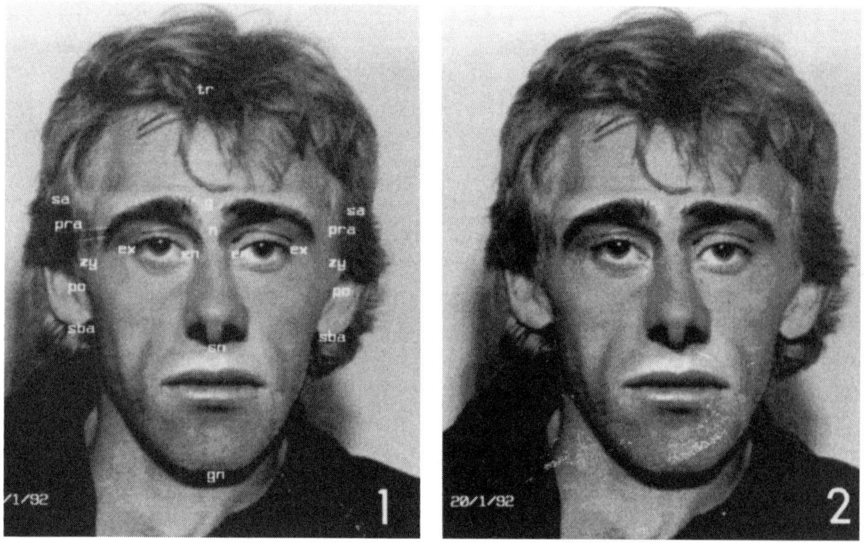

FIG. 14-9. 1: The reconstructed face with the landmarks. **2:** The fully reconstructed face.

matching the age group of the skull. By using this method of face reconstruction, there is a great deal of flexibility in adding weight and facial hair to the composite or doing two or three versions of the same face.

REFERENCES FOR PARTS I–III

1. Clark B. Computer assisted recovery enhancement system (CARES). *Gazette, Royal Canadian Mounted Police* 1989;51:20–21.
2. Farkas LG, Bryson W, Klotz J. Is photogrammetry of the face reliable? *Plast Reconstr Surg* 1980;66:346–355.
3. Farkas LG. *Anthropometry of the head and face in medicine.* New York: Elsevier, 1981.
4. Farkas LG, Munro IR, eds. *Anthropometric facial proportions in medicine.* Springfield IL: Charles C Thomas, 1987.
5. Farkas LG, Posnick JC. Growth and development of regional units in the head and face based on anthropometric measurements. *Cleft Palate Craniofac J* 1992;29:301–329.
6. Farkas LG, Posnick JC, Hreczko TM. Anthropometric growth study of the head. *Cleft Palate Craniofac J* 1992;29:303–308.
7. Farkas LG, Posnick JC, Hreczko TM. Growth patterns of the face: a morphometric study. *Cleft Palate Craniofac J* 1992;29:308–315.
8. Farkas LG, Posnick JC, Hreczko TM, Pron GE. Growth patterns in the orbital region: a morphometric study. *Cleft Palate Craniofac J* 1992;29:315–318.
9. Farkas LG, Posnick JC, Hreczko TM, Pron GE. Growth patterns of the nasolabial region: a morphometric study. *Cleft Palate Craniofac J* 1992;29:318–324.
10. Farkas LG, Posnick JC, Hreczko TM. Anthropometric growth study of the ear. *Cleft Palate Craniofac J* 1992;29:324–329.
11. Sills JD. Skull reconstruction identifies 1973 murder victim. *Blueline, Canada's National Law Enforcement Magazine* 1992;4:9.

Chapter 15

Medical Photography in Clinical Practice

Peter S. Reid and Leslie G. Farkas

Photographic documentation of the head and face provides a valuable resource for ongoing qualitative assessment of individuals with facial deformity (1–6). The photograph provides an important visual reference for monitoring growth and developmental changes in pediatric clients. It enables physicians to review health records, which can ultimately be used as an aid for planning surgery. The photograph creates a reliable means for qualitative evaluation of postoperative results, providing the client with a view of the changes. The photographer also provides the surgeon and other health professionals with credible visual material for teaching and research.

The process of producing good-quality clinical photographs can only be achieved by a thorough understanding of the individual components involved. The first component to consider is the technical aspect of photography. This includes the proper use of camera equipment and the techniques involved in film exposure and photographic lighting. The second component, a central focus in this chapter, involves the proper positioning of the patient with respect to the framing of the camera. Discrepancies in these areas diminish the value and effectiveness of the photograph, thereby creating the possibility of a visual misrepresentation of the patient's condition (7).

There are several other important questions that need to be considered as well (8). What degree of accuracy is required in the final image? Are the photographs required for teaching and medical records, and, therefore, needed to look comparable as a sequence? Are the photographs required for research purposes or as an aid for planning surgery where actual measurements from the photographs are required? Photographs produced for the latter context require a higher degree of accuracy and involve precise positioning and measuring of the client. In this case the final image is enlarged to standard size (life-size, 50% of life-size, or 25% of life-size) to enable actual measurements (photogrammetry) and drawings to be made directly from the photograph (9).

There is one further issue that needs consideration at

this point. This has to do with the practical question of availability of resources such as space, time, and trained personnel. In many busy hospital environments a compromise often has to be made between the amount of work involved in taking the photograph and the purpose of the final photograph.

ESTABLISHING A PROTOCOL

At the foundation of all clinical photography are the required elements of clarity, consistency, and technical proficiency. Clarity is simply getting the message across without creating confusion or visual clutter. Consistency is the concept that all photographs are carefully standardized so they are comparable to photographs taken at a later date (6). Technical proficiency is, as mentioned previously, knowledge and application of the technical aspects of photography.

What makes the system workable is the relationship or rapport between the photographer and the individual being photographed. The technical aspects at this point must be second nature; also, the photographer must be able to work confidently, helping the patient feel comfortable during the photo session. In turn the patient will be more cooperative, resulting in photographs of higher quality. This is especially important with younger clients. Children create additional problems because they rarely sit still for very long and are susceptible to mood changes. The photographer must work with a combination of patience and speed—that is, patience to wait until the young subject is sitting exactly in the right position, and speed to capture the moment once it has occurred.

If possible, the studio needs to have a quiet and friendly atmosphere. The subject usually sits on a rotating stool. The stool is adjustable and can be moved into different positions, depending on the required angle of view. Distracting clothing or jewelry are removed so as not to detract from the important elements in the photograph. The hair is pinned back when necessary, to make the forehead with the hairline and the ears visible. In some cases two separate sets of photographs are required, one with the hair pinned back and one without. The patient is asked to focus on an object to control for wandering eyes. Slight changes in the direction of the eyes can be distracting when photographs are taken for comparison purposes.

Judging the Quality of the Face

Before taking pictures, time should be reserved looking for major irregularities in each region of the craniofacial complex of the patient.

1. In the head, they would be the striking changes in the general shape of the skull, in horizontal or vertical direction.
2. In the face, the general contours of the facial frame and the symmetry between the two sides is judged. The quality of the facial midline is checked, whether the axis is a straight line or curved. Especially the position of the chin point in relation to the midaxis of the face should be observed.
3. In the orbits, we are carefully looking for changes in the mutual position of eyebrows and eye fissures (especially the inner commissures, endocanthions).
4. In the nose, attention is paid for any changes in the direction of the nasal bridge from the midfacial vertical axis.
5. In the mouth, we examine whether the labial fissure is located in the central portion of the face, symmetrical in its halves, and horizontal in the direction of its opening.
6. In the ears, symmetry or asymmetry between the levels of the ear lobes is checked.

Attention must be paid also to the rest position of the head of the patient. An abnormal head position can be photographed if required, but should not be accepted as rest position. For establishing the standard position of the head [Frankfort horizontal (FH)], the knowledge of the location of the orbitale point (on the lower rim of the orbit) and of the tragion landmark (on the upper edge of the tragus of the ear) (Fig. 15–1) is required. For establishing the standard position of the head, the line connecting the orbitale and tragion points must be kept in horizontal, with the help of a commercial angle meter

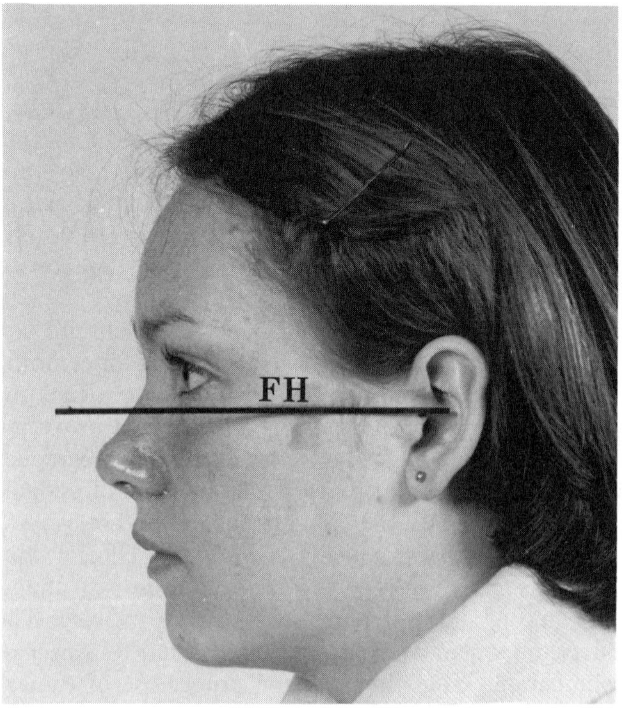

FIG. 15–1. Standard position of the head. In this position (FH) the line connecting the orbitale point and the tragion is kept horizontal.

(see Chapter 2). For determination of the FH, the healthier side of the face is chosen. If the ear canal is missing (atretic), the FH cannot be established.

The FH position is important when photographing patients for measurement purposes (photogrammetry). It also ensures that the client will be positioned in a consistent manner from visit to visit. For less critical documentation the head can be in the more natural "rest position." Many cameras have a grid system that can be fitted into the camera to aid in framing the patient.

FRAMING AND POSITIONING

Great care must be taken in the framing and positioning of the patient during photography. Any slight change in the angle of view can dramatically change the results (10).

Anatomical landmarks are available for the photographer to use for orientation purposes. First, the camera in almost all cases should be positioned at eye level and in a parallel plane to the client's face.

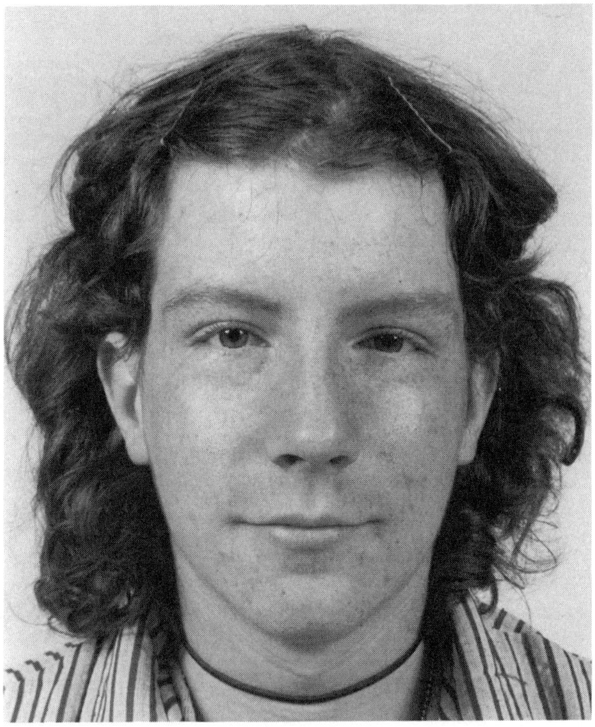

FIG. 15-2. Frontal view of the face. The head is in the rest position.

STANDARD VIEWS OF THE PHOTOGRAPHED FACE

As a rule there are four basic standardized views that are photographed of the head and face for a subject with a facial disfigurement. The four basic views include anteroposterior position, left and right lateral views, and nasal base view. These four views should remain the same for all clients. Depending on the diagnosis, area of specialization, or preference of individual medical staff, additional views can be added. It may be necessary to produce a close-up, detailed view of a particular area or to produce a view that is not a standard photograph but one that better illustrates the client's condition.

The Anteroposterior Position

In the anteroposterior and lateral position, the level and distance between the eyes and ears and the flat surfaces of the face can be used for orientation (Fig. 15–2). A frequent problem occurs when the camera placement in relation to the subject is just slightly out of position. This is a situation that is often not noticed by the untrained eye, until the final image is viewed. The difference in the angle of view appears slight, but the effect on perspective is considerable. The head can be in the rest position or in the standard position, if required. In optimal rest position of the head the frontal plane of the face is parallel with the lens. If the forehead is tilted toward the camera, the lower portion of the face seems shorter. If the chin is elevated, a long lower face will appear short.

The midsagittal plane (axis), which is the midline of the face, is used as a vertical reference when positioning the camera in the anteroposterior position.

Determination of the midaxis of the face may be difficult in a deformed face. The mutual position of the paired features of the face may offer some help. As mentioned already, the levels of the eyebrows, the eye fissures (or at least of the inner commissures, endocanthions), the wings (alae) of the nose, and the horizontal direction of the labial fissure (mouth) can be helpful if symmetrically located. Among the paired features, the levels of the ear lobes are the least reliable in a congenitally altered face. Among the landmarks located in the midaxis of the face, the point at the nasal root (nasion) is seldom dislocated from the midline. The landmark on the tip of the nose (pronasale) and the point at the base of the columella of the nose (subnasale) are usually not in the midline in a nose markedly deviated to the side. In faces with great asymmetry between the sides, usually the chin point is deviated from the midaxis of the face.

Lateral View of the Face

In the lateral view, only one side of the face should be visible in the frame of the camera. The vertical center of the frame in this position is lined up just behind the eye. If any portion of the far side of the face is visible (e.g., the eyebrow), the camera's position is too far forward (Fig. 15–3). Another technique for positioning the subject in

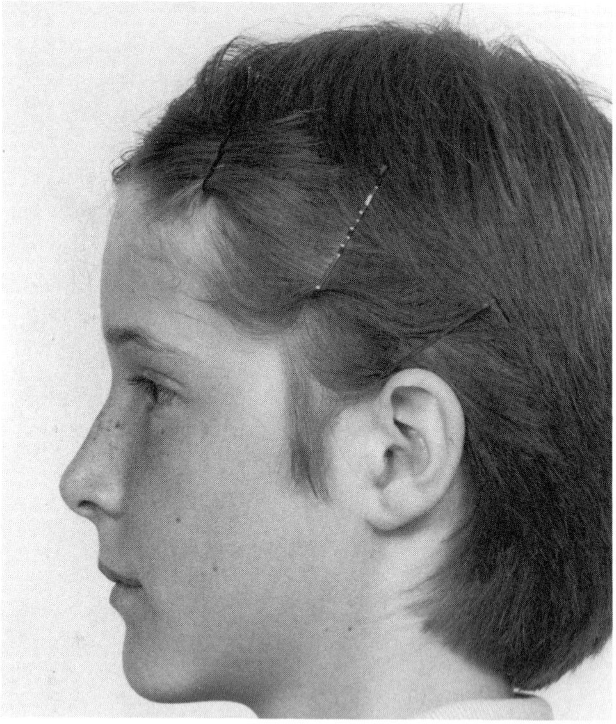

FIG. 15–3. Left lateral view of the face. The head is in the rest position.

the lateral view is by lining up the corners of the mouth while the mouth is in the open position. In this position the contours of the facial profile line are exposed from the forehead to the chin. The dominating sections are the forehead, the nasal bridge, and the chin. The "quality" of the inclinations in these three sections can differ greatly in various rest positions of the head. Thus, if the photograph is used for visual examination of the face, the impression can be misleading. In a head in a high rest position (10 degrees above the FH) the forehead appears more receding, a flat nose may show an "ideal" inclination of the bridge, and the chin—even if it is small—displays increased protrusion. In contrast, similar distortions in the opposite direction can be seen if the rest position of the head was very "low" (below the FH). In spite of these problems, there is no need to photograph the head in standard position. However, marking the orbitale and tragion landmarks on the skin prior to photographing will yield a possibility to measure the inclination in any section of the facial profile line in standard position of the head (FH) on a photograph with no requirements regarding the enlargement size of the print.

Base View of the Nose

For a longitudinal follow-up of the morphological changes in the structures of the soft nose, a standardized position of the head is indispensable. The key landmark for achieving this position is the most protruding point of the nasal tip (pronasale, prn), marked on the skin of the nose of the patient prior to photographing, looking at the face from the side. During photographing, the head is kept in a reclined position, vertical in midsagittal plane of the face, and showing the tip landmark on the top of the soft nose (Fig. 15–4).

Among the additional landmarks marked on the skin, the midpoint of the base of the columella (subnasale, sn) is the most important, followed by the points where the wings (alae) of the nose disappear in the skin of the upper lip right and left (subalare points, sbal). These points are marked on the skin also before photographing. The location of the subnasale landmark is seen from the front but must be checked also from the side of the face. The insertion points of the alar bases in the upper lip are marked from the front.

The skin markings allow us to estimate the size of the nasal tip protrusion (sn-prn), the size and asymmetry of the nostril floors (sbal-sn rt,lt), the width of the columella (sn′-sn′), the length of the right and left columella edges (sn-c′ rt,lt) (asymmetry), deviation of the columella, in-

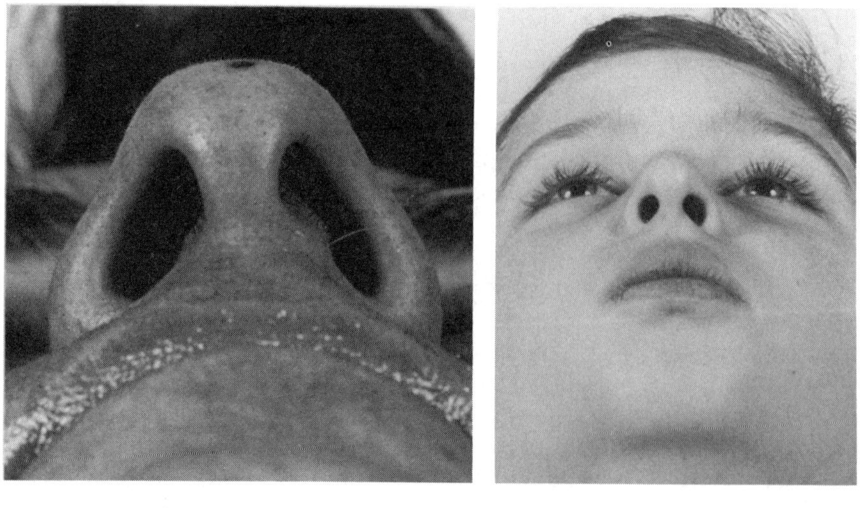

FIG. 15–4. Variations in the base views of the nose. **Left:** View of the soft nose only. The pronasale point (prn) on the tip of the nose was marked on the skin before photographing. In this position the pronasale is the highest point of the soft nose. **Right:** A modified base view of the nose. It is extended proximally to the bridge of the nose and the orbits, and distally to the lower face.

clinations of the longitudinal axis of the nostrils, and level differences between the alar base insertions (sbal rt,lt). For visual evaluation the nasal tip configuration and the shape of the nasal alae will be fully exposed.

It can happen that an "overhanging" nose would block—partly or completely—the view of the soft nose because of the decreased distance between the nasal tip and the upper lip surface. The same can happen in the presence of a marked protrusion of the upper lip. In such cases, two pictures are taken: one in standard base view and the other from a position which allows the nostrils to be seen.

ADDITIONAL VIEWS AT PHOTOGRAPHY

Depending on the diagnosis and the area of specialization or preference of the individual medical staff, additional views of the head and face are used.

The Ear

In a face photographed in standard lateral position, the length of the ear may be close to reality. The ear may appear narrower than in reality, depending on the degree of the protrusion from the head and on the position of the camera. The full shape of the ear and the true width of the ear can be seen only if the camera is parallel to the anterior surface of the auricle (Fig. 15–5). Inclination of the longitudinal medial axis of the ear can be determined only on the head maintained in FH (or if, prior to photographing, the orbitale and the tragion landmarks were marked on the skin of the patient).

The frontal view of the face, vertical in its midaxis, may reveal the difference in vertical level of the ear lobes. At visual examination, diagnosis of low-set ear can be made only in the case of an extremely low position of the ear(s).

Dorsal View of the Head

The dorsal view of the head helps to visualize the shape deformities in the head. It is easier to judge the vertical location of the ears in this position than from the front. During photographing, the head must be kept in vertical. The position of the camera must be adjusted to the general posterior plane of the head (Fig. 15–6).

Bird's-Eye View (Upper View) of the Head

The head is photographed from the top, maintaining the face in vertical and the head in rest position. The photograph reveals (a) the qualities of relationships between the width and the length of the skull and (b) any aberra-

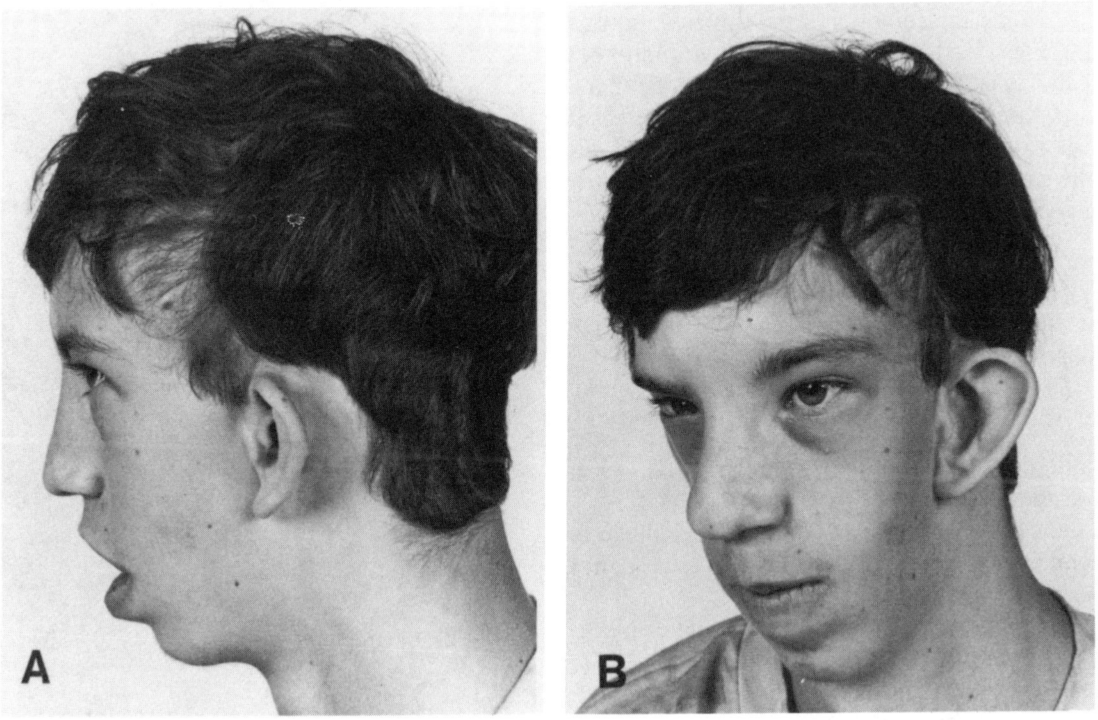

FIG. 15–5. In photographing the ears for *anthroposcopic* purposes, any position is good which shows the most striking disfigurement. Thus, the degree of the ear's protrusion from the head is well-demonstrated in the standard lateral position of the face (**A**). For *anthropometric* purposes (photogrammetry of standard size prints) the ear is photographed with its anterior surface parallel to the lens of the camera (**B**).

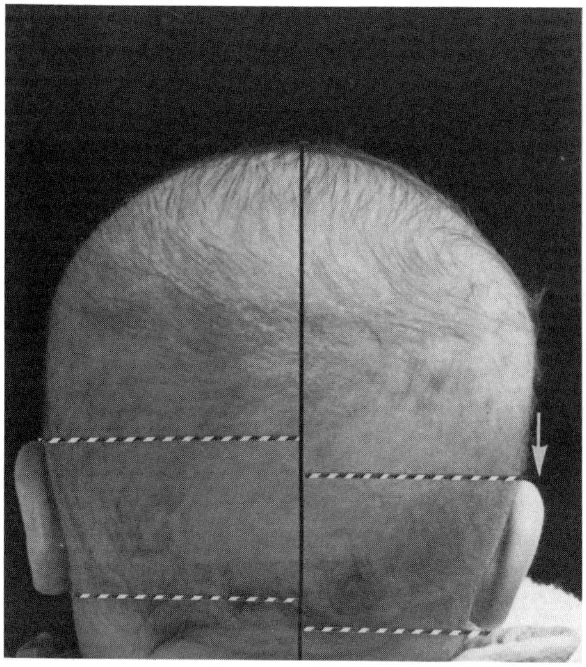

FIG. 15-6. Dorsal view of the head. The uneven level of the ear lobes is more discernible in this position than from the front.

tions from the usual contours of the head. The extent of the dislocation of the ear canals in anteroposterior direction is clearly visible in this position (Fig. 15–7). In modified upper view of the head the forehead is almost fully exposed, revealing the contours of the supraorbital rim (eyebrows) (Fig. 15–8). In this view, any asymmetry between the halves of the forehead is indicated by the shape differences between the contours of the right and left eyebrows (Fig. 15–9). This is the best position for demonstrating the striking disfigurement of the forehead in trigonocephaly (Fig. 15–10).

Special Views of the Face in the Orthodontic and Craniofacial Program at the Hospital for Sick Children in Toronto

Eleven views are required:

1. The anteroposterior position, photographed twice. Same as the basic frontal view.
2. With the subject smiling, showing teeth to illustrate the facial muscles; the smile must appear relaxed, not grimacing.
3. With the jaw relaxed, showing the natural displacement of the upper and lower jaw. This is especially noticeable when the head is in the lateral position. Self-conscious individuals may have a tendency to position the upper and lower jaw in a compensatory position to hide their degree of malocclusion. This position is to be avoided during photography.
4. The occlusal view with the subject's bite in occlusion to show the actual degree of malocclusion. This is achieved by having the subject bite down on their back teeth during photography. The occlusal view in most cases is a close-up of the immediate mouth area with dental retractor in position to separate the lips. These photographs include an anterior view, a left and right oblique view, and an optional right and left lateral view.
5. A final view using a dental mirror, to show the upper dental arch and palate.
6. Full face left and right lateral views.
7. Left and right 45-degree oblique views. The three-quarter view is more difficult to position because of the angular perspective of the face from this point of view. The three-quarter view can be lined up with those individuals with average facial features by placing the tip of the nose along the line of the edge of the cheek (Fig. 15–11).
8. The nasal base view. There is no standardized position for the nasal base view because the positioning of the head—by elevating the chin—is determined by

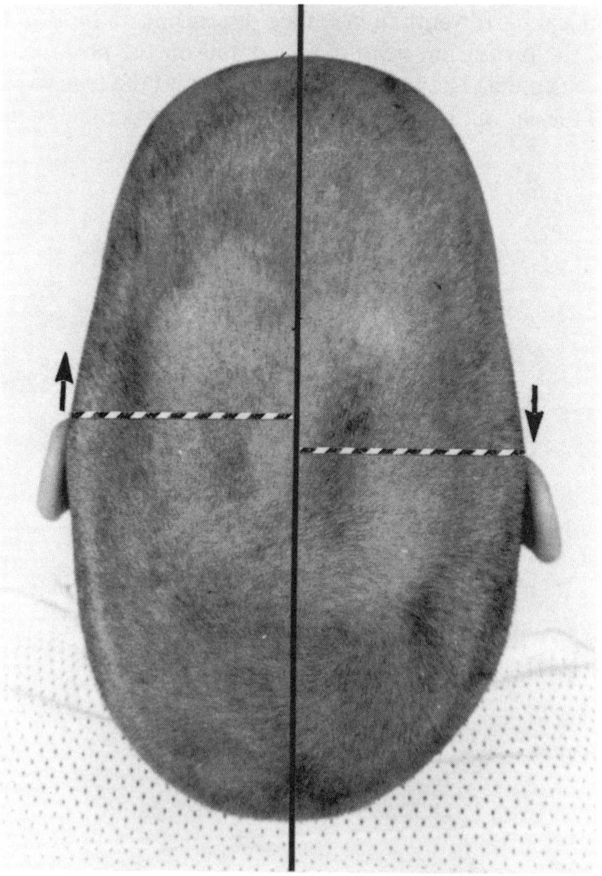

FIG. 15-7. Bird's-eye view (upper view) of the head. This view enables display of the asymmetry in location of the ears caused by horizontal dislocation of the ear canals (provided that the hair does not cover the ears). In this child the left ear canal is closer to the facial midline.

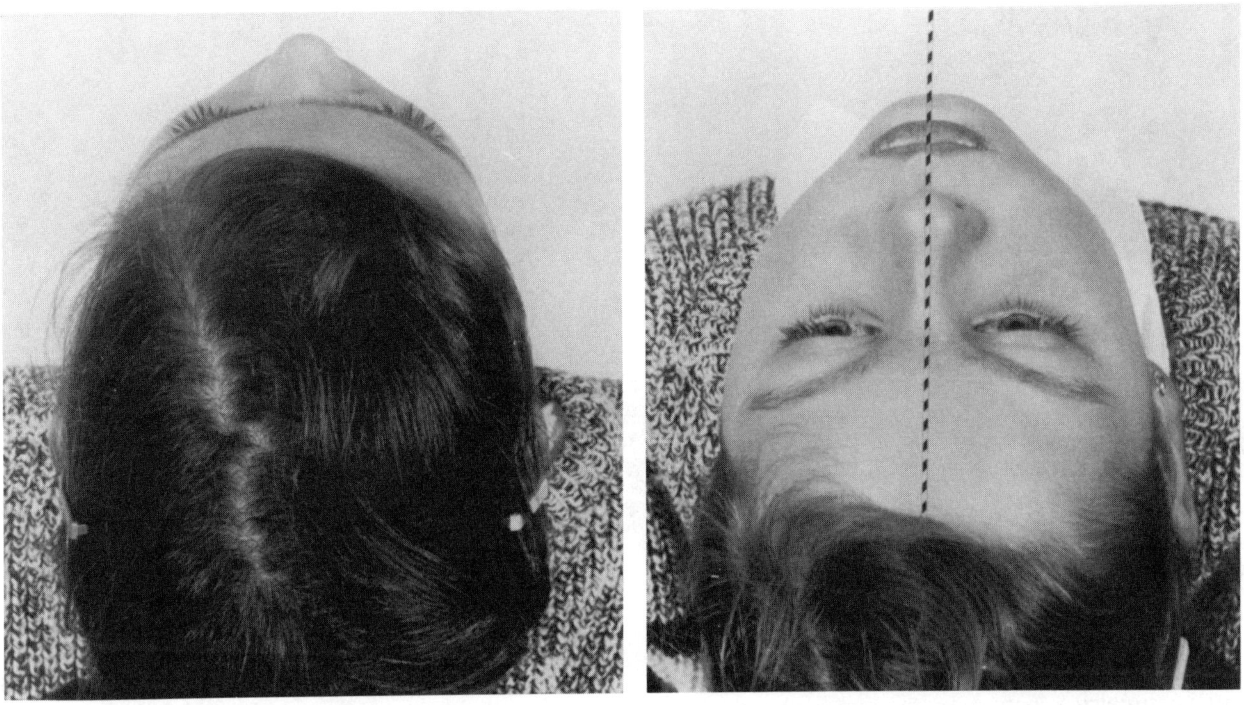

FIG. 15–8. Partial bird's-eye view of the head. **Left:** Position revealing the supraorbital contours of the forehead. **Right:** The view is extended to the face, indicating the position of the eyebrows, the direction of the medial axis (nasal bridge), and the position of the chin. In this view, any aberration from the normal would be more easily noticeable than in a front view of the face.

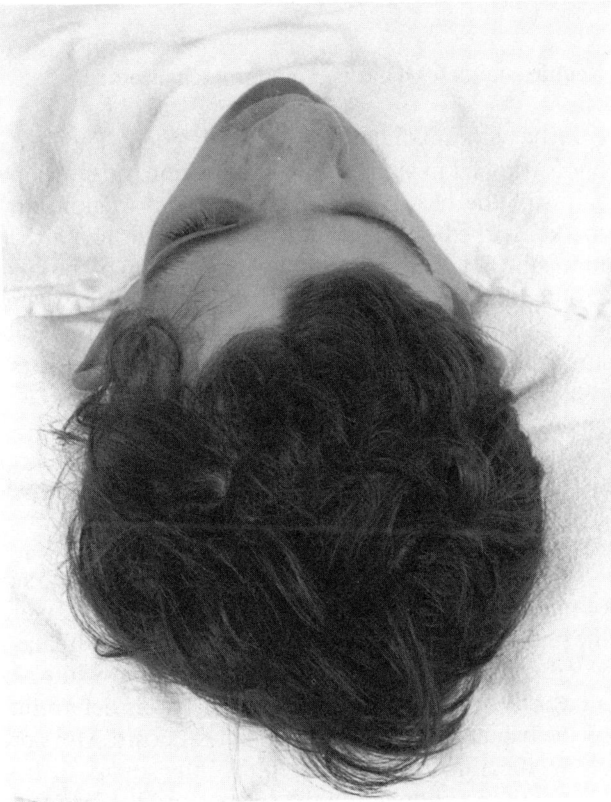

FIG. 15–9. Partial bird's-eye view of the head of a boy with Goldenhar syndrome. The asymmetry in the shape of the supraorbital contours is clearly exposed. The nasal bridge is deviated to the right, while the chin point maintains its axial position.

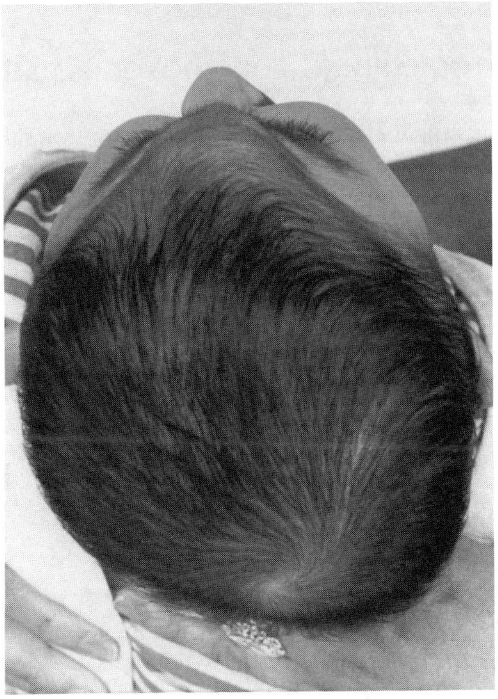

FIG. 15–10. Bird's-eye view of the head of a child with trigonocephaly. The triangular shape of the forehead is clearly visible.

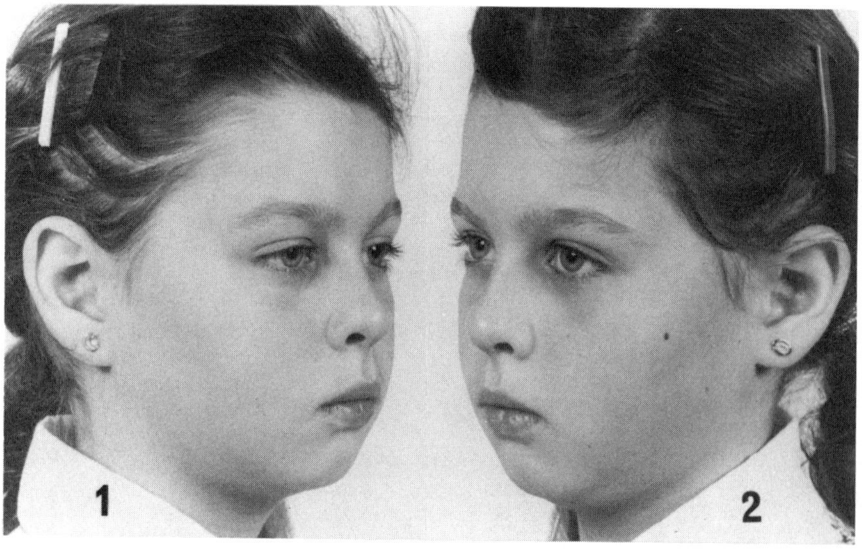

FIG. 15–11. Right and left oblique views of a healthy face.

what aspect of the face is of importance (e.g., the lip, the soft nose, or the forehead). A facial deformity of an asymmetrical nature or one that causes the anatomical landmarks to be out of position makes the task of positioning more challenging.

An important aspect of maintaining scale with the lens magnification preset at 1–10 is to standardize the locations on the face to visually focus the camera. With the exception of the nasal base view where the camera is generally focused on the nasal area, the medial aspect of the eye is used as a point of focus for the other positions.

PHOTOGRAPHING FOR PHOTOGRAMMETRY

Adjusting the clinical photograph both for qualitative and quantitative analysis has a twofold advantage. Making the clinical photographs measurable (photogrammetry) greatly decreases the time needed for completing a routine direct anthropometric examination. This is very important especially in small children, who generally do not cooperate well during the clinical examination. The adjusted photographs offer a number of measurements which can be taken only in standard position of the head, a requirement almost impossible to achieve in younger age groups.

Requirements for Measuring Linear Projective Distances

The enlargement size of the print should be standard (life-size, 25% of life-size, or 50% of life-size). The head and face are photographed in standard anteroposterior and lateral positions. The position of the head should be standard (in FH). The following landmarks should be marked on the face:

1. *On the frontal surface of the face:* the hairline point, the glabella, the nasion, the orbitale, the subnasale, the subalare right and left, the sublabiale, and the gnathion.
2. *On the lateral surface of the face:* the tragion, the nasal tip point (pronasale), the facial insertion point of the ala (ac), and the lower insertion point of the ear (otobasion inferius).

Requirements for Taking Angular Measurements

Print with any enlargement size is good. The face is photographed in standard lateral position, or in rest position of the head with the orbitale and tragion landmarks marked on the skin of the patient which are defining the FH. The inclination of any section of the facial profile line can be determined by a commercial angle meter (with relatively small errors). For measuring the angles in the profile, our multipurpose facial angle meter is used (see also Chapters 2 and 5). Measuring the facial angles does not require standard position of the head.

LIGHTING AND EXPOSURE

There are two components regarding the technical side of photography that are crucial to the final outcome of the photographic image. First, the quality of the image—whether it is printed from a color or black-and-white negative, or exists as an original color slide—is determined by the balance between lighting and exposure. More precisely, the difference is between the intensity of the flash being used to photograph the subject and the size of the aperture opening of the camera lens. Shutter speed is also an important component of exposure; however, it is not a factor in clinical photography with the use of electronic flash. Lens openings are measured in F-stops. As the F-stop number increases or decreases one full stop (e.g., F8, F11, or F16), the amount of light reaching the film is doubled in intensity or cut in half. Proper exposure should be calculated by using a trial-and-error "test" roll before any client photography begins. Another consideration when making preliminary tests is the accuracy of skin color in the final image. This is affected by a combination of the lighting system, type of film used, and processing. Finding a photo-finishing lab that provides consistent processing and color balance is another important factor.

The second consideration of photographic lighting is that it must best illustrate the condition of the client. Here again the point of importance is the element of consistency. There are several different lighting possibilities from which to choose. The important factor is to keep any lighting system relatively simple so it can be reproduced by more than one person over a long period of time and still maintain a standardized look.

There are two common problems that occur with improper use of lighting. The first is the use of harsh lighting that may obscure a portion of the subject's anatomy. This is usually a result of direct undiffused flash which also has the tendency to flatten the tonal range of the subject. The second problem occurs when the lighting is positioned in a way that alters or distorts the true appearance of the patient (7). Without proper lighting and exposure, all other efforts of the photographer are minimized.

Two electronic flashes on stands, positioned at 45-degree angles and slightly above the subject, provide equal illumination for both sides of the face. The appropriate position of the flash in relationship to the subject and photographer should be permanently marked in the photographic area. This is a straightforward lighting system that is easily standardized and provides good lighting quality.

An off-white seamless paper placed a minimum of three feet behind the client provides a clear neutral background. The use of a colored background (a popular choice is blue) causes a subtle but noticeable shift in the skin tone of the subject. This occurs because the light of the flash reflects off the colored background onto the subject being photographed (8).

THE CAMERA

Facial views are photographed with a 105-mm lens at a magnification scale of 1–10 (11). This is exactly four feet from camera to subject using a 35 mm single lens reflex camera. (Focal-length lenses between 90 mm and 105 mm are acceptable as long as the focal length remains consistent.) Keeping the magnification standardized is critical in order to maintain size relationships as patients are photographed over time. The magnification is set beforehand on the barrel of the lens, and then final focusing is done by moving the camera back and forth while viewing the subject through the viewfinder. Magnification or size ratio between the subject being photographed and the recorded image on film is expressed as a fraction such as 1/10. The size of the image on the film is one-tenth the size of whatever is being photographed. An occlusal view of a patient's bite, where a closer detailed view is required, would be photographed at a magnification of 1–2.

The 105-mm lens provides a good representation of what is normally considered an appropriate "true-to-life" perspective of the human face. Four feet is also a comfortable working distance between photographer and client. In contrast, the 55-mm lens, which has a wider angle of view, is suitable for photographing the upper and lower extremities. This lens distorts the facial features at a magnification of 1–10 because the lens would be focused at a distance too close to the subject's face (50- to 55-mm lens is actually classified photographically as a "normal" lens because it has the same approximate angle of view as the lens of the human eye; however, the 55-mm lens presents an image of the face that is too angular in perspective at a 1–10 magnification.)

SUMMARY

Medical photography carried out in an organized way increases the documentary value of the pictures. Photography of carefully planned positions is more economical, thereby eliminating pictures of no clinical importance. Standardization of medical photography means that the pictures are taken from the same distance, in as nearly identical positions as possible; this ensures nearly identical photographs by other photographers (12). However, photographs produced in standard sizes (not necessarily life-size), and in special cases in standard positions of the head (FH), would offer additional valuable measurements of the craniofacial complex. By standardization they could become valuable additions to the clinical charts.

REFERENCES

1. Skopec R. *Photography in medicine* [in Czech]. Prague: School of Photography, College of Art, 1952.
2. Gavan JA, Washburn SL, Lewis PH. Photography: an anthropometric tool. *Am J Phys Anthrop* 1952;10:331–354.
3. Neger M. A quantitative method for the evaluation of the soft-tissue facial profile. *Am J Orthod* 1959;45:738–751.
4. Sperli AE. Photographic standardization in plastic surgery. *Rev Latinoam Cir Plast* 1968;12:106–114.
5. Edgerton MT, McKnelly LO, Wolfort FG. Operating room photography for the plastic surgeon. *Plast Reconstr Surg* 1970;46:93–95.
6. Morello DC, Converse JM, Allen D. Making uniform pho-

tographic records in plastic surgery. *Plast Reconstr Surg* 1977;59: 366–367.

7. Chapple JG, Stephenson KL. Photographic misrepresentation. *Plast Reconstr Surg* 1970;45:135–140.
8. Williams AR. *Medical photography study guide,* 4th ed. Lancaster, England: MTP Press, 1984.
9. Farkas LG, Bryson W, Klotz J. Is photogrammetry of the face reliable? *Plast Reconstr Surg* 1980;66:346–355.
10. Dickason WL, Hanna DC. Pitfalls of comparative photography in plastic and reconstructive surgery. *Plast Reconstr Surg* 1976;58: 166–175.
11. Davidson TM. Photography in facial plastic and reconstructive surgery. *J Biol Photogr Assoc* 1979;47:59–67.
12. Jorgenson RJ. The real problems in dysmorphology. *Dysmorphol Clin Genet* 1988;2:54–57.

Chapter 16

Racial and Ethnic Morphometric Differences in the Craniofacial Complex

Karel Hajniš, Leslie G. Farkas, Rexon C. K. Ngim, S. T. Lee, and Govindasarma Venkatadri

Morphologic, quantitative, and proportional differences among various races and ethnic groups have been well known since the last century (1–4).

The main signs that distinguish one race from another develop from the action of adoptive mechanisms, which are influenced by the environment over time and are preserved by the evolutionary process in humans. However, the wide range of adaptability greatly narrows during the developmental period. Fewer outstanding differences within the individual basic races still occur; these are caused by microgenetic changes. These changes contribute to the differences between subgroups within the races, called in the past and especially in Europe "anthropologic types" (5–8).

The existence of the long-lasting microevolutionary changes have been observed also in our time—for example, the relative elongation of the head in relation to its width, described recently in some populations (9–12). Similar changes have also been reported in the Czech population (13).

One of the most varying parts of the human body is the craniofacial complex—in particular, the face. Because of its faultless mimic muscle apparatus, which is an important communication device, the face can reflect the most diverse psychological conditions. It is the most important part of the body because it provides the closest contact with the outside world (14). The head, and especially the face, is often deformed by injuries and congenital anomalies; therefore knowledge of the anatomy and morphology, as well as knowledge of the quantitative data of the craniofacial complex, is essential for those involved in restoring the normal proportions of the head and face (15–17).

The purpose of this study is to contribute to the knowledge of the differences in the morphology of the head and face in young adult males and females belonging to three races: Caucasians, blacks, and Asians, represented by North American Caucasian, African-Americans, and Singapore Chinese, respectively. These differences are also determined in three Caucasian ethnic groups: the North American Caucasians, the Germans (Germany), and the Czechs (Czech Republic).

SUBJECTS

The study group consisted of two subgroups. In the first subgroup, the morphological differences of the head and face in 103 North American Caucasians (52 males and 51 females), 100 African-Americans (50 males and 50 females), and 60 Chinese (30 males and 30 females) were analyzed. The age of the subjects ranged from 18 to 25 years. In the second subgroup, the morphological differences in the head and face were determined in three Caucasian ethnic groups—103 North American Caucasians, 120 Germans (Germany), and 109 Czechs (Czech Republic)—divided almost equally between the sexes, ranging in age between 18 and 25 years. The subjects of each subgroup were randomly selected. The members of the North American Caucasian sample were examined between 1973 and 1976 in Ontario and Quebec (Canada) (18).

The multiethnic character of the North American Caucasians was evident in its composition: 40% British, 32% Latin, 12% Germanic, 8% Slavic, and 8% other ethnic origin subjects.

The data for the African-American group, which consisted of university students from Buffalo (New York), were collected by one of the authors (LGF) in 1991.

The data for the Chinese population were collected in Singapore in 1988 (19). The data for the German sample were collected between 1967 and 1968 in Mainz am Rhein (Rhineland in Middle Germany) and Kiel (Schleswig-Holstein in Northern Germany) (20).

The data for the Czech sample came from Prague (Czech Republic). The examination was carried out between 1966 and 1967 (21).

METHODS

All investigators applied the same examination techniques (18,22). Twenty-five measurements were taken: four for the head (eu-eu, g-op, v-n, and circumference), four for the face (zy-zy, n-gn, n-sto, and sn-gn), five for the orbits (en-en, ex-ex, ex-en lt, ps-pi lt, and eye fissure inclination lt), four for the nose (n-sn, al-al, sn-prn, and nasal bridge inclination), six for the lips and mouth (ch-ch, sn-sto, sn-sl, ls-sto, sto-li, and sto-sl), and two for the ears (pra-pa lt and sa-sba lt) (see Chapter 2).

The relationship between the measurements was determined with the help of 16 proportion indices: one for the head, three for the face, two for the orbits, two for the nose, six for the lips and mouth, and two for the ears. For evaluation of the major indices of the head, face, and nose, the categories established by Martin and Saller (22) and the norms developed by Farkas and Munro (23) were used. The quality of proportions in the Chinese and black samples was evaluated from norms developed for them (*unpublished data*).

The morphological differences in the craniofacial complex were analyzed separately for the three groups representing North American Caucasians (whites, NAC), African-Americans (blacks, AA), and the Chinese population (CP), and then for the three ethnically distinguished Caucasian groups: North American Caucasians (NAC), Germans (GP), and Czechs (CzP).

Statistical Analysis

Student's *t*-test was used to analyze the mean values of the measurements collected. The degree of significance was regarded as *mild* (M) ($p = 0.05$), *moderate* (MOD) ($p = 0.01–0.02$), or *severe* (also high) (SEV) ($p \leq 0.001$).

RESULTS

Racial Differences

Head

Width of the head (eu-eu). Among the three races, the largest head width was seen in both sexes of CP (158.3 mm in males and 151.6 mm in females), and the smallest width was observed in AA (148.8 mm in males and 141.1 mm in females). In both sexes of the Chinese group, the head was highly (SEV) significantly wider than in the black or white groups. The difference in the width of the head in the males of the NAC compared with that of AA males was highly significant; in the females the difference was moderately significant.

Length of the head (g-op). The most elongated head was found in AA (199.2 mm in males and 186.6 mm in females), and the least elongated head was seen in CP (182 mm in males and 172.5 mm in females). The head of the blacks was highly significantly longer than the head of the Chinese in both sexes; in the head of the Whites, this occurred only in males. The head of black females was only slightly longer than that of white females. Length measurements in both sexes of whites were highly significantly larger than those in the Chinese.

The height of the head (v-n). The highest head was seen in NAC (117.7 mm in males and 112.6 mm in females), and the lowest was found in CP males (116.9 mm) and AA females (107.8 mm). Height measurements of the head were very similar in the males of the three races. In the females, only a moderate statistical difference (higher heads in white females) between NAC and AA was found; the findings between NAC and CP and between AA and CP were similar.

Head circumference. The AA group had the largest head circumference for both sexes (573.6 mm in males and 547.0 mm in females), and the CP group had the smallest (559.6 mm in males and 535.3 mm in females). The circumference was slightly larger for blacks than for whites for both sexes. However, the difference between the blacks and Chinese was highly significant.

Cephalic index (I = eu-eu × 100/g-op). The highest cephalic index values were observed in CP (I = 87.1 for males and 87.9 for females), and the lowest values were found in AA (I = 74.8 for males and 75.8 for females). The larger index values of the Chinese differed highly significantly from those of whites or blacks. In AA males, the smaller index value was highly significantly different in comparison with the findings in NAC males; in females the difference was moderately significant.

Conclusions. The cephalic index indicated a mesocephalic or medial head in NAC, brachycephalic or relatively short head in CP, and dolichocephalic or relatively elongated head in AA in both sexes. The greatest uniformity in the three races was found in the height of the head. The head of AA was the largest in length and circumference and was the smallest in width. In CP, the head was the largest in width and the smallest in length and circumference. The head of NAC was wider than that of AA, but narrower than that of CP. The NAC head was longer and had a greater circumference than that of CP, but these measurements, compared with the findings for AA, were smaller (Table 16–1).

Face

Width of the face (zy-zy). Among the three races the widest face was noted in CP (144.6 mm in males and 136.2 mm in females), and the narrowest was found in AA (138.7 mm in males and 130.5 mm in females). Chinese faces were highly significantly wider than those of blacks or whites. In whites and blacks the width measurements of the face were similar in both sexes.

Height of the face (n-gn). The highest face was found in AA (125.9 mm in males and 116.5 mm in females), and the shortest was noted in NAC (121.3 mm in males and 112 mm in females). These differences between the two racial groups were highly significant. The face of the blacks was slightly longer than that of the Chinese for both sexes. The face of the Chinese was longer than that of the whites, slightly in males and moderately significantly in females.

Height of the upper face (n-sto). The largest upper face height was recorded in males in CP (78.2 mm) and in females in AA (72.7 mm). The smallest measurements were noted in NAC (74 mm in males and 68.9 mm in females). In both sexes of Chinese and blacks, the almost

TABLE 16–1. *Anthropometric differences in the head of three racial groups*[a]

	NAC			AA			CP			Statistical difference[b]		
	N	$\bar{X}$	SD	*N*	$\bar{X}$	SD	*N*	$\bar{X}$	SD	NAC vs. AA	NAC vs. CP	AA vs. CP
						Head width (eu-eu) (mm)						
M	49	153.3	5.9	50	148.8	6.7	30	158.3	5.1	SEV	SEV	SEV
F	51	144.4	4.6	50	141.4	6.0	30	151.6	4.7	MOD	SEV	SEV
						Head length (g-op) (mm)						
M	49	193.7	7.6	50	199.2	6.0	30	182.0	6.2	SEV	SEV	SEV
F	51	184.9	7.0	50	186.6	6.5	30	172.5	6.1	NS	SEV	SEV
						Head height (v-n) (mm)						
M	48	117.7	8.0	50	117.1	9.5	30	116.9	8.0	NS	NS	NS
F	38	112.6	7.1	50	107.8	10.1	30	109.8	7.8	MOD	NS	NS
						Head circumference (mm)						
M	49	568.6	16.1	50	573.6	15.8	30	559.6	14.2	NS	MOD	SEV
F	51	544.4	14.8	50	547.0	16.2	30	535.3	15.5	NS	MOD	SEV
						Cephalic index (eu-eu × 100/g-op)						
M	49	79.2	4.6	50	74.8	3.7	30	87.1	3.6	SEV	SEV	SEV
F	51	78.2	4.2	50	75.8	3.9	30	87.9	3.6	MOD	SEV	SEV

[a] NAC, North American Caucasians; AA, African-Americans; CP, Chinese; M, male; F, female.
[b] Significancy: M, mild (p = 0.05); MOD, moderate (p = 0.01–0.02); SEV, severe ($p < 0.001$); NS, not significant.

identical upper face heights showed a moderately to highly significant difference when compared with those of NAC males and females.

Height of the lower face (sn-gn). The longest lower faces were found in AA (78.9 mm in males and 71.5 mm in females), and the shortest were noted in NAC (71.9 mm in males and 65.5 mm in females). These differences were highly significant, similar to those between AA and CP. The lower face measurements of whites and Chinese were almost identical.

Facial index (I = n-gn × 100/zy-zy). The largest facial index was recorded for AA (I = 90.9 for males and 89.4 for females), and the smallest was recorded for CP (I = 85.5 for males and 84.4 for females). The difference between the index values for AA males and NAC males was mildly significant; for AA males and CP males it was highly significant. The index value for black females was highly significantly larger than that for NAC and CP females. The index was larger in NAC than in CP, slightly for females and moderately significantly for males.

Upper face index (I = n-sto × 100/zy-zy). The highest upper face index values were seen in AA (I = 56.4 for males and 55.8 for females), and the lowest were noted in NAC (I = 54.0 for males and 52.6 for females), showing highly significant differences. The index values were almost identical for whites and Chinese in both sexes. The index was significantly smaller in Chinese than in blacks: moderately for males and highly for females.

Lower face–face height index (I = sn-gn × 100/n-gn). The AA exhibited the highest lower face–face height index values (I = 62.7 for males and 61.4 for females). The smallest indices were established for CP males (I = 58.8) and females (I = 57.8). The index values for both sexes of blacks were highly significantly larger than those for whites or Chinese. Almost identical was the index for white and Chinese males. The index value for NAC females was moderately significantly larger than that for Chinese females.

Conclusion. The facial index indicated a mesoprosop-type face in NAC males and in both sexes of CP, when a *medial face* is defined according to classical categories (24). The facial indices for NAC females and for both sexes of AA indicated a leptoprosop or relatively *long face.* The upper face index for NAC females and for both sexes of CP defined a mesen or *medial type* upper face. For NAC males and for both sexes of AA, the index indicated a lepten or relatively *long upper face.* The lower face–face height index defined a relatively *long lower face* for AA compared to the more balanced relationships between the two measurements for NAC and CP. In AA, the face was characterized by longer vertical profile dimensions (face height, upper face and lower face heights) and small face width. In CP, the face was the widest among the races. NAC had the smallest vertical profile measurements (face, upper face and lower face heights) among the races; the face of NAC was slightly wider than that of AA, but markedly narrower than that of CP (Table 16–2).

Orbits

Intercanthal width (en-en). Among the three races, the widest space between the orbits was noted in CP (37.6 mm in males and 36.5 mm in females), and the narrowest space was found in NAC (32.9 mm in males and 32.5 mm in females). The intercanthal width measurement in NAC showed highly significant differences compared with those in AA or CP. The differences between AA and CP were only moderately significant.

Biocular width (ex-ex). In AA, the biocular width measurement was highly significantly larger (96.8 mm in males and 92.9 mm in females) than that in NAC and CP in both sexes. The findings for males and females of NAC and CP were almost identical.

Eye fissure length left (ex-en). In AA, the eye fissure length left measurement was the largest (32.9 mm in males and 32.2 mm in females) and highly significantly greater than in NAC or CP, who had the smallest eye fissure lengths (29.4 mm in males and 28.4 mm in females). Compared with the smallest eye fissure lengths in the Chinese, the difference in these measurements in whites was highly significant.

Eye fissure height left (ps-pi). The highest eye fissure was seen in NAC (10.4 mm in males and 11.1 mm in females). They highly significantly differed from the smallest findings in CP (9.4 mm in males and 9.5 mm in females). The measurements in NAC and AA males were identical. The heights of the eye fissures of white females were larger than those of black females. The difference was moderately significant. A similar relationship was observed between the eye fissure heights of AA and CP males.

Eye fissure inclination left. Highly significant was the difference between the eye fissure inclinations in CP [which were the largest measured among the three races (12.1 degrees in males and 12.5 degrees in females)] and those in AA or in NAC [who exhibited the smallest inclinations (2.3 degrees in males and 5 degrees in females)]. The inclinations measured in blacks were significantly higher than those in NAC.

Intercanthal index (I = en-en × 100/ex-ex). The intercanthal indices for CP (I = 41.0 for males and 41.8 for females) were the highest measured among the three races; those for AA (I = 37.0 for both males and females) were the smallest. These indices showed a highly significant difference between the two races, as well as between the NAC and CP. The index values for whites and blacks were almost identical.

TABLE 16–2. *Anthropometric differences in the face of three racial groups*[a]

	NAC			AA			CP			Statistical difference[b]		
	N	$\bar{X}$	SD	*N*	$\bar{X}$	SD	*N*	$\bar{X}$	SD	NAC vs. AA	NAC vs. CP	AA vs. CP
							Face width (zy-zy) (mm)					
M	49	139.1	6.3	50	138.7	5.6	30	144.6	5.6	NS	SEV	SEV
F	51	131.1	5.3	50	130.5	4.8	30	136.2	4.0	NS	SEV	SEV
							Face height (n-gn) (mm)					
M	52	121.3	6.8	50	125.9	8.2	30	123.6	5.3	SEV	NS	NS
F	51	112.0	4.7	50	116.5	6.1	30	114.9	4.9	SEV	MOD	NS
							Upper face height (n-sto) (mm)					
M	52	74.0	4.2	50	78.0	4.8	30	78.2	4.0	SEV	SEV	NS
F	51	68.9	3.6	50	72.7	4.5	30	71.8	5.5	SEV	MOD	NS
							Facial index (n-gn × 100/zy-zy)					
M	52	88.5	5.1	50	90.9	6.5	30	85.5	4.0	M	MOD	SEV
F	51	86.2	4.6	50	89.4	5.5	30	84.4	3.4	SEV	NS	SEV
							Upper face index (n-sto × 100/zy-zy)					
M	52	54.0	3.1	50	56.4	4.1	30	54.1	2.9	SEV	NS	MOD
F	51	52.6	3.4	50	55.8	4.0	30	52.7	2.9	SEV	NS	SEV
							Lower face height (sn-gn) (mm)					
M	52	71.9	6.0	50	78.9	6.7	30	72.7	5.2	SEV	NS	SEV
F	51	65.5	4.6	50	71.5	5.2	30	66.4	5.6	SEV	NS	SEV
							Lower face–face height index (sn-gn × 100/n-gn) (mm)					
M	52	59.2	2.7	50	62.7	4.0	30	58.8	2.5	SEV	NS	SEV
F	51	58.6	2.9	50	61.4	2.8	30	57.8	3.9	SEV	MOD	SEV

[a] NAC, North American Caucasians; AA, African-American; CP, Chinese; M, male; F, female.
[b] Significancy: M, mild (p = 0.05); MOD, moderate (p = 0.01–0.02); SEV, severe (p < 0.001); NS, not significant.

Eye fissure index left (I = ps-pi × 100/ex-en). The eye fissure indices for NAC (*I* = 33.4 for males and 36.2 for females) were the largest recorded among the three races and highly significantly differed from the index values for AA (*I* = 30.4 for males and 32.5 for females), which were the lowest. The index values of whites were larger than those of the Chinese and differed highly significantly for females and mildly significantly for males. The index was slightly larger for Chinese females than that for black females. The index for males was moderately significantly larger than for black counterpart.

Conclusions. The main characteristics of the orbits of the Chinese group were the largest intercanthal width and intercanthal index, and the greatest inclination in the shortest eye fissure and the smallest eye fissure height. Blacks, had the largest biocular distance. Whites had the largest eye fissure heights, which contributed to the highest values of the eye fissure index among the races (Table 16–3).

Nose

Nose height (n-sn). In NAC males the height (53.2 mm) of the nose was identical to that of CP (53.5 mm). The nose height of black males (51.9 mm), which was the smallest observed among the three races, mildly significantly differed from that of white males and moderately significantly differed from that of Chinese males. The nose height of AA females (48.8 mm), which was the smallest recorded, slightly differed from that of NAC females (49.2 mm) but highly significantly differed from the that of CP females, which was greater.

Nose width (al-al). The widest noses were found in AA (44.1 mm in males and 40.1 mm in females), and the narrowest were noted in NAC (34.8 mm in males and 31.9 mm in females). The nose of blacks was highly significantly wider than that of whites or Chinese in both sexes. The difference between the wider Chinese and narrower white noses was also highly significant.

TABLE 16–3. *Anthropometric differences in the orbits of three racial groups*[a]

	NAC			AA			CP			Statistical difference[b]		
	N	$\bar{X}$	SD	*N*	$\bar{X}$	SD	*N*	$\bar{X}$	SD	NAC vs. AA	NAC vs. CP	AA vs. CP
							Intercanthal width (en-en) (mm)					
M	52	32.9	2.7	50	35.8	2.8	30	37.6	3.3	SEV	SEV	MOD
F	51	32.5	2.1	50	34.4	3.4	30	36.5	3.2	SEV	SEV	MOD
							Biocular width (ex-ex) (mm)					
M	49	90.7	3.8	50	96.8	4.6	30	91.7	4.0	SEV	NS	SEV
F	51	87.6	4.0	50	92.9	5.3	30	87.3	5.2	SEV	NS	SEV
							Eye fissure length (ex-en lt) (mm)					
M	49	31.3	1.4	50	32.9	1.6	30	29.4	1.3	SEV	SEV	SEV
F	51	30.7	1.8	50	32.2	2.0	30	28.4	1.7	SEV	SEV	SEV
							Eye fissure height (ps-pi lt) (mm)					
M	52	10.4	1.1	50	10.0	1.1	30	9.4	0.7	NS	SEV	MOD
F	51	11.1	1.2	50	10.4	1.2	30	9.5	1.2	MOD	SEV	SEV
							Eye fissure inclination lt (mm)					
M	49	2.3	1.7	50	7.0	3.3	30	12.1	2.1	SEV	SEV	SEV
F	51	5.0	2.8	50	9.0	2.6	30	12.5	2.0	SEV	SEV	SEV
							Intercanthal index (en-en × 100/ex-ex)					
M	49	37.4	2.2	50	37.0	2.6	30	41.0	2.6	NS	SEV	SEV
F	51	37.2	2.3	50	37.0	2.6	30	41.8	2.4	NS	SEV	SEV
							Eye fissure index (ps-pi × 100/ex-en lt)					
M	52	33.4	3.5	50	30.4	3.6	30	31.9	2.2	SEV	M	MOD
F	51	36.2	3.1	50	32.5	4.4	30	33.4	3.6	SEV	SEV	NS

[a] NAC, North American Caucasians; AA, African-American; CP, Chinese; M, male; F, female.
[b] Significancy: M, mild (p = 0.05); MOD, moderate (p = 0.01–0.02); SEV, severe (p < 0.001); NS, not significant.

Protrusion of the nasal tip (sn-prn). The nasal tip protrusions found in NAC (20.6 mm in males and 19.4 mm in females), which were the largest measured among the three races, differed highly significantly from those in CP (16.1 mm in males and 15.4 mm in females), which were the smallest, as well as from the protrusions in AA. In black females the tip protrusion was slightly larger, but in black males it was highly significantly larger than in the Chinese.

Nasal bridge inclination. The nasal bridge inclinations found in AA (32.2 degrees in males and 33.4 degrees in females) were the highest among the three races and differed highly significantly from those in CP (27.2 degrees in males and 24.5 degrees in females), which were the smallest. Highly significant also was the difference between the flat and more protruding noses in whites of both sexes. In comparison with NAC, the bridge inclination of black males was slightly larger, but in black females the difference became highly significant.

Nasal index (I = al-al × 100/n-sn). The highest nasal index was recorded for AA (I = 85.2 for males and 82.4 for females) and differed highly significantly from that for CP, who had the second largest index, and that for NAC, who had the smallest (I = 65.8 for males and 65.1 in females). The differences between the smallest indices for whites and the larger Chinese indices were highly significant.

Nasal tip protrusion width index (I = sn-prn × 100/al-al). The nasal tip protrusion width indices were the largest for NAC (I = 59.8 for males and 61.7 for females), exhibiting highly significant differences compared with the smaller CP and smallest AA indices (I = 40.1 for males and 40.5 for females). The indices for the Chinese were slightly larger than those for blacks.

Conclusions. According to the nasal proportion index, the NAC nose is leptorrhin or narrow, the CP nose is mesorrhin or medial, and the AA nose is chamerrhin or wide (24). The values of the soft nose index (tip protrusion/soft nose width) indicate a relatively protruding, narrow soft nose in NAC. In CP the soft nose is less protruding and wider, and in the AA the soft nose has the smallest protrusion in relation to its width (Table 16–4).

Lips and Mouth

Width of the mouth (ch-ch). The widest mouth was observed in both sexes of blacks (54.6 mm in males and 53.6 mm in females), and the smallest was noted in the Chinese (48.3 mm in males and 47.3 mm in females). The mouth of the blacks was larger than that of the NAC; these findings were not significant in males but were highly significant in females. Highly significant were the differences between the smallest mouth widths in CP and those in NAC.

Upper lip height (sn-sto). Upper lip heights were highly significantly larger in AA (26.1 mm in males and 24.5 mm in females) than in NAC or CP in both sexes. The upper lip height was smaller in NAC than in CP, with high significance in males and moderate significance in females.

Cutaneous upper lip height (sn-ls). The measurement of the height of the skin portion of the upper lip was largest in both sexes of AA (16.4 mm in males and 14 mm in females). The measurement in black males highly significantly differed from that in Chinese males, which was the smallest overall. The measurement in Chinese males was slightly smaller than that in white males. In females, the measurements of the cutaneous portion of the upper lips of the three races were almost identical.

Upper vermilion height (ls-sto). The largest measurements of upper vermilion height were found in AA (13.6 mm in males and 13.3 mm in females); these highly significantly differed from those in NAC and CP, which were much smaller in both sexes. The upper vermilion heights in NAC, which were the smallest overall, exhibited highly significant differences from those in CP.

Lower vermilion height (sto-li). The measurements of lower vermilion height in blacks were largest (13.8 mm in males and 13.2 mm in females), differing with high significance from the findings in NAC and CP. The differences between the measurements in NAC and CP were negligible in both sexes.

Lower lip height (sto-sl). The measurement of the

TABLE 16–4. *Anthropometric differences in the nose of three racial groups*[a]

	NAC			AA			CP			Statistical difference[b]		
	N	$\bar{X}$	SD	*N*	$\bar{X}$	SD	*N*	$\bar{X}$	SD	NAC vs. AA	NAC vs. CP	AA vs. CP
					Nose height (n-sn) (mm)							
M	49	53.2	3.3	50	51.9	3.0	30	53.5	2.8	M	NS	MOD
F	51	49.2	2.9	50	48.8	3.5	30	51.7	3.3	NS	SEV	SEV
					Nose width (al-al) (mm)							
M	49	34.8	2.7	50	44.1	3.4	30	39.2	2.9	SEV	SEV	SEV
F	51	31.9	1.9	50	40.1	3.2	30	37.2	2.1	SEV	SEV	SEV
					Nasal tip protrusion (sn-prn) (mm)							
M	52	20.6	2.2	50	17.6	2.0	30	16.1	1.5	SEV	SEV	SEV
F	51	19.4	1.7	50	16.1	2.1	30	15.4	1.8	SEV	SEV	NS
					Nasal bridge inclination (mm)							
M	51	31.6	4.6	50	32.2	5.0	30	27.2	3.5	NS	SEV	SEV
F	50	30.0	5.3	50	33.4	5.7	30	24.5	3.6	SEV	SEV	SEV
					Nasal index (al-al × 100/n-sn)							
M	52	65.8	6.8	50	85.2	7.9	30	73.5	6.2	SEV	SEV	SEV
F	51	65.1	5.8	50	82.4	8.0	30	72.3	6.6	SEV	SEV	SEV
					Nasal tip protrusion width index (sn-prn × 100/al-al)							
M	52	59.8	7.9	50	40.1	5.2	30	41.2	4.1	SEV	SEV	NS
F	51	61.7	6.0	50	40.5	6.3	30	41.6	5.7	SEV	SEV	NS

[a] NAC, North American Caucasians; AA, African-American; CP, Chinese; M, male; F, female.
[b] Significancy: M, mild ($p = 0.05$); MOD, moderate ($p = 0.01–0.02$); SEV, severe ($p < 0.001$); NS, not significant.

lower lip height in AA (22.1 mm in males and 20.2 mm in females), the largest among the three races, highly significantly differed from findings in NAC and CP in both sexes. In NAC males the measurement revealed a nonsignificant difference from that in CP males, whereas in NAC and CP females the findings were moderately significant.

Upper lip height–mouth width index (I = sn-sto × 100/ch-ch). The CP findings were the largest among the three races in both sexes (male I = 51.1; female I = 45.9) and significantly larger only in comparison with the smallest indices in NAC (male I = 41.1; female I = 39.5). In males the indices for CP were moderately significantly larger than for AA, but the values were almost the same in females. The index values in AA were highly different compared with NAC.

Skin portion upper lip height–upper lip height index (I = sn-ls × 100/sn-sto). The skin portion upper lip height–upper lip height indices for both sexes of NAC (*I* = 67.3 for males and 69.4 for females) were the largest among the three races and indicated a relatively high skin portion of the upper lip in relation to the entire upper lip height. These proportion indices for blacks differed with high significance from those for CP males or females. Compared with the AA, the index value for NAC males showed a mildly significant difference, whereas the value for NAC females showed a highly significant difference. The index of the AA females compared with that of CP females was moderately significantly smaller, whereas for AA males it was mildly significantly larger than for CP males.

Upper vermilion height–upper lip height index (I = ls-sto × 100/sn-sto). Blacks had the largest upper vermilion height–upper lip height indices (*I* = 52.3 for males and 54.7 for females), which indicated a relatively high upper vermilion in the upper lip and differed highly significantly from the proportion indices for both sexes of NAC, which were the smallest recorded. Significance was mild to moderate between the NAC and CP in both sexes. The significant difference between AA and CP males was mild, but there was a high significant difference between AA and CP females.

Upper vermilion height–upper lip skin portion height index (I = ls-sto × 100/sn-ls). The heights of the upper vermilion in relation to the skin portion height of the upper lip were largest in both sexes of AA (*I* = 84.1 for males and 96.7 for females), followed by CP and NAC, who had the smallest index values. The black and Chinese males differed mildly to moderately significantly from NAC males. The difference between the AA and CP males was not significant. The index values for AA females highly significantly differed from the lower index values for NAC and CP females. The index for CP females was moderately larger than that for NAC females.

Upper vermilion–lower vermilion height index (I = ls-sto × 100/sto-li). The largest upper vermilion height in relation to the lower vermilion height was observed in Chinese men (*I* = 104.8), followed by black men (*I* = 99.2) and NAC men (*I* = 88.6). The index values in white men highly significantly differed from those in Chinese or black men. The difference between the AA and CP males was mildly significant. Black females had the highest index (*I* = 101.1), followed by Chinese females (*I* = 97.0) and white females (*I* = 87.7). The difference between the indices was highly significant only between NAC and AA females. The difference was moderately significant between the NAC and CP females; black and CP females did not differ significantly.

Lower–upper lip height index (I = sto-sl × 100/sn-sto). In males, the largest index indicating a high lower lip in relation to the height of the upper lip was recorded for AA (*I* = 85.2), followed by that for NAC (*I* = 83.4) and CP (*I* = 79.0). The differences between AA and NAC were small, and those between AA and CP were moderately significant. In females, the differences between the index for NAC (*I* = 85.7, which was highest for the three races) and that for AA (*I* = 83.2) and CP (*I* = 82.4) were small.

Conclusions. All measurements of the upper and lower lips in both sexes of AA were the largest among the three races. Most of these measurements significantly differed from those in CP or NAC. In CP, the width of the mouth and the skin portion of the upper lip was smallest; in NAC the upper lip and its vermilion height were smallest. The heights of the cutaneous portions of the upper lips in the three races revealed the greatest similarities, especially in females.

Analysis of the relationships between the measurements showed that CP had the relatively highest upper lip in relation to mouth width. In NAC, the height of the skin portion of the upper lip in relation to the entire upper lip height was the largest. In AA, the upper vermilion height in relation to the upper lip skin portion height and the lower vermilion height in relation to the upper vermilion height were the largest. The smallest differences in the proportions in the three races were recorded between the relationships of the upper and lower lip heights (Table 16–5).

Ears

Ear width lt (pra-pa). The width of the ear in NAC males was almost the same as that in males in AA and CP. In black males the measurement was moderately significantly larger than that in Chinese males. In females, the ear width was identical in NAC and AA. The smallest measurement in CP females differed moderately significantly from that in NAC and AA.

TABLE 16–5. *Anthropometric differences in the lips and mouth of three racial groups*[a]

	NAC			AA			CP			Statistical difference[b]		
	N	$\bar{X}$	SD	*N*	$\bar{X}$	SD	*N*	$\bar{X}$	SD	NAC vs. AA	NAC vs. CP	AA vs. CP
Mouth width (ch-ch) (mm)												
M	49	53.5	3.6	50	54.6	4.1	30	48.3	6.8	NS	SEV	SEV
F	51	49.8	3.2	50	53.6	4.0	30	47.3	3.3	SEV	SEV	SEV
Upper lip height (sn-sto) (mm)												
M	52	21.8	2.2	50	26.1	2.5	30	23.5	2.2	SEV	SEV	SEV
F	51	20.1	2.3	50	24.5	3.0	30	21.6	2.1	SEV	MOD	SEV
Cutaneous upper lip height (sn-ls) (mm)												
M	13	14.8	2.6	50	16.4	2.0	30	14.2	2.0	MOD	NS	SEV
F	12	13.5	2.2	50	14.0	2.0	30	13.4	2.0	NS	NS	NS
Upper vermilion height (ls-sto) (mm)												
M	49	9.5	1.5	50	13.6	2.1	30	11.2	1.2	SEV	SEV	SEV
F	51	8.6	1.6	50	13.3	1.9	30	10.1	1.4	SEV	SEV	SEV
Lower vermilion height (sto-li) (mm)												
M	49	11.0	2.2	50	13.8	2.1	30	10.8	1.3	SEV	NS	SEV
F	51	10.0	1.5	50	13.2	1.9	30	10.5	1.3	SEV	NS	SEV
Lower lip height (sto-sl) (mm)												
M	52	18.8	2.5	50	22.1	2.4	30	18.5	2.0	SEV	NS	SEV
F	51	16.7	2.3	50	20.2	2.4	30	17.8	2.1	SEV	MOD	SEV
Upper lip height–mouth width index (sn-sto × 100/ch-ch)												
M	52	41.1	5.4	50	47.9	4.9	30	51.1	8.5	SEV	SEV	MOD
F	51	39.5	5.1	50	45.8	5.6	30	45.9	5.7	SEV	SEV	NS
Cutaneous–total upper lip height index (sn-ls × 100/sn-sto)												
M	16	67.3	5.1	50	63.3	7.3	30	60.2	5.8	M	SEV	M
F	12	69.4	6.3	50	57.8	8.0	30	62.0	5.2	SEV	SEV	MOD
Vermilion–total upper lip height index (ls-sto × 100/sn-sto)												
M	49	43.9	6.7	50	52.3	7.7	30	47.9	5.7	SEV	MOD	M
F	51	43.4	8.5	50	54.7	7.9	30	47.1	6.7	SEV	M	SEV
Vermilion–cutaneous upper lip height index (ls-sto × 100/sn-ls)												
M	16	71.8	13.7	50	84.1	18.1	30	80.8	15.6	MOD	M	NS
F	18	71.8	14.7	50	96.7	20.1	30	77.1	16.3	SEV	NS	SEV
Vermilion height index (ls-sto × 100/sto-li)												
M	49	88.6	17.3	50	99.2	12.3	30	104.8	12.5	SEV	SEV	M
F	51	87.7	17.5	50	101.1	10.0	30	97.0	14.3	SEV	MOD	NS
Lower–upper lip height index (sto-sl × 100/sn-sto)												
M	50	83.4	11.5	50	85.2	11.2	30	79.0	8.3	NS	NS	MOD
F	39	85.7	10.6	50	83.2	10.6	30	82.4	5.9	NS	NS	NS

[a] NAC, North American Caucasians; AA, African-American; CP, Chinese; M, male; F, female.
[b] Significancy: M, mild (p = 0.05); MOD, moderate (p = 0.01–0.02); SEV, severe ($p < 0.001$); NS, not significant.

Ear length lt (sa-sba). The greatest ear lengths were found in both sexes of NAC (62.4 mm in males and 59 mm in females), followed by CP (60.7 mm in males and 57.6 mm in females) and AA (59.8 mm in males and 57 mm in females). The difference between the ear lengths of NAC and AA was highly significant in males and moderately significant in females. The findings in AA and CP males and females showed great similarities. The ears were longer in NAC than in CP, with moderately significantly longer ears in NAC males than in CP males; the difference between females was not significant.

Ear index lt (I = pra-pa × 100/sa-sba). The largest indices, indicating relatively wide-short ears, were seen in AA (*I* = 60.7 for males and 60.2 for females). The smallest indices were noted in CP (*I* = 57.4 for males and 56.5 for females). The index values were very close for both sexes of NAC and CP. The indices for AA were highly significantly larger than those for CP. The larger ear index for black males was moderately significantly different compared with that for white males; in females the difference was highly significant.

Ear length–face height index lt (I = sa-sba × 100/n-gn). The length of the ear in relation to the face height was largest in NAC (*I* = 51.5 for males and 52.7 for females) and smallest in AA (*I* = 47.7 for males and 49.1 for females). The indices for AA and CP did not differ significantly. The differences between NAC and AA were highly significant for both sexes. For Chinese males the index value was slightly smaller than that for NAC males; for females the difference was moderately significant.

Conclusions. The shortest and widest ears of all three races were most characteristic of both sexes of AA. The longest ears were found in NAC. In both sexes of CP the ears were the narrowest, and their ear index revealed the smallest values, which indicates a relatively narrow-long ear. The ear proportion index, largest among the three races in AA, showed a relatively wide-short ear. The largest ear length–face height index was seen in NAC and designated a relatively long ear in relation to the face height. The same index, smallest in AA, disclosed a relatively short ear in relation to the height of the face (Table 16–6).

Ethnic Differences

Head

Width of the head (eu-eu). Among the ethnic groups, the widest head was seen in CzP (159.7 mm in males and 155.2 mm in females), whereas the narrowest was found in NAC (151.1 mm in males and 144.4 mm in females). These differences were highly significant, as were those between the wider CzP and the narrower GP head. The difference between the narrowest NAC and wider GP heads was small in females but moderately significant in males.

Length of the head (g-op). The longest heads were measured in GP (193.8 mm in males and 186.1 mm in females), whereas the shortest were found in CzP (188.1 mm in males and 180.7 mm in females). This difference

TABLE 16–6. *Anthropometric differences in the ears of three racial groups*[a]

	NAC			AA			CP			Statistical difference[b]		
	N	$\bar{X}$	SD	*N*	$\bar{X}$	SD	*N*	$\bar{X}$	SD	NAC vs. AA	NAC vs. CP	AA vs. CP
Ear width (pra-pa lt) (mm)												
M	49	35.9	2.2	50	36.2	3.2	30	34.7	2.4	NS	NS	MOD
F	50	34.1	2.6	50	34.2	2.9	30	32.4	2.4	NS	MOD	MOD
Ear length (sa-sba lt) (mm)												
M	52	62.4	3.7	50	59.8	4.0	30	60.7	3.8	SEV	MOD	NS
F	51	59.0	3.6	50	57.0	3.3	30	57.6	3.9	MOD	NS	NS
Ear index (pra-pa × 100/sa-sba lt)												
M	49	58.1	3.9	50	60.7	5.2	30	57.4	3.7	MOD	NS	SEV
F	51	57.4	3.8	50	60.2	5.3	30	56.5	4.8	SEV	NS	SEV
Ear–face height index (sa-sba lt × 100/n-gn)												
M	52	51.5	3.4	50	47.7	4.3	30	49.2	3.6	SEV	NS	NS
F	51	52.7	3.8	50	49.1	3.5	30	50.2	3.9	SEV	MOD	NS

[a] NAC, North American Caucasians; AA, African-American; CP, Chinese; M, male; F, female.
[b] Significancy: M, mild (p = 0.05); MOD, moderate (p = 0.01–0.02); SEV, severe ($p < 0.001$); NS, not significant.

was highly significant. The shorter heads found in NAC differed minimally from the longer heads in GP. Highly significant was the difference between CzP and NAC heads.

Circumference of the head. The smallest circumference was measured in the heads of NAC males and females. The largest circumference was noted in GP males and CzP females. Highly significant were the differences between the measurements of the NAC and those in GP or CzP. The differences between the findings for the GP and CzP males were minimal; in females they were moderately significant.

Cephalic index (I = eu-eu × 100/g-op). The index was the largest in CzP (*I* = 84.9 for males and 85.3 for females) among the three ethnic groups. This finding differed highly significantly from the index values of the NAC males (*I* = 78.3) and females (*I* = 78.2) (which were the smallest) as well as from those of GP (*I* = 78.8 for males and 77.9 for females). Almost identical were the findings for NAC and GP.

Conclusions. The widest head measurements were found in the Czech population, and the longest were noted in the German population. The index values indicated (a) a mesocephalic (medial) head type for both sexes of NAC and GP and (b) a brachycephalic (short-wide) head type for CzP (Table 16–7).

Face

Width of the face (zy-zy). The CzP males (140.4 mm) and females (135.2 mm) had the widest faces of the three ethnic groups. These findings were highly significantly different from those for NAC males (137.1 mm) and females (129.9 mm), which were the narrowest. The face of NAC females was only slightly narrower than the face of GP females; the face of NAC males revealed a moderately significant difference from the faces of GP males, which were wider.

Height of the face (n-gn). The face of NAC males was highly significantly higher (121.3 mm) than that of GP males [which was the smallest of the three groups (116.5 mm)] and moderately significantly higher than that of the CzP. In NAC, GP, and CzP females, the face heights were very similar.

Height of the upper face (n-sto). The measurements of the height of the upper face did not disclose highly significant differences among the three ethnic groups. The upper face heights of NAC and CzP females were identical. The differences between the findings of NAC and GP males, as well as between GP and CzP males, were minimal. The difference between the upper face height of GP and CzP females, and of NAC and GP females, was moderately significant.

Facial index (I = n-gn × 100/zy-zy). The facial index value of the males in NAC (*I* = 88.5) was highly significantly larger than that of CzP (*I* = 84.6) or GP (*I* = 83.2). In females, the differences among the indices in the three ethnic groups were nonsignificant to moderately significant: The highest index values were recorded for NAC (*I* = 86.2), and the smallest values were found for CzP (*I* = 81.8).

Upper face index (I = sn-sto × 100/zy-zy). In males,

TABLE 16–7. *Anthropometric differences in the head of three Caucasian ethnic groups*[a]

	NAC			GP			CzP			Statistical difference[b]		
	N	$\bar{X}$	SD	*N*	$\bar{X}$	SD	*N*	$\bar{X}$	SD	NAC vs. GP	NAC vs. CzP	GP vs. CzP
							Head width (eu-eu) (mm)					
M	52	151.1	5.8	60	153.7	5.9	57	159.7	5.0	MOD	SEV	SEV
F	51	144.4	4.6	60	145.7	4.8	52	155.2	5.6	NS	SEV	SEV
							Head length (g-op) (mm)					
M	52	192.7	6.7	60	193.8	7.1	57	188.1	5.1	NS	SEV	SEV
F	51	184.9	7.0	60	186.1	5.8	52	180.7	6.0	NS	SEV	SEV
							Head circumference (mm)					
M	52	562.5	14.4	60	574.3	14.9	57	572.6	15.7	SEV	SEV	NS
F	51	542.1	14.8	60	552.0	13.4	52	558.3	15.4	SEV	SEV	MOD
							Cephalic index (eu-eu × 100/g-op)					
M	52	78.3	3.8	60	78.8	4.1	57	84.9	2.7	NS	SEV	SEV
F	51	78.2	4.2	60	77.9	3.6	52	85.3	3.1	NS	SEV	SEV

[a] NAC, North American Caucasians; GP, German population; CzP, Czech population; M, male; F, female.

[b] Significancy: M, mild (p = 0.05); MOD, moderate (p = 0.01–0.02); SEV, severe ($p < 0.001$); NS, not significant.

the highest upper face index for the three ethnic groups occurred in NAC (I = 54.0). This values was highly significantly different from that for the CzP [who had the smallest index value (I = 51.6)] and was moderately significantly different from that for the GP (I = 52.4). In females, the difference between the NAC and GP was negligible (I = 52.4–52.6), but their indices revealed high significances compared with those for the Czechs (I = 50.4).

Conclusions. Among the three ethnic groups, the faces of CzP males and females were the widest and the NAC faces were the narrowest. The heights of female faces showed slight differences among the three ethnic groups. The face heights of NAC males, which were highest, significantly differed from those of GP and CzP males.

The highest facial index values in both sexes of NAC characterized a long-narrow face. They differed from the shorter-wider faces of GP and CzP. In NAC the high upper face related to the relatively narrow face in both sexes differed significantly from the proportion of the CzP face. The smallest upper face height–face width index of CzP indicated a markedly short-wide upper face. This face quality of CzP females significantly differed from that of the GP females, who had a more elongated upper face (Table 16–8).

Orbits

Intercanthal width (en-en). The widest space between the orbits was observed in both sexes of NAC (32.9 mm in males and 31.6 mm in females). It differed only slightly from that of the CzP (32.6 mm in males and 31 mm in females) but was highly significantly larger than that of the GP, who had the smallest measurements (29 mm in males and 28.1 mm in females) of the three ethnic groups. The differences between the measurements in CzP and GP were also highly significant.

Biocular width (ex-ex). The greatest projective distance between the exocanthions was recorded in the CzP (94.5 mm in males and 92.5 mm in females), followed by the GP and then the NAC (89.4 mm in males and 86.8 mm in females). In NAC males the measurements were highly significantly smaller than those of the other two ethnic groups. The measurement in NAC females (the smallest among the three groups) was highly significantly different from that of CzP females (the largest), but moderately significantly different from that of GP females. The differences between the GP and CzP were also significant: moderately in males and highly significantly in females.

Intercanthal index (I = en-en × 100/ex-ex). The largest intercanthal indices were found in NAC (I = 36.8

TABLE 16–8. *Anthropometric differences in the face of three Caucasian ethnic groups*[a]

	NAC			GP			CzP			Statistical difference[b]		
	N	$\bar{X}$	SD	*N*	$\bar{X}$	SD	*N*	$\bar{X}$	SD	NAC vs. GP	NAC vs. CzP	GP vs. CzP
Face width (zy-zy) (mm)												
M	52	137.1	4.3	60	139.6	5.2	57	140.4	5.6	MOD	SEV	NS
F	51	129.9	5.3	60	131.2	4.4	52	135.2	4.8	NS	SEV	SEV
Face height (n-gn) (mm)												
M	52	121.3	6.8	60	116.5	6.1	57	118.5	6.3	SEV	MOD	NS
F	51	111.8	5.2	60	110.3	5.2	52	111.8	5.7	NS	NS	NS
Upper face height (n-sto) (mm)												
M	52	74.0	4.2	60	73.4	4.4	57	72.5	3.9	NS	MOD	NS
F	51	68.1	3.4	60	69.5	3.1	52	68.1	3.3	MOD	NS	MOD
Facial index (n-gn × 100/zy-zy)												
M	52	88.5	5.1	60	83.2	4.4	57	84.6	5.5	SEV	SEV	NS
F	51	86.2	4.6	60	83.6	4.7	52	81.8	4.5	MOD	NS	MOD
Upper face index (n-sto × 100/zy-zy)												
F	52	54.0	3.1	60	52.4	3.1	57	51.6	2.3	MOD	SEV	NS
M	51	52.4	3.1	60	52.6	2.7	52	50.4	2.1	NS	SEV	SEV

[a] NAC, North American Caucasians; GP, German population; CzP, Czech population; M, male; F, female.
[b] Significancy: M, mild (p = 0.05); MOD, moderate (p = 0.01–0.02); SEV, severe (p < 0.001); NS, not significant.

for males and 36.4 for females), which differed highly significantly from those of the CzP (I = 34.6 for males and 33.8 for females) and from those of the GP [who had the smallest proportion values (I = 31 for males and 31.1 for females)]. The differences in the indices for CzP and GP were also highly significant.

Conclusions. North American Caucasians had the widest space between the orbits and the smallest distance between the lateral commissures of the eye fissures. These differences produced the largest intercanthal index of the three ethnic groups, a statistically significant finding. The smallest intercanthal index for both sexes of GP was produced by the smallest en-en value in combination with the second widest distance (after CzP) between the exocanthions (Table 16–9).

Nose

Height of the nose (n-sn). Among the three ethnic groups, the nose was highest in NAC males (53 mm), a finding that differs mildly significantly from that for the CzP nose (51.5 mm), which was smallest. The CzP and GP nose (52.1 mm) measurements differed slightly. In females, the differences in the heights of the noses were minimal in the three ethnic groups.

Width of the nose (al-al). The nose was the widest in GP (36 mm for males and 32.2 mm for females), moderately significantly differing from the width of the nose in both sexes of NAC. The nose of GP males was highly significantly wider than that of Czechs (34.4 mm), but not significantly differing in females. The width of the nose was almost identical in NAC and CzP males and females.

Protrusion of the nasal tip (sn-prn). The nasal tip of the NAC protruded more than that of GP. These differences were highly significant in females and slightly significant in males.

Nasal index (I = al-al $\times$ 100/n-sn). Among the three ethnic groups the nasal index for males was the largest for GP (I = 68.6). This value differed mildly significantly from that for NAC, who had the smallest index (I = 65.8), and nonsignificantly from that for CzP (I = 67.1). In females, the CzP had the largest index (I = 66.8), which differed mildly to moderately significantly from that for the NAC (I = 64.4) or GP (I = 64.9).

Nasal tip protrusion width index (I = sn-prn $\times$ 100/al-al). The nasal tip protrusion width index for NAC was highly significantly larger (I = 59.8 in males and 61.7 in females) than that for GP (I = 55.1 in males and 55.8 in females).

Conclusions. Among the three ethnic groups, the height of the nose revealed smaller differences than did the width of the nose, which differed highly significantly between GP and CzP males only; GP had the widest noses, whereas CzP the narrowest ones. In GP the highest nasal index was produced by the widest noses in both sexes. The greatly higher nasal tip protrusion in the NAC males and females influenced the quality of relationship between the tip protrusion and nose width, and it formed a significantly higher index than that for GP. North American Caucasians have a more protruding nose in relation to the width of the nose (Table 16–10).

Lips and Mouth

Width of the mouth (ch-ch). Among the three ethnic groups, the widest mouth in both sexes was seen in NAC. In males it differed moderately significantly from that in

TABLE 16–9. *Anthropometric differences in the orbits of three Caucasian ethnic groups*[a]

	NAC			GP			CzP			Statistical difference[b]		
	N	$\bar{X}$	SD	*N*	$\bar{X}$	SD	*N*	$\bar{X}$	SD	NAC vs. GP	NAC vs. CzP	GP vs. CzP
					Intercanthal width (en-en) (mm)							
M	52	32.9	2.7	60	29.0	2.4	57	32.6	2.4	SEV	NS	SEV
F	51	31.6	2.4	60	28.1	2.8	52	31.0	2.3	SEV	NS	SEV
					Biocular width (ex-ex) (mm)							
M	52	89.4	3.6	60	92.2	5.0	57	94.5	3.9	SEV	SEV	MOD
F	51	86.8	4.0	60	88.8	4.5	52	92.5	4.2	MOD	SEV	SEV
					Intercanthal index (en-en $\times$ 100/ex-ex)							
M	52	36.8	2.1	60	31.0	2.5	57	34.6	2.3	SEV	SEV	SEV
F	51	36.4	2.0	60	31.1	2.7	52	33.8	2.0	SEV	SEV	SEV

[a] NAC, North American Caucasians; GP, German population; CzP, Czech population; M, male; F, female.

[b] Significancy: M, mild (p = 0.05); MOD, moderate (p = 0.01–0.02); SEV, severe ($p < 0.001$); NS, not significant.

TABLE 16–10. *Anthropometric differences in the nose of three Caucasian ethnic groups*[a]

	NAC			GP			CzP			Statistical difference[b]		
	N	$\bar{X}$	SD	*N*	$\bar{X}$	SD	*N*	$\bar{X}$	SD	NAC vs. GP	NAC vs. CzP	GP vs. CzP
					Nose height (n-sn) (mm)							
M	52	53.0	3.5	60	52.1	3.9	57	51.5	3.8	NS	M	NS
F	51	48.9	2.6	60	49.4	2.9	52	48.9	2.6	NS	NS	NS
					Nose width (al-al) (mm)							
M	52	34.7	2.6	60	36.0	2.2	57	34.4	2.0	MOD	NS	SEV
F	51	31.4	1.9	60	32.2	1.6	52	31.9	1.8	MOD	NS	NS
					Nasal tip protrusion (sn-prn) (mm)							
M	52	20.6	2.2	60	19.8	1.7				M		
F	51	19.3	1.9	60	18.0	1.9				SEV		
					Nasal index (al-al × 100/n-sn)							
M	52	65.8	6.8	60	68.6	6.7	57	67.1	6.8	M	NS	NS
F	51	64.4	5.0	60	64.9	4.4	52	66.8	5.3	NS	MOD	M
					Nasal tip protrusion width index (sn-prn × 100/al-al)							
F	52	59.8	7.9	60	55.1	3.8				SEV		
M	51	61.7	6.0	60	55.8	4.4				SEV		

[a] NAC, North American Caucasians; GP, German population; CzP, Czech population; M, male; F, female.
[b] Significancy: M, mild (p = 0.05); MOD, moderate (p = 0.01–0.02); SEV, severe (p < 0.001); NS, not significant.

GP (the smallest measurement), but only a little from that in CzP males. The CzP mouth was slightly wider than that of the GP. In females the width of the mouth of NAC differed mildly significantly from that of GP and CzP. The sizes in GP and CzP females were almost the same.

Height of the upper lip (sn-sto). Male NAC (21.8 mm) and female GP (20.1 mm) had the largest lip heights. The differences among three ethnic groups were negligible.

Height of the upper vermilion (ls-sto). The height of the vermilion of the upper lip was highly significantly larger in both sexes of NAC (8.9 mm in males and 8.4 mm in females) than in GP or CzP (range between 5.9 and 5.4 mm), which differed minimally from each other.

Height of the lower vermilion (sto-li). The highest vermilion was found in both sexes of the NAC (10.4 mm in males and 9.7 mm in females), followed by CzP and then GP, who had the smallest measurements (8.9 mm in males and 8.7 mm in females). The differences were highly significant only between NAC and GP in both sexes. The findings in CzP and NAC females were almost the same; in CzP males the difference was moderately significantly smaller.

Upper lip height–mouth width index (I = sn-sto × 100/ch-ch). Upper lip height–mouth width indices were identical for NAC and GP males, with slightly smaller index value in CzP males. In females the closest index values were found between NAC and CzP and between GP and CzP. Statistical evaluation revealed mild significancy between the GP and CzP males, and moderate between the NAC and GP females.

Vermilion–total upper lip height index (I = ls-sto × 100/sn-sto). The vermilion–total upper lip height index was highly significantly larger in both sexes of NAC (I = 41.1 for males and 43.1 for females) than in the other two ethnic groups. Differences between the GP and CzP indices were minimal in both sexes.

Vermilion height index (I = ls-sto × 100/sto-li). In both sexes of the NAC the vermilion height index values were highly significantly larger (I = 87.7 for males and 87.4 for females) than the values for the other two ethnic groups. The differences were moderately significant when the higher index values of GP males and females were compared with those of the CzP.

Conclusions. The mouth width, upper lip height, and proportion index showing the relation of upper lip height to mouth width exhibited the lowest statistical differences or showed the most similar findings in the three ethnic groups. Heights of the vermilions and their indices demonstrated the greatest differences. The upper vermilion heights of NAC males and females, which were the largest, differed highly significantly from those of the GP and CzP; the largest lower vermilion heights of NAC males and females differed highly significantly only from the GP. The vermilion measurements in the GP and CzP were close to each other (Table 16–11).

TABLE 16–11. *Anthropometric differences in the lips and mouth of three Caucasian ethnic groups*[a]

	NAC			GP			CzP			Statistical difference[b]		
	N	$\bar{X}$	SD	*N*	$\bar{X}$	SD	*N*	$\bar{X}$	SD	NAC vs. GP	NAC vs. CzP	GP vs. CzP
Mouth width (ch-ch) (mm)												
M	52	53.3	3.3	60	51.6	3.2	57	52.5	3.6	MOD	NS	NS
F	51	49.8	3.2	60	48.7	2.7	52	48.6	2.7	M	M	NS
Upper lip height (sn-sto) (mm)												
M	52	21.8	2.2	60	21.3	1.4	57	21.0	2.9	NS	NS	NS
F	51	19.6	2.4	60	20.1	1.1	52	19.8	2.7	NS	NS	NS
Upper vermilion height (ls-sto) (mm)												
M	52	8.9	1.5	60	5.9	1.6	57	5.4	1.8	SEV	SEV	NS
F	51	8.4	1.3	60	5.6	1.4	52	5.4	1.7	SEV	SEV	NS
Lower vermilion height (sto-li) (mm)												
M	52	10.4	1.9	60	8.9	1.7	57	9.4	2.5	SEV	MOD	NS
F	51	9.7	1.6	60	8.7	1.5	52	9.3	2.1	SEV	NS	NS
Upper lip height–mouth width index (sn-sto × 100/ch-ch)												
M	52	41.1	5.4	60	41.3	2.6	57	40.0	4.2	NS	NS	M
F	51	39.5	5.1	60	41.2	2.3	52	40.7	3.9	MOD	NS	NS
Vermilion–total upper lip height index (ls-sto × 100/sn-sto)												
M	52	41.1	7.5	60	27.5	4.6	57	25.7	6.1	SEV	SEV	NS
F	51	43.1	7.6	60	28.0	4.2	52	27.3	6.2	SEV	SEV	NS
Vermilion height index (ls-sto × 100/sto-li)												
M	52	87.8	18.5	60	65.7	15.2	57	57.4	17.2	SEV	SEV	MOD
F	51	87.4	15.8	60	65.0	13.7	52	58.1	15.7	SEV	SEV	MOD

[a] NAC, North American Caucasians; GP, German population; CzP, Czech population; M, male; F, female.
[b] Significancy: M, mild (p = 0.05); MOD, moderate (p = 0.01–0.02); SEV, severe ($p < 0.001$); NS, not significant.

Ears

Width of the ear lt (pra-pa). The differences in the width of the ear were minimal or mildly to moderately significant among the males of the three ethnic groups. In females the ear was the widest in NAC (33.5 mm). This finding differed highly significantly from that in GP (31.3 mm) or CzP (30.2 mm). The difference between the GP and CzP was mildly significant.

Length of the ear lt (sa-sba). In males the longest ear was seen in CzP (65.6 mm). This finding differed highly significantly from that for NAC (62.4 mm), who had the shortest ears. In females the length of the ears differed slightly among the three ethnic groups.

Ear index lt (I = pra-pa × 100/sa-sba). The largest ear index value was found in both sexes of NAC (I = 57 for males and 57.4 for females). This value differs highly significantly from the CzP, who had the smallest index (I = 53.8 for males and 51.8 for females). The index of the NAC was highly significantly larger than that of GP only in females; for males the difference was moderately significant.

Ear–face height index lt (I = sa-sba × 100/n-gn). The ear–face height index value in males was highest in CzP (I = 55.4). This value was not significantly different from that for GP (I = 54.9) but was highly significantly larger than that for NAC, who had the smallest index value (I = 51.5). In females the index values exhibited slight differences among the three ethnic groups.

Conclusions. The differences in the width of the ears were much greater between NAC females (who had the widest ears) and the other two ethnic groups than between GP and CzP females. In males the differences were not significant or mildly to moderately significant. The ear length in CzP males revealed the largest measurements of the three groups. In females the length of the ears only slightly differed among the three ethnic groups. The largest ear index values of NAC males and females indicated a relatively wide-short ear which differed from the relatively narrower-longer CzP and GP ears. The lon-

TABLE 16–12. *Anthropometric differences in the ears of three Caucasian ethnic groups*[a]

	NAC			GP			CzP			Statistical difference[b]		
	N	$\bar{X}$	SD	*N*	$\bar{X}$	SD	*N*	$\bar{X}$	SD	NAC vs. GP	NAC vs. CzP	GP vs. CzP
						Ear width (pra-pa lt) (mm)						
M	52	35.4	2.2	60	35.4	3.1	57	34.4	2.4	NS	MOD	M
F	51	33.5	2.1	60	31.3	2.6	52	30.2	3.0	SEV	SEV	M
						Ear length (sa-sba lt) (mm)						
M	52	62.4	3.7	60	64.0	3.3	57	65.6	3.1	MOD	SEV	MOD
F	51	58.5	3.4	60	59.0	4.3	52	59.3	4.1	NS	NS	NS
						Ear index (pra-pa × 100/sa-sba lt)						
M	52	57.0	4.0	60	54.9	3.5	57	53.8	3.5	MOD	SEV	NS
F	51	57.4	3.8	60	52.7	3.8	52	51.8	4.4	SEV	SEV	NS
						Ear–face height index (sa-sba lt × 100/n-gn)						
M	52	51.5	3.4	60	54.9	2.0	57	55.4	2.8	SEV	SEV	NS
F	51	52.4	3.4	60	53.5	3.2	52	53.0	3.2	NS	NS	NS

[a] NAC, North American Caucasians; GP, German population; CzP, Czech population; M, male; F, female.
[b] Significancy: M, mild (p = 0.05); MOD, moderate (p = 0.01–0.02); SEV, severe (p < 0.001); NS, not significant.

gest ear in relation to face height was identified in CzP males, and the smallest was found in NAC males. Ear length–face height index values were smaller in females than in males. In females the ear length–face height index values showed small differences among the three ethnic groups (Table 16–12).

DISCUSSION AND CONCLUSIONS

A comparison of the major anthropometric signs of the head and face among various races and major ethnic subgroups within the races is becoming an urgent requirement in medicine because of the continuous migration of large groups of people from one continent to another after the Second World War. The morphological changes caused by facial syndromes in the craniofacial complex may be influenced by special racial or ethnic characteristics. The aim of this study was to discover some of the differences and similarities among the morphologic signs of the head and face in healthy subjects belonging to three racial groups and three Caucasian ethnic subgroups.

Morphologic Categories in the Three Races

The cephalic index indicated that North American Caucasians have a mesocephalic (medium wide-long) head, whereas African-Americans have a dolichocephalic (long-narrow) head and the Chinese have a hyperbrachycephalic (short-wide) head. No noteworthy differences were found in the height of the head in the three races.

The facial framework expressed by the facial index was largest in blacks; the face is long (leptoprosop) in relation to its width—in contrast to the faces of NAC and CP, which have a proportionally more balanced facial frame (mesoprosop). Although the upper face height was almost identical in blacks and Chinese and markedly shorter in North American Caucasians, the height of the lower face in blacks was the largest, occupying close to two-thirds of the height of the face. The width of the face in the Chinese group greatly exceeded the findings in North American Caucasians or African-Americans in both sexes.

The most striking differences between the three races were found in the orbits, the nose, and the lips and mouth.

The intercanthal width and the intercanthal index were the largest in CP and were the smallest in NAC (en-en) and AA (intercanthal index) (Table 16-3). The eye fissure length and height was the smallest in the Chinese group, but the inclination was largest (12.1 degrees in males and 12.5 degrees in females) compared with that of AA (7 degrees in males and 9 degrees in females) or NAC, who had the smallest inclination (2.3 degrees in males and 5 degrees in females).

The nasal framework described by the nasal index was highly significantly different among the three races. Blacks had the largest index values. These indicated a relatively wide-short (chamaerrhin) nose that differed from the narrower nose of the CP or from the narrowest nose of the NAC. The differences in the height of the noses were less notable. The difference in the nasal tip protrusion was great: largest in NAC and smallest in CP. In NAC, this finding contributed to the narrowest soft nose shape, typical of Caucasians. The nasal bridge inclination in the CP (27.2 degrees in males and 24.5 degrees in females), the smallest among the races and associated with the smallest nasal tip protrusion, greatly enhanced the visual impression of the flatness of the nose. The nasal bridge inclination in AA was slightly larger than in NAC. The very close values of the nasal tip protrusion–nose width indices for the AA and CP indicated a similar build of the soft nose, which was wider and lower than that of the NAC.

In the orolabial region, the greatest differences observed between the races were in the heights of the entire upper lip and upper vermilion. The blacks had the highest upper lip and the highest lower lip with the highest upper and lower vermilions. These measurements differed markedly from those of the CP and NAC. In females of all three racial groups, the heights of the cutaneous portions of the upper lip were very close.

The mouths of black males and females were the largest, and those of the CP the smallest. Although the comparison of the heights of the entire upper and lower lips did not show highly significant differences among the races, the variations in the heights of the vermilions of the upper and lower lips revealed great differences: In NAC the upper vermilion was moderately smaller than the lower, in CP the upper vermilion was larger, and in AA the vermilions were well-balanced.

The widest ears were seen in AA, and the narrowest ears were found in the CP. In blacks the length of the ear was the shortest among the races, a highly significant difference compared with the length of the ear of NAC, which was the longest of the three races. In blacks, the highest ear index values indicated a wide-short ear. Concordantly, the relatively short ear in relation to the long face of the AA produced the smallest ear length–face height index. This finding was highly significant when compared with the values for NAC, who had the largest index value among the races, indicating a relatively long ear in relation to the face height.

Morphologic Categories in the Three Caucasian Ethnic Groups

Analysis of anthropometric data collected from ethnic groups of Caucasians was restricted to the NAC and two European Caucasian groups because of the limited number of norms of anthropometric measurements and proportion indices available for other European Caucasian ethnic groups. In 10 of 28 measurements and proportion indices, NAC significantly differed from both European Caucasian groups. The orbits, lips and mouth, and ears revealed the greatest differences. The least differing were the measurements and proportions of the nose.

The head measurements of NAC males and females were the smallest in width and circumference among the three ethnic groups. The GP had the longest heads. The CzP had the largest head width and cephalic index and the smallest length of head among the three ethnic subgroups (Table 16–7). Between the NAC and the GP, more similarities were found in measurements of the head than between the GP and the CzP. Thus, in NAC and GP the head length and cephalic index (mesocephalic type) were close in both sexes; only the head width was similar in females. Cephalic index showed, on average, a brachycephalic head in CzP.

In NAC the facial index values (the highest) indicated a relatively long-narrow face, which differed from the wider face of CzP males and the relatively widest of GP males. The largest upper face height in NAC produced the largest upper face index values among the three ethnic subgroups (Table 16-8). Small differences were found between GP and CzP in the face width, upper face, and face index of males and in the height and upper face heights of both sexes.

The values of the intercanthal index revealed the greatest statistical differences: In both sexes of NAC the highest index (I = 36.8 for males and 36.4 for females) indicated a relatively wide space between the orbits in relation to the distance between the exocanthions. In GP the index was significantly smaller than in NAC (I = 31 for males and 31.3 for females), indicating the relatively smallest space between the orbits (29 mm for males and 28.1 mm for females). The differences among the intercanthal indices were caused less by differing sizes of intercanthal width (en-en) than by significant differences among biocular widths (ex-ex); these were largest in both sexes of CzP, followed by the GP and then NAC, who had the smallest measurements. For the orbits, only the intercanthal width of both sexes of NAC and CzP were close in values (Table 16–9).

Nasal measurements did not show highly significant differences in the height and width of the nose and, consequently, between the nasal indices of the NAC and two European Caucasian groups. The widest nose was found in GP males and females, whereas the narrowest nose was noted in CzP males and NAC females. In comparison with the GP, the tip of the nose of NAC protruded more. Narrower noses with more protruded tips produced a more balanced soft nose form.

For NAC and GP, similar findings were observed in height of the nose in both sexes and in the nasal index in females. Nose heights were close also in NAC and CzP females, as was the nose width of both sexes. Nose heights for GP and CzP were similar in both sexes. Nose widths were similar in females, and the nasal indices were similar in males.

The greatest difference between the North American and European Caucasians occurred in the upper lip vermilion height, which was significantly larger than that of GP or CzP. In NAC the high upper vermilion occupied almost half (mean 42%) of the entire upper lip height, in GP it occupied less than one-third (mean 28%), and in CzP it occupied more than one-quarter (mean 26.5%).

The males or females, or both, in the GP and CzP samples exhibited a greater number of similar measurements (mouth width, upper lip height, upper and lower vermilion heights, upper lip–mouth index, and the upper lip skin–upper lip height index) than did the NAC and GP (upper lip height and the upper lip height–mouth width index) or the NAC and CzP (mouth width, upper lip height, lower vermilion height, and the upper lip height–mouth width index).

The width of the ear in NAC and GP was the same in males and moderately smaller in CzP. The ear was significantly wider in NAC females than in GP or CzP females. The length of the ear of females in the three ethnic groups was almost the same. The ear index in males and females showed a relatively wide-short ear in NAC, compared with the narrow-long ears of CzP. The similar length of the ears in females of the three ethnic groups produced almost identical proportion indices of the length of the ear in relation to the height of the face. In NAC males, in relation to the height of the face, the length of the ear was markedly shorter than that of the GP or CzP.

In this study the primary intention was to demonstrate the major differences among three racial groups of North America. However, our sample populations may not be racially homogeneous. Our sample of NAC, because it included 40% Anglo-Saxons (18), may not represent the white population in all regions of North America. Similarly, findings based on a small sample of AA ($N = 100$) that included only 28% who had no other racial heritage may be not representative of the entire black population of North America. The Chinese population sample, the most homogeneous of the analyzed groups, was drawn from one of the South Chinese populations (Hong Kong, Canton, Singapore), who represent one of the largest group of Asians settled in North America. In North America the anthropometric differences in the head and face of Chinese subjects may be smaller than those of white or black populations because many may have immigrated from the same geographical areas of the southern mainland of Asia. The marked differences between the two ethnic groups of European Caucasians and between the North American and European Caucasians were clearly demonstrated. The determination of the genetic causes of these differences is beyond the scope of the present study.

Not all differences are immediately apparent. Some signs of the head and face that are mildly or moderately different among the races or ethnic subgroups may not be discernible during visual examination. In contrast, some mildly differing measurements or proportion indices, if they are located in the sensitive area of the face (see Chapter 7, Fig. 1) (25), may be noticeable during visual inspection. The search for craniofacial norms should be extended to all major ethnic groups within the races. Surgeons urgently require this information. The present study should serve as an inspiration for those who are willing to participate in this challenging program.

REFERENCES

1. Topinard P. *Eléments d'anthropologie générale.* Paris: A Delahaye et E Legrosivier, 1885.
2. Coon CS. *The origin of races.* New York: AA Knopf, 1962.
3. Biasutti R. *Le Razze e i Popoli della Terra.* Torino: Unione tipografico, 1959.
4. Roginskij JA, Levin MG, eds. *Osnovy antropologii* [in Russian]. Moskva: Izd Moskovsko Universiteta, 1955.
5. Deniker J. *Les races et peuples de la terre.* Paris: Masson et Cié, 1926.
6. Czekanowski J. Das Typenfrequenzgesetz. *Anthropol Anz* 1928;5:335–359.
7. Czekanowski J. Polska synteza antropologiczna w perspektywie historicznej [in Polish]. *Przegl Antropol* 1956;22:470–613.
8. Michalski I. *Struktura antropologiczna Polski* [In Polish]. Lódz: Acta Anthropol Univ Lodziensis, 1949.
9. Tobias PV. The negative secular trend. *J Hum Evol* 1985;14:347–356.
10. Billy G. Migration et évolution chez quelques populations actuelles. In: Ferembach D, ed. *Les processus de l'hominisation. L'évolution humaine, les faites, les modalités.* Paris: Colloque International du CNRS, no. 599, 1981;265–268.
11. Vercauteren M, Susanne C, Orban R. Évolution seculaire des dimensions céphaliques chez les enfants Belges, entre 1960 et 1980. *Bull Mem Soc Anthropol Paris* 1983;10:13–24.
12. Gast A. Über Veränderungen von Kopfmassen bei Kindern im Vorschulalter in den Jahren 1953 und 1976. *Ärztl Jugendkd* 1983;74:322–329.
13. Brůžek J, Hajniš K, Tláskal P, Blažek V, Krásničanová H. Temporal trends and debrachycephalization in Czech children in the first year of life [in Czech]. *Česk Pediatr* 1988;43:199–203.
14. Dahan J. Morphogenetische Tendenz und Umweltbedingtes Profilverhalten. *Fortschr Kieferorthop* 1984;45:198–216.
15. Richardson ER. Racial differences in dimensional traits of the human face. *Angle Orthod* 1980;50:301–311.
16. Chung CS, Kau MCW, Walker GF. Racial variation of cephalometric measurements in Hawaii. *J Craniofac Genet Dev Biol* 1982;2:99–106.
17. Chung CS, Runck DW, Bilben SE, Kau MCW. Effects of interracial crosses on cephalometric measurements. *Am J Phys Anthropol* 1986;69:465–472.
18. Farkas, LG. *Anthropometry of the head and face in medicine.* New York: Elsevier, 1981.
19. Farkas LG, Ngim RCK, Lee ST. The fourth dimension of the face: a preliminary report of growth potential in the face of the Chinese population of Singapore. *Ann Acad Med Singapore* 1988;17:319–345.
20. Hajniš K. *Kopf-, Ohrmuschel- und Händwachstum (Verwendung bei den Operationen der angeborenen Mißbildungen und Unfallsfolgen).* Praha: Acta Universitatis Carolinae (Biology), 1972;1974.
21. Hajnišová M. Growth of the face and basic characteristics of the skull in children and youth from 6 to 18 years [in Czech]. Praha: Chair of Anthropology, Charles University, Faculty of Natural Sciences, PhD thesis, 1968.
22. Martin R, Saller K, eds. *Lehrbuch der Anthropologie.* Stuttgart: G Fischer, 1957.
23. Farkas LG, Munro IR. eds. *Anthropometric facial proportions in medicine.* Springfield: Charles C Thomas, 1987.
24. Hajniš K. Categories in classical anthropometric proportion systems. In: Farkas LG, Munro IR, eds. *Anthropometric facial proportions in medicine.* Springfield, IL: Charles C Thomas, 1987;9–17.
25. Farkas LG, Kolar JC. Anthropometrics and art in the aesthetics of women's faces. *Clin Plast Surg* 1987;14:599–616.

Chapter 17

Computer-Assisted Acquisition of Facial Surface Topography

David E. Altobelli

The measurement and description of facial form continues to provide technical challenges. The subtle complexity of the face is extraordinary, in that such a small area can assume millions of distinctive appearances. And with the face providing the primary interface for human interaction and communication, description of its characteristics continues as a dominant theme.

Limitations in instrumentation made early approaches to measurement of the facial form almost exclusively distances, arc lengths, angles, and proportions between these parameters. Studies in the scientific literature have amassed a large database of these measurements, with data sources largely from cephalometric radiographs and direct anthropometric facial measurements. Only recently with hardware and software developments for digital acquisition of spatial data, such as registered and nondistorted medical studies i.e., computerized tomography (CT), magnetic resonance imaging (MRI), and structured light surface imaging techniques, has it become feasible to accurately and efficiently sample the three-dimensional spatial relationships of anatomic tissues with minimal distortion. However, many new problems arise, some of which include the following: (a) How do we accurately and in a tangible fashion describe the intricate contour and structural forms in way that is both statistically robust and usable in clinical practice? (b) How do we best integrate the large database of two-dimensional relationships with these new three-dimensional data formats? (c) How does one establish normative data for contour and form?

The potential applications for this three-dimensional facial surface information are numerous, including such disciplines as anthropology, medicine, ergonomics, art, the adaptation of military equipment, and especially the entertainment and fashion industries. In our case, the primary interest in facial measurement has been for the planning of facial surgery. Understanding of the response of soft tissues to the movements of underlying bone is still poorly understood. With advanced computer-assisted graphic techniques emerging to plan the detailed movements of the abnormal skeleton, the response of the facial surface can only be roughly approximated. The ideal approach to the design of the surgery most probably is from the *outside in* (determine the ideal position of the skin surface first, then move the bones to achieve this result) rather than from the *inside out* (reposition and augment the bones, then hope the skin response is acceptable). The ideal facial form will probably be based on (a) existing soft-tissue constraints, (b) culturally based facial aesthetic characteristics, and (c) hereditary features. The design of underlying bones would then be optimized to achieve both (a) the structural criteria for mastication, speech, and breathing and (b) the aesthetic concerns of the idealized facial form customized for the specific patient.

The objective of this chapter is to introduce recent developments in technology that can be applied to recording the three-dimensional anatomic form. This includes hardware advances for three-dimensional digitizing of single points and the scanning of the entire surface of objects. Approaches to automated anthropometric measurements, as well as the use of surface data for planning facial surgery, will be introduced.

DATA ACQUISITION

Conventional Techniques

Description of the head and face has relied primarily on the measurement of the distances, arc lengths, and angles between anatomic landmarks. Radiographically, homologous landmarks (anatomic landmarks with biological correspondence from subject to subject—for example, the nasion or the midline junction of the nasal and frontal bones) are frequently the edges and junctions of bones. Similarly with the facial surface, the landmarks are often found at the edges or boundaries of structure, or at the most prominent convex or concave regions. Based on the distance and angular relationships between two or more points, means and standard deviations are derived for these scalar (magnitudes without orientation) parameters (i.e., the distance between the inner corners of the eyes is a magnitude) without knowing whether one inner canthi is higher than, in front of, or symmetric to the midline with respect to the other. These relationships can be adjusted for subject size variation somewhat, by using proportional relationships between parameters.

Information in scalar form continues to be very useful in assessing normal growth and planning treatment. These parameters are often plotted as a function of age with the mean and standard deviations at each age, often with a curve that has been best fit to the data. The patient's measurement can then be directly compared to the normal population chosen, to determine the extent of their variation from average. Parameters outside a two-standard-deviation envelope are considered abnormal. Estimates of the rates of growth and the change in the rates of growth can be determined from the first and second derivative of the parameter–age curve.

This approach to measurement has had distinct advantages, including low cost, minimal requirements for specialized instrumentation, and a relatively short learning curve for the training of specialized technicians to record the data. Furthermore, comparison and statistical analysis of the data in this form are usually straightforward and have provided the basis for the majority of published studies that have compared morphologic differences in the craniofacial region. However, manual anthropometric measurement of the face can also be quite time-consuming, and in young children it can be extremely difficult to perform.

Conventional approaches of anthropometric measurement are also limited in their ability to comprehensively describe the three-dimensional characteristics of the craniofacial region. Although very important discrete relationships can be derived between anatomic points, these distances or angular measures only provide a minuscule description of the three-dimensional morphology (form and structure). For example, if one were provided with all of the conventional anthropometric facial measurements from a subject and then were to attempt to "backtrack" and reconstruct the spatial positions (x, y, z coordinates) of each of the anatomic landmarks in space, this task would probably not be accomplished. The deficiencies with the data in this form become even more clear when attempting to design a surgical procedure fully in three dimensions. With current methods in computer visualization of medical data, the problem in analysis and treatment planning is no longer the ability to view the anatomy in a three-dimensional form, but rather the paucity of normative data in a three-dimensional form to determine the extent of the deformity or the optimal new position of the anatomic units (1).

In the next section, we will briefly review some of the techniques that have been used to document the appearance of patients. Two-dimensional techniques, such as photography and video imaging, will be briefly discussed. Other techniques to record and measure three-dimensional morphology, which fall into the science of biostereometrics, will be reviewed, such as moulages, stereophotogrammetry, holography, and the instrumentation available to digitize the three-dimensional coordinates of individual landmarks. Finally, a more detailed discussion of structured light surface scanning techniques will be presented. This will include the specific characteristics of these data formats and how they can be applied to anthropometric measurement and analysis.

Photography

Photography is the most common method used clinically to document a patient's external appearance. Although standardized views are commonly used, subtle differences in orientation, magnification, and lighting often make it difficult to make both qualitative and quantitative comparisons at different time-points. More recently, techniques have become available using video cameras and high-resolution charged couple device (CCD) camera backs to capture the image in a digital form that allows immediate viewing of the image data on the computer monitor. Advantages of this method include the ability to instantly assess the quality of the captured image, and once the image is in a digital form, methods of image enhancement and measurement are possible. Measurements such as distances (components of distance in the plane perpendicular to the long axis of the camera) can be made if a fudicial of known dimension [we often tape a penny (19 mm or 0.75 in.) on the patient's neck] is in the view of the camera at the approximate distance of the structures to be measured. This allows a scale factor for the image to be approximated at a later time. In one study comparing photogrammetric to 104 direct anthropometric facial measurements, 62 measurements were possible from front and lateral photographs, and of these, 26 measurements were reliable (2).

Two-Dimensional Simulation of Facial Surgery Based on Anthropometry

Personal computer (PC)-based systems can be used to estimate surgical treatment options based on two-dimensional photographic or image data (3). Frontal and lateral photographs or color slides can be digitized with a video camera or a high-resolution slide digitizer, or they can be transferred directly from photographic film to Kodak photo CD and entered into the computer in a high-resolution digital format. Several image processing and manipulation commercial software packages are available—such as TIPS (Truevision, Indianapolis, IN) software on PC DOS systems, or Photoshop (Adobe, Inc.) on MacIntosh (MAC) and PC Window (Microsoft, Inc.)-based systems—that can manipulate sections of the image (such as translation, rotation, and scaling) in a controlled fashion. A scale factor is initially determined (pixels/cm) and a grid pattern with a 10-mm grid is centered on the patient's midline from the frontal view and from nasion on the lateral view. As patches of the image are moved, the grid provides reference to keep track of the cumulative movements.

Based on the patients' anthropometric measurements and their deviation from normal standards, the facial units (periorbital, nasal, forehead, midface, and mandible) can be repositioned, based on anticipated correction of hard and soft tissues, to provide more normalized anthropometric proportions. The altered appearance of the patient is then reevaluated both visually and by basic anthropometric measurements of the computer screen display. The final combination of movements that approximate the operative design are determined from the displaced grids, and further analysis and planning can be pursued in three dimensions using the surface-rendered three-dimensional reconstructions from surface scan and CT data sources.

Photogramometric Techniques

The basic concepts in stereophotogrammetry are analogous to the principles of binocular vision. The discrepancy in the two images, as viewed at differing vantage points by the eyes separated by 50–60 mm, is perceived in the brain where the difference is interpreted in terms of depth, as well as in terms of lateral and vertical position. This same principle can be mimicked with lateral separation of two cameras recording images of the same object from different vantage points. If the images are then played back to the corresponding eyes, the illusion of three dimensions is re-created.

This has been the basic approach used for mapping the topography of earth's terrain using Landsatt data from satellite pictures. The terrain is photographed by metric cameras at different time-points. The separation between the camera images is determined by the time between exposures and the speed of the satellite. The x and y positions on the film of identifiable landmarks in both views are recorded in each image. Based on the separation of the camera images, the focal length of the lens, and the relative positions of the corresponding landmark(s) in both views, the depth information can be determined from basic trigonometric relationships (4,5).

This approach allows rapid and inexpensive acquisition of data in a three-dimensional format. Although the approach works well at describing landmarks comprising the boundaries of a structure (e.g., the corners of the eyes and mouth), difficulties arise trying to identify areas of maximum and minimum convexity and concavity on the surfaces of the forehead, cheeks, and chin. These areas are characterized by a smooth, rather featureless surface, where a corresponding landmark cannot be clearly identified in both views.

This problem can be addressed in several ways. One approach plays back the stereophotographs, and then one manually plots the contours lines that intersect at the same height above a specified base using a stereoviewer and stereoplotter (6). Other approaches have used techniques to mark several points on the surface of the object (7) or to project light, in the form of lines, grids, or a rastered point, to provide easily identifiable, temporary landmarks on the surface of objects to be analyzed (8–12). Cyberware [Monterey, CA] has developed a commercial system in widespread use and this will be discussed in more detail later in the chapter.

The light projected onto the surface can be used in two basic ways. The most common approach is to allow the light pattern to provide temporary landmarks on the surface of the object when being viewed by cameras in offset positions. The challenge is to be able to provide a large number of distinguishable locations to provide a detailed representation of the surface. The other technique is for the location of the light on the objects surface to represent predefined spatial locations, such as a plane of light representing a constant range value from the camera. The challenge with these approaches is again to provide enough distinguishable lines to provide a high-resolution description of the surface. Difficulties arise in correcting for the optics in the systems. For example, it is easy to project a line of light that is planar using a laser and a cylindrical lens. However, for example, to project a grid onto an object—using, for example, a slide in a projector—the divergence of the beam has to be accounted for and corrected for different ranges. Telecentric photogramometric techniques are used to provide these corrections (13).

Moiré Diffraction

Moiré techniques have been used frequently in medicine to map three-dimensional contour, for example, to assess facial asymmetry and in the assessment of scoliosis. The standard moiré technique uses a transparent series of lines in the form of a grid (rhonchi ruling) placed in front of the object to be contour-mapped. A light source is offset from the viewing angle. The light passes through the diffraction grading twice: First it passes from the source, through the grading to the object, and then as the light is reflected off the object, it passes through the grading and is viewed by the film or investigator. An interference pattern of light and dark lines or fringes is created; each fringe represents a set of equidistant points from the grid. The fringes appear superimposed on the object as a series of contour plots of similar elevation (14).

The distance (in millimeters) of any fringe line to the grid is determined by the following equation:

$$\text{distance} = \frac{nl}{\frac{d}{g} - n}$$

where n is the number of the fringe, l is the distance from the point source of light to the grid, g is the number of grid lines per millimeter of the grading, and d is the lateral distance between the eye (camera) and the light source. For example, studies have used these techniques to study changes in facial contour after repair of zygomatic fractures (15). Techniques have been developed to automate the interpretation of the moiré patterns and convert them to a three-dimensional format (16). However, the method is limited by the vantage point of the system to the subject, with areas of rapid elevation change being difficult to characterize because of inability to distinguish the line separation. This requires evaluation of the objects from different vantage points.

Holography

A hologram can be thought of as a window in time, with the entire panoramic scene from any vantage point through the window, represented and stored in the glass. When the hologram is made, it captures the "instant" of the light passing through the glass in the form of both amplitude and phase. Later, when the hologram is reviewed and the holographic pattern is replayed, the image precisely replicates the object's three-dimensional appearance and thus the illusion that the object is still in its original position. Original holograms required coherent laser light and a very stable environment for exposure, with laser light also needed for playback. Newer techniques have been developed to make holograms without the need for lasers or real objects. Computer-generated representations of three-dimensional objects can be recorded in a holographic form as well (17). Although true, vivid three-dimensional views are possible, the current forms do not readily allow extraction of quantitative information.

Moulages

A moulage, or model of the facial form, is usually obtained with an alginate-based impression material applied to the facial surface in a semirecumbent position. This material, with a consistency something like that of gelatin, is picked up or attached to a more rigid framework to keep the impression from distorting. Plaster or stone is then used to fill the impression and make the master cast. Although small regions of the head and face (e.g., the ear) are fairly easy to obtain, the whole face is much more difficult, requiring the patient to remain fairly still, often breathing through just two straws from their nose (18). Obviously, this technique is not amenable to small children unless they are under general anesthesia. There are several advantages to this technique, most notably that a spatial model of the patient is available for archival measurement and for the fabrication of implants or prosthetic devices in facial reconstruction. The disadvantages include (a) the time-consuming and claustrophobic nature of the process, (b) storage and the brittle nature of the models, and (c) little data with regard to the facial distortion caused by the weight of the impression material.

Discrete Landmarks

An alternative to the conventional approach of measuring the relationships of facial landmarks with calipers, protractors, and measuring tapes is to record the spatial coordinates (x, y, z) of each landmark as referenced to the anatomic coordinate system. With this approach, the three-dimensional position of a landmark is recorded using a three-dimensional digitizing instrument (more about these devices later), and the various relationships between the recorded or digitized points can be determined at a later time using trigonometric formulas (19).

The landmark points can be acquired sequentially (such as the right medial canthi, the left medial canthi, nasion, glabella, etc.) and stored in a data file such as a spreadsheet. The distance between any of the two points (x_1, y_1, z_1) and (x_2, y_2, z_2) can then be calculated using the following relationship:

$$\text{distance} = \overline{12} = \sqrt{(x_1 - x_2)^2 + (y_1 - y_2)^2 + (z_1 - z_2)^2}$$

In addition, the angular relationship between three points (x_1, y_1, z_1), (x_2, y_2, z_2), and (x_3, y_3, z_3) can be determined by first calculating the distance between each of the points using the distance formula above ($\overline{12}$, $\overline{23}$, and $\overline{13}$) and then using the basic trigonometric relationship from the Law of Cosines:

$$a^2 = b^2 + c^2 - 2bc\cos\alpha$$

$$b^2 = a^2 + c^2 - 2ac\cos\beta$$

$$c^2 = a^2 + b^2 - 2ab\cos\gamma$$

where a, b, and c are the lengths of the sides of any triangle, and α, β, and γ are the vertices or angles opposite the corresponding sides, respectively.

The angle $\angle\overline{123}$ between points 1, 2, and 3 equals

$$\angle\overline{123} = \frac{\cos^{-1}(\overline{12}^2 + \overline{23}^2 - \overline{31}^2)}{2(\overline{12} * \overline{23})}$$

Geodesic or arc distances can be approximated by taking several points along the arc scribed on the face and summing the distances between each of the sets of points.

Often when assessing symmetry and planning surgery, it is important to know the relative position of the points with respect to the anatomic coordinate system: superior–inferior, anterior–posterior, medial–lateral. For our example with the medial canthi, knowing that one of the canthi is higher and anterior to the other would be important information for surgical planning. The distance alone between the canthi does not provide this information. If the three-dimensional digitizers coordinate system was aligned with the patient coordinate system, then simply the difference between each of the components provides this information. For example, in an adult, if the coordinate system (in millimeters) has its origin at soft-tissue nasion, with the following orientation ($+x$ is left, $+y$ is posterior, and $+z$ is superior), the left inner canthus has the coordinates (15, 13, −5.5), and the right inner canthus has the coordinates (−19, 11, −7), then the following relationships can be derived (Fig. 17–1):

1. *Anthropometric distance.* The distance between the right (−19, 11, −7) and left canthi (15, 13, −5.5) can be calculated from the distance formula above:

$$= \sqrt{((15 - -19)^2 + (13 - 11)^2 + (-5.5 - -7)^2}$$

$$= \sqrt{34^2 + 2^2 + 2.2^2}$$

$$= 34.13 \text{ mm}$$

In the adult, the average inner canthal distance is 32 mm with a standard deviation of about 2.4 mm. Thus, by conventional analysis, the relationship between the inner canthi is normal.

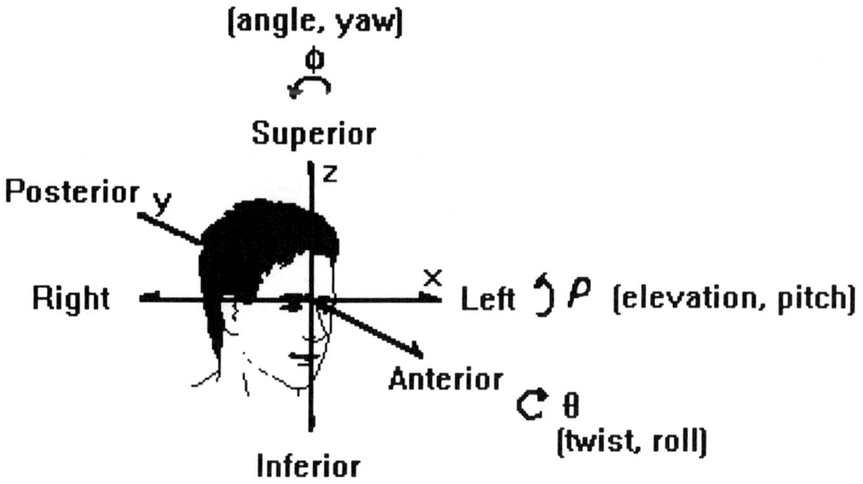

FIG. 17–1. Schematic representation of the anatomic coordinate system in relation to an arbitrarily assigned Cartesian coordinate system. Synonyms for rotation about the various axes are shown. In this example, the origin of the coordinate system is placed at soft-tissue nasion. This can be defined arbitrarily at other locations along the midline plane.

2. *Medial–lateral position.* Because the origin was at the midline, the relationship of the right and left canthi to the midline is directly related to the absolute value of their x coordinate. The left canthi (**15**, 13, −5.5) is 15 mm to the left of the midline, and the right canthi is (**−19**, 11, −7) 19 mm to the right of the midline. Thus, even though the distance between the canthi is normal, their positions are not symmetric: The right canthi is displaced laterally 4 mm greater than the left side, with the estimate being in the normal range (i.e., half the inner canthal distance of 32 ± 2.3 or 16 ± 1.2). There are published data available to provide an estimate of inner canthal lateral asymmetry in the general population (20) (see Appendix A-1).

3. *Anterior–posterior position.* The difference in the y values of the left canthi (15, **13**, −5.5) and the right canthi (−19, **11**, −7) [or (13 − 11) = 2 mm] demonstrates that the left canthi is 2 mm anterior to the right canthi. The average anterior–posterior range for the offset position of this structure in relation to the nasion is not known. The range of asymmetry of anterior–posterior position in relation to the sellion was reported earlier (20) (see Appendix A-1).

4. *Superior–inferior position.* The difference in the z values of the left canthi (15, 13, **−5.5**) and the right canthi (−19, 11, **−7**) [or (−5.5 − −7) = 2.5 mm] demonstrates that the left endocanthion is asymmetrically positioned 2.5 mm higher than the right endocanthion. The average superior–inferior offset for this structures' position in relation the nasion is not known.

Thus, preservation of the coordinates of the landmark provides the conventional scalar measurement of distance between any two points consistent with the measure found directly with the caliper. In addition, however, the vector components (magnitude and direction) in the anatomic coordinate system can also be determined. Furthermore, if an another anatomic coordinate system is chosen based on different criteria, the vector components can be recalculated. Or, if at a later time, some new relationship is required between two or more points, all the necessary information has been preserved for the new calculation.

INSTRUMENTATION FOR MEASUREMENT OF SPATIAL POSITIONS

Several devices are available for position (x, y, z) and orientation [elevation or pitch (ρ), roll or twist (θ), and angle or yaw (ϕ) tracking]. Full description of an object's location (6 degrees of freedom: x, y, z, ρ, θ, ϕ) requires both the position and orientation information of a specific point on the rigid body, or the position of at least three points on a rigid body. These devices fall into two basic categories: contact and noncontact. The simplest contact device uses three or more translating stages to position the tip of a stylus in the x, y, and z directions (3 degrees of freedom). The other degrees of freedom (orientation) are fixed at 90 degrees (i.e., orthogonal). The advantages of this approach are precision, accuracy, and low cost. The disadvantages are that data acquisition is slow, and often some points on the inside of the structures or under overhangs can be difficult to access. This can usually be accommodated by changing the shape and orientation of the stylus.

Articulated arms, similar to robots without active positioning, provide another mechanical approach to measurement. The arm will usually consist of six articulations with rotary transducers at each joint to allow determination of the position and orientation of the stylus tip. The transducers at each joint can be based on variable resistance (potentiometers), optically encoded discs, Hall effect magnetic sensors, and other techniques that involve variable capacitance or inductance can be used. High resolution for position and orientation are possible for these devices (FARO Technologies, Inc., Lake Mary, FL).

Electromagnetic devices are also available for tracking that were originally developed for military applications such as pilot head position. This approach relies on three orthogonally positioned wire coils embedded in a sensor that is attached to the object to be tracked. They can be fairly small in size with small wire connection. A source or reference creates a magnetic field gradient that allows the sensor to determine its position and orientation. This is based on the magnitude of the currents induced in the sensor's three orthogonally oriented coils with the oscillation of the magnetic field. Resolutions are on the order of about 0.4 mm and with positional accuracy of 0.8 mm and angular position of 0.1 degree (Polhemus, Colchester, VT, Ascension Technology Corp., Burlington, VT).

Sonic techniques can also be used to determine three-dimensional spatial coordinates. This technique, based on triangulation, uses three microphones separated by 1–2 feet in a triangular or L formation. The sound on the probe is usually generated by small spark gaps. The time for the sound to reach each of the three microphones is measured. The speed of sound multiplied by time determines the distance of each spark gap to each microphone. The three distances uniquely determine spatial position. Two spark gaps are required to determine with 5 degrees of freedom the position of the probe, and three sources are needed for full 6-degree-of-freedom position and orientation (Science Accessories Corp., Stratford, CT).

Optical techniques are also available for spatial measurements. Simple techniques use two or three video cameras to follow the position of fiducials or light-emitting diodes (LEDs). Using triangulation, the position of these points are tracked and their position is calculated. To track multiple objects or total rigid-body motion, multiple LEDs are mounted, with the sequence of illumination determining which LED was localized. Modified CCD in dense linear arrays allow greater resolution of position than do lower-resolution NTSC CCD video cameras (Origin Instruments, Grand Prairie, TX; Pixsys, Boulder, CO).

In summary, the measurement of the coordinates of each landmark preserves their spatial relationships. This allows us to perform conventional anthropometric measurements, and also the vector relationships between points can be determined. However, even this approach is relatively time-consuming, and it only provides an abbreviated representation of facial morphology. In the next section, instrumentation that samples the entire surface of the object, so that contour and form can be measured and analyzed, will be presented.

PATTERNED OR STRUCTURED LIGHT TECHNIQUES

Although triangulation with stereophotographs can provide the spatial position of landmarks identifiable from both views, it is difficult to measure regions which have a paucity of features. Furthermore, the single vantage points of the stereo cameras often have a limited view of the entire object, and although that 180 degrees may be visible, it is nearly impossible to find distinguishing features on surfaces because their normals (vectors perpendicular to the surface at a given location) slope away from the cameras. The use of projected patterns of light in the form of lines, grids, or rastered points temporally provides features to these regions and allows their positions to be mapped. Viewing the object from multiple vantage points enhances the surface detail and minimizes the chance of obstructed views.

The basic technique usually involves a projector (usually a laser with a cylindrical lens to project a line of light or a strobe light projected through a glass slide with a grid or dot pattern) and a camera positioned and oriented in a fixed manner to each other. The separation between the camera and projector provides the stereoptic baseline. The light pattern is projected as planes or rays of light in space that intersect with the surface of the object. A profile or contour line(s) is visible on the surface of the object, and its position is recorded by video camera. The offset of the line in the camera's field of view can be calculated, with this offset a function of depth information based on the relative position of the camera and the light source. The imaging apparatus (the light and camera) can be rotated or translated with respect to the object to be imaged, or the object can be rotated or translated with respect to the sensors. Usually, several hundred profile lines or views are acquired and used to build the geometry of the object (Fig. 17–2).

In addition to a single profile line, a sequential series of light patterns can be projected onto the surface of the object and viewed from multiple camera views covering the entire region of interest. This approach allows rapid acquisition of data and no physical movement of the imaging apparatus with respect to the subject. One device uses six cameras and six projectors to image the entire head. As the pattern of light is projected on the subject, the cameras detect, digitize, and transmit a sequence of 24 images each to a digital image processor for surface triangulation, decalibration, and fusion. The 144 total views are processed, and the overlapping regions are resampled in a cylindrical coordinate system. The data are then displayed as a shaded polygon model or is transformed into a voxel format for use in the biomedi-

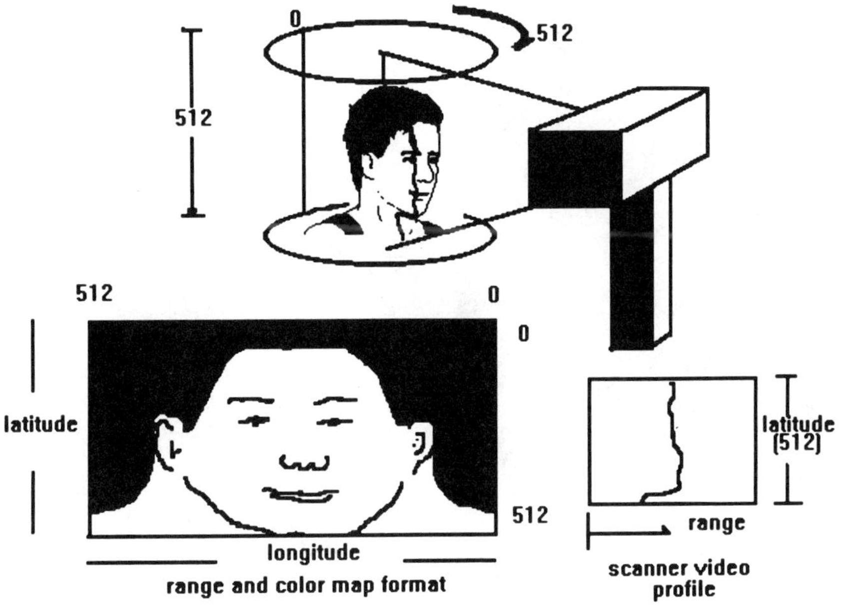

FIG. 17–2. Schematic representation of the Cyberware scanner, with its sensor that rotates around the subject as it projects a vertical plane of laser light onto the object surface. The contour of the line is sampled from an offset angular view by a video camera in the scanner head, imaging the curved profile line of the surface. The left-to-right shift of the line segment is a function of the range or depth from the scanners center of rotation. Five hundred twelve profile line increments are used to represent the full 360 degrees around the object. The data are stored initially in a cylindrical matrix of latitudes and longitudes, with the range values expressed in microns (note that the resolution of the scanner is not at the micron level). These 512 profiles become the basis of synthesizing the object's geometric surface. The 262,144 range values are acquired from top to bottom in a clockwise direction.

cal image processing program ANALYZE (Mayo Foundation, Rochester, MN) (21,22).

Cyberware Scanners

Our group has experience with the commercially available Cyberware ECHO 3030 surface scanner (Cyberware, Monterey, CA) that has been used for medical imaging applications (23). This scanner is based on the radial projection of a plane of low-intensity laser light onto the surface of the subject such that the laser light can be distinguished from the ambient light (Figs. 17–3 and 17–4).

Data Acquisition

A video camera with 512 lines of vertical resolution samples the object surface about 30 times per second from two symmetrically angled vantage points using a beam splitter (half-silvered mirror). This dual view enhances visualization of surfaces that may be obstructed from one of the viewing directions. The scanner head is rotated around the subject while digitizing 512 circumferential views (30 frames/second, 0.73 degrees per frame) of the entire head with a scanning time of about 17 seconds. The range values are calculated based on the offset of the laser line from the camera's center of view, and they are stored in a range or matrix of grid values

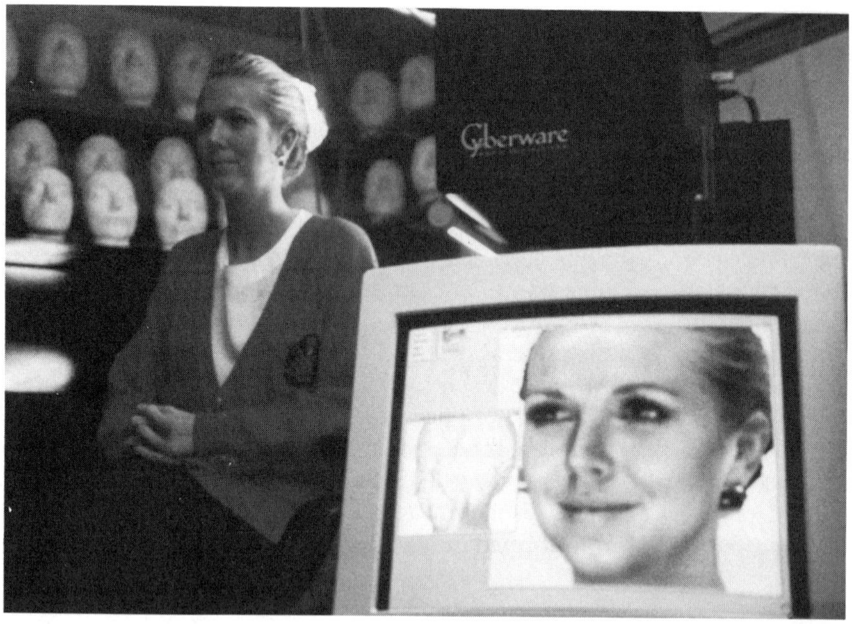

FIG. 17–3. Subject positioned in the scanner after acquiring image data. Images of the wire-frame model (low-resolution format) and the color texture-mapped surface model (converted to black and white for this illustration) can be seen on the computer monitor. The computer models can be rotated, translated, scaled, and rendered using various surface models from any user-defined vantage point interactively.

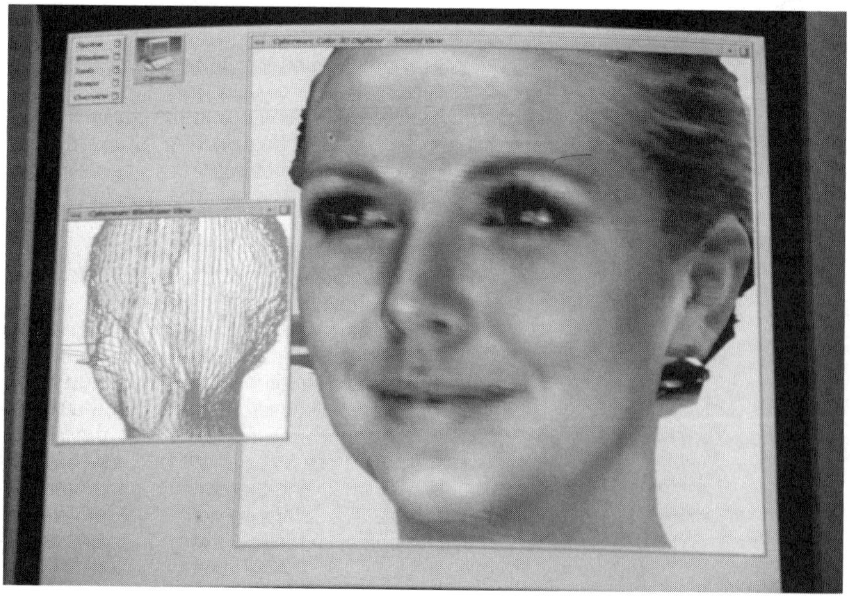

FIG. 17–4. Close-up view of computer monitor from Fig. 17–3. Texture-mapped surface model demonstrates the high resolutions possible and the life-like appearance.

(512 latitudes by 512 longitudes). If, at a given point, the digitizer receives no reflection (or is too weak to detect), or the surface is out of the allowable range, a special value is stored called *void.* With the color option, a second set of optics looks at the vertical line on the surface slightly offset from the place illuminated by the laser. The RGB value detected using ambient light is recorded for each point whose radius value was detected.

Subject Positioning

It is important that the chair allows the subject to remain still during the entire scan. A chair back, arm rests, or even an old-style dental head rest, cradling the mastoid region, helps stabilize the subject during the scan. It is also helpful to have the subject stare at a point somewhat above the scanner head, so that the gaze of the eyes remains conjugate. If the subject is looking straight forward, the right and left eye directions will be subtly different from shifts in gaze caused by the eyes viewing their reflection in mirrors of the scanner head.

Lighting

The lighting in the scanner room is also important. Light from outside windows, halogen lights, and uneven room lighting can affect the quality of the scan. We usually dim the room lights and have mounted two fluorescent lamps on either side of the scanner head. This provides uniform lighting throughout the scan. Areas of high reflection—metal objects or oily skin, and sometimes the hair surface—will not image well, so they must be dulled somewhat using powdered material such as cornstarch. Finally, eyeglasses cause some problems as well: The surface of the glasses is usually imaged, not the eyes underneath. To intentionally image the eyeglasses, jewelry, and so on, dull their reflective surface with cornstarch.

Data Format

The three-dimensional data, acquired as a series of contours or profiles, are initially stored in a cylindrical coordinate system, with a matrix of 512 longitudes and 512 latitudes. Thus, a total of 262,144 (512 × 512) range values are measured radially from the center of the cylindrical coordinate system which corresponds to the scanner heads' center of rotation. Typically, each contour is divided equally into 512 Y intervals or latitudes, each assigned a range of Z value. The range values are expressed in microns such that integer values can be used (note that the resolution of the scanner is not actually expressed in microns). Micron values are easily converted to millimeters by dividing by 1000. To describe the entire surface of the object, a series of contour lines (longitudes) are acquired side by side across the objects surface. This can be done using linear or straight-line motion of the object past the detector: With the PS base of the Cyberware scanner, the detector moves in a circular pattern around the object, producing equal angular intervals between successive contours. One convenient way to display the data is in the form of a range map, which is a 512 × 512 image derived by unwrapping the cylindrical data and displaying either the range values (with distance mapped to shades of gray, with the black values near the center of the object and the white values near the periphery; see Fig. 17–5) or color values (Fig. 17–6, converted to black and white) of the object surface in the lattice. This essentially provides a Mercator projection or flat image of the surface data, analogous to planar projections or wall maps of the earth.

The data from a color Cyberware scan is stored in two files: One of these contains the range information, and the other contains the surface color information. The range data file, usually about 600 kb in size, has an ASCII header that ends with *DATA=*. This is followed by the 262,144 range values in binary form, with two bytes per range value. The coordinates of each range value are not listed, but rather the order of presentation of the range values implies longitude and latitude position. The values are given in order by contour or profile, from top to bottom, clockwise around a top view of the object. Typically, the range values are given in microns (1000 microns = 1 mm) using 2 bytes. The vertical interval between the 512 latitude values in each contour is 0.8 mm, covering a total range of 41 cm or about 16 in. vertically. The angular spacing between each contour is 0.73 degrees, or 0.122718 radians (2π radians per 360 degrees).

Thus, the spatial resolution of the data is variable, with a smaller surface represented by the range value in regions closer to the center of the scan. Typically, for object surfaces 5 cm from the center of the scan, the area represented by each range value is 0.8 mm × 0.87 mm, whereas from a surface 10 cm from the center, the surface patch represented by the range value is 0.8 mm × 1.745 mm. The color information is stored in a SGI rgb file (Silicon Graphics—red, green, blue) that has a *file.color* filename. The parameters of the study can be varied, so that it is best to consult the header on each scan file for these values.

Other File Formats

The range data can be converted into a number of different formats, including ASCII, AutoCad® (.dxf), Wavefront, Dig.Arts, .byu, and IGES. This allows the data to be imported into a number of commercially available computer-aided design (CAD) and solid modeling programs. We also convert the Cyberware data into

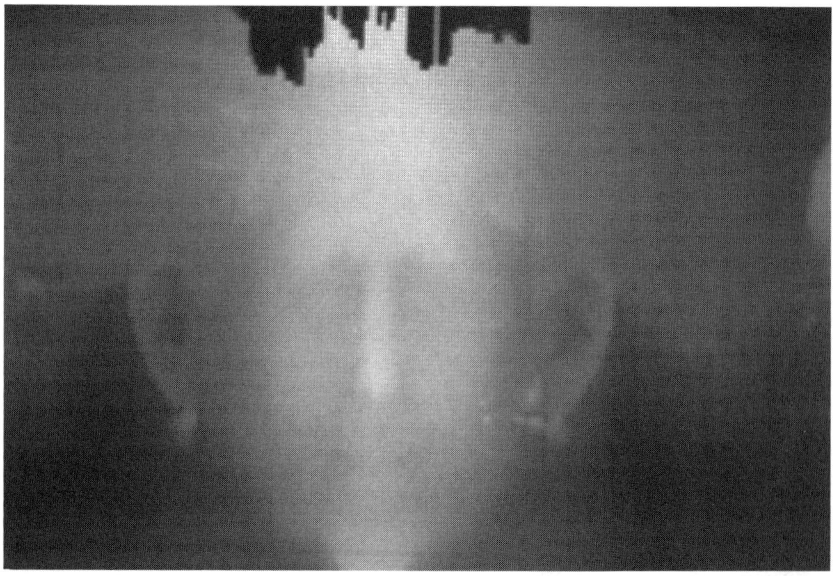

FIG. 17–5. Mercator representation of range data. The 512-latitude and 512-longitude array of range values in the original cylindrical coordiate system is modeled as a planar image, an unrolled cylindrical map with a vertical seam at the center of the back of the head. The range values are mapped to a gray scale look-up table; the range values closer to the scanner center are dark, and values farther away from the scanner center are light in color.

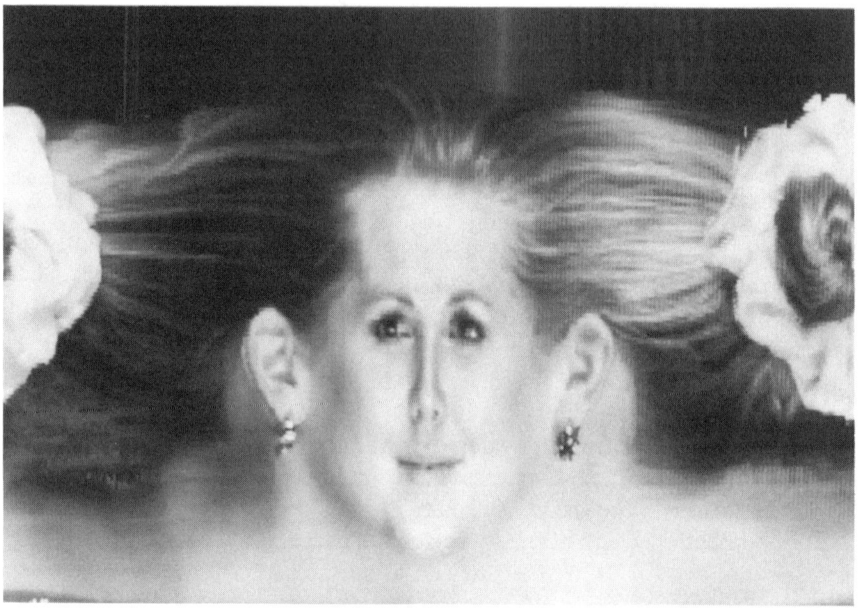

FIG. 17–6. Mercator representation of color texture data. The format is similar to that in Fig. 17–5. The color texture map (converted to black and white) is displayed.

formats that can be viewed in conjunction with CT or MRI data in the form of display lists (consistent with dividing cubes algorithm) or polygons (marching cubes algorithm) (24).

DATA PROCESSING

The creation of an image that represents an object of the three-dimensional world is called *modeling*. The addition of shading, highlights, textures, and shadowing to these images is termed *rendering* and is based on the position of object in world space. Below we will describe the basic techniques for modeling surface scan data and their visualization within a computer graphic environment.

Surface Rendering

Surface rendering, as the name implies, is the visualization of surfaces, in a manner familiar to viewing a solid object or a patient's anatomy. To represent a three-dimensional object, the boundaries of the object must first be determined to model the structure, and then the surface is rendered for viewing on the computer screen. The Cyberware range data represents an array of points. A surface between the points can be represented by a matrix of polygons in the form of triangles or quadrilaterals. The Cyberware models are quite large for real-time manipulation on low-end workstations. Algorithms that decimate or reduce the number of polygons to be ren-

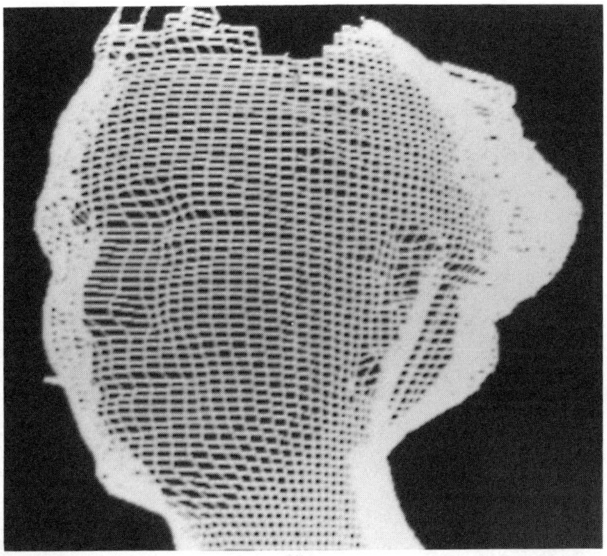

FIG. 17-7. Low-resolution wire-frame model, full face, three-quarter view. This is the lowest-resolution wire-frame model (approximately every fourth profile), used for rapid rotation, translation, and scaling before rendering the data as a surface shaded model. Higher-resolution wire-frame models (not shown) allow each profile line to be visualized if required.

dered by combining multiple polygons into one larger polygon in areas where there is little change in surface can reduce the number polygons nearly 90% and allow rapid rendering of the data with little loss in surface detail.

The object(s) can be represented on the computer screen in various formats. Wire-frame models in various resolutions are frequently used to perform initial manipulations on the data and can be rendered fairly quickly (Figs. 17–7 and 17–10). Surface shaded models are then used to provide a more realistic-appearing surface to the object (Fig. 17–8). Finally, texture mapping can be applied to give true realism. With the Cyberware data, the color data can be mapped back to the surface-rendered image to provide enhanced surface detail (Figs. 17–4 and 17–12).

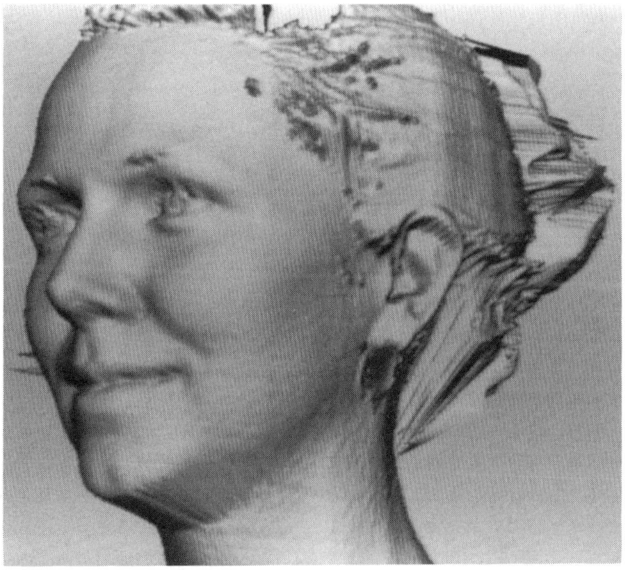

FIG. 17-8. Light-shaded surface model of full face, three-quarter view. Shading based on surface depth and orientation to simulated light source.

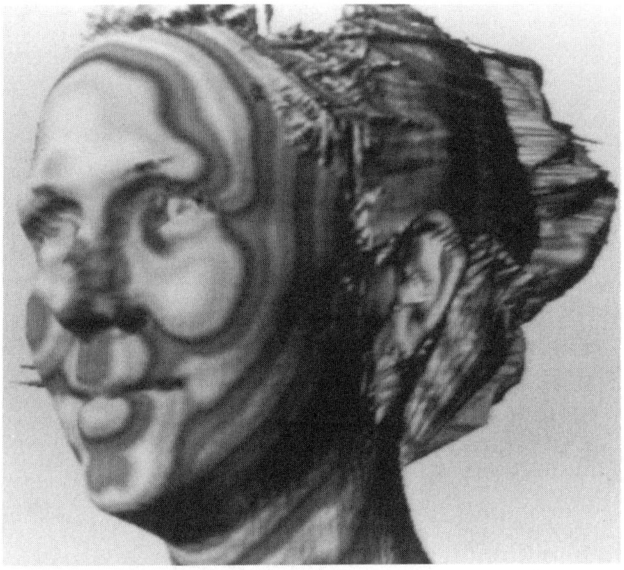

FIG. 17-9. Texture map of range data on surface model. Range values are mapped to color look-up table (converted to black and white for this illustration) and mapped to the corresponding surface. The surface appearance is similar to that of a terrain contour map, with isoradius surfaces with respect to the center of the scanner represented as contour lines. This provides an initial estimate of facial symmetry. However, this representation can also be deceiving, especially if the subject was not centered in the scanner or if the data were not transformed from the object to anatomic coordinate system.

To provide the illusion of three dimensions on the computer monitor, perspective and shading are applied. This brightness of the surface at a point is determined from the depth, the depth gradient (rate of change in depth), and the orientation of the surface normal (the direction perpendicular to the surface) with respect to the simulated light source. The more parallel the surface normal is to the light source, and the closer the object is to the viewing plane, the brighter the shading at that point.

DATA MANIPULATION

Once the data has been acquired and a computer-based model has been reconstructed, the model can then be studied in great detail. The first step is visualization, or the ability to interactively view the data from various vantage points. Next it is often necessary to see behind

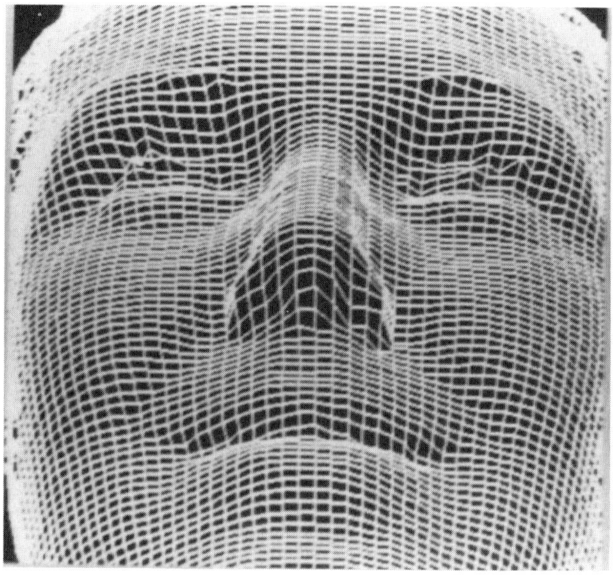

FIG. 17-10. Higher magnification, submental view, low-resolution wire frame for rapid positioning.

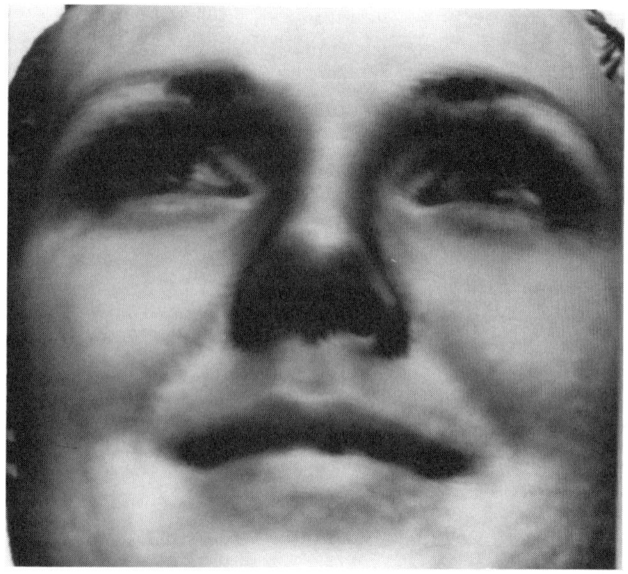

FIG. 17-12. Higher magnification, submental view, color texture map (converted to black and white) on surface of geometric model provides realistic appearance and assists in identification of anthropometric landmarks.

structures or to see inside objects, or to view them in cross section, so the equivalent of dissection tools are available for these inside looks. Next, the extraction of quantitative information may be needed in the analysis. This can include the *x, y, z* coordinates of specific landmarks, the distance and angle between point(s), the surface area of a defined patch of the objects surface, or the volume subtended by a region of the object. Finally, it is frequently necessary to compare the same object at different time-points, or different objects, so methods of registration and alignment and scaling are often used.

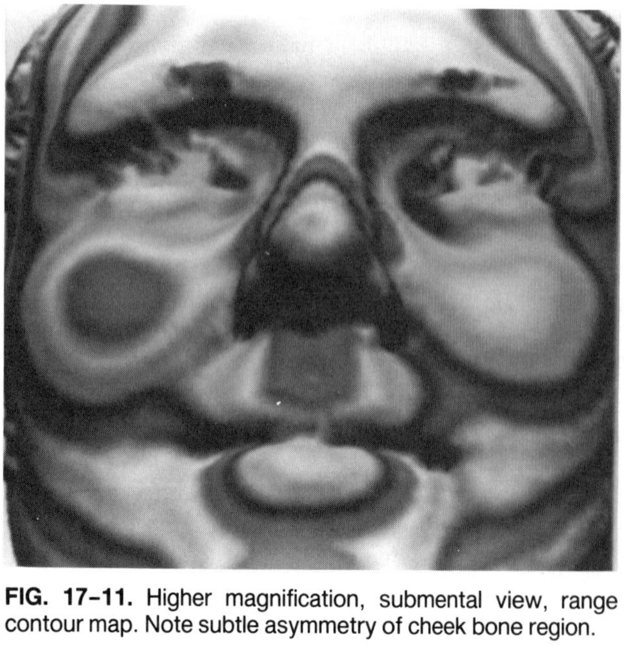

FIG. 17-11. Higher magnification, submental view, range contour map. Note subtle asymmetry of cheek bone region.

Graphic User Interface

The current graphic user interface (GUI) is based on a conventional mouse, keyboard, and color monitor. The tools for manipulating the software to view the data are via graphic buttons and slider bars. Changes in parameters can also be implemented directly by keyboard or in the form of script files (lists of instructions, similar to a macro). Multiple objects can be displayed and independently controlled. Each object can be altered with multiple parameters, including on/off, color, translation, rotation, center of rotation, clipping, and scaling. The object's position and orientation are described in at least four separate coordinate systems: (i) object, (ii) world, (iii) camera, and (iv) screen.

Graphic Coordinate Systems

The object or local coordinate system is anchored to the object as the original data were acquired, such as the original points from the Cyberware scan. Often these data are transformed to the anatomic coordinate system to correct for malpositions in head position during the original data acquisition (Fig. 17-1). The next level is the world, global, or absolute coordinate system, which is the coordinate system where multiple objects or models reside within a global scene. Translation and rotational changes of objects relative to each other occur in this system. Next, the camera, or eye, coordinate system is used to determine what an observer would perceive in world space from specific vantage points, such as through a camera or through the eye. Finally, the representation of the three-dimensional scene is projected into the display coordinate system, which is the two-dimensional computer monitor viewing plane. Multiple coordinate systems of variable hierarchy are required to represent complex motions—for example, the movement of the fingers with respect to the hand, the hand to the forearm, the forearm to the upper arm, the upper arm to the shoulder, and so on.

Anatomic Coordinate System

After the data have been acquired and reconstructed into a three-dimensional format, it must be aligned to an anatomic coordinate system to facilitate interpretation of the data in clinical terms. More often than not, the patient's position in the scanner does not correspond to the conventional anatomic coordinate system of superior–inferior, anterior–posterior, and medial–lateral. The midsagittal plane is estimated from midline and bilateral symmetric facial hard/soft-tissue landmarks of the face and the cranial base. This process is often subjective and nontrivial in patients with severe craniofacial asymmetries. The horizontal plane is based on the upper aspect porion and infra-orbitale. Again, different philosophies to establish the horizontal plane are available. Some investigators use the Frankfort horizontal (25), others use natural head position (26). The frontal plane is then orthogonally determined and can be centered at porion, nasion, or any arbitrary position. To determine the translation and rotation parameters to transform the data from the object to the anatomic coordinate system, we simply manually align the object in world space to a wire-frame orthogonal reference, based on the criteria listed above. The position and orientation parameters to reposition the object correctly in world space are used in the transformation matrix. Another technique for registration includes least-squares fitting between three or more landmark points.

DATA MEASUREMENT AND ANALYSIS

Basic measurements are possible from the surface models. These include cylindrical and/or Cartesian coordinates of discrete points, which can be stored in a data file and exported into a spreadsheet for further analysis and comparison with normative values. Other common scalar measurements include distances between points, the angle subtended by three points, and the surface distance between two points. Other measurements include the surface areas and volumes within user-defined boundaries.

Further information can be extracted about the surface contour. The simplest form of display is to map the radius values to color table and display the equivalent of a topographical map with respect to the center of the scanner coordinate system (see Figs. 17–9 and 17–11). This display is helpful in detecting areas of asymmetry. The measurement of the curvature and the rate of curvature also provide important information. Basic approaches to measuring curvature usually involve Gaussian and mean curvature. Investigators have used Cyberware-acquired data to measure facial curvature, texture-mapping these values back onto the computer-based model. These data can be used for registration of imaging studies done at different time-points or can be used to extract characteristic features of the face, such as ridges of maximum or minimum curvature, for facial recognition (27).

The data from the facial surface can be described globally by its generalized shape. Techniques to describe shape have included superquadrics, which are deformable solid modeling primitives, and also mesh-based surface models, where a physically motivated energy function is employed. More recently, closed-form solutions have been employed for physically based shape modeling and recognition. These techniques have been applied to Cyberware-based data to provide parametrically based models for analysis, comparison, and recognition (28).

Precision/Accuracy

Our goal is to fully exploit the information intrinsic to the image data. These measurements can be in the form of simple scalar measurements such as distances, angles, proportions, surface area, or volume, or they can involve more complex relationships related to morphological parameters such as contour and form. In any measurement system, the objective from the observation is to best estimate the true value of the quantity to be measured. Likewise, the risk associated with accepting the best measurement must also be determined.

Precision and accuracy are two terms commonly used to describe the quality of the measurement. The *precision* is a function of the ability to detect a change or is a measure of the dispersion and/or how the values cluster about the mean. For example, if a gun site is aligned to the target bull's-eye, the tightness of the grouping of multiple rounds fired is a function of the precision of the sight, gun, bullets, and so on. The *accuracy* is a function of the closeness of the mean of the values of the observation to the true measurement. With the gun site, the cross-hairs of the scope may be centered on the bull's-eye; however, the rounds hitting the target may consistently be about 5 cm to the right of the target center. Thus, the scope or site of the gun has been incorrectly mounted.

Errors can be classified as either accidental or systematic. Accidental errors are those which in a large number of measurements are as often negative as positive—for example, the error in identifying a specific anatomic landmark with inter and intraobserver identification. They affect the arithmetic mean very little. All other errors are systematic, where they do not tend to balance out with multiple measurements, but rather give a definite bias to the mean. For example, if the observer identifies an anatomic landmark, but the voxel dimension was assigned and scaled incorrectly in the three-dimensional rendered image, the aspect ratio of the computer model is incorrect. In cephalometric analysis, the measurement error may be related to not taking the magnification error if the cephalometric radiograph was not taken into account. In both cases the magnification factor was taken into account. In both cases the measurements would be fairly precise, but inaccurate.

Details of the precision and accuracy of the Cyberware data on moulages are presented in the next chapter. This provides a good indication of the resolution of the instrument without the complications of motion or landmark identification. The highest resolution possible of the system is limited by the optics of the camera, the density of their CCD arrays, and the number of points used to represent the data. Other sources of error are introduced when scanning animate objects. The subject will always have a slight tremor or sway, and this error can be significant. Systems that acquire the data more rapidly will help to reduce these errors and will make it possible to study infants and small children. Finally, the ability to find the classical facial landmarks can be difficult on the computer screen image. Addition methods such as stereoscopic displays and curvature information mapped to the texture-mapped color surface will facilitate accurate landmark identification.

SUMMARY

The development of instrumentation to efficiently map facial surface topography now allows investigators to characterize its morphology using detailed quantitative methods. Measurement can extend beyond scalar measurements of distance, angle, and proportion. The object is then to describe (a) the shape and size of the object globally and (b) the details of the surface contours locally. The applications of these data are numerous, and below a few examples in regard to the facial region are given with their references for the more inquisitive reader.

The data can be used as the basis for modeling facial animation. This physically based approach begins with the geometry and the color data from the surface scan. A three-dimensional model is constructed using a mesh-based geometry. The mesh can then be transformed into a physically based head model of the subject, incorporating the elastic features of the facial tissue, with a set of anatomically motivated facial muscle actuators. These components are integrated into the facial modeling, and animation allows the simulation of facial expression and the movements associated with speech (29).

Other work has utilized facial surface data to model the response of the skin to local excisions of tissue. These finite element-based models assist the surgeon in the design of the excision margins and local flaps, and they have excellent potential for teaching residents on computer-based surgery simulators (30).

The data acquired from the laser surface scans can also be used to construct physical models using a computer numerically controlled (CNC) milling machine. The fabrication of portrait sculpture ironically was one of the strong motivating factors in the development of this technology. Our personal experience using physical models has been for estimating soft-tissue deficiencies and designing implants, and we have also used these models in the design of facial prosthesis for either (a) reconstruction of facial structures lost secondary to trauma or neoplasm or (b) correction of a congenital deformity.

The description and modeling of the change in the facial surface in relation to underlying bone movements are still in their preliminary stages. The advantages of using structured light scanners versus computed tomography to follow the changes in facial surface form include no additional radiation exposure and reduced costs. The current approaches to visualize and quantitate facial changes with surgery basically register preoperative and postoperative scans and quantitate the differences volumetrically (1,31–33). We are working to extend this work to model the deformations of the soft-tissue elements that will enable prospective estimates of soft-tissue contour during the design of the osseous surgery.

In our own preliminary work, we have merged the laser scan image data with both three-dimensional CT and MRI data. The surface scan data provides a more detailed skin surface for planning. In preliminary work at our institution, these data were applied clinically in an 8-year-old girl with Crouzon syndrome characterized with hypertelorbitism and forehead retrusion. The data from her 12-year-old brother's facial surface scan was used as a template to estimate the optimal parameters for repositioning her eyes more medially and advancing her forehead.

We commonly hear descriptions of newborn babies and their siblings such as: "You have your mother's eyes and your father's nose." The human recognition system, somehow, is able to rapidly glean these recognizable features in an instant. However, when you ask someone if they could give more detail as to why they think these features are similar, they are often lost for words. Maybe technology will finally let us get a quantitative handle on what human nature has known from the beginning.

REFERENCES

1. Altobelli DE, Kikinis R, Mulliken JB, Cline H, Lorensen W, Jolesz F. Computer-assisted three-dimensional planning in craniofacial surgery. *Plast Reconstr Surg* 1993;92(4):576–585.
2. Farkas LG, Bryson B, Tech B, Klotz J. Is photogrammetry of the face reliable? *Plast Reconstr Surg* 1980;66(3):346–355.
3. Mattison RC. Facial video image processing: standard facial image capturing, software modification, development of a surgical plan, and comparison of presurgical and postsurgical results. *Ann Plast Surg* 1992;29(5):385–389.
4. Deacon AT, Anthony AG, Bhatia SN, Muller JP. Evaluation of a CCD-based facial measurement system. *Med Inform* 1991;16(2):213–228.
5. Berkowitz S, Cuzzi J. Biostereometric analysis of surgically corrected abnormal faces. *Am J Orthod* 1977;72(5):526–538.
6. Savara BS, Miller SH, Demuth RJ, Kawamoto HK. Biostereometrics and computer graphics for patients with craniofacial malformations: diagnosis and treatment planning. *Plast Reconstr Surg* 1985;75(4):495–501.
7. Kobayashi T, Ueda K, Honma K, Sasakura H, Hanada K, Nakauima T. Three-dimensional analysis of facial morphology before and after orthognathic surgery. *J Craniomaxillofac Surg* 1990;18:68–73.
8. Uesugi M. Three-dimensional curved shape measuring system using image encoder. *J Robot Mechatron* 1991;3(3):190–195.
9. Dunn SM, Keizer RL, Yu J. Measuring the area and volume of the human body with structured light. *IEEE Trans Syst Man Cybern* 1989;19(6):1350–1364.
10. Motoyoshi M, Namura S, Arai HY. A three-dimensional measuring system for the human face using three-directional photography. *Am J Orthod Dentofac Orthop* 1992;101(5):431–440.
11. Keizer RL. Recovering the shape of the human face. *Innov Tech Biol Med* 1989;10(3):315–329.
12. Arridge S, Moss JP, Linney AD, James DR. Three-dimensional digitization of the face and skull. *J Maxillofac Surg* 1985;13:136–143.
13. Robertson NRE, Volp CR. Telecentric photogrammetry: its development, testing, and application. *Am J Orthod* 1981;80(6):623–637.
14. Karlan MS, Madden M, Habal MB. Biostereometric analysis in plastic and reconstructive surgery: a one-step, on-line technique. *Plast Reconstr Surg* 1978;62(2):235–239.
15. Kawano Y. Three-dimensional analysis of the face in respect of zygomatic fractures and evaluation of the surgery with the aid of moiré topography. *J Craniomaxillofac Surg* 1987;15:68–74.
16. Cline HE, Lorensen WE, Holik AS. Automatic moiré contouring, *Appl Optics* 1984;23(10):1454–1459.
17. Benton SA. Alcove holograms for computer-aided design. *True three-dimensional imaging technologies and display technologies.* San Diego: SPIE, 1987;761.
18. Shatkin S, Stark DB. Cleft lip and palate moulages. *Plast Reconstr Surg* 1965;36(2):235–238.
19. Hildebolt CF, Vannier MW. Three-dimensional measurement accuracy of skull surface landmarks. *Am J Phys Anthropol* 1988;76:497–503.
20. Farkas LG. *Anthropometry of the head and face in medicine.* New York: Elsevier, 1981.
21. Vannier MW, Pilgram T., Bhatia G, Brunsden B, Commean P. Facial surface scanner. *IEEE Comput Graphics Appl* 1991;Nov:72–80.
22. Vannier MW, Pilgram TK, Bhatia G, Brunsden B, Nemecek JR, Young VL. Quantitative three-dimensional assessment of face-lift with an optical facial surface scanner. *Ann Plast Surg* 1993;30(3):204–211.
23. Cutting CB, McCarthy JG, Karron DB. Three-dimensional input of body surface data using a laser light scanner. *Ann Plast Surg* 1988;21(1):38–45.
24. Cline HE, Lorensen WE, Ludke S, Crawford CR, Teeter BC. Two algorithms for the three-dimensional reconstruction of tomograms. *Med Phys* 1988;15(3):320–327.
25. Steiner CC. The use of cephalometrics as an aid to planning and assessing orthodontic treatment. *Am J Orthod* 1960;46:721–735.
26. Moores CFA, van Venrooij ME, Lebret LML, et al. New norms for the mesh diagram analysis. *Am J Orthod* 1976;69(1):57–71.
27. Gordon GG. Face recognition based on depth maps and surface curvature. *Geometric methods in computer vision.* San Diego: SPIE, 1991;1570.
28. Pentland A, Sclaroff S. Closed-form solutions for physically based shape modeling and recognition. *IEEE Trans Pattern Anal Machine Intell* 1991;13(7):715–729.
29. Waters K, Terzopoulos D. Modelling and animating faces using scanned data. *J Visual Comput Anim* 1991;2:123–128.
30. Pieper SD. CAPS: Computer-aided plastic surgery. Ph.D. thesis, Media Arts and Sciences, MIT, Cambridge, MA, 1992.
31. Coombes AM, Moss JP, Linney AD, Richards R, James DR. A mathematical method for the comparison of three-dimensional changes in the facial surface. *Eur J Orthod* 1991;13:95–110.
32. Linney AD, Grindrod SR, Arridge SR, Moss JP. Three-dimensional visualization of computerized tomography and laser scan data for the simulation of maxillofacial surgery. *Med Inform* 1989;14(2):109–121.
33. McCance AM, Moss JP, Wright WR, Linney AD, James DR. A three-dimensional soft tissue analysis of 16 skeletal class III patients following bimaxillary surgery. *Brit J Oral Maxillofac Surg* 1992;30:221–232.

Chapter 18

Correspondence Between Direct Anthropometry and Structured Light Digital Measurement

Daniel B. Bača, Curtis K. Deutsch, and Ralph B. D'Agostino, Jr.

The surface scanning methods described in the previous chapter hold considerable promise for assessing craniofacial morphology (1,2; also see Chapter 17, *this volume*). There are advantages to using surface scanning techniques over direct anthropometry. Direct measurement is highly reliable, but painstaking and laborious. In contrast, surface scanning is exceedingly rapid. Though not as costly a procedure in terms of human resources, surface scanning does require a substantial investment in equipment and software.

Surface scanning also permits the recording of an extraordinary volume of data (the image is essentially a three-dimensional photocopy), and it archives these data for future use. The digital image of the face and head appears resolute at 512 × 512 pixels; with color mapping, the image on the computer screen is startlingly real. But do measurements from surface scans correspond closely to the direct anthropometric assessments described in this book? We performed a validation study to address this question.

We contrasted linear measurements derived from on-line computer assessment of digitized craniofacial scans to direct anthropometric examinations. To remove the potential effect of movement artifact in this study, we performed the measurements on moulages molded from alginate (a material used for dental impressions). These moulages bear considerable resemblance to "life masks" used to memorialize individuals since the Middle Ages. By using these casts, we removed sources of variability that would have been present in facial measurements *in situ*. Among these sources are:

1. *Facial animation.* We have observed that some areas (especially the eyes and mouth) can vary considerably with facial expression. By using moulages, facial expression was held constant.

2. *Landmark identification.* Landmarks were drawn on the moulage, and thus were identical for both techniques.
3. *Facial pliability.* The surface of the face can be slightly distorted when touched by an instrument (e.g., spreading calipers), especially in measurements involving soft landmarks on the skin. This factor was held constant by measuring the fixed contours of the moulage.
4. *Interobserver variability.* Interobserver variability was eliminated by using a single rater (D.B.B.), who measured all subjects using both techniques.

Because so many sources of potential error were eliminated, this validation study is biased in the direction of achieving good agreement between the two techniques. We are currently performing validation and reliability studies on the structured light images to estimate the magnitude of these potential sources of variability (Deutsch, et al., *in preparation*).

METHODS

We prepared moulages (alginate casts) for 30 normal adults (N = 15 males, N = 15 females) ranging in age from 19 to 41 years (mean 28 years). Direct anthopometric measurements were made on the moulages using calipers. Landmarks were identified in both techniques by application of this books' operational definitions, and they were marked in pencil directly on the moulages. The 21 dependent variables (measured on a mm scale) studied are listed in Table 18–1.

Each moulage was scanned using the Cyberware 3030/PS Color 3D Laser Light Scanner in combination with a Silicon Graphics Iris Indigo XS24 computer. The computer renders the data and allows manipulation of the image, allowing one to identify landmarks X, Y, and Z coordinates on the digital array. We compared the linear measurements from the two techniques by performing a correlation analysis. The Pearson product–moment correlation coefficients (r) are presented in Table 18–1.

RESULTS

The agreement between the two methods was excellent, yielding r values between 0.93 and 1.00. Furthermore, the magnitude of the linear measurements were highly comparable using the two methods. For each measurement, we computed the mean of the difference scores (direct measurement, digital measurement). The mean differences (in millimeters) for these measurements were very small, ranging from 1.9 (for facial height) to 0.0 (for upper lip vermilion height, cutaneous upper lip height, and left orbital width). For most of the dependent variables, it seems that the two techniques yielded practically identical results.

The range of mean differences was −0.65 to 1.89 mm, and the range of standard deviations for these mean differences was 0.25 to 1.22 mm. These difference scores are shown graphically in Fig. 18–1. Approximately 83% of the surface scan measurements made on the moulages are within 1.0 mm of the direct caliper measurements, and 98% are within 2.0 mm.

TABLE 18-1. *Interrater reliability coefficients*

Variable	Measurement	Correlation coefficient (r)	Average difference (mm)
Interocular diameter	en-en	.929	−0.5
Supraorbital diameter	fs-fs	.995	−0.4
Nasal root width	mf-mf	.940	−0.5
Biocular diameter	ex-ex	.975	−0.1
Right orbital width	fs-mf	.990	−0.4
Right orbital height	os-or	.991	0.2
Left orbital width	fs-mf	.995	0.0
Left orbital weight	os-or	.993	0.2
Forehead width	ft-ft	.996	−0.7
Face height	n-gn	.993	1.9
Upper face height	n-sto	.997	0.6
Mandible height	sto-gn	.978	0.3
Face width	zy-zy	.996	−0.1
Lower face width	go-go	.993	−0.2
Nose length	n-sn	.994	0.5
Nose width	al-al	.987	−0.2
Nasal protrusion	sn-prn	.989	0.1
Medial height of cutaneous upper lip	sn-ls	.981	0.0
Medial vermilion height of upper lip	ls-sto	.986	0.0
Medial vertical upper lip height	sn-sto	.980	0.1
Mouth width	ch-ch	.994	−0.4

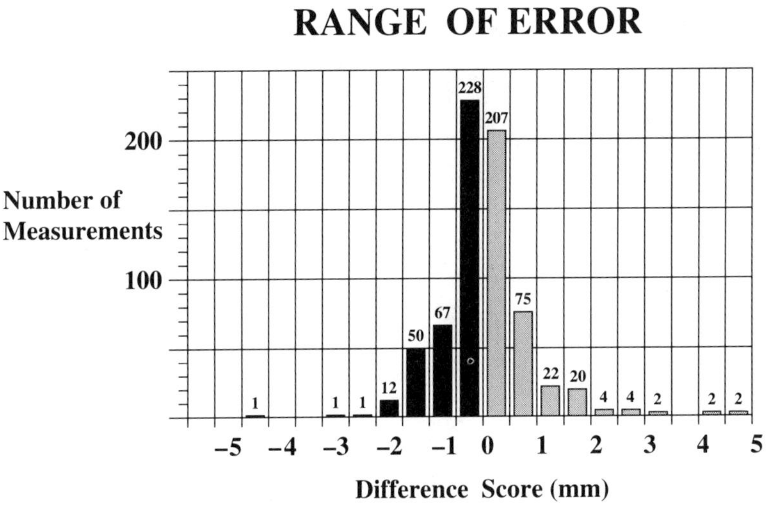

FIG. 18-1. Frequency distribution of difference scores (structured light scanner digital measurement versus direct anthropometric measurement).

DISCUSSION

We are encouraged by these findings; with respect to anthropometric measurements, the structured light scanner provides veridical images of the face and head. Across all measurements in the study, the average of mean difference scores approximated zero (0.03 mm). In other words, there is no systematic bias between the direct and digital measurements.

Several sources of variance are introduced by scanning live subjects rather than moulages. For instance, how does one identify the eu, po, and v landmarks on the cranium, which are usually obscured by a thick hair cover? A surface approximating the cranial contour may be achieved using a tight elastic cap, but this introduces a new source of variability. Another methodological challenge is found in identifying projective surface analogs of bony landmarks (e.g., ft, zy, go, and gn) used in linear and arc measurements. Further, it will be necessary to standardize subject positioning during scanning to optimize test-retest reliability.

We must bear in mind that this test was biased in the direction of obtaining good agreement between methods. Various effects of measurement error must be independently estimated and incorporated into models of validity and reliability.

REFERENCES

1. Arridge SR, Moss, JP, Linney, AD, James DR. Three dimensional digitization of the face and skull. *J Maxillofac Surg* 1985;13:136–143.
2. Cutting CB, McCarthy JG, Karron DB. Three-dimensional input of body surface data using a laser light scanner. *Ann Plast Surg* 1988;21:38–45.

Chapter 19

Instructional Video for Anthropometric Methods

Curtis K. Deutsch, David E. Altobelli, and Leslie G. Farkas

Over the years, we have been happy to provide training in anthropometry to many of our colleagues. These colleagues have come from a variety of backgrounds. Most have been craniofacial plastic surgeons, who have used these methods in planning procedures and to assess pre- versus postoperative morphology. Also numbered among them have been medical anthropologists, with interests ranging from physical attractiveness to forensic applications. Recently, we have been approached by increasing numbers of medical geneticists, hoping to render the description of dysmorphology more objective. Requests for training have become increasingly frequent in the last few years.

INSTRUCTIONAL METHODS

What was needed was an efficient means of providing hands-on training. We concluded that the most convenient and cost-effective way to provide this instruction was in video format.

We created a video entitled *The Craniofacial Examination in Medicine* (1). It is structured to be a "Masters' Class". In this program, we describe (a) the instruments used in the craniofacial examination, (b) landmark identification, (c) measurement techniques, and (d) potential sources of measurement error and how to minimize them. If you would like to order a copy of this video for your laboratory or clinic, please write Dr. Deutsch at the Eunice Kennedy Shriver Center.

STANDARDIZATION OF MEASUREMENT

Our aim is to standardize the anthropometric examination, facilitating comparability of collected data.

It is crucial for each laboratory or clinic to convince itself that its measurement techniques are reliable, *prior* to their research or clinical use. This requires statistical tests of test–retest (*intrarater*) and between-rater (*interrater*) reliability. Though there is no objective answer to the question "How many subjects must we see in our reliability study?", we recommend a minimum sample size (N) of 20. Generally, institutions require that proposed reliability studies be institutionally reviewed before their human experimentation committees.

A reliability study is painstaking. Prior to performing the study, it is necessary to practice test–retest and between-rater procedures, periodically comparing figures on small groups of subjects (often, patient lab or clinic personnel). At this training stage, it is possible to develop consistent procedures and to correct rater biases.

During the interrater component of the reliability study, it is important that raters be "blind" to each others' measurements. Also, intrarater measurements must be "blind"—that is, structured so that the second value is separated by passage of time from the first, with numerous intervening measurements to be made. Obviously, if two measurements were made consecutively, considerable bias would be introduced, artificially optimizing intrarater agreement.

Upon completion of the reliability study, it is helpful to plot each score in a scatterplot, which depicts the direction, form, and strength of relationships between two values. This plot is illustrated in all introductory statistics books (e.g., see refs. 2 and 3). It is also important to compute the correlation coefficient (r) for each measure. For small samples (e.g., $N = 20$), this can be accomplished by hand using computational formulas (see ref. 2, pages 162–168). Computation of correlation coefficients is also a common feature of statistical packages, such as SAS, BMDP, SPSS, SPSSx, and JUMP.

The value of r always falls between -1.00 and $+1.00$. Your values will lie between 0.00 (no correlation) and 1.00 (a perfect correlation). The correlation measures the strength of linear (not curved) relationships between two variables. As the r approaches 1.00, the points in the scatterplot will lie closer to a straight line. The value of r has no unit of measurement, being a "dimensionless" number (2).

How large must a correlation coefficient be? (Here, we discuss Pearson product-moment correlations, for continuous, normal distributions.) The rule-of-thumb most laboratories use is that the coefficient r must be greater than .70. For this application, we would rate r values as follows: between .70 and .80, "fair"; between .80 and .90, "good"; and between .90 and 1.00, "excellent."

When comparing raters' values, do not average multiple measurements. This will bias your estimate of interrater reliability in the direction of increasing correlation (see ref. 2, page 175).

QUALITY CONTROL

It is straightforward to compare your measurements to those of your colleagues and to test the reproducibility of your own data, but how do your values compare to those of an expert? We have devised a means of testing your values. At your request, we will send you a set of craniofacial moulages (plaster casts) on which we have performed the anthropometric measurements described in this atlas. The moulage is a *tabula rasa,* a blank slate upon which you can draw your landmarks and perform your measurements. Send us your data (care of Dr. Deutsch), and we will provide in return the data collected by the "experts." If our measurements are in close agreement, we will also send you a certificate attesting to that fact.

REFERENCES

1. Deutsch CK, Altobelli D, Farkas LG. *The Craniofacial examination in medicine: Anthropometric methods* [video]. Boston: Medical Video Associates, 1994.
2. Moore DS, McCabe GP. *Introduction to the practice of statistics,* 2nd ed. New York: WH Freeman, 1993.
3. Winkler RL, Hays WL. *Statistics: Probablility, inference, and decision,* 2nd ed. New York: Holt, Rinehart, & Winston, 1971.

Appendix A

Craniofacial Norms in North American Caucasians from Birth (One Year) to Young Adulthood

Leslie G. Farkas, Tania A. Hreczko, and Marko J. Katic

The North American Caucasian population norms are reported for 132 anthropometric measurements of the head and face. In addition, the norms for the body height and weight were also developed. Of the total of 132 measurements, 70 were single and 62 were paired. There were 103 linear and 29 angular measurements (inclinations and angles): 15 cranial, 31 facial, 21 orbital, 28 nasal, 18 labio-oral, and 19 aural.

In 111 of the 132 measurements (84.1%) the norms (mean and standard deviations) were developed by direct examination of the subjects. In 66 of 111 measurements (59.5%) the norms were obtained in 20 (21) age groups, from birth (1 year) to 18 years (19–25 years). Nineteen measurements (17.1% of 111) were limited to 13 age groups, between 6 and 18 years. In 17 of 111 (15.3%) measurements the data were recorded only in young adults (19–25 years). In 9 of 111 measurements (8.1%), norms were established for ages 1 through 5 and then in young adulthood (19–25 years), but not for ages 6 through 18.

In a group of 21 of 132 measurements (15.9%), in which the data were obtained by examination of the subjects between 1 and 5 years of age and also in young adulthood (19–25 years), the missing intermediate norms for one cranial, seven facial, seven nasal, and six labio-oral measurements were subsequently *derived,* three directly and 18 indirectly.

Among the 18 indirectly derived measurements, 14 were linear (one cranial, five facial, four nasal, and four labial) and four angular (two in the face and two in the lower lip). In order to estimate the growth curve of the missing data in the 14 linear measurements, a regression method of proportional adjustment was used (1). This method is based on the requirement that a related and fully measured facial feature is available as well as estimates of the youngest (1–5 years) and oldest (19–25 years) age groups of the required measurement. This method assumes that the growth characteristics of the reference measure is a reasonable base to build estimates of the required or target measure. Mathematically, deviations from the fitted age slope of the reference measure is proportionally applied to the slope of the target measure. In effect, this method fits an adjusted growth curve of the reference measure between the given youngest and oldest age means of the target measure. Norms developed by this method are marked with an asterisk (*) in the tables.

For four angular measurements (labiomental angle and the inclinations of the mandible, chin, and lower lip) with missing norms between the youngest and oldest age groups, no dependable reference measure could be used. In these cases the difference between the means of the youngest and oldest age categories was evenly distributed across the intermediate age groups in which the norms were missing. This simple method provides only a first approximation of the unmeasured norms and as such should be used with caution. These data are marked in the tables with a "#" sign. The standard deviations of the derived norms were estimated by pooling the standard deviations in the youngest and oldest age categories.

In a group of three angular measurements (inclination of the columella, the glabellonasal angle, and the nasal tip angle) the norms (mean and standard deviations) were not developed by direct measurements of the subjects. In these cases the required measurements were reconstructed *geometrically* from the recorded component inclination and angles of the nose. In the tables the norms developed by this method are marked with a "+" sign.

Tables reporting the craniofacial and other norms in the Caucasian population are marked with a capital A, and the Roman numeral after it identifies the anatomical region: I for the head, II for the face, III for the orbits, IV for the nose, V for the lips and mouth, VI for the ears, and VII for height and weight. The Arabic numeral denotes the order of measurements. In tables reporting the data obtained by direct measurements, the size of the sample (*N*) and the mean values of the measurements with the standard deviations (SD) are given.

REFERENCE

1. Daniel C, Wood FS, eds. *Fitting equations to data.* New York: John Wiley & Sons, 1980.

Head

HEAD

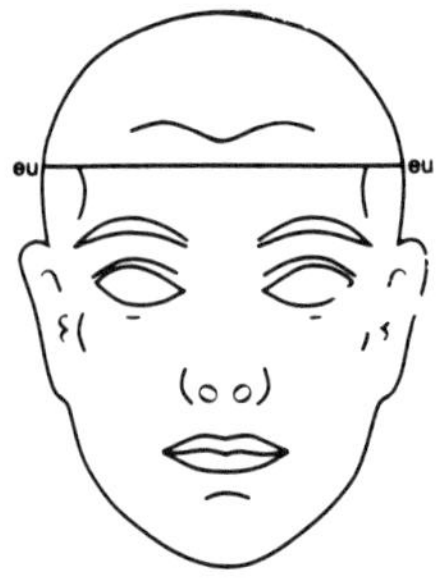

TABLE A–I-1. *Width of the head (eu-eu) (mm)*

Age (years)	Male			Female		
	N	Mean	SD	*N*	Mean	SD
0–5 months	8	110.1	2.0	5	108.2	3.8
6–12 months	20	118.8	6.2	8	115.7	4.9
1	18	125.5	5.6	28	122.0	6.0
2	30	130.5	5.5	32	127.8	3.7
3	30	133.7	4.0	30	130.8	4.0
4	30	136.4	4.7	30	135.8	3.8
5	30	138.2	4.0	30	135.4	3.8
6	50	139.8	4.8	50	136.8	4.6
7	50	140.8	5.3	50	137.6	4.6
8	51	142.6	4.4	51	138.6	4.8
9	51	142.5	5.4	50	139.4	4.9
10	50	141.5	5.2	49	139.7	4.6
11	50	145.0	5.4	51	141.4	4.2
12	52	145.5	5.3	52	141.0	5.3
13	50	146.7	5.3	49	142.4	4.9
14	49	147.2	4.8	51	142.7	5.1
15	50	148.7	6.3	51	144.5	4.8
16	50	149.4	6.0	51	145.2	4.9
17	49	153.3	5.9	51	144.0	5.1
18	52	151.1	5.8	51	144.4	4.6
19–25	109	151.1	5.7	200	144.1	5.1

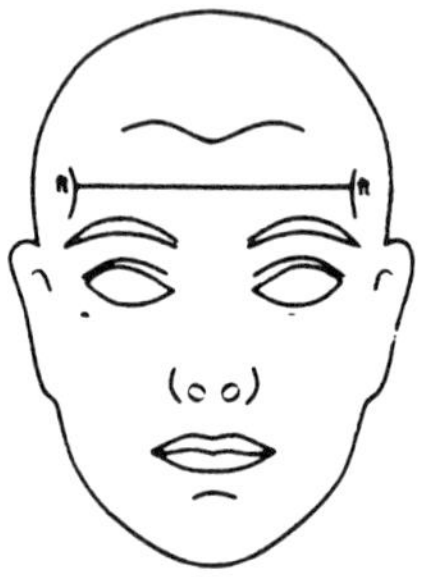

TABLE A–I-2. *Width of the forehead (ft-ft) (mm)*

Age (years)	Male			Female		
	N	Mean	SD	*N*	Mean	SD
0–5 months	8	76.2	3.6	5	74.2	4.4
6–12 months	20	80.2	3.5	8	79.0	3.0
1	18	83.3	3.5	28	82.5	4.6
2	30	87.8	4.8	32	85.5	4.3
3	30	89.8	5.6	30	92.1	4.4
4	30	96.2	4.0	30	93.6	3.7
5	30	98.2	4.3	30	95.5	4.1
6	35	104.4	4.9	39	103.3	5.8
7	33	105.8	4.9	42	105.1	4.4
8	45	106.4	4.4	41	105.4	5.0
9	47	109.1	5.9	38	107.3	5.3
10	37	107.2	5.5	37	106.7	5.0
11	39	110.2	5.7	41	108.5	4.4
12	50	109.9	6.6	50	109.3	5.5
13	47	113.1	6.2	49	111.6	5.4
14	46	114.2	5.6	43	112.5	5.7
15	44	115.0	5.9	43	111.9	4.5
16	49	115.8	5.9	35	113.3	4.9
17	48	118.8	6.0	36	114.4	4.7
18	50	117.5	5.2	38	114.0	4.3
19–25	109	115.9	5.2	199	111.5	4.4

HEAD

TABLE A-I-3. *Skull-base width (t-t) (mm)*

Age (years)	Male			Female		
	N	Mean	SD	*N*	Mean	SD
0–5 months	8	95.5	6.7	5	94.2	3.6
6–12 months	20	99.1	4.7	8	94.7	4.1
1	18	103.3	3.9	28	102.9	4.9
2	30	109.1	4.7	32	105.7	3.7
3	30	110.8	4.1	30	110.1	3.4
4	30	120.7	5.4	30	116.0	4.4
5	30	123.4	3.9	30	116.9	4.8
6	50	123.1	5.4	50	120.3	4.6
7	50	125.3	5.4	50	122.1	4.4
8	51	127.7	4.5	51	122.7	4.9
9	51	129.0	5.2	50	124.9	5.7
10	50	129.5	5.2	49	126.8	4.6
11	50	133.5	5.9	51	127.8	5.4
12	52	135.4	5.6	53	129.7	5.3
13	50	137.3	5.7	49	132.2	4.6
14	49	138.1	5.7	51	132.8	4.3
15	50	140.4	5.9	51	135.5	4.7
16	50	142.6	5.3	51	136.1	5.2
17	49	146.4	6.4	51	135.3	5.9
18	52	143.9	5.5	51	135.9	5.1
19–25	109	146.8	5.6	200	138.3	4.9

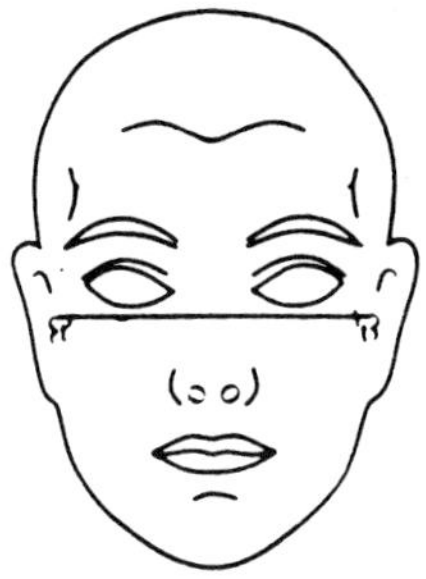

TABLE A-I-4. *Height of the calvarium (v-tr) (mm)*

Age (years)	Male			Female		
	N	Mean	SD	*N*	Mean	SD
1	17	32.4	6.0	17	31.8	5.3
2	31	32.6	5.3	32	31.6	5.7
3	30	33.9	5.9	30	32.1	5.1
4	30	41.8	7.1	30	43.9	8.3
5	30	45.8	7.6	30	45.3	5.9
6*	35	46.6	7.7	38	45.8	6.1
7*	33	47.3	7.7	42	47.0	6.2
8*	45	47.3	7.8	41	46.8	6.4
9*	47	47.2	7.9	38	46.9	6.5
10*	37	47.2	8.0	37	47.4	6.7
11*	38	47.7	8.0	41	47.3	6.8
12*	50	46.6	8.1	50	49.4	7.0
13*	47	46.8	8.2	49	48.0	7.1
14*	46	46.3	8.2	43	48.0	7.3
15*	44	47.3	8.3	43	48.9	7.4
16*	49	46.3	8.4	35	48.3	7.6
17*	48	46.9	8.4	36	48.5	7.7
18	50	46.5	8.5	38	49.0	7.9
19–25	109	46.3	9.0	198	47.4	8.0

* Indirectly derived norms (for details see pages 141–142).

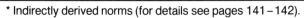

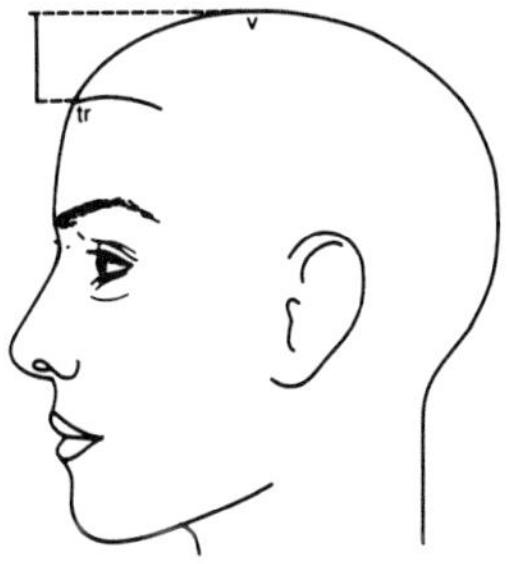

HEAD

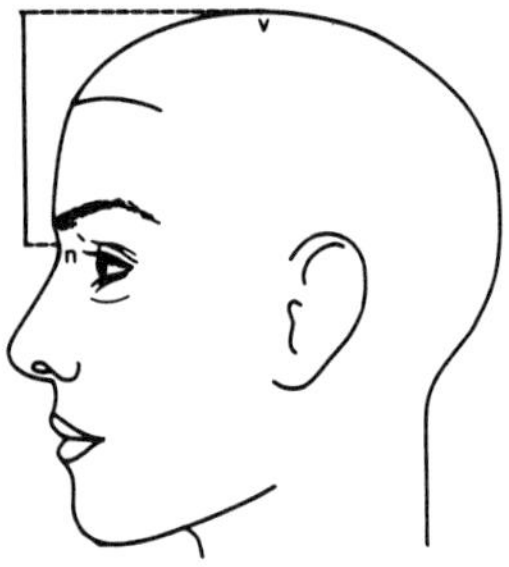

TABLE A–I-5. *Anterior height of the head (v-n) (mm)*

Age (years)	Male			Female		
	N	Mean	SD	*N*	Mean	SD
1	17	97.5	6.4	28	95.8	8.8
2	31	99.2	7.2	32	96.9	7.0
3	30	103.5	7.0	30	96.9	7.1
4	30	103.6	6.4	30	101.0	6.4
5	30	102.9	6.6	30	102.8	6.2
6	35	105.7	7.7	38	104.0	7.4
7	33	108.3	6.1	42	106.9	5.4
8	45	109.3	6.0	41	106.5	6.7
9	47	110.2	5.7	38	106.8	6.8
10	37	111.2	5.9	37	108.2	6.3
11	38	113.5	9.3	41	108.0	6.3
12	50	111.8	6.5	50	112.8	8.0
13	47	113.3	7.4	49	109.8	6.8
14	46	113.1	7.5	43	109.8	5.2
15	44	116.5	8.2	43	112.1	6.4
16	49	115.2	5.8	35	110.8	6.4
17	48	117.7	8.0	36	111.4	6.9
18	50	117.6	9.7	38	112.6	7.1
19–25	109	111.3	6.9	199	108.9	6.3

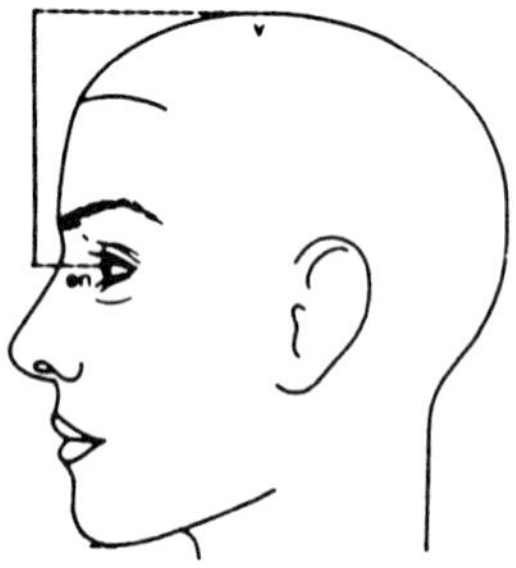

TABLE A–I-6. *Special height of the head (v-en) (mm)*

Age (years)	Male			Female		
	N	Mean	SD	*N*	Mean	SD
18	40	120.5	6.3	40	117.3	7.7
19–25	109	121.3	6.8	198	118.7	6.1

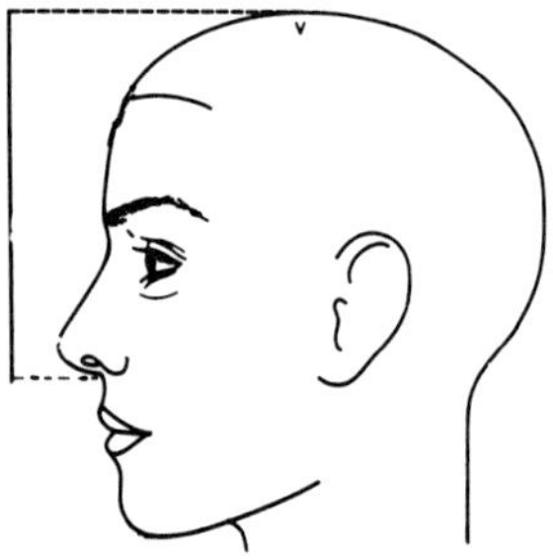

TABLE A–I-7. *Height of the head and nose (v-sn) (mm)*

Age	Male			Female		
	N	Mean	SD	*N*	Mean	SD
18	40	163.6	7.4	40	159.8	7.9
19–25	109	164.3	7.7	198	159.4	7.2

TABLE A–I-8. *Craniofacial height (v-gn) (mm)*

Age (years)	Male			Female		
	N	Mean	SD	*N*	Mean	SD
1	17	177.5	7.1	28	173.8	6.2
2	31	182.5	8.6	32	179.3	6.5
3	30	187.4	7.1	30	181.6	7.0
4	30	193.0	7.1	30	188.1	5.9
5	30	193.7	6.1	30	190.9	6.6
6	50	198.2	9.9	50	194.0	9.8
7	50	201.1	10.7	50	199.0	9.4
8	51	207.3	7.9	50	200.9	8.5
9	51	208.7	9.7	49	202.8	9.7
10	50	211.0	8.2	49	204.9	10.4
11	49	214.7	10.4	51	206.7	11.1
12	52	218.2	8.3	53	216.1	11.3
13	50	220.7	10.8	49	216.3	8.1
14	49	224.1	11.0	51	214.6	9.6
15	50	231.6	11.0	51	218.7	9.8
16	50	232.5	8.0	51	217.4	10.1
17	49	236.1	9.3	51	218.0	9.9
18	52	234.3	12.3	51	216.2	12.6
19–25	109	229.4	7.3	200	215.0	7.9

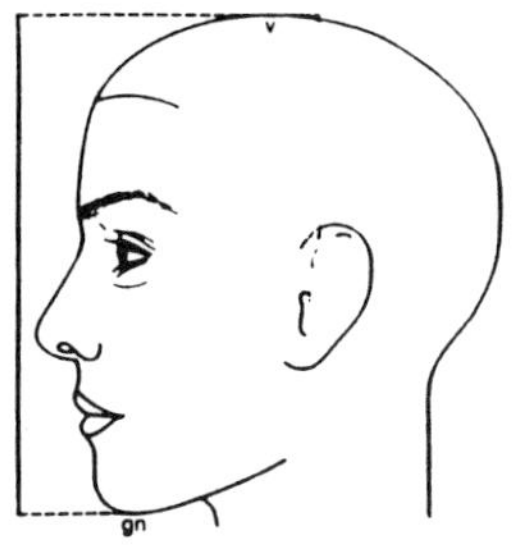

TABLE A–I-9. *Height of the forehead I (tr-g) (mm)*

Age (years)	Male			Female		
	N	Mean	SD	*N*	Mean	SD
6	50	49.3	6.5	50	50.1	6.2
7	50	50.4	5.5	50	50.5	6.3
8	51	50.8	6.5	51	50.8	5.2
9	51	51.1	5.3	50	50.9	6.5
10	50	50.9	6.1	49	50.2	6.7
11	50	51.7	6.6	51	50.0	7.7
12	52	54.0	6.8	53	49.7	5.7
13	50	53.8	6.5	49	49.9	6.0
14	49	52.8	6.5	51	50.7	5.8
15	50	57.6	6.1	51	51.7	6.8
16	50	56.9	8.7	51	51.0	5.9
17	49	55.5	7.0	51	50.7	6.0
18	52	58.4	6.5	51	51.3	6.3
19–25	109	57.0	7.4	200	52.7	6.0

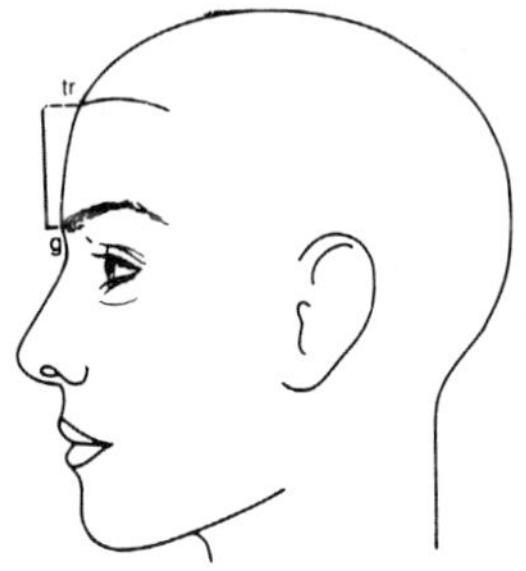

HEAD

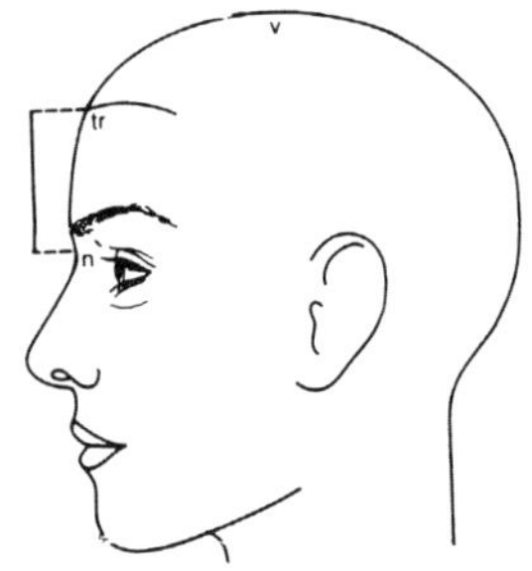

TABLE A–I-10. *Height of the forehead II (tr-n) (mm)*

Age (years)	Male *N*	Male Mean	Male SD	Female *N*	Female Mean	Female SD
1	18	62.7	5.9	28	64.0	5.2
2	31	62.9	4.4	31	62.1	4.9
3	30	65.4	4.5	30	61.1	3.8
4	30	67.0	4.6	30	60.9	6.5
5	30	64.2	4.9	30	61.2	5.2
6	50	61.0	7.0	50	62.0	6.3
7	50	63.0	7.0	50	61.7	6.7
8	51	62.2	7.1	51	62.9	6.4
9	51	62.2	5.6	50	62.1	6.6
10	50	62.5	6.8	49	61.7	6.7
11	50	64.2	7.5	51	62.1	7.5
12	52	66.8	6.7	53	62.6	6.3
13	50	65.3	6.7	49	61.3	6.6
14	49	63.5	6.5	51	61.9	6.5
15	50	68.7	6.3	51	62.0	6.7
16	50	67.1	9.3	51	61.5	6.5
17	49	65.9	8.0	51	62.8	7.0
18	52	70.1	6.4	51	63.5	6.5
19–25	109	67.1	7.5	200	63.0	6.0

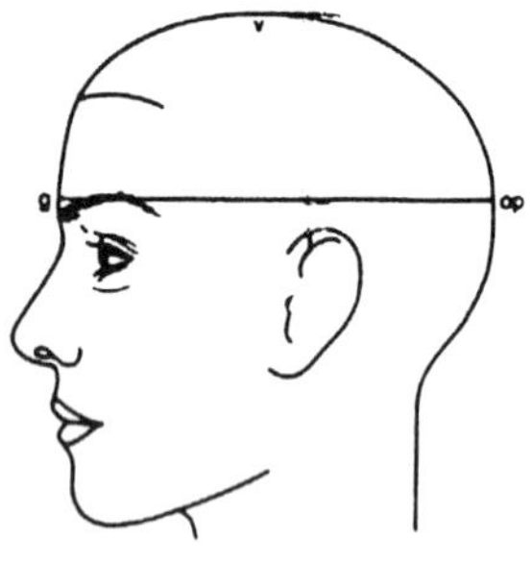

TABLE A–I-11. *Length of the head (g-op) (mm)*

Age (years)	Male *N*	Male Mean	Male SD	Female *N*	Female Mean	Female SD
0–5 months	8	149.0	5.9	5	141.6	3.0
6–12 months	20	151.9	5.3	8	158.3	3.9
1	18	166.7	6.2	28	162.0	7.9
2	30	170.5	12.4	32	168.6	5.7
3	30	177.5	6.7	30	173.7	6.3
4	30	181.5	6.2	30	175.2	5.2
5	30	180.5	6.2	30	178.8	5.2
6	50	183.2	7.6	50	177.7	5.8
7	50	184.0	7.7	50	180.8	6.4
8	51	185.9	7.5	51	181.1	7.0
9	51	185.8	5.7	50	181.2	6.7
10	50	187.8	6.0	49	182.7	6.5
11	50	187.0	7.8	51	183.3	6.7
12	52	188.8	7.6	53	184.2	6.3
13	50	188.3	7.3	49	183.9	6.5
14	49	189.2	7.4	51	184.3	6.2
15	50	194.1	6.8	51	184.9	6.0
16	50	193.3	5.6	51	184.5	6.1
17	49	193.7	7.6	51	184.5	6.8
18	52	192.7	6.7	51	184.9	7.0
19–25	109	197.4	6.7	199	186.8	6.8

HEAD

TABLE A–I-12. *Circumference of the head (mm)*

Age (years)	Male *N*	Male Mean	Male SD	Female *N*	Female Mean	Female SD
0–5 months	8	433.8	12.4	5	418.0	9.5
6–12 months	20	452.6	14.1	8	451.6	16.3
1	17	490.9	11.1	28	475.5	16.8
2	31	500.8	14.5	30	490.7	10.5
3	30	508.8	12.9	30	502.2	12.8
4	30	518.4	14.7	30	508.9	10.3
5	30	520.0	11.9	30	516.7	9.4
6	50	518.6	14.3	49	507.4	12.1
7	50	521.2	14.2	50	515.4	14.4
8	51	529.1	15.9	51	517.8	14.3
9	51	528.7	13.9	50	522.0	13.4
10	50	534.7	13.9	49	525.9	13.1
11	50	537.5	17.4	51	530.2	14.6
12	52	542.8	13.6	53	534.8	13.5
13	50	545.7	15.3	49	537.8	13.6
14	49	547.8	16.2	51	538.7	12.7
15	50	562.8	14.9	50	542.0	15.3
16	50	563.2	11.5	51	546.4	13.6
17	49	568.6	16.1	50	544.4	14.6
18	52	562.5	14.4	51	542.1	14.8
19–25	109	579.0	14.4	200	549.0	14.8

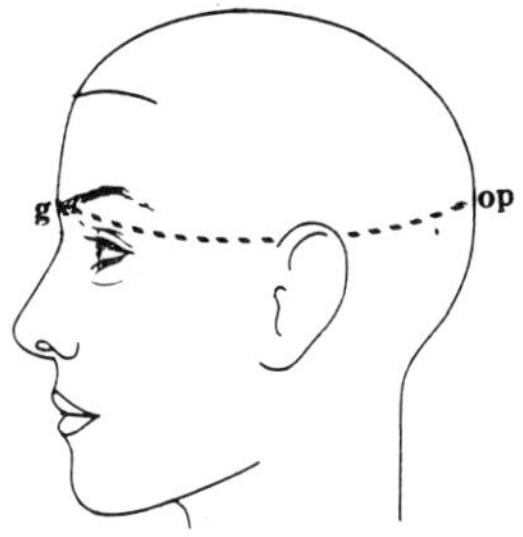

TABLE A–I-13. *Inclination of the anterior surface of the forehead (degrees)*

Age (years)	Male *N*	Male Mean	Male SD	Female *N*	Female Mean	Female SD
1	18	4.9	2.8	19	5.7	3.5
2	31	4.8	3.1	34	5.8	3.6
3	30	5.5	2.9	30	5.3	3.4
4	30	7.2	3.5	30	7.8	4.1
5	30	5.6	8.0	30	8.1	4.5
6	50	4.2	5.3	49	7.0	5.1
7	50	4.7	4.8	50	5.6	4.8
8	51	4.1	5.1	51	4.9	5.5
9	51	1.4	5.5	50	3.0	4.4
10	50	0.3	5.6	49	1.1	5.3
11	50	−0.3	5.2	51	0.6	5.1
12	52	−3.3	5.8	53	−0.5	5.2
13	50	−3.3	5.8	49	−2.1	5.0
14	49	−5.2	5.8	51	−2.1	5.9
15	50	−7.5	5.7	51	−2.2	5.2
16	50	−7.0	5.4	51	−3.4	5.5
17	49	−8.7	4.5	51	−4.3	6.5
18	52	−10.1	4.8	51	−6.6	6.0
19–25	109	−9.8	4.4	200	−5.9	5.2

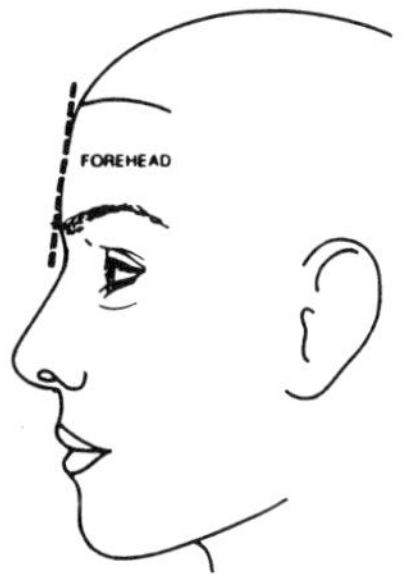

HEAD

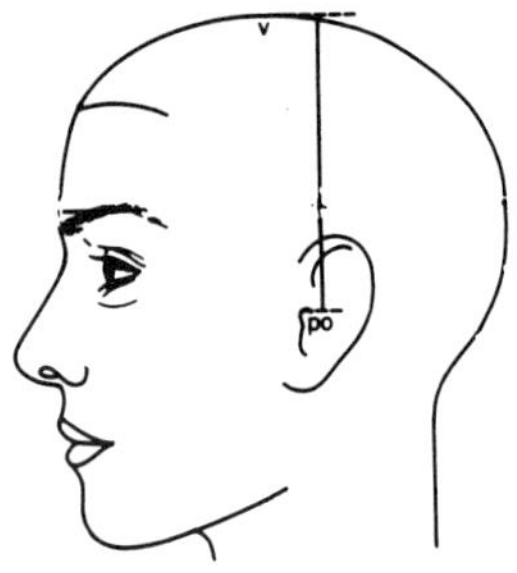

TABLE A–I-14. *Auricular height of the head (v-po) (mm)*

Age (years)	Male, right side			Female, right side		
	N	Mean	SD	*N*	Mean	SD
6	50	128.7	5.9	50	125.3	4.7
7	50	129.2	6.0	50	127.5	4.4
8	51	130.3	5.3	51	127.8	4.5
9	51	131.9	5.1	50	128.0	4.8
10	50	133.7	6.1	49	130.1	5.7
11	48	133.3	6.4	51	130.3	5.7
12	52	133.1	5.4	53	130.2	5.8
13	50	134.8	5.8	49	132.2	4.5
14	49	136.2	5.8	51	132.2	5.2
15	50	138.5	5.4	51	133.9	5.1
16	50	138.1	4.3	50	133.3	5.1
17	49	140.4	4.4	51	132.8	5.4
18	52	137.3	6.0	51	130.4	5.2
19–25	1	132.0	—	200	132.4	5.0

Age (years)	Male, left side			Female, left side		
	N	Mean	SD	*N*	Mean	SD
1	17	116.8	5.7	17	114.2	5.8
2	31	119.9	5.6	32	115.6	5.8
3	30	123.1	6.4	30	118.6	6.1
4	30	123.0	5.7	30	119.2	3.8
5	30	123.7	5.0	30	121.5	5.3
6	50	128.6	5.9	50	125.4	4.9
7	50	129.1	6.2	50	127.5	4.5
8	51	130.5	5.4	51	127.7	4.5
9	51	131.9	4.8	50	127.7	4.9
10	50	133.3	6.2	49	129.8	5.8
11	48	132.9	6.2	51	129.5	5.6
12	52	132.8	5.2	53	130.1	5.8
13	50	134.7	5.9	49	131.9	5.5
14	49	136.1	5.9	51	132.0	5.1
15	50	138.5	5.5	51	133.7	5.1
16	50	138.1	4.8	50	133.5	5.0
17	49	140.2	4.5	51	132.9	5.3
18	52	136.9	5.9	51	130.1	5.4
19–25	109	132.7	5.7	199	132.3	5.0

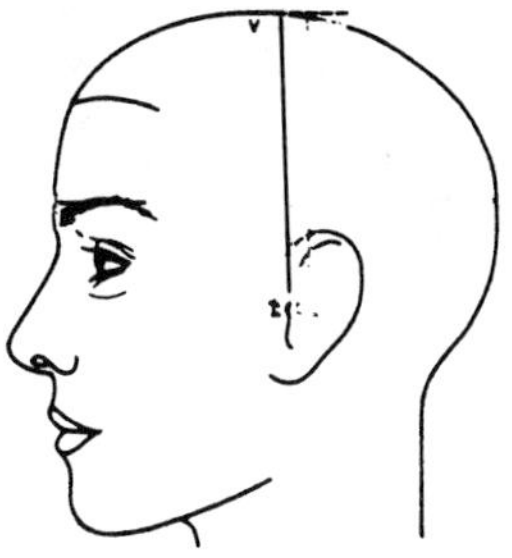

TABLE A–I-15. *Distance between the vertex and the tragion (v-t) (mm)*

Age (years)	Male, right side			Female, right side		
	N	Mean	SD	*N*	Mean	SD
6	22	136.4	5.0	22	131.8	4.4
7	21	138.7	6.9	36	133.9	4.0
8	41	137.6	4.7	38	133.9	4.4
9	47	139.6	5.4	38	135.6	5.0
10	37	140.2	4.5	35	137.1	5.9
11	36	141.3	6.3	40	137.9	5.7
12	34	141.5	4.5	28	137.3	5.3
13	44	142.8	5.9	46	139.6	4.4
14	46	144.2	5.9	43	140.2	5.0
15	44	147.3	5.5	43	141.9	5.4
16	48	146.9	4.4	34	140.9	5.4
17	44	148.8	4.9	33	140.0	4.3
18	26	146.2	6.5	17	140.7	3.7

Age (years)	Male, left side			Female, left side		
	N	Mean	SD	*N*	Mean	SD
6	22	136.1	4.6	22	131.5	4.5
7	21	138.3	6.7	36	133.6	4.5
8	41	137.5	4.9	38	133.7	4.5
9	47	139.3	5.2	38	135.1	5.0
10	37	139.4	4.5	35	136.4	6.0
11	36	140.6	6.3	40	136.5	5.7
12	34	141.2	4.1	28	137.2	5.3
13	44	142.2	6.3	46	139.2	5.7
14	46	143.8	6.2	43	139.6	4.9
15	44	147.1	5.4	43	141.5	5.4
16	48	146.6	5.2	34	141.3	5.3
17	44	148.2	5.0	33	139.9	4.3
18	26	145.3	6.1	17	139.7	4.6

FACE

FACE

TABLE A–II-1. *Width of the face (zy-zy) (mm)*

Age (years)	Male			Female		
	N	Mean	SD	*N*	Mean	SD
0–5 months	8	92.5	3.0	5	93.2	4.1
6–12 months	20	97.8	5.2	8	94.6	4.6
1	18	96.7	3.3	27	95.6	4.3
2	31	98.9	4.9	32	97.9	3.0
3	30	101.4	5.0	30	101.2	4.2
4	30	110.2	5.4	30	106.8	4.6
5	30	111.8	5.1	30	109.4	3.6
6	50	114.9	5.3	50	113.4	5.1
7	50	116.0	5.8	50	115.8	4.6
8	51	120.5	4.1	51	117.3	4.6
9	51	121.8	5.8	50	119.4	6.0
10	50	121.9	6.1	49	120.7	4.7
11	50	125.7	6.0	51	122.5	5.2
12	52	125.5	5.6	53	123.6	5.8
13	50	128.5	5.9	49	126.8	5.2
14	49	130.9	5.7	51	128.0	4.5
15	50	133.5	6.2	51	129.7	4.5
16	50	134.9	5.5	51	130.6	5.3
17	49	139.1	6.3	51	131.1	5.3
18	52	137.1	4.3	51	129.9	5.3
19–25	109	139.1	5.3	200	130.0	4.6

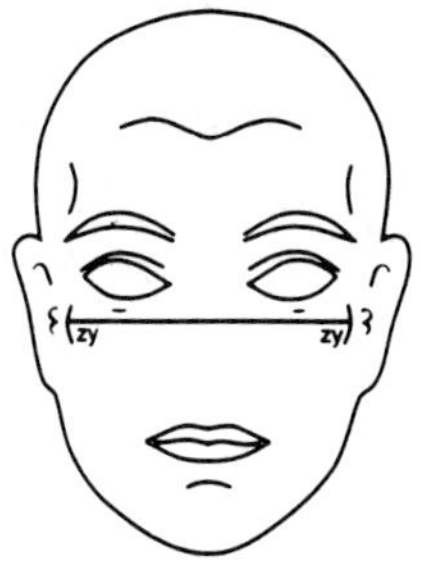

TABLE A–II-2. *Width of the mandible (go-go) (mm)*

Age (years)	Male			Female		
	N	Mean	SD	*N*	Mean	SD
0–5 months	8	70.5	3.9	5	69.0	2.3
6–12 months	20	75.1	4.7	8	69.9	4.2
1	18	76.2	6.0	24	74.6	5.2
2	30	79.8	4.2	32	78.2	4.5
3	30	81.6	6.0	30	80.9	4.5
4	30	86.0	5.5	30	83.5	4.4
5	30	87.2	3.9	30	85.7	3.5
6	49	82.2	5.3	50	79.1	5.4
7	50	83.0	5.8	50	83.5	5.0
8	51	87.5	4.1	50	83.8	5.4
9	51	89.3	4.9	50	87.0	5.5
10	50	89.2	5.4	48	86.4	4.9
11	49	90.0	6.1	51	87.1	5.4
12	51	89.3	6.8	52	87.1	6.6
13	48	94.1	5.8	49	89.3	6.2
14	48	96.3	5.2	51	92.5	4.0
15	48	99.5	5.5	50	93.8	4.3
16	49	99.4	4.8	50	93.3	4.8
17	48	99.5	6.3	49	92.6	4.9
18	52	97.1	5.8	49	91.1	5.9
19–25	109	105.6	6.7	200	94.5	5.0

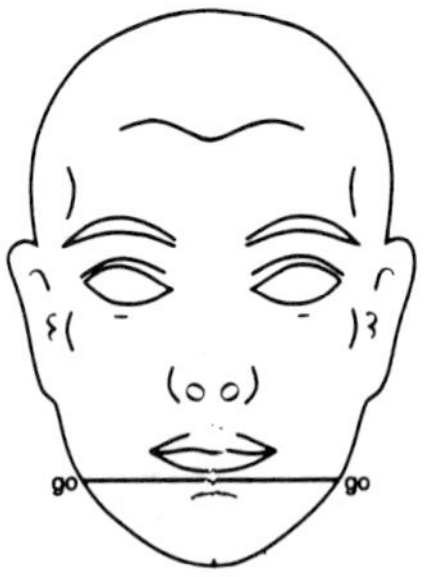

FACE

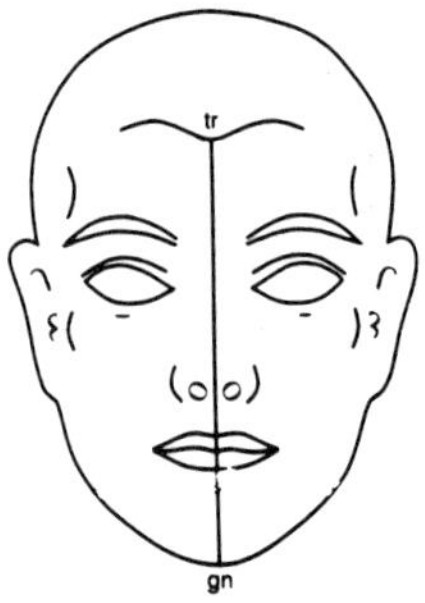

TABLE A–II-3. *Physiognomical height of the face (tr-gn) (mm)*

Age (years)	Male			Female		
	N	Mean	SD	*N*	Mean	SD
1	18	143.6	6.7	21	141.1	6.6
2	31	150.1	6.6	32	145.8	6.9
3	30	153.4	6.4	30	148.0	4.7
4	30	157.5	6.5	30	145.2	18.0
5	30	155.4	6.4	30	151.9	6.7
6	50	157.6	8.0	50	155.0	6.5
7	50	161.0	8.0	50	158.9	6.2
8	51	163.4	7.4	51	159.3	6.5
9	51	163.8	7.3	50	161.6	8.2
10	50	166.1	7.5	49	164.1	7.5
11	50	169.5	8.6	51	164.3	8.2
12	52	173.5	8.8	53	168.0	7.3
13	50	175.4	9.4	49	168.6	6.8
14	49	176.4	8.9	51	170.8	7.7
15	50	184.8	7.8	51	170.7	7.1
16	50	185.0	8.9	51	172.1	7.7
17	49	184.1	9.4	51	172.7	8.3
18	52	187.5	8.1	51	172.5	7.5
19–25	109	187.2	12.1	200	173.3	7.8

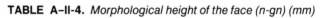

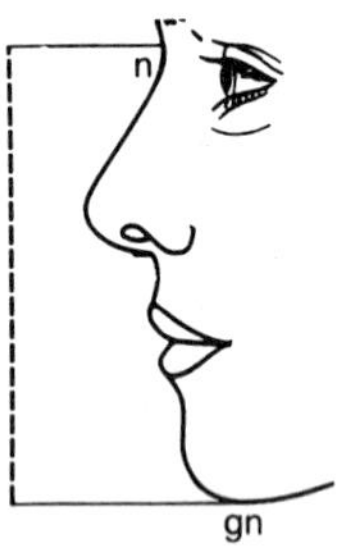

TABLE A–II-4. *Morphological height of the face (n-gn) (mm)*

Age (years)	Male			Female		
	N	Mean	SD	*N*	Mean	SD
0–5 months	8	70.0	4.5	5	68.0	4.5
6–12 months	20	70.5	4.8	8	72.7	4.4
1	18	80.6	4.8	20	77.2	4.9
2	31	87.5	3.5	31	83.8	3.9
3	30	88.5	3.5	30	86.9	2.4
4	30	96.4	4.3	30	92.6	3.6
5	30	96.7	3.5	30	96.5	4.6
6	50	98.5	5.0	50	95.7	4.4
7	50	99.5	5.0	50	98.3	3.7
8	51	101.8	4.9	51	98.1	5.4
9	51	102.7	5.3	50	101.3	5.3
10	50	105.2	4.5	49	103.9	5.0
11	50	107.1	6.0	51	104.7	5.0
12	52	109.1	5.4	53	108.2	4.6
13	50	111.6	5.7	49	109.1	5.0
14	49	114.1	6.5	51	110.7	5.3
15	50	119.1	5.7	51	111.0	5.1
16	50	120.9	4.6	51	113.5	6.0
17	49	120.9	7.1	51	112.0	4.7
18	52	121.3	6.8	51	111.8	5.2
19–25	109	124.7	5.7	200	111.4	4.8

TABLE A–II-5. *Physiognomical height of the upper face (n-sto) (mm)*

Age (years)	Male			Female		
	N	Mean	SD	*N*	Mean	SD
0–5 months	8	41.4	2.3	5	40.8	4.4
6–12 months	20	44.0	2.8	8	44.0	2.6
1	18	49.0	4.3	20	46.5	4.0
2	31	52.5	3.0	31	50.7	3.5
3	30	54.3	2.7	30	53.4	2.6
4	30	58.9	2.3	30	56.1	2.6
5	30	58.6	2.1	30	58.0	2.6
6	50	60.0	3.5	50	57.9	3.5
7	50	60.4	3.4	50	59.7	2.6
8	51	61.8	3.2	51	60.4	3.4
9	51	62.3	3.5	50	62.3	3.4
10	50	64.5	3.2	49	63.2	3.4
11	50	65.4	3.3	51	64.4	2.9
12	52	67.3	4.0	53	66.3	3.2
13	50	68.3	3.2	49	67.3	3.2
14	49	70.0	4.2	51	68.2	4.0
15	50	73.3	3.9	51	68.8	3.4
16	50	74.1	3.2	51	70.4	3.8
17	49	74.0	4.4	51	68.9	3.6
18	52	74.0	4.2	51	68.1	3.4
19–25	109	76.6	4.0	200	69.4	3.2

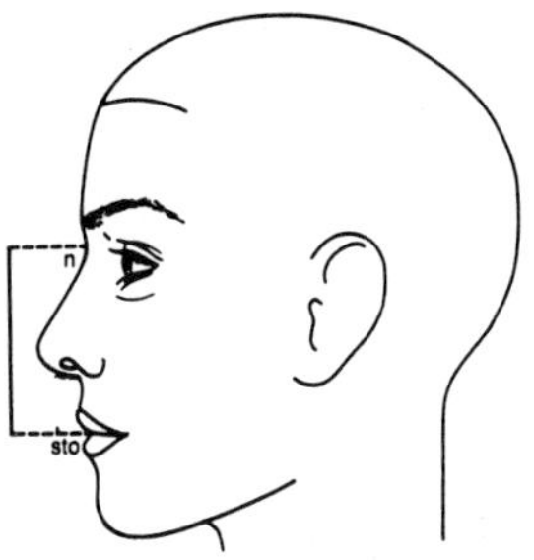

TABLE A–II-6. *Height of the lower face (sn-gn) (mm)*

Age (years)	Male			Female		
	N	Mean	SD	*N*	Mean	SD
1	18	49.9	2.8	19	47.3	3.7
2	31	54.5	3.1	31	51.7	3.2
3	30	55.2	3.9	30	54.3	2.7
4	30	60.1	3.5	30	57.8	3.7
5	30	60.3	3.2	30	59.4	3.6
6	50	61.4	3.8	50	58.8	4.3
7	50	61.1	4.2	50	59.7	2.7
8	51	61.9	4.0	51	59.3	3.8
9	51	61.7	4.3	50	59.9	3.7
10	50	63.5	3.8	49	62.2	4.3
11	50	65.3	4.5	51	62.1	3.9
12	52	64.8	4.7	53	64.6	4.3
13	50	66.5	4.3	49	63.9	4.0
14	49	67.8	4.5	51	64.8	4.3
15	50	70.6	4.5	51	64.1	4.3
16	50	71.3	3.9	51	65.9	4.8
17	49	70.8	5.6	51	65.3	4.6
18	52	71.9	6.0	51	65.5	4.5
19–25	109	72.6	4.5	200	64.3	4.0

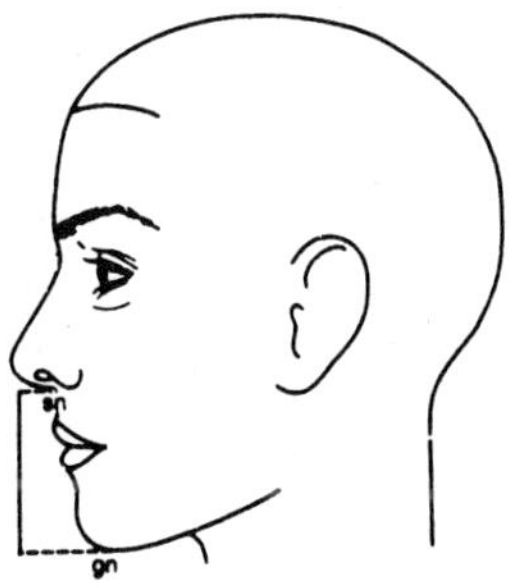

FACE

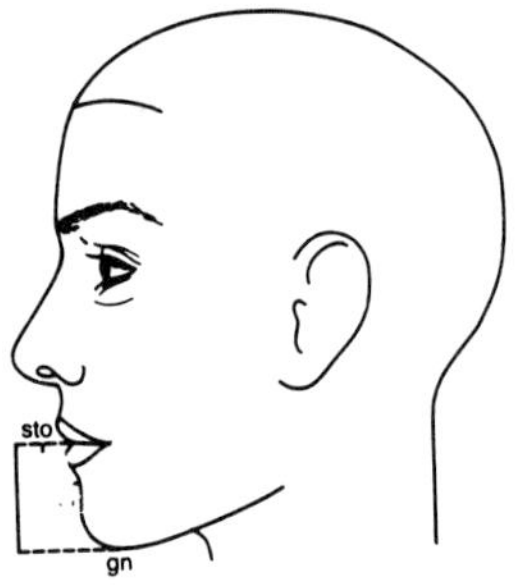

TABLE A–II-7. *Height of the mandible (sto-gn) (mm)*

Age (years)	Male			Female		
	N	Mean	SD	*N*	Mean	SD
1	18	31.9	2.2	19	31.4	3.1
2	31	36.1	2.8	31	34.4	2.4
3	30	35.6	2.0	30	35.5	2.5
4	30	41.1	3.7	30	40.2	3.4
5	30	42.2	2.7	30	41.3	2.8
6	50	41.4	3.2	50	40.3	3.3
7	50	42.4	3.1	50	40.7	2.4
8	51	42.2	3.4	51	40.6	3.0
9	51	42.4	3.6	50	40.9	3.1
10	50	43.3	3.5	49	42.5	3.5
11	50	44.0	3.2	51	42.2	3.0
12	52	44.1	3.9	53	44.1	3.6
13	50	45.7	3.2	49	43.5	3.2
14	49	46.4	3.5	51	44.2	3.3
15	50	47.8	3.5	51	43.5	3.1
16	50	48.9	3.4	51	44.7	4.0
17	49	48.5	3.8	51	44.7	2.9
18	52	50.1	4.4	51	45.2	2.9
19–25	109	50.7	4.0	200	43.4	3.1

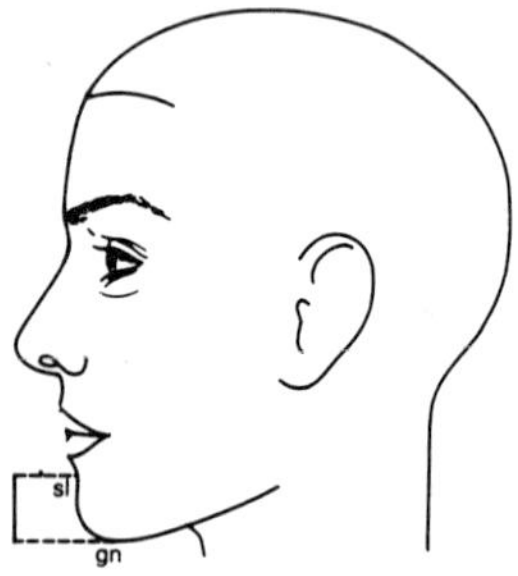

TABLE A–II-8. *Height of the chin (sl-gn) (mm)*

Age (years)	Male			Female		
	N	Mean	SD	*N*	Mean	SD
0–5 months	8	17.3	1.9	5	15.6	0.9
6–12 months	20	18.4	1.8	8	18.7	2.1
1	18	20.4	1.6	19	19.8	1.5
2	31	24.0	2.3	31	22.1	2.2
3	30	23.9	2.0	30	24.1	2.5
4	30	27.0	2.7	30	26.6	2.2
5	30	27.4	2.3	30	27.5	2.2
6	*	26.9	2.4	*	26.6	2.2
7	*	27.5	2.4	*	26.7	2.2
8	*	27.3	2.5	*	26.4	2.2
9	*	27.5	2.5	*	26.4	2.2
10	*	28.0	2.6	*	27.2	2.3
11	*	28.5	2.7	*	26.8	2.3
12	*	28.5	2.7	*	27.8	2.3
13	*	29.5	2.8	*	27.2	2.3
14	*	30.0	2.8	*	27.4	2.3
15	*	30.9	2.9	*	26.8	2.3
16	*	31.6	3.0	*	27.3	2.3
17	*	31.3	3.0	*	27.1	2.3
18	*	32.3	3.1	*	27.2	2.3
19–25	109	33.1	3.0	200	27.0	2.5

TABLE A–II-9. *Height of the upper profile (tr-prn) (mm)*

Age (years)	Male *N*	Male Mean	Male SD	Female *N*	Female Mean	Female SD
18	40	114.4	9.3	40	107.2	7.4
19–25	109	115.3	9.3	197	106.4	8.1

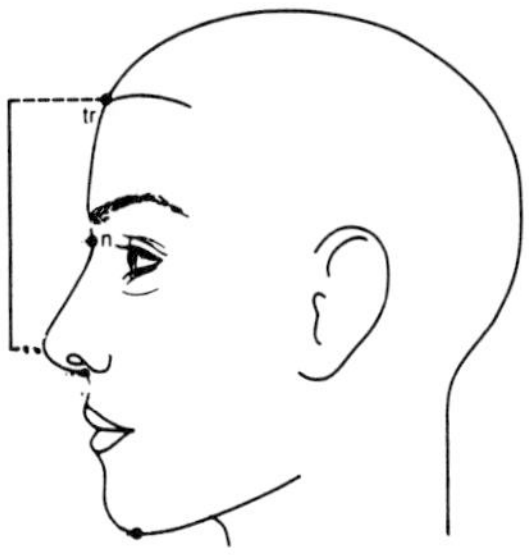

TABLE A–II-10. *Height of the lower profile (prn-gn) (mm)*

Age (years)	Male *N*	Male Mean	Male SD	Female *N*	Female Mean	Female SD
18	50	92.2	5.5	50	78.5	5.7
19–25	109	91.7	5.6	200	81.4	4.6

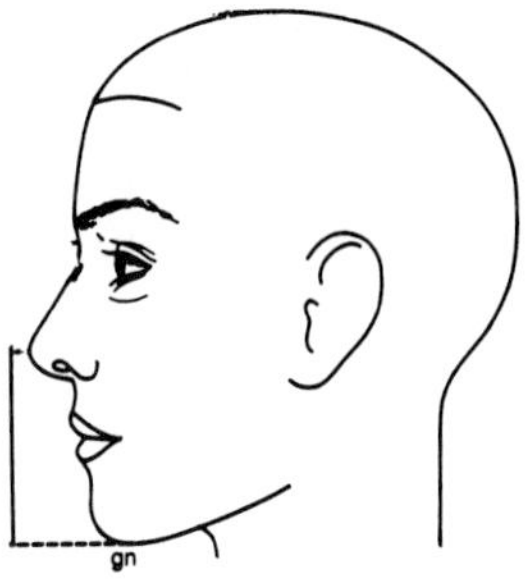

TABLE A–II-11. *Lower half of the craniofacial height (en-gn) (mm)*

Age (years)	Male *N*	Male Mean	Male SD	Female *N*	Female Mean	Female SD
19–25	109	117.7	5.6	200	102.7	5.1

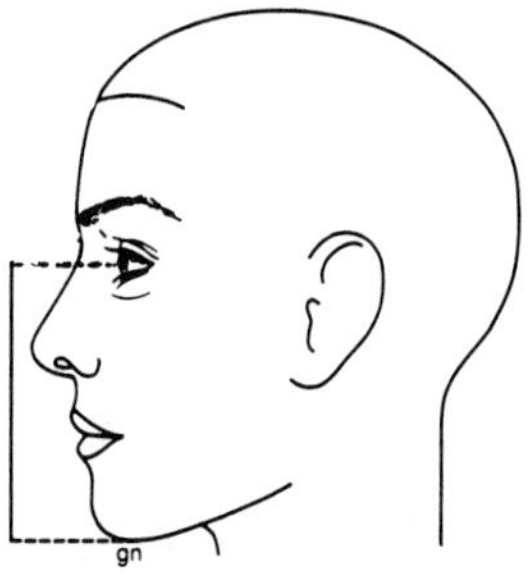

TABLE A–II-12. *Distance between the glabella and the subnasale (g-sn) (mm)*

Age (years)	Male *N*	Male Mean	Male SD	Female *N*	Female Mean	Female SD
19–25	109	67.2	4.9	200	63.1	4.4

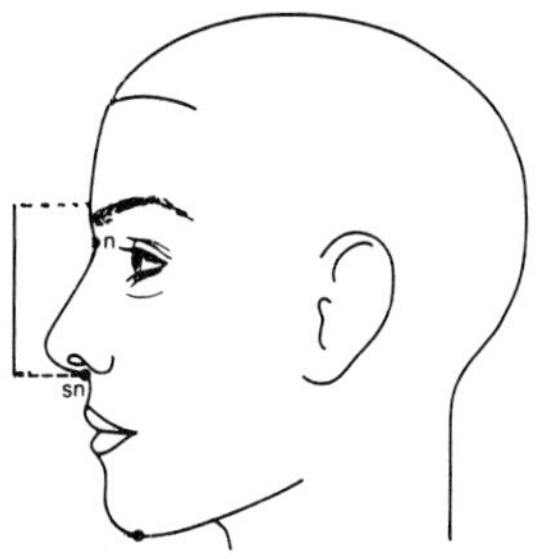

FACE

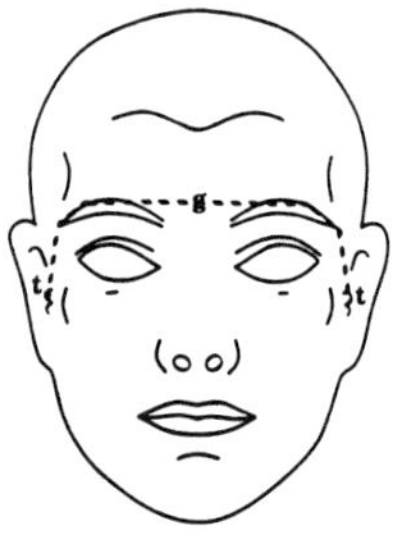

TABLE A-II-13. *Supraorbital arc (t-g-t) (mm)*

Age (years)	Male *N*	Male Mean	Male SD	Female *N*	Female Mean	Female SD
1	17	252.1	8.0	15	244.7	8.5
2	31	255.6	7.4	30	252.1	3.1
3	30	261.6	9.8	30	259.0	9.4
4	30	265.0	9.0	30	258.1	9.7
5	30	268.8	11.2	30	263.3	8.6
6	*	277.2	11.6	*	270.7	8.8
7	*	280.7	12.0	*	276.3	9.1
8	*	285.6	12.4	*	275.6	9.3
9	*	288.6	12.8	*	281.1	9.5
10	*	292.3	13.2	*	283.5	9.8
11	*	299.2	13.5	*	287.1	10.0
12	*	301.4	13.9	*	289.7	10.2
13	*	304.8	14.3	*	292.6	10.4
14	*	310.6	14.7	*	292.3	10.7
15	*	317.5	15.1	*	296.2	10.9
16	*	320.1	15.5	*	293.8	11.1
17	*	323.2	15.9	*	292.2	11.4
18	50	336.7	16.3	51	294.8	11.6
19–25	109	336.7	16.7	193	294.8	11.8

* Indirectly derived norms (for details see pages 141–142).

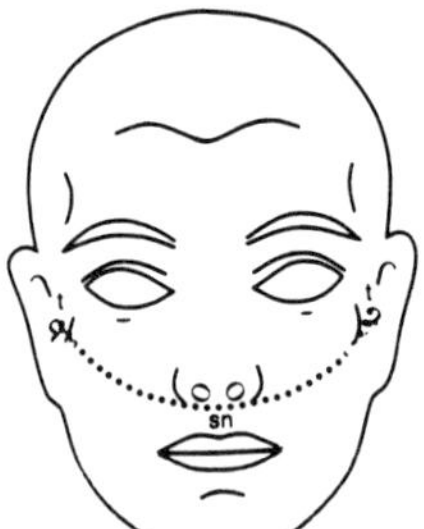

TABLE A-II-14. *Maxillary arc (t-sn-t) (mm)*

Age (years)	Male *N*	Male Mean	Male SD	Female *N*	Female Mean	Female SD
1	17	226.4	9.0	15	221.3	8.0
2	31	232.0	7.8	30	227.1	8.5
3	30	235.8	9.6	30	231.8	9.7
4	30	244.5	10.3	30	237.4	6.8
5	30	247.1	9.1	30	241.8	7.5
6	50	241.4	9.8	50	238.3	6.9
7	50	243.3	8.9	50	241.9	8.0
8	51	250.3	7.6	51	244.3	9.0
9	51	253.6	10.4	50	249.0	9.7
10	50	257.3	11.3	49	252.6	7.8
11	50	261.7	10.4	51	255.5	9.8
12	52	263.8	9.7	53	259.0	9.9
13	50	266.9	11.1	49	263.1	10.2
14	49	270.2	13.2	51	264.3	8.9
15	50	278.2	11.7	51	266.9	9.8
16	50	278.6	10.4	51	267.5	10.7
17	49	284.0	9.5	51	268.4	8.7
18	52	279.3	10.8	51	267.5	11.3
19–25	109	302.2	9.9	200	280.2	9.4

TABLE A–II-15. *Mandibular arc (t-gn-t) (mm)*

Age (years)	Male			Female		
	N	Mean	SD	*N*	Mean	SD
1	17	226.3	8.7	15	225.0	8.5
2	30	235.3	8.0	30	230.5	10.0
3	30	242.3	10.1	30	237.8	8.8
4	30	255.0	10.2	30	245.1	7.9
5	30	258.0	11.0	30	254.2	7.3
6	50	246.9	9.3	50	242.7	7.7
7	50	250.6	10.6	50	249.7	7.9
8	51	260.3	8.2	51	251.7	9.7
9	51	263.5	10.7	50	258.3	10.7
10	50	267.7	12.3	49	261.1	8.9
11	50	273.6	11.6	51	266.2	11.1
12	52	277.4	11.0	53	269.3	11.5
13	50	283.5	12.5	49	275.3	11.1
14	49	288.2	15.5	51	278.9	10.4
15	50	297.9	14.7	51	281.5	11.2
16	50	298.7	11.2	51	284.1	11.3
17	49	306.2	13.0	51	285.6	9.7
18	52	304.1	13.2	51	284.7	11.3
19–25	109	336.7	16.7	193	294.8	11.8

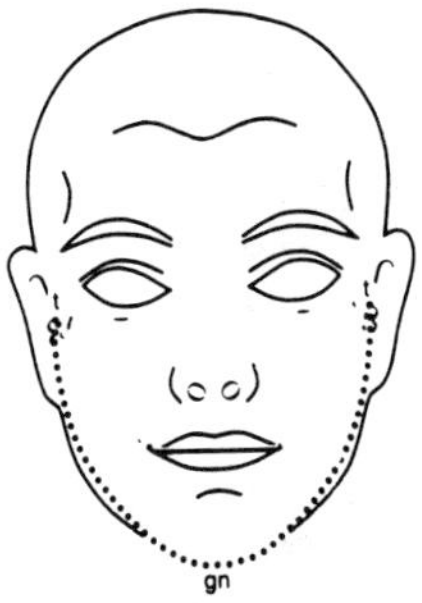

TABLE A–II-16. *Inclination of the upper face profile (g-sn) (degrees)*

Age (years)	Male			Female		
	N	Mean	SD	*N*	Mean	SD
1	18	−5.2	3.2	18	−5.3	3.5
2	31	−5.7	3.0	34	−5.1	3.3
3	30	−6.3	2.6	30	−5.9	3.2
4	30	−3.1	3.0	30	−0.9	3.2
5	30	−3.2	4.0	30	−1.6	2.9
6	21	−2.0	2.9	20	−3.2	4.0
7	19	−2.2	3.7	29	−2.6	3.3
8	29	−0.7	3.3	21	−2.5	3.3
9	34	−0.3	2.8	24	−1.0	2.6
10	21	−0.1	3.2	18	−1.6	2.0
11	15	−2.3	4.4	9	−1.4	2.6
12	10	2.3	3.7	5	1.0	4.5
13	25	1.3	3.7	12	0.2	3.6
14	34	0.5	3.1	23	0.5	3.5
15	23	2.3	3.9	9	0.1	1.5
16	24	0.6	3.2	5	0.4	2.1
17	16	0.6	3.3	12	0.3	4.1
18	13	2.1	3.7	8	−0.8	4.1
19–25	109	2.4	3.3	200	1.7	3.1

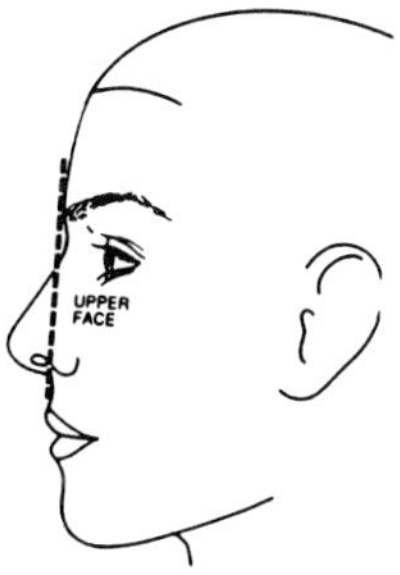

FACE

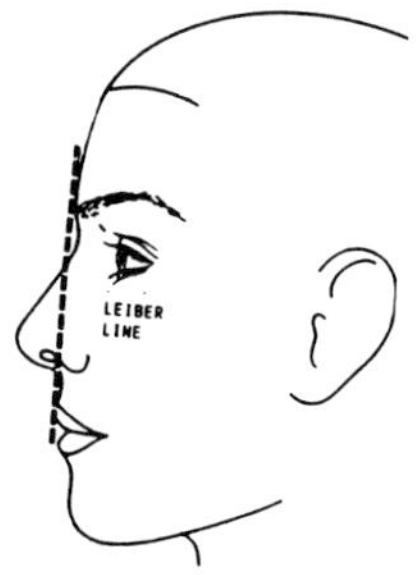

TABLE A–II-17. *Inclination of the Leiber line (g-ls) (degrees)*

Age (years)	Male			Female		
	N	Mean	SD	*N*	Mean	SD
1	17	3.6	2.6	18	1.6	3.0
2	31	2.0	3.3	34	1.9	3.1
3	30	0.3	3.8	30	2.3	3.0
4	30	−0.3	3.8	30	1.3	2.8
5	30	−0.7	3.6	30	0.9	3.0
19–25	40	1.3	3.5	40	1.6	2.5

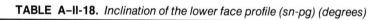

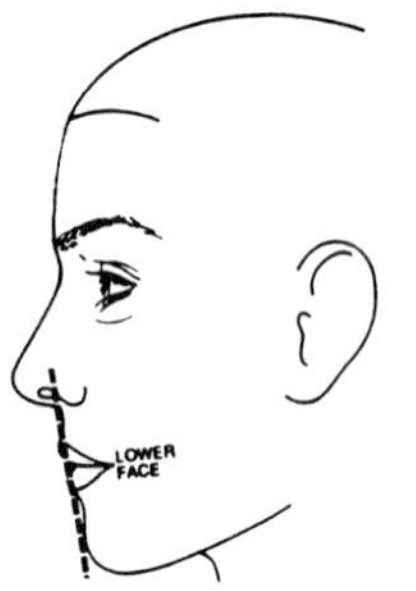

TABLE A–II-18. *Inclination of the lower face profile (sn-pg) (degrees)*

Age (years)	Male			Female		
	N	Mean	SD	*N*	Mean	SD
1	17	−16.8	3.3	18	−17.3	3.3
2	31	−19.3	3.2	34	−17.9	3.1
3	30	−19.3	4.2	30	−19.3	3.7
4	30	−21.6	4.0	30	−17.8	3.9
5	30	−18.1	5.3	30	−18.0	4.4
6	21	−21.0	3.5	20	−19.8	2.0
7	19	−20.7	2.9	29	−19.7	4.8
8	29	−19.8	4.1	21	−20.1	4.3
9	34	−18.2	4.0	24	−17.9	5.2
10	21	−18.4	5.1	18	−17.4	4.1
11	15	−18.9	3.3	9	−16.4	5.0
12	10	−15.3	4.6	5	−22.2	9.4
13	25	−17.2	4.3	12	−17.4	5.3
14	34	−16.7	5.2	26	−15.9	5.4
15	23	−15.8	6.0	9	−15.2	4.7
16	24	−17.3	5.4	5	−14.6	6.7
17	16	−17.3	3.7	12	−12.3	3.5
18	13	−14.5	5.7	8	−14.1	7.9
19–25	109	−10.6	5.3	200	−13.3	4.5

TABLE A–II-19. *Inclination of the mandible (li-pg) (degrees)*

Age (years)	Male *N*	Male Mean	Male SD	Female *N*	Female Mean	Female SD
1	17	−24.8	3.1	18	−24.4	3.9
2	31	−26.4	4.2	34	−26.5	3.2
3	30	−26.5	4.0	30	−27.0	4.9
4	30	−30.6	4.7	30	−28.7	5.0
5	30	−27.8	6.5	30	−28.3	5.3
6	#	−26.9	6.5	#	−27.7	5.4
7	#	−26.0	6.5	#	−27.0	5.5
8	#	−25.1	6.5	#	−26.4	5.6
9	#	−24.2	6.5	#	−25.7	5.7
10	#	−23.3	6.6	#	−25.1	5.9
11	#	−22.4	6.6	#	−24.5	6.0
12	#	−21.4	6.6	#	−23.8	6.1
13	#	−20.5	6.6	#	−23.2	6.2
14	#	−19.6	6.6	#	−22.5	6.3
15	#	−18.7	6.6	#	−21.9	6.4
16	#	−17.8	6.6	#	−21.3	6.5
17	#	−16.9	6.6	#	−20.6	6.6
18	#	−16.0	6.6	#	−20.0	6.7
19–25	109	−15.0	6.6	200	−19.4	6.9

Indirectly derived norms (for details see pages 141–142).

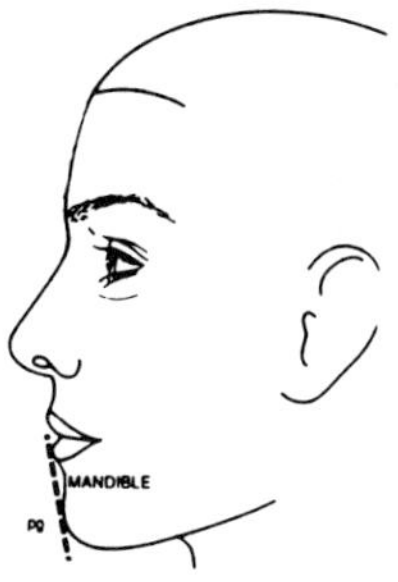

TABLE A–II-20. *Inclination of the chin (degrees)*

Age (years)	Male *N*	Male Mean	Male SD	Female *N*	Female Mean	Female SD
1	17	−5.9	7.7	18	−3.3	8.9
2	31	−7.2	7.7	34	−5.9	6.9
3	30	−7.7	6.9	30	−5.8	8.4
4	30	−7.6	6.4	30	−8.5	7.0
5	30	−7.3	8.8	30	−8.2	9.2
6	#	−5.6	8.9	#	−7.0	9.2
7	#	−3.9	9.0	#	−5.7	9.3
8	#	−2.2	9.1	#	−4.5	9.3
9	#	−0.5	9.2	#	−3.2	9.3
10	#	1.2	9.4	#	−2.0	9.4
11	#	3.0	9.5	#	−0.8	9.4
12	#	4.7	9.6	#	0.5	9.4
13	#	6.4	9.7	#	1.7	9.4
14	#	8.1	9.8	#	3.0	9.5
15	#	9.8	9.9	#	4.2	9.5
16	#	11.5	10.0	#	5.4	9.5
17	#	13.2	10.1	#	6.7	9.6
18	#	14.9	10.2	#	7.9	9.6
19–25	109	16.7	10.4	45	9.1	9.6

Indirectly derived norms (for details see pages 141–142).

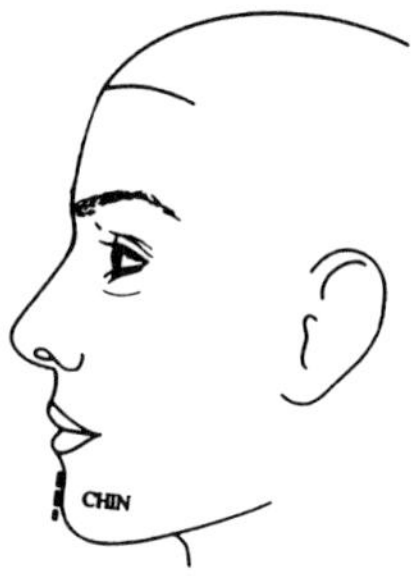

FACE

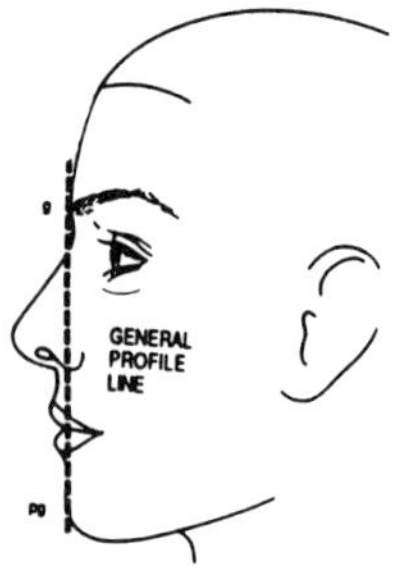

TABLE A–II-21. *Inclination of the general profile line (g-pg) (degrees)*

Age (years)	Male			Female		
	N	Mean	SD	*N*	Mean	SD
1	17	−10.2	2.5	18	−8.6	2.4
2	31	−9.8	2.3	34	−9.7	2.7
3	30	−11.2	2.2	30	−9.6	2.6
4	30	−11.7	2.5	30	−10.0	2.8
5	30	−11.2	2.8	30	−9.6	2.9
18	101	−4.7	3.1	131	−4.9	3.8
19–25	109	−3.0	3.4	200	−4.1	3.0

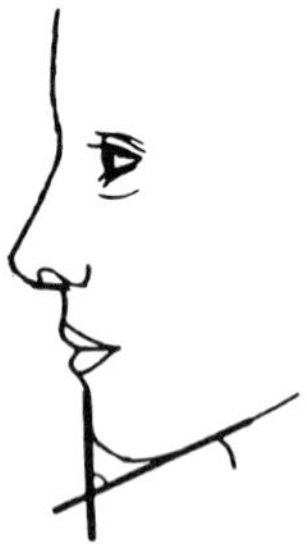

TABLE A–II-22. *Mentocervical angle (degrees)*

Age (years)	Male			Female		
	N	Mean	SD	*N*	Mean	SD
1	18	94.2	7.5	20	94.6	6.5
2	31	95.8	5.8	34	94.4	6.6
3	30	96.6	5.8	30	94.9	6.9
4	30	90.3	6.9	30	87.3	5.2
5	30	86.8	7.3	30	91.0	8.8
19–25	40	78.3	7.9	42	83.9	9.3

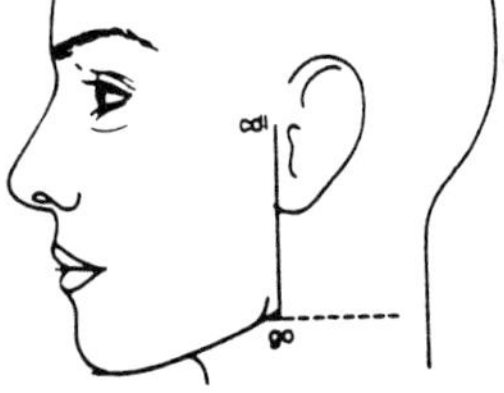

TABLE A–II-23. *Height of the mandibular ramus (go-cdl), right and left (mm)*

Age (years)	Male			Female		
	N	Mean	SD	*N*	Mean	SD
Right						
19–25	35	67.1	5.3	40	61.6	5.0
Left						
19–25	35	67.7	5.3	40	62.2	4.7

FACE

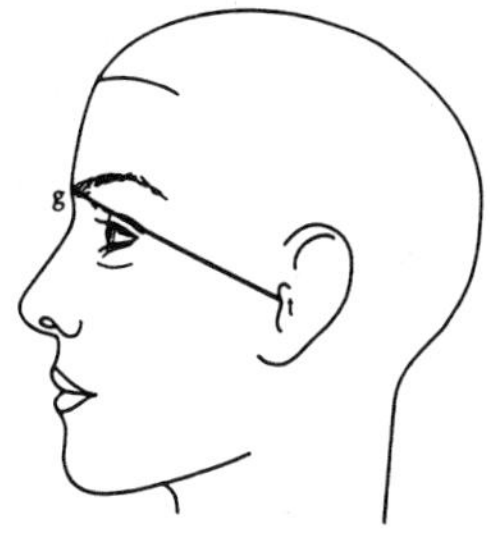

TABLE A-II-24. *Depth of the supraorbital rim (t-g), right and left (mm)*

Age (years)	Male, right			Female, right		
	N	Mean	SD	*N*	Mean	SD
1	18	101.3	3.6	20	96.9	3.3
2	30	104.4	4.2	31	102.5	2.9
3	30	108.1	5.0	30	106.2	4.3
4	30	108.1	4.2	30	102.8	3.1
5	30	108.5	3.4	30	105.7	2.9
6	*	111.9	3.6	*	108.5	3.0
7	*	113.1	3.7	*	110.5	3.1
8	*	115.5	3.9	*	110.8	3.2
9	*	116.6	4.1	*	112.8	3.3
10	*	118.3	4.3	*	114.4	3.4
11	*	121.1	4.4	*	115.8	3.4
12	*	122.3	4.6	*	117.1	3.5
13	*	123.6	4.8	*	118.0	3.6
14	*	126.0	4.9	*	118.1	3.7
15	*	129.0	5.1	*	119.7	3.8
16	*	130.4	5.3	*	119.2	3.9
17	*	131.8	5.4	*	118.4	4.0
18	*	129.5	5.6	*	119.7	4.1
19–25	109	129.5	5.8	193	119.7	4.2

Age (years)	Male, left			Female, left		
	N	Mean	SD	*N*	Mean	SD
1	18	101.2	3.8	20	96.8	3.3
2	30	104.3	4.3	31	102.7	3.0
3	30	108.2	5.4	30	106.2	4.3
4	30	107.5	4.0	30	102.6	3.0
5	30	108.3	4.0	30	105.3	3.3
6	*	111.9	4.1	*	108.6	3.5
7	*	113.7	4.2	*	111.3	3.6
8	*	115.5	4.4	*	110.6	3.8
9	*	117.0	4.5	*	113.2	3.9
10	*	118.5	4.6	*	113.7	4.1
11	*	121.5	4.7	*	115.4	4.3
12	*	122.3	4.8	*	116.4	4.4
13	*	124.0	5.0	*	118.0	4.6
14	*	126.5	5.1	*	117.8	4.7
15	*	129.3	5.2	*	119.5	4.9
16	*	130.2	5.3	*	118.3	5.1
17	*	131.5	5.4	*	117.9	5.2
18	*	129.5	5.6	*	118.9	5.4
19–25	109	129.2	5.7	193	118.9	5.6

* Indirectly derived norms (for details see pages 141–142).

FACE

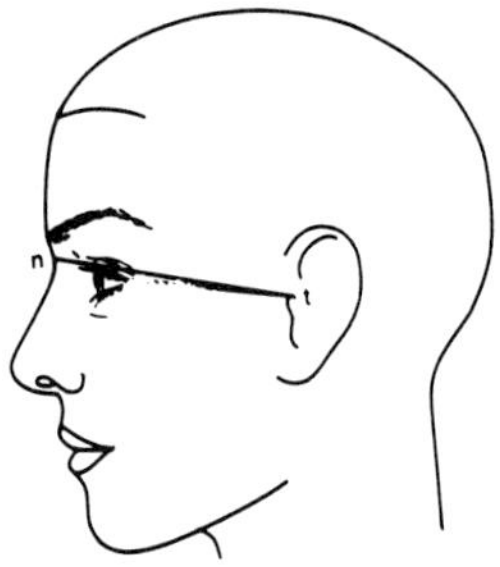

TABLE A–II-25. *Depth of the upper third of the face (n-t), right and left (mm)*

Age (years)	Male, right side			Female, right side		
	N	Mean	SD	*N*	Mean	SD
1	18	92.6	2.1	20	89.8	3.1
2	30	95.9	3.6	32	93.6	2.8
3	30	97.1	3.5	30	96.0	2.9
4	30	102.8	4.0	30	98.3	3.0
5	30	104.0	3.5	30	101.1	3.2
6	50	107.1	3.8	50	104.0	3.2
7	50	108.2	4.1	50	106.1	3.4
8	51	110.4	3.7	51	106.6	3.7
9	51	111.3	4.2	50	108.7	4.1
10	50	112.8	4.2	49	110.4	3.3
11	50	115.4	4.8	51	112.0	3.7
12	52	116.4	4.6	53	113.4	4.4
13	50	117.6	4.7	49	114.5	4.3
14	49	119.8	5.1	51	114.7	3.6
15	50	122.5	4.5	51	116.5	4.6
16	50	123.7	3.9	51	116.2	3.8
17	49	125.0	4.6	51	115.6	3.7
18	52	122.7	4.3	51	117.0	4.7
19–25	109	126.6	4.2	200	116.5	4.5

Age (years)	Male, left side			Female, left side		
	N	Mean	SD	*N*	Mean	SD
1	18	92.5	2.1	20	89.8	3.2
2	30	96.0	3.5	32	93.6	2.8
3	30	97.2	3.7	30	96.0	2.9
4	30	102.6	3.9	30	98.5	3.4
5	30	103.5	3.7	30	101.0	3.2
6	50	106.9	4.4	50	104.4	3.3
7	50	108.6	3.7	50	107.3	3.7
8	51	110.3	3.9	51	106.9	3.8
9	51	111.7	3.7	50	109.6	4.2
10	50	113.1	4.2	49	110.4	3.1
11	50	115.9	4.5	51	112.3	3.8
12	52	116.7	4.5	53	113.5	4.2
13	50	118.3	4.8	49	115.3	4.0
14	49	120.6	5.4	51	115.4	3.7
15	50	123.3	4.0	51	117.3	3.8
16	50	124.1	4.0	51	116.3	4.0
17	49	125.3	4.5	51	116.2	3.5
18	52	123.4	4.3	51	117.4	5.3
19–25	109	126.2	4.3	200	116.0	4.5

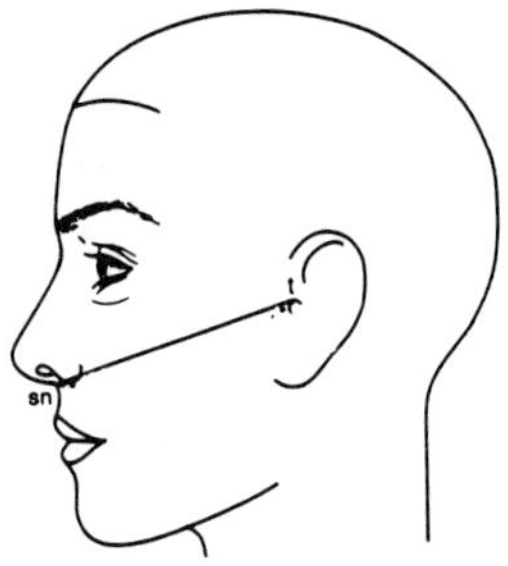

TABLE A-II-26. *Depth in the maxillary region (sn-t), right and left (mm)*

Age (years)	Male, right side			Female, right side		
	N	Mean	SD	*N*	Mean	SD
1	18	95.3	2.0	20	91.3	2.6
2	30	96.4	3.1	32	95.1	3.0
3	30	98.7	3.7	30	97.0	2.2
4	30	105.0	4.0	30	100.3	2.8
5	30	106.1	3.2	30	103.1	3.0
6	50	107.5	4.3	50	104.3	3.2
7	50	109.3	4.0	50	106.2	3.1
8	51	111.2	3.7	51	107.2	3.7
9	51	112.4	4.5	50	109.5	4.2
10	50	114.2	4.6	49	111.4	3.2
11	50	116.4	4.3	51	113.1	4.0
12	52	118.4	4.8	53	115.1	4.2
13	50	118.9	5.0	49	116.3	4.6
14	49	121.4	5.6	51	116.3	3.7
15	50	124.5	5.0	51	117.9	4.8
16	50	125.8	4.5	51	118.2	4.4
17	49	127.6	4.1	51	117.5	4.1
18	52	125.3	5.0	51	117.9	4.3
19–25	109	133.0	4.6	200	120.5	3.8

Age (years)	Male, left side			Female, left side		
	N	Mean	SD	*N*	Mean	SD
1	18	95.2	1.9	20	91.4	2.8
2	30	96.7	3.2	32	95.2	3.1
3	30	98.6	3.2	30	97.1	2.3
4	30	104.6	4.0	30	100.4	2.9
5	30	106.0	3.5	30	102.9	3.2
6	50	106.6	4.5	50	104.5	3.4
7	50	109.1	3.6	50	106.5	3.2
8	51	111.0	3.8	51	107.5	4.0
9	51	111.9	4.2	50	110.0	4.2
10	50	113.2	5.0	49	110.6	3.2
11	50	116.0	4.3	51	112.8	4.7
12	52	118.2	4.8	53	114.9	4.0
13	50	119.2	5.0	49	116.0	4.5
14	49	121.4	6.0	51	116.3	3.6
15	50	124.7	5.0	51	117.9	3.6
16	50	125.1	4.6	51	117.1	4.5
17	49	126.7	4.3	51	117.6	4.0
18	52	125.6	4.9	51	118.3	5.2
19–25	109	131.8	4.3	200	119.3	4.0

FACE

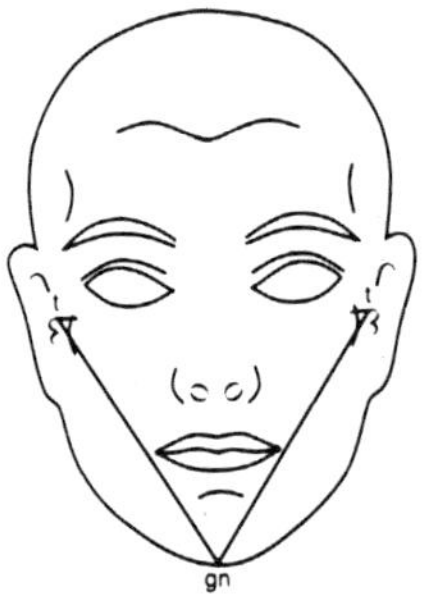

TABLE A–II-27. *Depth in the mandibular region (gn-t), right and left (mm)*

Age (years)	Male, right side			Female, right side		
	N	Mean	SD	*N*	Mean	SD
1	18	100.2	3.1	19	97.9	2.6
2	30	103.9	3.9	31	102.5	2.9
3	30	106.9	4.7	30	105.9	3.6
4	30	111.8	4.4	30	107.4	3.9
5	30	114.4	4.9	30	111.2	4.2
6	50	113.3	4.9	50	111.3	4.1
7	50	115.3	5.6	50	114.4	3.8
8	51	119.1	3.8	51	115.4	4.6
9	51	120.9	4.6	50	117.5	4.7
10	50	122.6	5.5	49	119.6	4.7
11	50	125.5	5.1	51	121.9	5.3
12	52	127.9	5.4	53	123.6	6.0
13	50	129.5	6.2	49	126.3	5.2
14	49	132.6	6.7	51	127.5	4.9
15	50	136.6	6.7	51	128.9	5.8
16	50	137.6	5.2	51	128.8	5.8
17	49	140.9	6.1	51	129.9	4.7
18	52	139.6	6.4	51	129.7	5.2
19–25	109	148.5	5.0	200	134.0	5.0

Age (years)	Male, left side			Female, left side		
	N	Mean	SD	*N*	Mean	SD
1	18	100.1	3.2	19	98.1	2.6
2	30	104.0	3.9	31	102.7	3.0
3	30	107.2	4.8	30	106.3	3.5
4	30	112.3	4.2	30	107.9	4.4
5	30	115.0	5.3	30	111.5	4.1
6	50	112.4	4.4	50	110.5	4.2
7	50	114.7	5.4	50	113.4	3.8
8	51	118.6	4.0	51	113.9	4.7
9	51	119.5	4.5	50	116.9	5.0
10	50	121.1	5.4	49	118.2	4.4
11	50	124.4	5.1	51	120.7	5.9
12	52	126.8	5.8	53	123.0	5.0
13	50	129.1	6.0	49	125.4	5.4
14	49	132.5	7.6	51	126.8	4.8
15	50	136.0	6.6	51	128.1	5.3
16	50	136.1	5.9	51	127.7	5.8
17	49	139.1	6.2	51	129.3	4.7
18	52	138.5	6.8	51	129.1	5.7
19–25	109	148.2	5.2	200	133.9	5.1

FACE

TABLE A–II-28. *Depth of the lower jaw (go-gn), right and left (mm)*

Age (years)	Male, right side *N*	Mean	SD	Female, right side *N*	Mean	SD
1	18	67.3	3.4	18	66.3	2.9
2	30	70.6	3.1	30	69.5	3.7
3	30	72.4	4.1	30	71.1	3.7
4	30	75.3	3.9	30	73.0	3.9
5	30	77.1	4.6	30	76.5	4.1
6	*	76.8	4.6	*	76.2	4.1
7	*	77.7	4.7	*	77.8	4.0
8	*	79.3	4.7	*	77.2	4.0
9	*	79.3	4.8	*	78.5	4.0
10	*	80.3	4.8	*	79.4	4.0
11	*	81.2	4.8	*	80.3	3.9
12	*	82.9	4.9	*	81.2	3.9
13	*	82.7	4.9	*	80.9	3.9
14	*	84.1	5.0	*	81.7	3.8
15	*	86.2	5.0	*	82.0	3.8
16	*	86.5	5.0	*	81.9	3.8
17	*	88.5	5.1	*	81.3	3.7
18	*	86.6	5.1	*	81.3	3.7
19–25	35	86.6	5.2	40	81.3	3.7

Age (years)	Male, left side *N*	Mean	SD	Female, left side *N*	Mean	SD
1	18	67.3	3.4	18	66.2	2.9
2	30	70.7	3.2	30	69.5	3.7
3	30	72.5	4.3	30	71.2	3.8
4	30	75.5	3.9	30	73.1	3.8
5	30	77.1	4.1	30	76.6	4.1
6	*	76.3	4.2	*	77.0	4.1
7	*	78.3	4.3	*	78.0	4.0
8	*	78.9	4.3	*	77.8	4.0
9	*	79.4	4.4	*	79.1	4.0
10	*	79.7	4.5	*	79.6	4.0
11	*	81.3	4.6	*	80.3	3.9
12	*	83.2	4.7	*	81.5	3.9
13	*	82.6	4.7	*	81.4	3.9
14	*	84.3	4.8	*	82.3	3.8
15	*	86.1	4.9	*	82.7	3.8
16	*	85.8	5.0	*	82.3	3.8
17	*	87.5	5.1	*	81.7	3.7
18	*	86.8	5.1	*	81.8	3.7
19–20	35	86.8	5.2	40	81.8	3.7

* Indirectly derived norms (for details see pages 141–142).

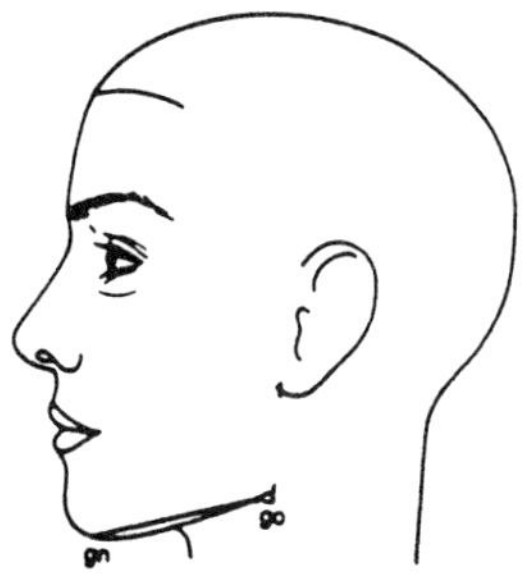

FACE

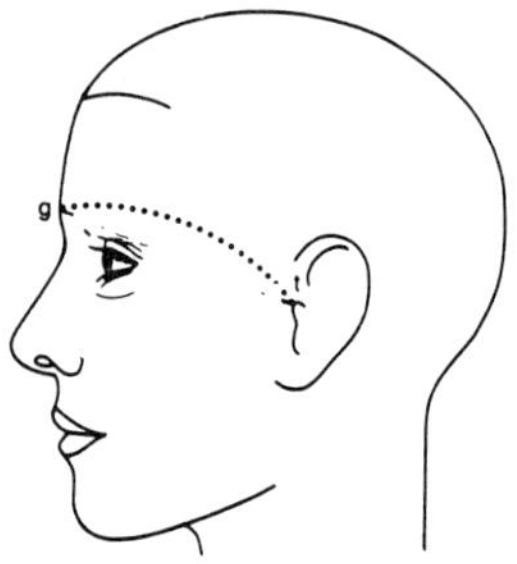

TABLE A–II-29. *Lateral surface half-arc in the upper third of the face (t-g surf), right and left (mm)*

Age (years)	Male, right side			Female, right side		
	N	Mean	SD	*N*	Mean	SD
1	17	126.1	3.9	15	122.5	4.2
2	30	127.7	3.6	30	126.0	4.2
3	30	131.0	4.9	30	129.5	4.8
4	30	132.8	4.7	30	129.2	4.9
5	30	134.5	5.1	30	131.8	4.3
6	*	138.6	5.1	*	135.2	4.4
7	*	140.0	5.1	*	137.6	4.4
8	*	142.9	5.2	*	137.9	4.5
9	*	144.2	5.2	*	140.3	4.5
10	*	146.2	5.2	*	142.2	4.6
11	*	149.5	5.2	*	143.8	4.7
12	*	150.9	5.2	*	145.4	4.7
13	*	152.4	5.3	*	146.4	4.8
14	*	155.3	5.3	*	146.4	4.8
15	*	158.9	5.3	*	148.3	4.9
16	*	160.6	5.3	*	147.6	5.0
17	*	162.2	5.3	*	146.6	5.0
18	*	159.3	5.4	*	148.1	5.1
19–25	109	159.3	5.4	193	148.1	5.1

Age (years)	Male, left side			Female, left side		
	N	Mean	SD	*N*	Mean	SD
1	17	125.9	4.1	15	122.2	4.4
2	30	127.9	3.9	30	125.9	4.0
3	30	130.6	5.0	30	129.6	4.8
4	30	132.2	4.7	30	128.9	5.1
5	30	134.3	6.5	30	131.5	4.7
6	*	138.6	4.7	*	135.5	4.7
7	*	140.7	4.7	*	138.8	4.7
8	*	142.7	4.8	*	137.9	4.8
9	*	144.4	4.8	*	141.0	4.8
10	*	146.1	4.8	*	141.6	4.8
11	*	149.7	4.8	*	143.6	4.8
12	*	150.5	4.8	*	144.8	4.8
13	*	152.5	4.9	*	146.7	4.9
14	*	155.4	4.9	*	146.3	4.9
15	*	158.7	4.9	*	148.4	4.9
16	*	159.6	4.9	*	146.8	4.9
17	*	161.1	4.9	*	146.2	4.9
18	*	158.5	4.9	*	147.4	4.9
19–25	109	158.5	4.9	193	147.4	4.9

* Indirectly derived norms (for details see pages 141–142).

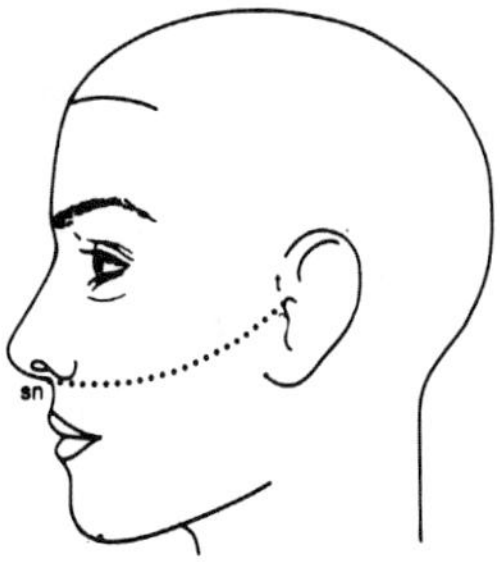

TABLE A–II-30. *Maxillary half-arc (sn-t surf), right and left (mm)*

Age (years)	Male, right side			Female, right side		
	N	Mean	SD	*N*	Mean	SD
1	17	113.2	4.2	15	110.6	4.3
2	30	115.8	4.0	30	113.8	4.4
3	30	118.0	4.7	30	115.9	4.7
4	30	121.4	5.3	30	117.9	3.5
5	30	122.6	4.6	30	120.7	3.8
6	50	121.2	5.0	50	119.6	3.5
7	50	121.7	4.6	50	120.8	4.2
8	51	125.7	3.8	51	122.4	4.7
9	51	127.1	5.6	50	124.7	4.7
10	50	129.4	5.8	49	127.1	3.7
11	50	131.4	5.5	51	128.2	5.1
12	52	132.3	4.7	53	129.9	5.2
13	50	133.8	5.3	49	132.2	5.4
14	49	135.7	6.5	51	132.6	4.5
15	50	139.7	5.9	51	134.3	5.4
16	50	140.5	5.2	51	134.4	5.7
17	49	143.1	4.9	51	134.8	4.7
18	52	140.2	5.6	51	134.1	5.7
19–25	109	151.7	5.3	200	140.7	4.9

Age (years)	Male, left side			Female, left side		
	N	Mean	SD	*N*	Mean	SD
1	17	112.8	4.7	15	110.9	4.0
2	30	115.8	3.9	30	113.6	4.3
3	30	117.8	5.0	30	115.8	5.1
4	30	123.1	5.3	30	119.5	3.6
5	30	124.5	4.9	30	121.4	3.4
6	50	120.2	5.1	50	118.7	3.7
7	50	121.5	4.6	50	121.1	4.0
8	51	124.6	4.3	51	121.9	4.6
9	51	126.5	5.1	50	124.3	5.2
10	50	127.9	5.9	49	125.4	4.4
11	50	130.3	5.4	51	127.2	5.0
12	52	131.6	5.3	53	129.1	5.0
13	50	133.1	6.1	49	130.9	5.1
14	49	134.4	7.0	51	131.6	4.8
15	50	138.5	6.3	51	132.5	4.8
16	50	138.1	5.6	51	133.1	5.3
17	49	140.9	5.2	51	133.6	4.4
18	52	139.2	5.6	51	133.5	6.0
19–25	108	150.5	5.0	200	140.0	8.6

FACE

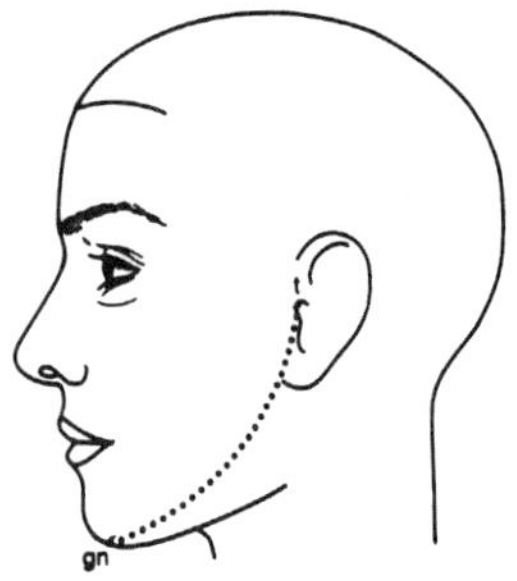

TABLE A–II-31. *Mandibular half-arc (gn-t surf), right and left (mm)*

Age (years)	Male, right side			Female, right side		
	N	Mean	SD	*N*	Mean	SD
1	17	113.2	4.4	15	112.5	4.1
2	30	117.9	4.0	30	115.3	5.0
3	30	121.3	5.2	30	119.1	4.4
4	30	127.2	5.3	30	122.5	4.0
5	30	128.8	5.7	30	127.3	3.8
6	50	124.6	5.0	50	122.4	4.4
7	50	126.0	5.5	50	125.8	3.8
8	51	131.0	4.1	51	126.7	5.2
9	51	132.8	5.5	50	130.0	5.5
10	50	135.2	6.6	49	131.9	4.8
11	50	137.8	5.9	51	134.5	5.9
12	52	139.8	5.5	53	135.6	6.2
13	50	142.8	6.6	49	139.1	5.6
14	49	145.0	7.9	51	140.6	5.3
15	50	150.1	7.7	51	141.6	5.9
16	50	150.8	5.4	51	143.2	5.7
17	49	155.1	6.7	51	144.2	5.3
18	52	153.2	6.8	51	143.4	5.5
19–25	109	170.4	6.4	200	153.9	6.3

Age (years)	Male, left side			Female, left side		
	N	Mean	SD	*N*	Mean	SD
1	17	113.1	4.3	15	112.5	4.4
2	30	117.6	4.0	30	115.1	4.9
3	30	121.1	5.1	30	118.7	4.4
4	30	127.9	5.3	30	122.6	4.2
5	30	129.2	5.5	30	126.9	3.8
6	50	122.3	4.7	50	120.3	3.8
7	50	124.6	5.6	50	123.9	4.6
8	51	129.3	4.5	51	125.0	5.0
9	51	130.7	5.6	50	128.3	5.5
10	50	132.5	6.0	49	129.3	4.5
11	50	135.8	6.0	51	131.7	5.6
12	52	137.6	5.8	53	133.7	5.7
13	50	140.7	6.3	49	136.3	5.7
14	49	143.2	8.0	51	138.3	5.5
15	50	147.9	7.3	51	139.9	5.6
16	50	147.9	6.1	51	140.9	6.0
17	49	151.3	7.1	51	141.4	4.8
18	52	150.9	6.8	51	141.3	6.2
19–25	109	168.3	6.3	200	151.5	6.0

Orbits

ORBITS

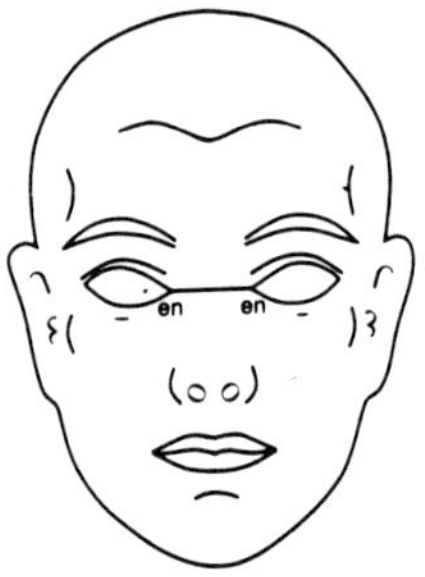

TABLE A–III-1. *Intercanthal width (en-en) (mm)*

Age (years)	Male			Female		
	N	Mean	SD	*N*	Mean	SD
0–5 months	8	25.6	1.1	5	25.0	2.5
6–12 months	8	25.6	2.1	8	25.6	2.1
1	18	27.3	1.8	22	26.9	2.3
2	31	26.5	2.0	31	26.6	1.7
3	30	27.2	1.9	30	27.0	2.0
4	30	30.3	1.9	30	29.0	2.0
5	30	30.8	2.1	30	29.4	2.2
6	50	30.6	2.3	50	29.8	2.0
7	50	30.2	2.5	50	30.1	1.9
8	51	31.2	2.2	51	30.5	1.9
9	51	31.7	2.4	50	31.1	2.2
10	50	31.2	2.0	49	31.2	2.6
11	50	32.6	2.3	51	31.6	2.2
12	52	32.0	2.1	53	31.6	2.6
13	50	32.8	2.8	49	32.2	2.3
14	49	33.1	2.6	51	32.4	2.1
15	50	33.7	2.3	51	32.7	2.3
16	50	33.4	3.2	51	31.8	2.0
17	49	33.9	2.6	51	32.5	2.1
18	52	32.9	2.7	51	31.6	2.4
19–25	109	33.3	2.7	200	31.8	2.3

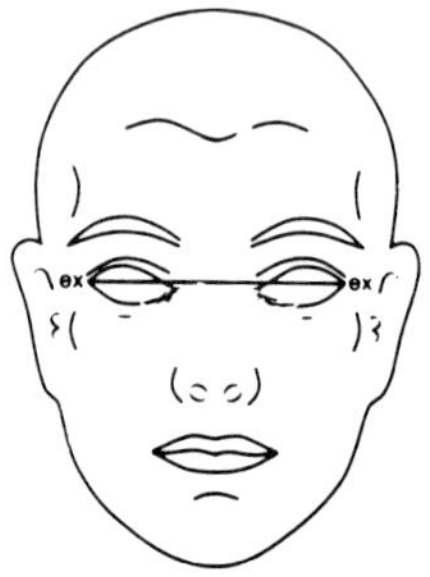

TABLE A–III-2. *Biocular width (ex-ex) (mm)*

Age (years)	Male			Female		
	N	Mean	SD	*N*	Mean	SD
0–5 months	8	74.1	3.5	5	74.8	2.6
6–12 months	8	75.1	4.2	8	75.1	4.2
1	18	76.0	2.3	22	75.3	4.3
2	31	76.2	2.9	31	75.5	2.6
3	30	77.5	3.4	30	77.3	3.7
4	30	77.2	3.3	30	75.3	2.4
5	30	78.7	4.2	30	76.5	2.5
6	50	80.0	3.6	50	77.8	3.2
7	50	79.2	3.5	50	79.4	3.5
8	51	81.5	2.9	51	79.2	3.2
9	51	82.9	4.1	50	81.4	3.9
10	50	82.8	3.3	49	81.8	3.8
11	50	85.2	3.2	51	82.8	3.0
12	52	85.6	3.0	53	83.6	3.4
13	50	86.8	3.5	49	85.4	3.2
14	49	86.9	4.2	51	85.3	3.2
15	50	89.4	3.5	51	87.0	3.6
16	50	89.7	3.8	51	86.9	3.8
17	49	90.7	3.8	51	87.6	4.0
18	52	89.4	3.6	51	86.8	4.0
19–25	109	91.2	3.0	200	87.8	3.2

ORBITS

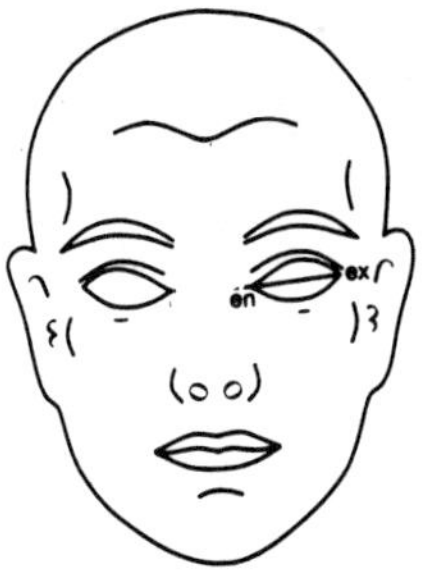

TABLE A–III-3. *Length of the eye fissure (ex-en), right and left (mm)*

Age (years)	Male, right side			Female, right side		
	N	Mean	SD	*N*	Mean	SD
0–5 months	8	24.3	1.5	5	25.0	1.0
6–12 months	20	25.6	1.2	8	24.8	1.3
1	18	25.9	1.0	22	25.4	1.4
2	31	26.1	1.5	31	25.8	1.5
3	30	26.9	1.5	30	26.3	1.2
4	30	26.7	1.4	30	26.0	1.1
5	30	27.2	1.5	30	26.8	1.4
6	50	27.4	1.4	50	27.0	1.2
7	50	27.4	1.1	50	27.4	1.3
8	51	27.9	1.0	51	27.3	1.0
9	51	28.4	1.4	50	28.1	1.7
10	50	28.9	1.2	49	28.3	1.4
11	50	29.1	1.3	51	28.7	1.7
12	52	29.7	1.4	53	29.2	1.5
13	50	30.0	0.9	49	29.8	1.3
14	49	30.0	1.3	51	29.6	1.4
15	50	31.0	1.4	51	30.3	1.3
16	50	31.2	1.2	51	30.3	1.5
17	49	31.3	1.4	51	30.4	1.3
18	52	31.2	1.3	51	30.7	1.8
19–25	109	31.3	1.2	200	30.7	1.2

Age (years)	Male, left side			Female, left side		
	N	Mean	SD	*N*	Mean	SD
0–5 months	8	24.3	1.5	5	25.0	1.0
6–12 months	20	25.6	1.2	8	24.8	1.3
1	18	25.9	1.0	22	25.4	1.4
2	31	26.1	1.5	31	25.8	1.5
3	30	26.9	1.6	30	26.3	1.2
4	30	26.7	1.3	30	26.1	1.0
5	30	27.3	1.5	30	26.8	1.4
6	50	27.4	1.4	50	27.0	1.3
7	50	27.4	1.1	50	27.4	1.3
8	51	27.9	1.0	51	27.3	1.0
9	51	28.4	1.4	50	28.1	1.7
10	50	28.9	1.2	49	28.3	1.4
11	50	29.1	1.3	51	28.7	1.7
12	52	29.6	1.3	53	29.2	1.4
13	50	30.0	0.9	49	29.8	1.3
14	49	29.9	1.2	51	29.6	1.4
15	50	31.0	1.3	51	30.3	1.3
16	50	31.2	1.2	51	30.3	1.5
17	49	31.3	1.4	51	30.4	1.3
18	52	31.2	1.3	51	30.7	1.8
19–25	109	31.3	1.2	200	30.7	1.2

ORBITS

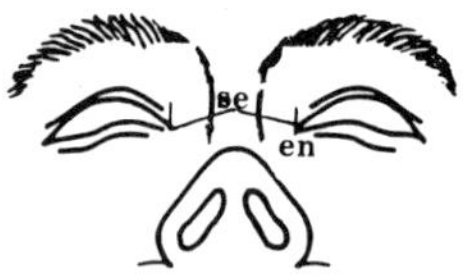

TABLE A–III-4. *Endocanthion–facial midline distance (en-se), right and left (mm)*

Age (years)	Male, right side			Female, right side		
	N	Mean	SD	*N*	Mean	SD
1	17	16.7	1.7	15	16.5	2.2
2	30	16.9	1.3	30	17.1	1.8
3	30	17.6	1.7	30	17.7	1.4
4	30	20.5	1.9	30	19.4	1.4
5	30	20.4	1.4	30	19.6	1.3
6	50	20.3	1.4	50	19.4	1.3
7	50	20.7	1.4	50	19.8	1.4
8	51	20.7	1.5	51	19.7	1.3
9	51	21.2	1.4	50	20.5	1.6
10	50	20.9	1.5	49	20.7	1.3
11	50	22.0	1.8	51	21.3	1.4
12	52	21.8	1.6	53	21.5	1.7
13	50	22.3	1.9	49	21.7	1.7
14	49	23.1	1.9	51	21.5	1.5
15	50	23.9	1.8	51	21.9	1.9
16	50	23.9	2.0	51	21.5	1.3
17	49	24.2	2.0	51	22.0	1.7
18	52	24.0	2.1	51	21.9	1.8
19–25	109	25.6	2.1	199	21.9	1.6

Age (years)	Male, left side			Female, left side		
	N	Mean	SD	*N*	Mean	SD
1	17	16.7	1.7	15	16.5	2.2
2	30	16.9	1.3	30	17.1	1.8
3	30	17.6	1.7	30	17.7	1.4
4	30	20.3	1.9	30	19.3	1.4
5	30	20.3	1.4	30	19.5	1.2
6	50	20.3	1.4	50	19.4	1.4
7	50	20.6	1.5	50	19.7	1.4
8	51	20.5	1.4	51	19.6	1.4
9	51	21.1	1.4	50	20.7	1.6
10	50	20.9	1.5	49	20.7	1.3
11	50	22.1	1.8	51	21.1	1.5
12	52	21.8	1.6	53	21.2	1.5
13	50	22.4	1.9	49	21.7	1.7
14	49	23.1	2.0	51	21.8	1.3
15	50	23.8	1.8	51	22.1	1.7
16	50	23.8	2.1	51	21.9	1.5
17	49	24.2	2.1	51	22.1	1.9
18	52	23.9	1.9	51	21.9	1.8
19–25	109	25.3	2.1	199	22.0	1.5

ORBITS

TABLE A–III-5. *Pupil–facial midline distance (pupil-se), right and left (mm)*

Age (years)	Male, right side			Female, right side		
	N	Mean	SD	*N*	Mean	SD
19–25	40	33.5	2.0	40	31.2	1.8

Age (years)	Male, left side			Female, left side		
	N	Mean	SD	*N*	Mean	SD
19–25	40	33.4	2.0	40	31.4	1.8

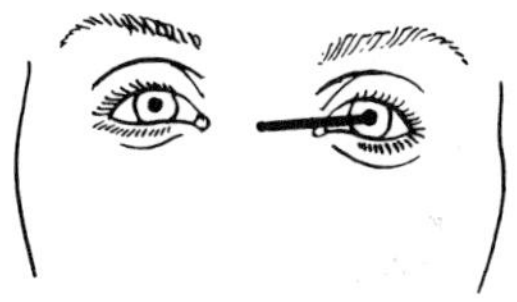

TABLE A–III-6. *Orbito-aural distance (ex-obs), right and left (mm)*

Age (years)	Male, right side			Female, right side		
	N	Mean	SD	*N*	Mean	SD
6	35	70.4	3.6	39	68.7	4.7
7	33	72.6	4.5	42	70.9	3.7
8	45	74.0	3.5	41	72.2	3.5
9	47	75.1	4.2	38	73.9	3.6
10	37	75.8	4.6	37	74.0	3.0
11	39	76.9	5.1	41	75.0	3.2
12	50	76.7	4.9	50	74.1	4.2
13	47	77.7	3.6	49	75.9	4.6
14	46	79.5	4.5	43	77.0	4.2
15	44	81.3	4.0	43	78.6	3.6
16	49	81.5	3.7	35	78.2	3.7
17	48	82.5	4.5	36	76.4	4.4
18	50	79.8	5.4	38	76.1	3.7

Age (years)	Male, left side			Female, left side		
	N	Mean	SD	*N*	Mean	SD
6	35	70.5	4.0	39	68.9	5.0
7	33	73.0	4.1	42	71.5	3.8
8	45	74.3	3.7	41	72.1	3.7
9	47	75.2	4.6	38	74.6	4.1
10	37	76.2	4.3	37	74.1	2.9
11	39	76.8	5.2	41	75.3	3.3
12	50	76.5	5.0	50	74.0	4.5
13	47	79.3	3.9	49	76.2	4.2
14	46	80.5	4.3	43	77.4	4.1
15	44	82.0	3.8	43	78.4	3.1
16	49	80.9	4.2	35	77.9	3.6
17	48	83.0	4.6	36	76.6	4.4
18	50	79.8	5.1	38	76.1	4.2

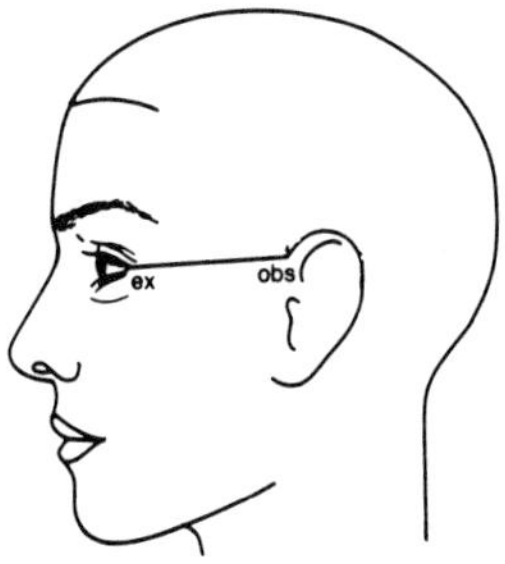

ORBITS

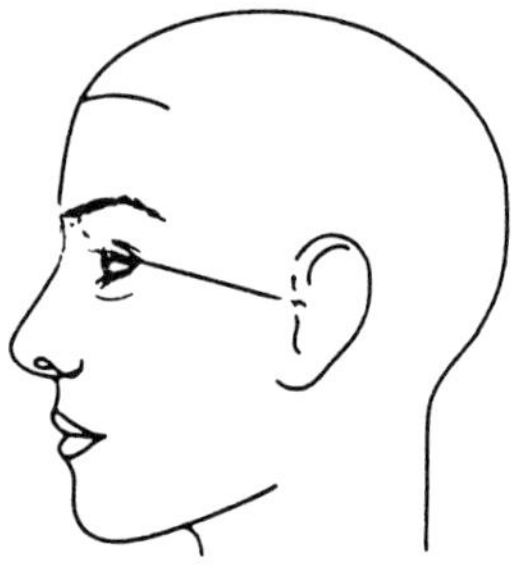

TABLE A-III-7. *Orbitotragial distance (ex-t), right and left (mm)*

Age (years)	Male, right side			Female, right side		
	N	Mean	SD	*N*	Mean	SD
1	18	60.9	1.7	18	59.8	2.9
2	30	65.1	2.9	30	63.0	2.2
3	30	65.8	2.2	30	64.8	2.4
4	30	71.1	3.2	30	68.8	2.3
5	30	71.5	2.8	30	70.3	2.3
6	50	72.8	3.4	50	71.8	2.8
7	50	74.2	3.5	50	73.1	3.0
8	51	75.6	3.0	51	73.5	2.5
9	51	76.2	3.6	50	74.7	3.8
10	50	77.6	3.5	49	76.0	2.9
11	50	78.2	4.0	51	76.0	3.5
12	52	79.3	3.2	53	77.8	3.5
13	50	79.2	3.6	49	78.0	3.6
14	49	81.3	4.0	51	78.3	3.4
15	50	82.6	3.6	51	79.7	3.7
16	50	83.2	3.2	51	79.0	3.4
17	49	84.5	3.7	51	78.3	2.9
18	52	82.6	3.4	51	79.4	3.6
19–25	109	85.3	3.2	199	78.9	3.7

Age (years)	Male, left side			Female, left side		
	N	Mean	SD	*N*	Mean	SD
1	18	61.0	1.8	18	59.8	2.9
2	30	65.1	3.1	30	63.1	2.2
3	30	65.7	2.4	30	64.7	2.5
4	30	71.1	2.9	30	69.0	2.4
5	30	71.5	3.0	30	70.3	2.5
6	50	72.6	3.6	50	71.8	3.1
7	50	74.4	3.7	50	73.7	3.0
8	51	75.7	2.5	51	73.5	3.2
9	51	76.5	3.6	50	74.9	3.8
10	50	77.7	3.7	49	75.9	3.0
11	50	78.4	3.9	51	75.9	3.5
12	52	79.3	3.6	53	77.6	3.7
13	50	80.3	3.7	49	78.4	3.3
14	49	81.9	4.2	51	78.6	3.4
15	50	83.3	3.2	51	79.5	3.5
16	50	83.4	3.4	51	79.0	3.6
17	49	84.7	3.6	51	78.3	2.9
18	52	82.8	3.5	51	79.5	3.8
19–25	109	84.8	3.3	199	78.2	3.5

TABLE A–III-8. *Orbitogonial distance (ex-go), right and left (mm)*

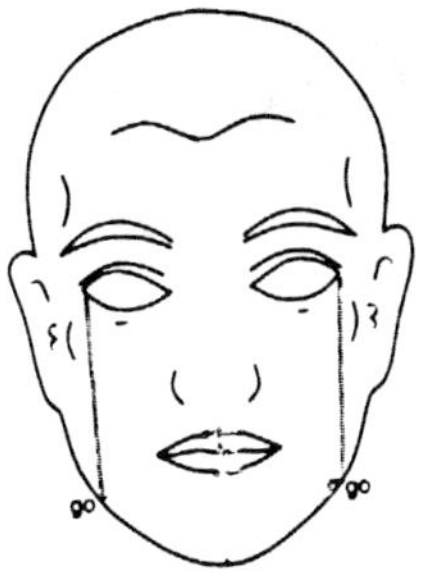

Age (years)	Male, right side			Female, right side		
	N	Mean	SD	*N*	Mean	SD
6	50	81.4	4.5	50	79.9	4.0
7	50	83.3	3.7	50	82.3	3.7
8	51	86.0	3.5	51	82.7	3.2
9	51	87.1	4.1	50	83.9	4.0
10	50	88.1	4.1	49	87.2	3.7
11	50	90.3	3.6	51	87.8	4.3
12	52	91.2	3.8	53	89.9	4.9
13	50	93.5	4.2	49	90.6	4.4
14	49	94.7	4.6	51	92.1	4.7
15	50	99.0	5.1	51	93.3	4.3
16	50	99.3	4.7	51	93.4	6.1
17	49	102.2	5.8	51	93.2	3.9
18	52	101.5	5.3	51	93.2	5.0
19–25	109	103.9	4.7	198	93.0	4.5

Age (years)	Male, left side			Female, left side		
	N	Mean	SD	*N*	Mean	SD
6	50	81.5	4.6	50	80.0	3.8
7	50	83.5	3.9	50	82.8	3.7
8	51	85.9	3.1	51	83.1	3.5
9	51	87.6	4.0	50	84.2	4.0
10	50	88.3	3.7	49	87.4	4.0
11	50	90.5	3.7	51	87.4	4.2
12	52	91.3	4.2	53	89.7	4.6
13	50	93.7	4.4	49	90.5	4.0
14	49	95.0	4.8	51	92.0	4.5
15	50	98.7	4.9	51	93.0	4.6
16	50	99.0	4.8	51	93.2	5.7
17	49	101.9	6.0	51	93.1	4.5
18	52	101.6	5.5	51	93.8	5.0
19–25	109	103.6	4.5	198	92.4	4.4

TABLE A–III-9. *Orbitoglabellar distance (ex-g), right and left (mm)*

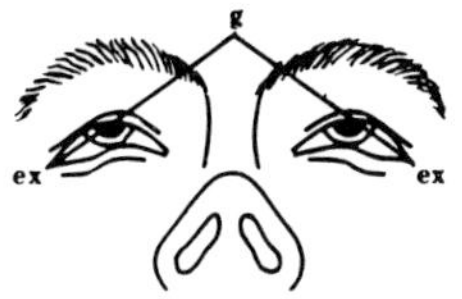

Age (years)	Male, right side			Female, right side		
	N	Mean	SD	*N*	Mean	SD
19–25	35	52.3	2.8	40	48.9	2.8

Age (years)	Male, left side			Female, left side		
	N	Mean	SD	*N*	Mean	SD
19–25	35	52.2	2.9	40	48.9	2.7

ORBITS

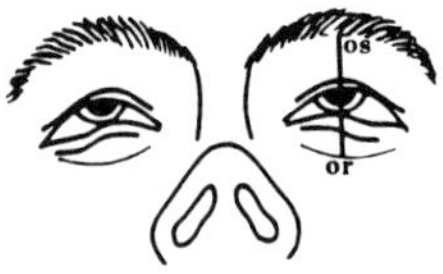

TABLE A–III-10. *Height of the orbit (or-os), right and left (mm)*

Age (years)	Male, right side			Female, right side		
	N	Mean	SD	*N*	Mean	SD
1	18	19.5	1.8	19	19.2	1.6
2	31	20.6	1.6	31	19.7	2.1
3	30	21.2	1.8	30	21.8	2.1
4	30	23.7	2.3	30	23.0	2.0
5	30	23.1	2.6	30	23.6	2.3
19–25	40	29.5	5.3	40	29.4	3.1

Age (years)	Male, left side			Female, left side		
	N	Mean	SD	*N*	Mean	SD
1	18	19.5	1.8	19	19.2	1.6
2	31	20.6	1.6	31	19.6	2.0
3	30	21.2	1.8	30	21.8	2.2
4	30	23.8	2.4	30	22.9	2.1
5	30	23.1	2.6	30	23.7	2.3
19–25	40	29.8	4.2	40	29.4	3.2

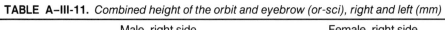

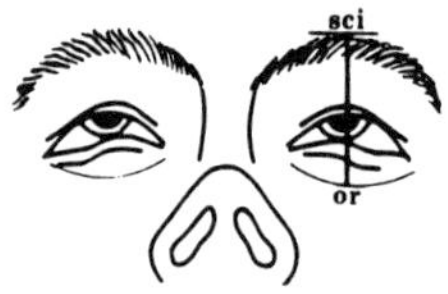

TABLE A–III-11. *Combined height of the orbit and eyebrow (or-sci), right and left (mm)*

Age (years)	Male, right side			Female, right side		
	N	Mean	SD	*N*	Mean	SD
1	18	24.3	1.4	19	24.3	2.1
2	31	25.8	2.2	31	25.4	1.9
3	30	27.1	2.8	30	26.8	3.0
4	30	31.0	2.8	30	29.8	1.7
5	30	30.7	2.7	30	30.9	2.0
19–25	109	38.0	3.5	195	38.4	3.4

Age (years)	Male, left side			Female, left side		
	N	Mean	SD	*N*	Mean	SD
1	18	24.3	1.4	19	24.3	2.1
2	31	25.8	2.2	31	25.5	1.9
3	30	27.1	2.7	30	26.8	3.0
4	30	31.1	2.7	30	29.8	1.7
5	30	30.7	2.7	30	30.9	2.0
19–25	109	38.0	3.5	195	38.3	3.3

ORBITS

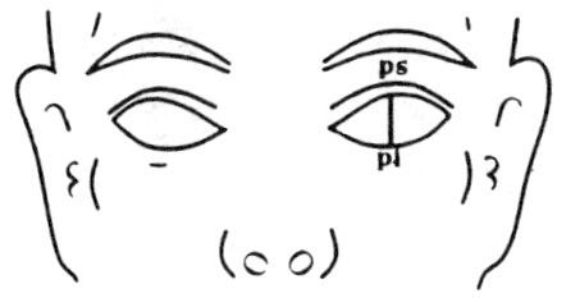

TABLE A–III-12. *Height of the eye fissure (ps-pi), right and left (mm)*

Age (years)	Male, right side			Female, right side		
	N	Mean	SD	*N*	Mean	SD
0–5 months	8	9.5	1.4	5	9.6	0.5
6–12 months	8	10.1	1.4	8	10.1	1.4
1	18	9.4	1.0	22	9.7	1.1
2	31	9.6	0.8	31	9.5	1.0
3	30	10.0	1.0	30	9.8	1.0
4	30	9.8	0.7	30	10.0	1.0
5	30	9.8	0.7	30	10.0	0.9
6	50	9.5	1.0	50	9.4	0.8
7	50	9.2	1.0	50	9.3	0.7
8	51	9.4	1.0	51	9.3	1.0
9	51	9.5	0.8	50	9.7	1.0
10	50	9.6	0.9	49	9.8	1.0
11	50	9.6	1.1	51	9.9	1.1
12	52	9.8	0.9	53	10.2	1.1
13	50	9.9	1.1	49	10.1	1.1
14	49	9.7	1.1	51	10.5	1.0
15	50	10.0	0.9	51	10.7	1.1
16	50	10.2	0.8	51	10.9	1.3
17	49	10.3	1.2	51	10.7	1.3
18	52	10.4	1.1	51	11.1	1.2
19–25	109	10.8	0.9	200	10.9	1.2

Age (years)	Male, left side			Female, left side		
	N	Mean	SD	*N*	Mean	SD
0–5 months	8	9.5	1.4	5	9.6	0.5
6–12 months	8	10.1	1.4	8	10.1	1.4
1	18	9.4	1.0	22	9.7	1.1
2	31	9.6	0.8	31	9.5	1.1
3	30	10.0	1.0	30	9.7	1.0
4	30	9.9	0.7	30	10.0	1.0
5	30	9.8	0.7	30	10.0	0.9
6	50	9.5	1.1	50	9.4	0.8
7	50	9.2	1.0	50	9.3	0.7
8	51	9.4	1.0	51	9.3	1.0
9	51	9.5	0.8	50	9.7	1.0
10	50	9.6	0.9	49	9.8	1.0
11	50	9.6	1.1	51	9.9	1.1
12	52	9.8	0.9	53	10.2	1.1
13	50	9.9	1.0	49	10.1	1.1
14	49	9.7	1.1	51	10.5	1.0
15	50	10.0	0.9	51	10.7	1.1
16	50	10.2	0.8	51	10.9	1.3
17	49	10.3	1.1	51	10.7	1.3
18	52	10.4	1.1	51	11.1	1.2
19–25	109	10.8	0.9	200	10.9	1.2

ORBITS

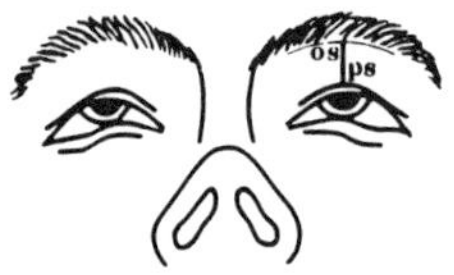

TABLE A–III-13. *Height of the upper lid (ps-os), right and left (mm)*

Age (years)	Male, right side			Female, right side		
	N	Mean	SD	*N*	Mean	SD
1	18	6.3	0.9	19	5.7	1.0
2	31	6.4	0.9	31	6.3	1.0
3	30	6.5	0.9	30	6.6	1.0
4	30	7.4	0.7	30	7.2	0.9
5	30	7.3	0.6	30	7.8	1.2
19–25	40	11.2	2.0	40	12.6	2.6

Age (years)	Male, left side			Female, left side		
	N	Mean	SD	*N*	Mean	SD
1	18	6.3	0.9	19	5.7	1.0
2	31	6.4	0.9	31	6.3	1.0
3	30	6.5	0.9	30	6.6	1.0
4	30	7.4	0.6	30	7.2	0.9
5	30	7.3	0.6	30	7.8	1.2
19–25	40	11.2	2.0	40	12.6	2.6

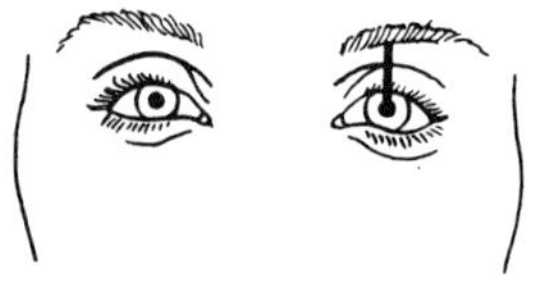

TABLE A–III-14. *Pupil–upper lid height (pupil-os), right and left (mm)*

Age (years)	Male, right side			Female, right side		
	N	Mean	SD	*N*	Mean	SD
19–25	40	24.2	3.1	40	23.0	3.0

Age (years)	Male, left side			Female, left side		
	N	Mean	SD	*N*	Mean	SD
19–25	40	24.4	3.3	40	22.9	3.3

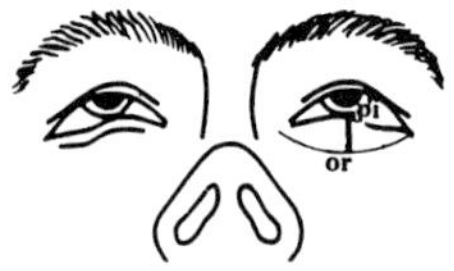

TABLE A–III-15. *Height of the lower lid (pi-or), right and left (mm)*

Age (years)	Male, right side			Female, right side		
	N	Mean	SD	*N*	Mean	SD
1	18	5.6	1.0	19	4.9	0.7
2	31	5.3	0.8	31	5.0	0.9
3	30	5.1	0.6	30	5.2	0.7
4	30	5.8	0.8	30	5.5	0.7
5	30	5.7	0.8	30	6.0	0.9
19–25	40	8.1	1.6	40	7.5	1.7

Age (years)	Male, left side			Female, left side		
	N	Mean	SD	*N*	Mean	SD
1	18	5.6	1.0	19	4.9	0.7
2	31	5.3	0.8	31	5.0	0.9
3	30	5.1	0.6	30	5.2	0.7
4	30	5.8	0.7	30	5.5	0.7
5	30	5.7	0.8	30	6.0	0.9
19–25	40	8.1	1.5	40	7.5	1.7

TABLE A–III-16. *Pupil–lower lid height (pupil-or), right and left (mm)*

Age (years)	Male, right side			Female, right side		
	N	Mean	SD	*N*	Mean	SD
19–25	40	12.6	2.1	40	12.0	1.3

Age (years)	Male, left side			Female, left side		
	N	Mean	SD	*N*	Mean	SD
19–25	40	12.6	2.1	40	12.0	1.3

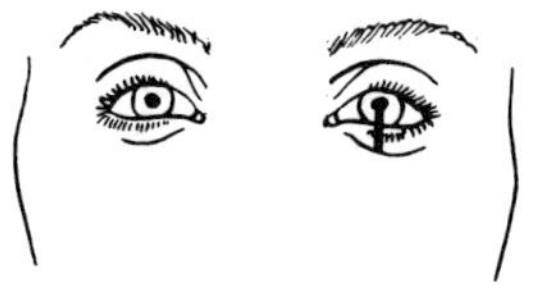

TABLE A–III-17. *Difference between the sagittal levels of the upper and lower orbital rims (os′/or′), right and left (mm)*

Age (years)	Male, right side			Female, right side		
	N	Mean	SD	*N*	Mean	SD
19–25	40	14.1	4.4	40	11.1	3.2

Age (years)	Male, left side			Female, left side		
	N	Mean	SD	*N*	Mean	SD
19–25	40	14.1	4.4	40	11.3	3.1

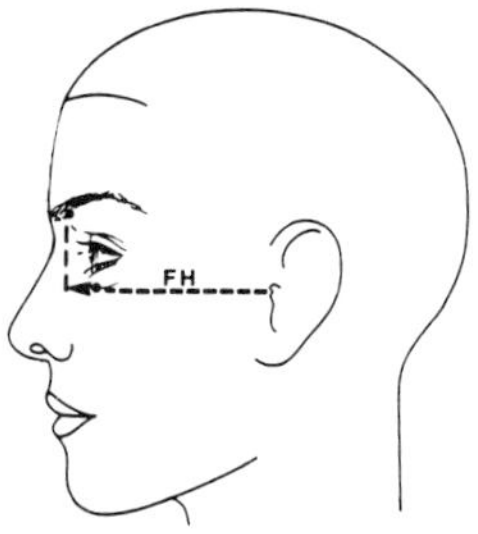

TABLE A–III-18. *Difference between the sagittal levels of the endocanthion and the exocanthion landmarks (en/ex), right and left (mm)*

Age (years)	Male, right side			Female, right side		
	N	Mean	SD	*N*	Mean	SD
19–25	41	5.1	1.6	45	3.6	1.7

Age (years)	Male, left side			Female, left side		
	N	Mean	SD	*N*	Mean	SD
19–25	41	5.1	1.6	45	3.6	1.7

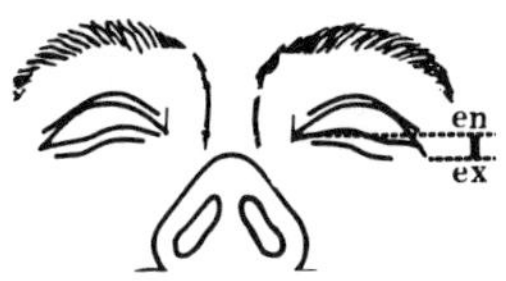

ORBITS

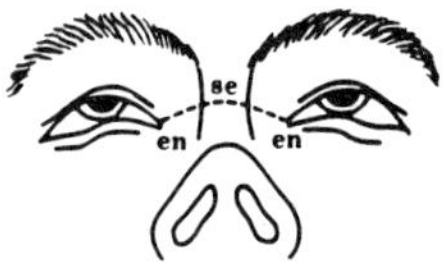

TABLE A-III-19. *Endocanthion–facial midline surface distance (en-se surf), right and left (mm)*

Age (years)	Male, right side			Female, right side		
	N	Mean	SD	*N*	Mean	SD
6	50	21.5	1.6	50	20.6	1.6
7	50	22.0	1.6	50	21.2	1.3
8	51	22.2	1.6	51	21.5	1.3
9	51	22.6	1.3	50	22.1	1.5
10	50	22.8	1.6	49	22.4	1.4
11	49	23.7	1.9	51	23.0	1.6
12	52	23.4	1.7	52	23.2	1.8
13	50	24.0	2.1	49	23.5	1.8
14	49	24.8	2.1	51	23.2	1.7
15	50	25.6	1.9	51	23.8	2.1
16	50	26.0	2.2	51	23.8	1.8
17	49	26.4	2.4	51	23.9	1.7
18	52	26.4	2.2	51	24.3	1.8

Age (years)	Male, left side			Female, left side		
	N	Mean	SD	*N*	Mean	SD
6	50	21.5	1.6	50	20.5	1.6
7	50	21.8	1.8	50	21.0	1.4
8	51	22.0	1.5	51	21.3	1.3
9	51	22.5	1.3	50	22.2	1.6
10	50	22.7	1.6	49	22.3	1.4
11	49	23.6	1.9	51	22.8	1.7
12	52	23.3	1.7	52	22.8	1.5
13	50	24.0	2.2	49	23.4	1.9
14	49	24.7	2.2	51	23.5	1.4
15	50	25.4	1.8	51	23.9	2.0
16	50	25.8	2.2	51	23.7	1.7
17	49	26.5	2.5	51	23.8	1.9
18	52	26.2	2.1	51	24.1	1.9

ORBITS

TABLE A–III-20. *Inclination of the eye fissure (en-ex), right and left (degrees)*

Age (years)	Male, right side			Female, right side		
	N	Mean	SD	*N*	Mean	SD
0–5 months	8	5.9	2.5	5	7.4	2.1
6–12 months	20	5.3	2.2	8	4.7	1.6
1	18	5.2	2.5	22	5.3	2.2
2	31	3.9	1.9	34	4.3	1.4
3	30	3.4	1.6	30	4.3	1.7
4	30	5.8	2.2	30	6.9	2.1
5	30	5.8	2.3	30	6.6	2.0
6	50	2.9	1.7	49	4.1	2.1
7	50	3.2	2.1	50	4.7	2.4
8	51	3.6	2.5	51	3.4	2.3
9	51	3.3	1.7	49	3.7	2.1
10	50	3.3	2.0	49	3.8	1.9
11	50	2.7	1.8	50	3.5	2.4
12	52	2.8	2.0	53	3.4	1.8
13	50	3.1	2.0	49	3.2	2.0
14	48	3.0	2.2	51	4.9	2.7
15	50	3.0	2.0	51	4.1	2.2
16	48	2.7	1.5	51	4.0	2.5
17	49	2.3	1.8	51	5.0	2.8
18	50	2.1	1.9	50	4.1	2.2

Age (years)	Male, left side			Female, left side		
	N	Mean	SD	*N*	Mean	SD
0–5 months	8	5.9	2.5	5	7.4	2.1
6–12 months	20	5.3	2.2	8	4.7	1.6
1	18	5.2	2.5	22	5.4	2.2
2	31	3.8	1.8	34	4.2	1.5
3	30	3.3	1.6	30	4.1	1.8
4	30	5.8	2.2	30	6.8	2.2
5	30	5.8	2.3	30	6.5	2.1
6	50	2.9	1.7	49	4.0	1.9
7	50	3.2	2.1	50	4.7	2.4
8	51	3.6	2.5	51	3.5	2.3
9	51	3.3	1.7	49	3.7	2.1
10	50	3.3	2.0	49	3.9	1.9
11	50	2.7	1.8	50	3.6	2.4
12	52	2.9	1.9	53	3.4	1.8
13	50	3.1	2.0	49	3.2	2.0
14	48	3.0	2.2	51	4.9	2.7
15	50	3.0	2.0	51	4.1	2.2
16	48	2.8	1.4	51	4.0	2.5
17	49	2.3	1.7	51	5.0	2.8
18	50	2.1	1.9	50	4.1	2.2

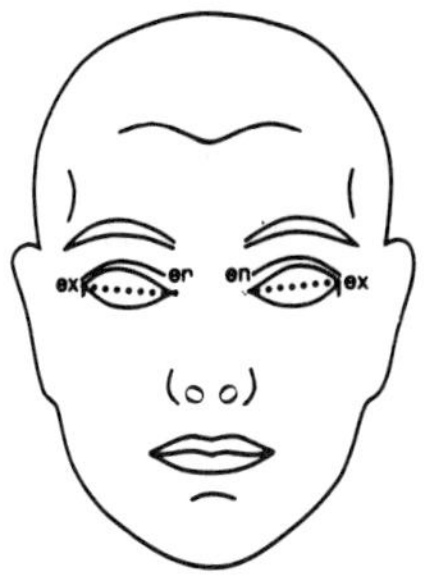

ORBITS

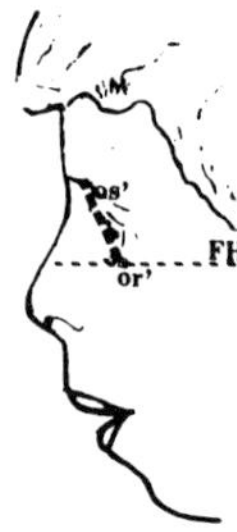

TABLE A-III-21. *Inclination of the line placed over the upper and the lower orbital rims (os'/or'), right and left (degrees)*

Age (years)	Male, right side			Female, right side		
	N	Mean	SD	*N*	Mean	SD
1	17	10.1	3.2	15	9.7	2.0
2	31	11.8	7.1	33	10.7	3.5
3	30	10.3	3.1	30	9.5	3.6
4	30	13.9	3.5	30	14.6	3.5
5	30	13.8	3.7	30	13.7	2.9
19–20	40	18.0	4.5	40	14.4	3.2
21–25	109	18.9	4.0	187	13.2	3.4

Age (years)	Male, left side			Female, left side		
	N	Mean	SD	*N*	Mean	SD
1	17	10.1	3.2	15	9.7	2.0
2	31	11.9	7.1	33	10.7	3.5
3	30	10.3	3.1	30	9.4	3.7
4	30	13.8	3.3	30	14.5	3.5
5	30	13.9	3.7	30	13.9	3.0
19–20	40	18.0	4.5	40	14.2	3.8
21–25	109	18.9	4.0	187	13.2	3.2

Nose

NOSE

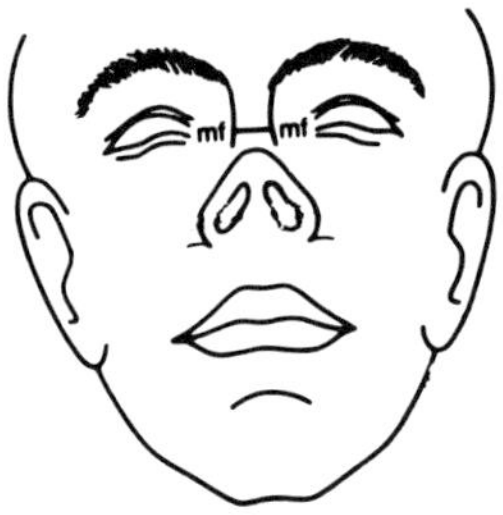

TABLE A–IV-1. *Width of the nasal root (mf-mf) (mm)*

Age (years)	Male			Female		
	N	Mean	SD	*N*	Mean	SD
0–5 months	8	15.3	1.5	5	15.6	0.5
6–12 months	20	17.6	1.6	8	17.3	1.8
1	18	17.1	1.6	22	17.1	2.1
2	31	16.3	1.4	31	16.4	1.8
3	30	16.2	1.4	30	16.2	1.4
4	30	17.2	1.3	30	17.3	0.9
5	30	18.0	1.4	30	17.4	1.1
6	50	16.3	2.6	50	15.7	2.1
7	50	16.3	2.5	50	16.7	2.3
8	51	17.2	2.1	51	16.6	1.9
9	51	18.2	2.5	49	17.3	2.4
10	50	17.2	2.2	49	17.2	2.4
11	50	18.1	2.5	51	16.9	2.1
12	52	17.4	2.0	53	16.8	1.8
13	50	18.9	2.0	49	17.3	2.3
14	49	18.8	2.0	51	17.6	2.3
15	50	19.1	2.1	51	18.7	2.2
16	50	18.8	2.4	51	16.9	2.5
17	49	18.1	2.2	51	17.1	2.4
18	52	16.2	2.2	51	15.4	2.1
19–25	109	19.6	1.9	200	18.4	1.9

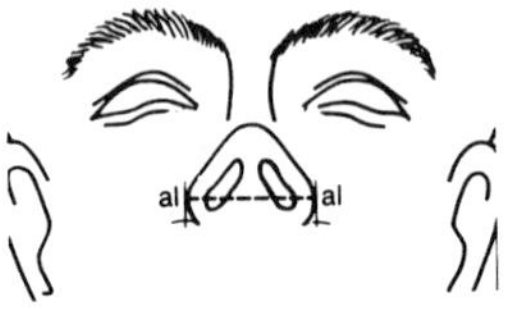

TABLE A–IV-2. *Width of the nose (al-al) (mm)*

Age (years)	Male			Female		
	N	Mean	SD	*N*	Mean	SD
0–5 months	8	25.6	1.1	5	24.4	1.5
6–12 months	20	26.5	1.4	8	25.4	1.5
1	18	26.5	1.5	21	25.9	1.4
2	31	25.6	1.4	31	26.1	1.2
3	30	26.1	1.5	30	25.9	1.1
4	30	28.4	1.7	30	27.8	1.3
5	30	28.9	1.5	30	28.5	1.5
6	50	28.6	1.6	50	27.8	1.3
7	50	28.8	1.9	50	28.6	1.7
8	51	29.8	1.5	51	28.5	1.8
9	51	29.4	1.8	50	29.2	2.3
10	50	30.2	1.9	49	29.6	1.9
11	50	30.1	1.7	51	29.9	2.4
12	52	31.6	1.9	53	30.9	2.1
13	50	32.4	2.3	49	31.0	2.0
14	49	33.1	2.5	51	31.0	1.6
15	50	34.2	2.2	51	31.7	1.9
16	50	34.0	2.0	51	31.6	2.0
17	49	34.8	2.7	51	31.9	1.9
18	52	34.7	2.6	51	31.4	1.9
19–25	109	34.9	2.1	200	31.4	2.0

NOSE

TABLE A-IV-3. *Width between the facial insertion points of the alar base (ac-ac) (mm)*

Age (years)	Male			Female		
	N	Mean	SD	*N*	Mean	SD
0–5 months	8	24.0	1.1	5	23.2	1.5
6–12 months	20	24.9	1.5	8	23.6	1.3
1	18	25.3	1.5	19	23.5	4.0
2	31	25.2	1.3	31	25.0	1.1
3	30	25.1	1.1	30	25.0	1.0
4	30	28.3	1.5	30	27.7	1.5
5	30	28.3	1.4	30	28.3	1.3
6	*	27.2	1.5	*	25.8	1.4
7	*	27.4	1.5	*	26.7	1.4
8	*	28.3	1.6	*	26.7	1.5
9	*	27.9	1.6	*	27.5	1.5
10	*	28.7	1.7	*	28.0	1.6
11	*	28.6	1.8	*	28.3	1.7
12	*	30.0	1.8	*	29.4	1.7
13	*	30.7	1.9	*	29.6	1.8
14	*	31.3	1.9	*	29.7	1.8
15	*	32.4	2.0	*	30.5	1.9
16	*	32.2	2.1	*	30.5	2.0
17	*	32.9	2.1	*	30.9	2.0
18	*	32.8	2.2	*	30.5	2.1
19–25	86	32.8	2.3	45	30.5	2.2

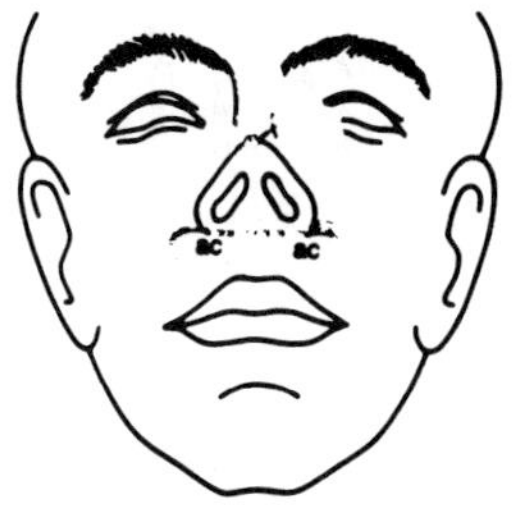

TABLE A-IV-4. *Width between the labial insertions of the alar base (sbal-sbal) (mm)*

Age (years)	Male			Female		
	N	Mean	SD	*N*	Mean	SD
1	18	13.8	1.4	18	12.7	1.4
2	31	14.5	1.2	30	14.1	1.6
3	30	15.3	1.5	30	14.7	1.0
4	30	16.7	1.3	30	16.1	1.0
5	30	17.4	1.2	30	16.8	1.3
6	*	17.5	1.3	*	16.9	1.3
7	*	18.4	1.4	*	17.6	1.4
8	*	18.3	1.5	*	18.0	1.4
9	*	18.5	1.6	*	18.1	1.5
10	*	18.6	1.7	*	18.6	1.5
11	*	18.6	1.7	*	18.6	1.5
12	*	19.8	1.8	*	19.3	1.6
13	*	19.6	1.9	*	19.3	1.6
14	*	19.7	2.0	*	19.0	1.7
15	*	19.9	2.1	*	19.8	1.7
16	*	20.2	2.2	*	20.2	1.7
17	*	20.5	2.3	*	20.4	1.8
18	*	21.0	2.4	*	19.9	1.8
19–25	85	21.0	2.4	45	19.9	1.9

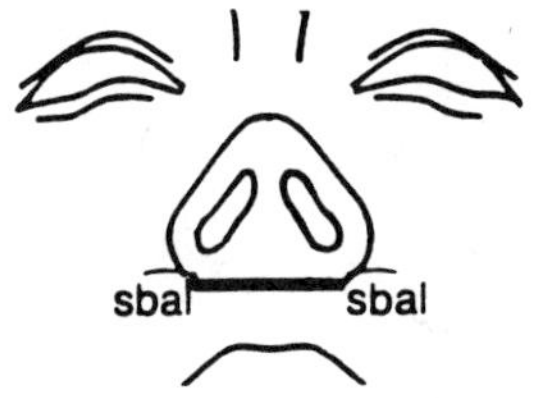

NOSE

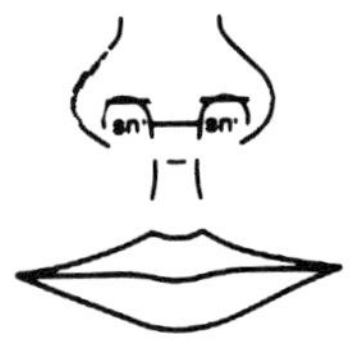

TABLE A–IV-5. *Width of the columella (sn'-sn') (mm)*

Age (years)	Male			Female		
	N	Mean	SD	*N*	Mean	SD
1	18	5.5	0.8	18	5.1	0.8
2	31	5.3	0.5	31	5.2	0.6
3	30	5.3	0.7	30	5.4	0.8
4	30	6.7	0.6	30	6.7	0.5
5	30	7.0	0.6	30	6.9	0.6
6	50	6.7	0.7	50	6.5	0.5
7	50	6.7	0.4	50	6.7	0.7
8	51	7.1	0.7	51	6.5	0.8
9	51	6.8	0.6	50	6.9	0.8
10	50	7.2	0.7	49	7.0	0.8
11	50	7.1	0.8	51	7.3	0.8
12	52	7.7	0.6	53	7.5	0.6
13	50	7.6	0.8	49	7.7	0.8
14	49	7.8	0.9	51	7.4	0.7
15	50	8.0	0.7	51	7.5	0.7
16	50	7.9	0.7	51	7.6	0.7
17	49	8.0	0.9	51	7.5	0.8
18	52	8.1	0.6	51	7.6	0.7
19–25	109	6.9	0.7	200	6.8	0.7

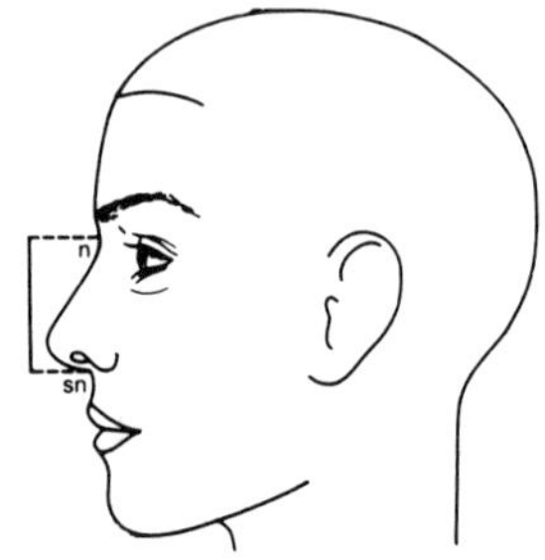

TABLE A–IV-6. *Height of the nose (n-sn) (mm)*

Age (years)	Male			Female		
	N	Mean	SD	*N*	Mean	SD
0–5 months	8	26.5	1.7	5	24.6	1.8
6–12 months	20	27.0	1.7	8	26.9	1.6
1	18	30.9	1.9	22	29.2	2.6
2	31	33.7	2.7	31	32.6	2.6
3	30	35.3	2.6	30	34.6	2.3
4	30	39.5	1.9	30	37.8	1.9
5	30	38.9	2.7	30	39.3	2.1
6	50	40.1	3.1	50	39.3	2.7
7	50	41.4	3.0	50	40.7	2.7
8	51	42.1	2.4	51	41.5	2.8
9	51	43.7	2.9	50	43.6	3.1
10	50	45.0	2.9	49	44.5	3.1
11	50	45.0	2.3	51	45.7	2.6
12	52	47.5	3.4	53	47.2	3.3
13	50	48.8	3.2	49	48.2	3.4
14	49	49.7	3.8	51	49.1	3.2
15	50	51.9	3.4	51	49.2	3.3
16	50	53.0	3.1	51	50.4	3.0
17	49	53.2	3.3	51	49.2	2.9
18	52	53.0	3.5	51	48.9	2.6
19–25	109	54.8	3.3	200	50.6	3.1

NOSE

TABLE A-IV-7. *Length of the nasal bridge (n-prn) (mm)*

Age (years)	Male *N*	Male Mean	Male SD	Female *N*	Female Mean	Female SD
0–5 months	8	22.8	1.8	5	21.2	1.5
6–12 months	20	24.1	1.6	8	23.4	1.7
1	18	27.6	2.4	22	25.9	2.0
2	31	29.9	2.5	31	29.2	2.5
3	30	31.4	1.9	30	30.2	2.8
4	30	33.8	2.0	30	31.6	2.3
5	30	33.5	1.8	30	32.8	2.1
6	*	34.8	2.0	*	33.1	2.2
7	*	36.2	2.3	*	34.7	2.4
8	*	37.0	2.5	*	35.7	2.5
9	*	38.7	2.7	*	37.8	2.7
10	*	40.1	2.9	*	39.0	2.8
11	*	40.3	3.1	*	40.4	2.9
12	*	42.8	3.2	*	42.0	3.1
13	*	44.1	3.4	*	43.2	3.2
14	*	45.2	3.6	*	44.4	3.4
15	*	47.4	3.8	*	44.8	3.5
16	*	48.6	4.0	*	46.2	3.6
17	*	49.0	4.1	*	45.4	3.8
18	40	49.0	4.2	40	45.4	3.9
19–25	109	50.0	3.6	200	44.7	3.4

* Indirectly derived norms (for details see pages 141–142).

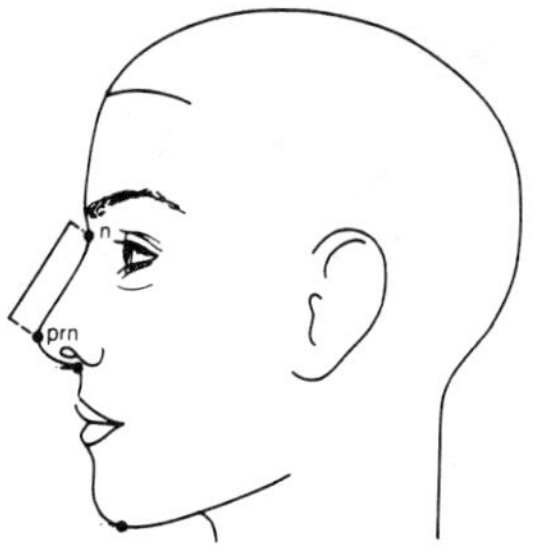

TABLE A-IV-8. *Nasal tip protrusion (sn-prn) (mm)*

Age (years)	Male *N*	Male Mean	Male SD	Female *N*	Female Mean	Female SD
0–5 months	8	9.5	1.4	5	8.4	1.3
6–12 months	20	9.1	1.2	8	9.7	0.8
1	18	10.1	1.5	20	10.2	1.4
2	31	11.3	1.5	31	11.5	1.4
3	30	12.1	1.4	30	12.4	1.8
4	30	13.0	1.1	30	12.3	1.1
5	30	13.3	0.8	30	13.1	1.2
6	50	15.1	1.5	50	14.8	1.2
7	50	15.3	1.3	50	15.5	1.1
8	51	15.9	1.3	51	15.5	1.2
9	51	16.0	1.4	50	16.1	1.3
10	50	16.5	1.2	49	16.6	1.3
11	50	16.7	1.4	51	16.8	1.5
12	52	17.7	1.5	53	17.4	1.5
13	50	18.1	1.5	49	18.1	1.6
14	49	18.9	2.2	51	18.6	1.5
15	50	18.6	1.5	51	18.7	1.4
16	50	20.1	1.7	51	19.3	1.6
17	49	20.5	1.9	51	19.4	1.7
18	52	20.6	2.2	51	19.3	1.9
19–25	109	19.5	1.9	200	19.7	1.6

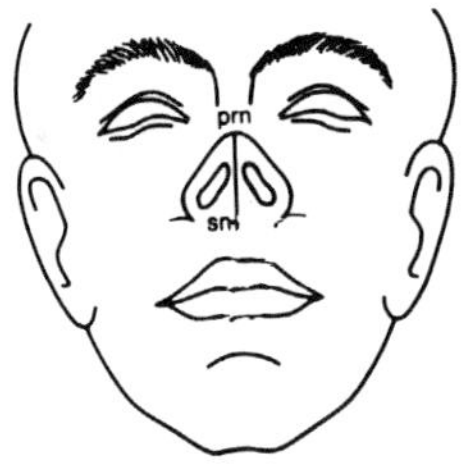

NOSE

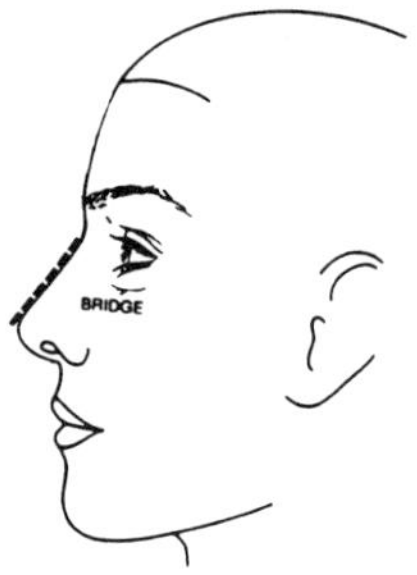

TABLE A–IV-9. *Inclination of the nasal bridge (degrees)*

Age (years)	Male			Female		
	N	Mean	SD	*N*	Mean	SD
0–5 months	8	33.1	3.9	5	31.8	2.9
6–12 months	20	32.5	2.9	8	32.6	3.7
1	18	29.9	3.3	22	29.7	4.0
2	31	28.6	4.7	34	28.0	3.5
3	30	27.8	4.4	30	28.3	4.0
4	30	27.6	4.1	30	29.0	4.2
5	30	28.2	4.5	30	28.9	3.0
6	49	31.4	4.6	49	29.7	4.5
7	50	29.2	4.2	50	28.8	4.3
8	51	29.5	4.7	51	29.0	4.1
9	51	30.0	4.1	50	28.7	4.2
10	50	28.0	4.0	49	28.3	4.3
11	49	28.1	4.9	51	27.9	4.3
12	52	30.0	3.4	51	28.7	4.3
13	50	29.7	4.5	49	29.4	4.3
14	49	30.7	4.3	51	29.3	4.4
15	50	29.8	4.4	50	29.1	4.2
16	50	29.5	5.1	51	29.5	4.8
17	49	29.9	4.6	51	29.4	5.2
18	51	31.6	4.6	50	30.0	5.3
19–25	109	30.4	3.6	200	29.9	3.9

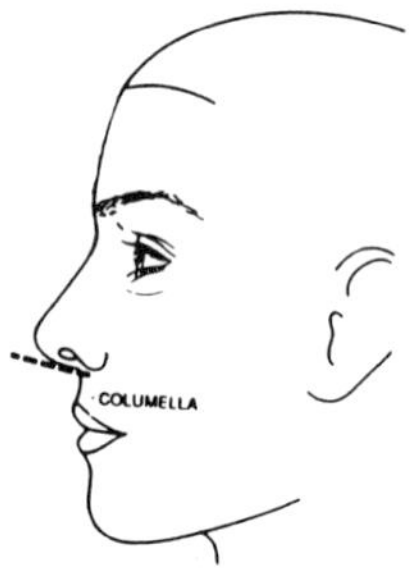

TABLE A–IV-10. *Inclination of the columella (degrees)*

Age (years)	Male			Female		
	N	Mean	SD	*N*	Mean	SD
1	17	63.6	4.6	18	63.8	5.0
2	31	63.6	5.8	34	64.7	5.4
3	30	65.6	6.0	30	62.7	4.5
4	30	68.6	5.7	30	65.1	7.1
5	30	68.3	6.1	30	64.4	7.4
6	+	66.9	8.0	+	63.4	8.6
7	+	66.4	10.7	+	70.5	10.3
8	+	71.0	8.4	+	72.4	5.7
9	+	68.4	9.9	+	72.2	11.7
10	+	71.8	8.9	+	68.4	8.9
11	+	74.5	7.6	+	75.3	4.9
12	+	69.7	5.1	+	73.4	19.8
13	+	72.4	9.6	+	75.7	9.6
14	+	71.9	9.1	+	70.6	8.7
15	+	75.0	8.3	+	74.3	5.3
16	+	73.5	8.6	+	78.0	13.8
17	+	80.4	7.1	+	76.6	9.9
18	+	74.3	7.4	+	82.9	5.7
19–25	109	77.9	8.9	45	78.2	7.5

$^{+}$ Indirectly derived norms (for details see pages 141–142).

NOSE

TABLE A-IV-11. *Inclination of the nasal tip portion (degrees)*

Age (years)	Male			Female		
	N	Mean	SD	*N*	Mean	SD
19–25	67	61.7	7.1	45	34.3	11.2

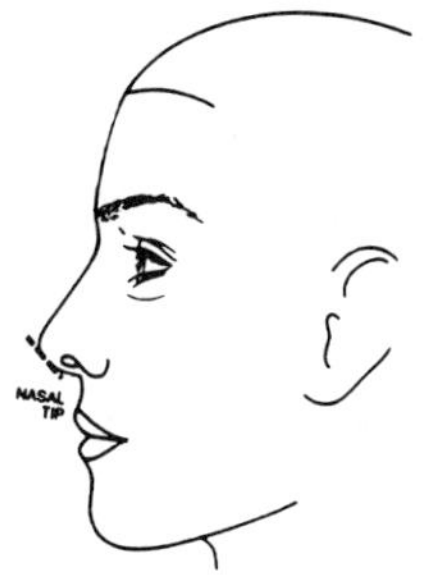

TABLE A-IV-12. *Glabellonasal angle (degrees)*

Age (years)	Male			Female		
	N	Mean	SD	*N*	Mean	SD
1	18	166.6	7.1	21	166.6	7.9
2	31	158.1	6.8	34	163.4	7.2
3	30	159.3	6.6	30	160.8	6.3
4	30	162.2	6.3	30	166.3	5.2
5	30	162.6	5.4	30	164.3	8.8
6	+	156.4	9.7	+	152.3	9.5
7	+	154.8	10.1	+	153.6	8.9
8	+	156.8	6.9	+	155.2	8.7
9	+	160.1	9.1	+	158.9	9.0
10	+	159.5	8.5	+	160.5	8.5
11	+	161.1	11.6	+	163.1	8.4
12	+	166.3	9.5	+	161.6	10.2
13	+	166.6	8.9	+	165.1	9.1
14	+	169.6	10.1	+	164.7	11.2
15	+	169.2	11.3	+	162.6	8.7
16	+	168.5	8.5	+	165.4	10.1
17	+	168.1	9.9	+	165.3	12.6
18	+	172.1	9.7	+	170.6	11.2
19–25	44	146.0	11.4	45	164.9	5.6

+ Indirectly derived norms (for details see pages 141–142).

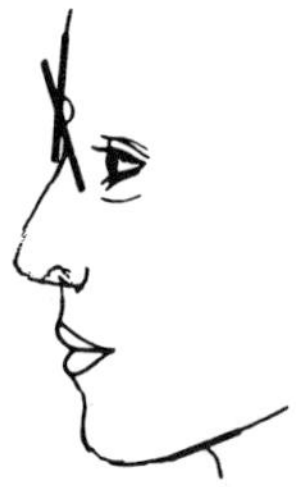

NOSE

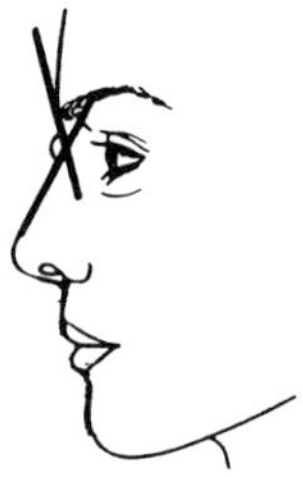

TABLE A-IV-13. *Nasofrontal angle (degrees)*

Age (years)	Male			Female		
	N	Mean	SD	*N*	Mean	SD
0–5 months	8	123.4	8.5	5	123.4	3.8
6–12 months	20	125.5	11.1	8	124.7	5.2
1	18	125.8	6.8	24	123.0	24.6
2	31	128.6	8.7	34	127.0	8.1
3	30	131.9	6.4	30	129.6	5.9
4	30	128.5	6.0	30	126.7	7.8
5	30	128.0	6.3	30	129.4	5.5
6	50	129.2	7.5	50	129.6	6.3
7	50	130.3	7.4	49	130.3	7.1
8	51	131.4	7.4	51	131.1	6.3
9	51	131.6	6.5	50	133.0	7.3
10	50	131.8	7.0	49	133.3	6.9
11	49	132.8	9.2	51	135.8	6.4
12	52	132.9	7.3	53	132.3	8.4
13	50	133.6	7.3	49	133.6	8.1
14	48	133.8	7.4	50	133.2	7.7
15	50	131.8	8.0	51	131.5	7.0
16	50	132.0	7.2	50	132.3	7.5
17	48	129.5	9.3	51	131.6	8.2
18	48	130.5	8.1	51	134.0	7.4
19–25	109	130.3	7.4	200	134.3	7.0

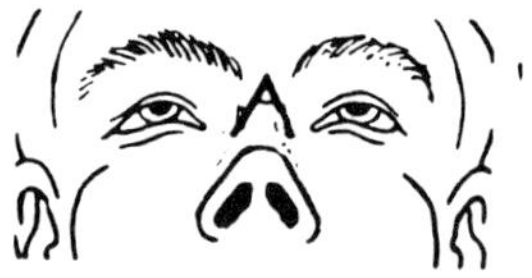

TABLE A-IV-14. *Nasal root–slope angle (degrees)*

Age (years)	Male			Female		
	N	Mean	SD	*N*	Mean	SD
19–25	42	44.8	6.9	45	43.9	7.7

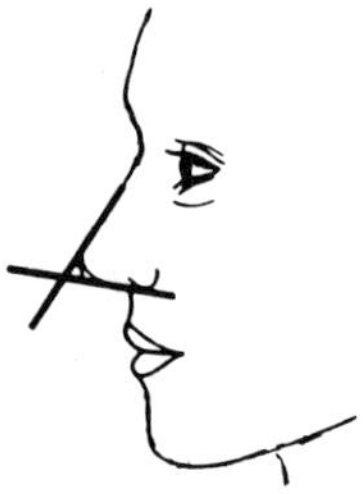

TABLE A-IV-15. *Nasal tip angle (degrees)*

Age (years)	Male			Female		
	N	Mean	SD	*N*	Mean	SD
1	18	82.6	5.9	21	81.5	3.7
2	31	83.2	5.6	34	82.9	4.6
3	30	84.9	4.3	30	83.2	5.3
4	30	81.3	6.4	30	81.1	5.8
5	30	80.2	6.7	30	83.4	4.9
6	+	81.0	6.5	+	88.7	7.1
7	+	83.8	8.7	+	81.6	9.3
8	+	79.1	9.9	+	80.0	6.1
9	+	81.7	9.4	+	79.3	10.3
10	+	80.5	7.9	+	83.2	8.3
11	+	79.5	8.4	+	76.8	5.2
12	+	81.2	5.3	+	78.0	15.9
13	+	78.5	9.7	+	76.4	9.9
14	+	78.1	8.9	+	79.7	9.1
15	+	74.8	8.8	+	77.8	5.1
16	+	77.5	7.7	+	76.0	9.8
17	+	71.4	7.8	+	75.6	12.0
18	+	74.6	6.0	+	69.9	7.8
19–25	109	71.7	7.4	45	67.4	7.4

+ Indirectly derived norms (for details see pages 141–142).

NOSE

TABLE A-IV-16. *Nasal ala-slope angle (degrees)*

Age (years)	Male			Female		
	N	Mean	SD	*N*	Mean	SD
19–25	42	63.9	5.8	45	59.4	5.3

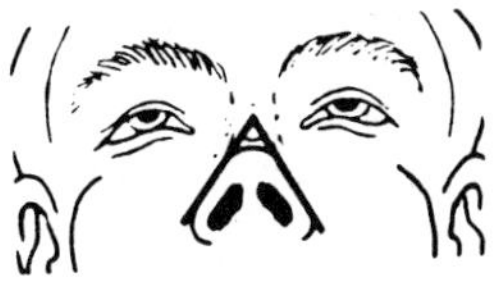

TABLE A-IV-17. *Nasolabial angle (degrees)*

Age (years)	Male			Female		
	N	Mean	SD	*N*	Mean	SD
0–5 months	8	98.1	7.6	5	102.2	7.3
6–12 months	20	98.7	10.6	8	101.9	9.2
1	18	94.9	6.8	24	95.7	8.5
2	31	100.8	9.1	34	98.9	10.0
3	30	103.8	7.5	30	104.3	8.3
4	30	106.0	10.4	30	109.2	10.0
5	30	104.6	11.9	30	108.6	9.2
6	50	103.7	9.3	50	106.1	9.4
7	50	101.0	8.3	49	100.4	8.2
8	51	101.0	9.5	51	102.0	8.2
9	51	100.8	9.6	50	101.0	8.2
10	50	100.6	8.6	49	100.7	7.2
11	50	102.3	7.8	51	102.3	8.9
12	52	100.6	9.1	53	101.4	9.5
13	50	98.6	9.4	49	100.9	9.3
14	48	102.3	9.8	51	102.3	8.5
15	50	99.8	8.9	51	100.8	8.9
16	50	100.8	7.8	50	101.1	9.1
17	49	96.2	9.4	51	99.3	8.6
18	49	98.9	8.0	51	99.1	8.7
19–25	109	99.8	11.8	200	104.2	9.8

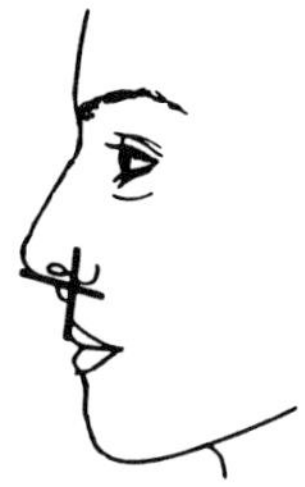

TABLE A-IV-18. *Deviation of the nasal bridge (degrees)*

Age (years)	Male			Female		
	N	Mean	SD	*N*	Mean	SD
1	18	0.0	0.0	20	0.0	0.0
2	31	0.0	0.2	34	0.0	0.0
3	30	0.1	0.4	30	0.0	0.3
4	30	0.1	0.5	30	0.0	0.0
5	30	0.0	0.0	30	0.0	0.0
6	50	2.3	1.5	50	2.5	0.7
7	50	2.3	0.8	49	2.7	1.1
8	51	2.0	—	51	3.0	1.4
9	51	3.7	2.7	49	3.0	0.6
10	50	3.5	1.1	49	3.3	6.5
11	50	2.3	0.6	50	3.2	1.2
12	52	5.0	2.4	53	3.4	1.2
13	49	4.0	1.1	49	3.4	1.1
14	49	4.3	2.3	51	3.7	1.5
15	50	3.7	1.1	51	4.6	2.4
16	50	3.9	1.3	51	2.8	1.3
17	49	3.8	1.4	51	3.1	1.1
18	52	1.2	0.4	51	3.4	1.3
19–25	40	4.6	2.5	40	3.4	1.3

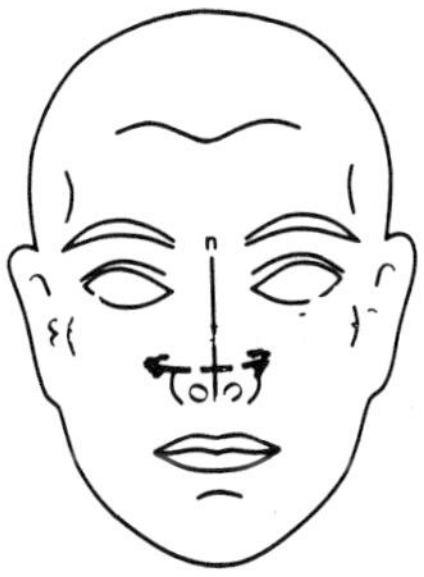

NOSE

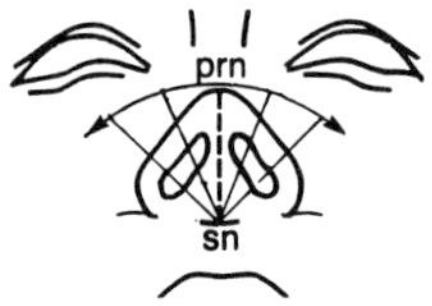

TABLE A-IV-19. *Deviation of the columella (degrees)*

Age (years)	Male			Female		
	N	Mean	SD	*N*	Mean	SD
1	18	0.0	0.0	20	0.1	0.2
2	31	0.0	0.0	34	0.1	0.4
3	30	0.1	0.3	30	0.0	0.0
4	30	0.2	1.3	30	0.0	0.0
5	30	0.0	0.0	30	0.1	0.7
6	50	0.0	0.0	50	0.0	0.0
7	50	3.3	1.4	50	3.7	2.4
8	51	4.2	2.3	51	5.4	2.9
9	51	3.4	1.1	49	2.0	—
10	50	3.4	1.3	49	3.7	1.5
11	50	3.8	1.7	50	5.0	2.4
12	52	6.0	2.0	53	3.5	1.3
13	49	3.6	1.1	49	4.8	2.6
14	49	4.5	1.9	51	3.4	1.5
15	50	4.5	2.1	51	4.5	1.7
16	50	3.7	1.5	51	4.2	1.2
17	49	4.5	2.1	51	2.6	0.9
18	52	4.7	3.1	51	4.4	1.5
19–25	108	0.5	1.6	199	0.4	1.7

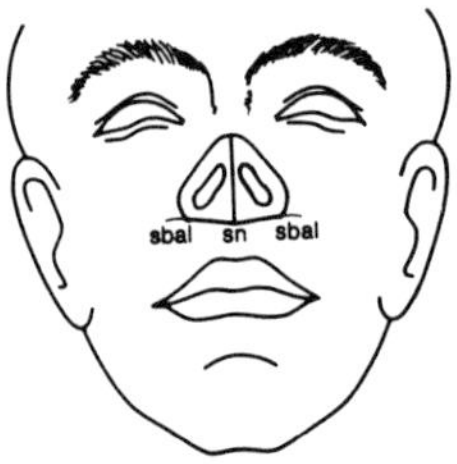

TABLE A-IV-20. *Width of the nostril floor (sbal-sn), right and left (mm)*

Age (years)	Male, right side			Female, right side		
	N	Mean	SD	*N*	Mean	SD
1	18	6.6	1.1	18	6.1	0.7
2	31	7.2	1.3	30	7.1	1.1
3	30	7.6	1.4	30	7.3	1.0
4	30	9.7	1.8	30	9.0	1.1
5	30	9.4	1.1	30	9.2	1.1
6	50	9.5	1.2	50	9.3	1.0
7	50	10.2	1.0	50	9.7	1.1
8	51	10.3	1.2	51	10.0	1.1
9	51	10.4	1.2	50	10.1	1.4
10	50	10.5	1.3	49	10.4	1.3
11	50	10.6	1.2	51	10.5	1.3
12	52	11.5	1.5	53	10.9	1.3
13	50	11.5	1.3	49	10.9	1.2
14	49	11.7	1.3	51	10.7	1.3
15	50	11.8	1.1	51	11.3	1.4
16	50	12.1	1.5	51	11.5	1.3
17	49	12.4	1.3	51	11.7	1.5
18	52	12.8	1.7	51	11.3	1.7
19–25	109	12.9	1.7	200	10.9	1.4

TABLE A–IV-20. *Continued.*

Age (years)	Male, left side			Female, left side		
	N	Mean	SD	*N*	Mean	SD
1	18	6.6	1.1	18	6.1	0.7
2	31	7.2	1.4	30	7.1	1.1
3	30	7.6	1.4	30	7.4	1.1
4	30	9.8	1.7	30	9.0	1.1
5	30	9.3	1.0	30	9.2	1.1
6	50	9.5	1.2	50	9.3	1.0
7	50	10.0	0.9	50	9.7	1.0
8	51	10.0	1.1	51	9.9	1.1
9	51	10.3	1.1	50	10.0	1.3
10	50	10.5	1.3	49	10.3	1.3
11	50	10.6	1.3	51	10.3	1.4
12	52	11.3	1.3	53	10.8	1.3
13	50	11.3	1.3	49	10.8	1.2
14	49	11.4	1.3	51	10.7	1.3
15	50	11.7	1.2	51	11.1	1.3
16	50	11.9	1.6	51	11.4	1.3
17	49	12.1	1.4	51	11.5	1.5
18	52	12.5	1.6	51	11.4	1.7
19–25	109	12.8	1.7	200	10.9	1.5

TABLE A–IV-21. *Thickness of the ala (al′-al′), right and left (mm)*

Age (years)	Male, right side			Female, right side		
	N	Mean	SD	*N*	Mean	SD
1	18	4.4	0.6	18	4.0	0.8
2	31	4.7	0.6	31	4.4	0.6
3	30	4.6	0.6	30	4.4	0.6
4	30	5.6	0.7	30	5.7	0.7
5	30	5.4	0.6	30	5.3	0.7
6	50	4.6	0.8	50	4.7	0.6
7	50	4.8	0.8	50	4.9	0.6
8	51	5.1	0.7	51	4.9	0.7
9	51	5.0	0.6	50	5.1	0.7
10	50	5.2	0.7	49	5.2	0.8
11	50	5.1	0.6	51	5.3	0.9
12	52	5.5	0.7	53	5.6	0.8
13	50	5.7	0.7	49	5.7	0.8
14	49	5.7	0.8	51	5.4	0.6
15	50	5.7	0.6	51	5.6	0.8
16	50	5.8	0.5	51	5.5	0.8
17	49	6.0	0.8	51	5.6	0.9
18	52	5.9	0.9	51	5.3	0.8
19–25	109	5.9	0.7	198	5.3	0.7

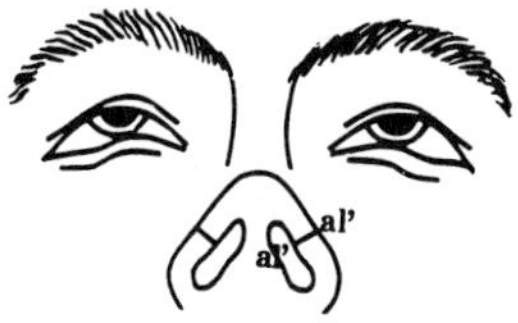

NOSE

TABLE A–IV-21. *Continued.*

Age (years)	Male, left side			Female, left side		
	N	Mean	SD	*N*	Mean	SD
1	18	4.4	0.6	18	4.0	0.8
2	31	4.7	0.6	31	4.4	0.6
3	30	4.6	0.6	30	4.4	0.6
4	30	5.6	0.6	30	5.6	0.7
5	30	5.3	0.5	30	5.3	0.8
6	50	4.6	0.8	50	4.7	0.6
7	50	4.8	0.8	50	4.9	0.6
8	51	5.1	0.7	51	4.9	0.7
9	51	5.0	0.6	50	5.1	0.7
10	50	5.2	0.8	49	5.2	0.9
11	50	5.1	0.6	51	5.3	0.9
12	52	5.4	0.7	53	5.5	0.9
13	50	5.7	0.7	49	5.6	0.7
14	49	5.7	0.8	51	5.4	0.6
15	50	5.7	0.6	51	5.6	0.8
16	50	5.7	0.5	51	5.4	0.9
17	49	6.0	0.8	51	5.6	0.9
18	52	5.9	0.9	51	5.3	0.8
19–25	109	5.8	0.7	198	5.3	0.7

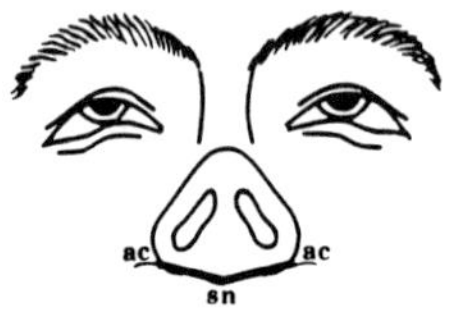

TABLE A–IV-22. *Distance between the facial insertions of the alar base and the subnasale point (ac-sn), right and left (mm)*

Age (years)	Male, right side			Female, right side		
	N	Mean	SD	*N*	Mean	SD
0–5 months	8	12.0	0.5	5	10.4	2.1
6–12 months	20	12.5	1.0	8	11.7	0.5
1	18	13.6	1.2	18	13.2	1.4
2	31	14.3	1.0	31	14.0	1.0
3	30	14.8	1.2	30	14.6	0.9
4	30	17.1	1.0	30	17.0	0.9
5	30	17.2	0.8	30	16.9	1.1
6	*	18.5	0.9	*	18.9	1.1
7	*	19.7	0.9	*	19.5	1.2
8	*	19.8	1.0	*	19.9	1.2
9	*	19.8	1.1	*	19.9	1.2
10	*	19.8	1.2	*	20.3	1.3
11	*	19.9	1.2	*	20.3	1.3
12	*	21.5	1.3	*	20.9	1.3
13	*	21.3	1.4	*	20.8	1.3
14	*	21.6	1.4	*	20.3	1.4
15	*	21.7	1.5	*	21.3	1.4
16	*	22.1	1.6	*	21.5	1.4
17	*	22.6	1.6	*	21.8	1.5
18	*	23.2	1.7	*	20.9	1.5
19–25	86	23.2	1.8	45	20.9	1.5

TABLE A–IV-22. *Continued.*

Age (years)	Male, left side			Female, left side		
	N	Mean	SD	*N*	Mean	SD
0–5 months	8	12.0	0.5	5	10.4	2.1
6–12 months	20	12.5	0.5	8	11.7	0.5
1	18	13.6	1.2	18	13.2	1.4
2	31	14.4	1.1	31	14.0	1.0
3	30	14.9	1.1	30	14.7	0.9
4	30	17.1	1.1	30	16.9	0.9
5	30	17.2	0.8	30	16.9	1.1
6	*	18.7	0.9	*	18.8	1.2
7	*	19.6	0.9	*	19.4	1.2
8	*	19.4	1.0	*	19.6	1.3
9	*	19.9	1.1	*	19.6	1.3
10	*	20.2	1.2	*	20.0	1.4
11	*	20.3	1.2	*	19.8	1.4
12	*	21.5	1.3	*	20.6	1.5
13	*	21.4	1.4	*	20.5	1.5
14	*	21.5	1.4	*	20.1	1.6
15	*	22.0	1.5	*	20.7	1.6
16	*	22.2	1.6	*	21.2	1.7
17	*	22.5	1.6	*	21.2	1.7
18	*	23.2	1.7	*	20.9	1.8
19–25	86	23.2	1.8	45	20.9	1.8

+ Indirectly derived norms (for details see pages 141–142).

TABLE A–IV-23. *Length of the ala (ac-prn), right and left (mm)*

Age (years)	Male, right side			Female, right side		
	N	Mean	SD	*N*	Mean	SD
0–5 months	8	16.5	0.9	5	15.8	1.3
6–12 months	20	17.9	1.6	8	16.9	1.2
1	18	19.7	1.5	18	19.3	1.3
2	31	21.0	1.5	31	20.5	1.2
3	30	21.9	1.3	30	22.0	1.7
4	30	23.3	1.3	30	22.7	1.1
5	30	23.6	0.9	30	23.2	1.5
6	50	24.9	1.6	50	24.4	1.3
7	50	25.6	1.4	50	25.1	1.1
8	51	26.5	1.4	51	25.4	1.6
9	51	26.8	1.4	50	26.3	1.5
10	50	27.3	1.3	49	27.1	1.5
11	50	27.8	1.7	51	27.7	1.8
12	52	29.1	1.8	53	28.6	1.9
13	50	30.5	2.3	49	29.5	1.6
14	49	31.8	2.4	51	29.9	1.7
15	50	33.2	2.1	51	30.5	1.6
16	50	33.7	1.8	51	31.2	1.5
17	49	34.7	1.9	51	30.7	2.1
18	52	34.5	1.8	51	30.7	1.7
19–25	109	35.0	1.6	199	31.5	1.8

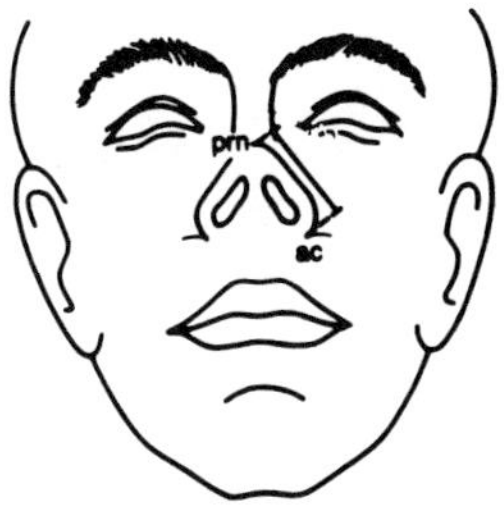

NOSE

TABLE A–IV-23. *Continued.*

Age (years)	Male, left side			Female, left side		
	N	Mean	SD	*N*	Mean	SD
0–5 months	8	16.5	0.9	5	15.8	1.3
6–12 months	20	17.9	1.6	8	16.9	1.2
1	18	19.7	1.5	18	19.3	1.3
2	31	21.0	1.5	31	20.7	1.2
3	30	21.9	1.3	30	22.0	1.7
4	30	23.3	1.3	30	22.7	1.1
5	30	23.6	0.9	30	23.2	1.5
6	50	24.9	1.6	50	24.4	1.3
7	50	25.6	1.4	50	25.1	1.1
8	51	26.4	1.3	51	25.4	1.6
9	51	26.7	1.4	50	26.4	1.6
10	50	27.3	1.4	49	27.1	1.5
11	50	27.9	1.9	51	27.7	1.8
12	52	29.0	1.8	53	28.6	1.9
13	50	30.5	2.3	49	29.5	1.6
14	49	31.8	2.4	51	29.8	1.6
15	50	33.3	2.1	51	30.5	1.6
16	50	33.7	1.9	51	31.2	1.5
17	49	34.8	1.9	51	30.7	2.1
18	52	34.5	1.7	51	30.8	1.6
19–25	109	35.0	1.7	198	31.4	1.8

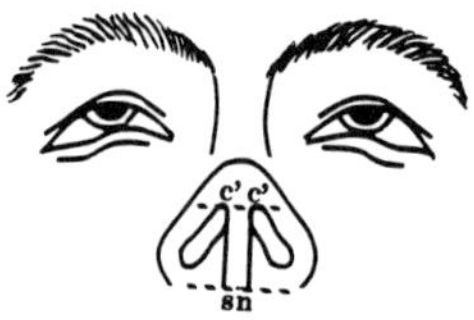

TABLE A–IV-24. *Length of the columella (sn-c'), right and left (mm)*

Age (years)	Male, right side			Female, right side		
	N	Mean	SD	*N*	Mean	SD
0–5 months	8	3.2	0.7	5	3.2	0.4
6–12 months	20	4.3	0.9	8	4.7	0.8
1	18	5.1	0.8	21	5.1	0.9
2	31	6.0	0.8	30	6.6	1.1
3	30	6.3	1.2	30	6.5	1.2
4	30	6.9	0.9	30	6.7	0.7
5	30	6.6	0.9	30	7.1	0.9
6	50	8.1	1.2	50	7.5	1.0
7	50	8.1	1.0	50	7.8	1.0
8	51	8.0	1.1	51	7.8	1.2
9	51	8.3	1.0	50	8.4	1.2
10	50	8.4	0.8	49	8.6	1.1
11	50	8.8	1.3	51	8.6	1.4
12	52	9.0	1.2	53	8.9	1.3
13	50	9.5	1.3	49	9.6	1.1
14	49	10.1	1.6	51	10.0	1.3
15	50	10.4	1.3	51	10.0	1.5
16	50	10.8	1.5	51	10.5	1.4
17	49	10.9	1.7	51	10.4	1.6
18	52	11.4	1.8	51	10.7	1.5
19–25	109	11.6	1.7	200	11.5	1.7

TABLE A-IV-24. *Continued.*

Age (years)	Male, left side			Female, left side		
	N	Mean	SD	*N*	Mean	SD
0–5 months	8	3.2	0.7	5	3.2	0.4
6–12 months	20	4.3	0.9	8	4.7	0.8
1	18	5.1	0.8	21	5.1	0.9
2	31	5.9	0.9	30	6.5	1.1
3	30	6.2	1.2	30	6.5	1.0
4	30	6.9	0.9	30	6.7	0.7
5	30	6.7	0.8	30	7.1	0.9
6	50	8.1	1.2	50	7.5	1.0
7	50	8.1	1.1	50	7.8	0.9
8	51	8.0	1.1	51	7.8	1.2
9	51	8.3	1.0	50	8.3	1.1
10	50	8.4	0.8	49	8.6	1.1
11	50	8.7	1.2	51	8.6	1.4
12	52	9.0	1.1	53	9.0	1.3
13	50	9.5	1.2	49	9.6	1.1
14	49	10.1	1.6	51	10.0	1.4
15	50	10.3	1.3	51	10.1	1.5
16	50	10.7	1.6	51	10.5	1.3
17	49	11.0	1.7	51	10.4	1.6
18	52	11.3	1.7	51	10.7	1.5
19–25	109	11.5	1.7	200	11.4	1.7

TABLE A-IV-25. *Depth of the nasal root (en-se sag), right and left (mm)*

Age (years)	Male, right side			Female, right side		
	N	Mean	SD	*N*	Mean	SD
1	17	8.6	1.5	15	8.5	1.4
2	31	9.5	1.3	32	9.1	1.6
3	30	9.8	1.2	30	9.8	1.3
4	30	9.4	1.4	30	9.3	1.5
5	30	9.7	1.3	30	9.5	1.6
6	34	11.0	1.5	39	11.1	1.8
7	33	12.1	1.2	42	11.4	1.8
8	45	11.8	1.7	41	11.3	1.6
9	47	12.6	1.8	38	11.9	1.8
10	37	12.3	1.9	37	12.2	1.9
11	39	12.7	1.6	41	12.5	1.9
12	50	12.8	1.8	50	12.3	1.6
13	47	13.9	2.1	49	13.2	1.9
14	46	15.0	2.4	43	13.7	1.8
15	44	15.2	2.3	43	14.0	1.8
16	48	15.8	2.1	35	13.6	1.3
17	48	16.5	2.4	36	13.6	2.3
18	48	15.3	2.3	37	13.6	1.9
19–25	107	15.9	2.1	200	14.0	1.7

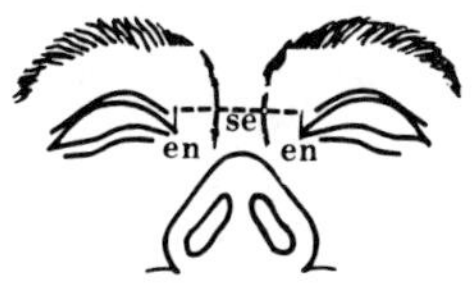

NOSE

TABLE A–IV-25. *Continued.*

Age (years)	Male, left side			Female, left side		
	N	Mean	SD	*N*	Mean	SD
1	17	8.6	1.5	15	8.5	1.4
2	31	9.5	1.3	32	9.2	1.4
3	30	9.7	1.2	30	9.8	1.2
4	30	9.3	1.3	30	9.2	1.5
5	30	9.7	1.3	30	9.5	1.6
6	34	11.1	1.5	39	11.1	1.8
7	33	12.1	1.2	42	11.4	1.8
8	45	11.8	1.7	41	11.3	1.6
9	47	12.6	1.8	38	11.9	1.8
10	37	12.3	1.9	37	12.3	1.9
11	39	12.7	1.6	41	12.5	1.9
12	50	12.7	1.8	50	12.3	1.6
13	47	14.0	2.1	49	13.2	1.9
14	46	15.0	2.4	43	13.7	1.8
15	44	15.3	2.2	43	14.0	1.8
16	48	15.8	2.1	35	13.6	1.3
17	48	16.5	2.2	36	13.6	2.4
18	48	15.3	2.3	37	13.6	2.0
19–25	108	15.9	2.1	200	14.0	1.7

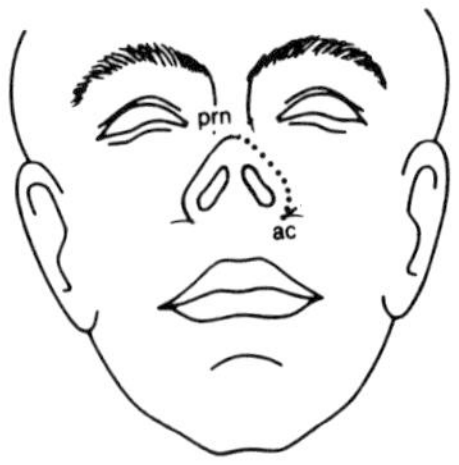

TABLE A–IV-26. *Surface length of the ala (ac-prn surf), right and left (mm)*

Age (years)	Male, right side			Female, right side		
	N	Mean	SD	*N*	Mean	SD
1	17	23.8	1.3	15	22.1	1.9
2	30	24.9	1.4	30	24.1	1.7
3	30	25.0	2.0	30	25.4	1.8
4	30	28.1	1.4	30	27.1	1.4
5	30	28.5	1.2	30	27.9	1.6
6	50	28.5	2.1	50	27.8	1.6
7	50	29.1	1.6	50	28.6	1.3
8	51	29.9	1.6	51	28.7	2.0
9	51	30.4	1.6	50	30.2	1.9
10	50	31.2	1.6	49	30.7	1.6
11	50	31.7	2.0	51	31.7	2.0
12	52	33.3	2.0	53	32.9	2.2
13	50	35.2	3.0	49	34.0	2.0
14	49	36.2	2.8	51	34.5	1.5
15	50	38.4	2.5	51	35.1	1.7
16	50	38.9	2.1	51	35.6	1.7
17	49	39.9	2.3	51	35.4	2.3
18	52	39.7	2.3	51	35.2	2.0
19–25	108	40.5	3.3	197	36.4	2.1

TABLE A-IV-26. *Continued.*

Age (years)	Male, left side			Female, left side		
	N	Mean	SD	*N*	Mean	SD
1	17	23.8	1.3	15	22.1	1.9
2	30	24.8	1.3	30	24.1	1.7
3	30	25.0	2.0	30	25.3	1.8
4	30	28.1	1.4	30	27.1	1.4
5	30	28.5	1.2	30	27.9	1.6
6	50	28.6	2.1	50	27.8	1.6
7	50	29.2	1.6	50	28.6	1.3
8	51	29.9	1.6	51	28.7	2.0
9	51	30.4	1.6	50	30.3	1.9
10	50	31.1	1.6	49	30.7	1.5
11	50	31.8	2.0	51	31.8	2.1
12	52	33.2	2.1	53	33.0	2.2
13	50	35.2	2.9	49	34.0	2.0
14	49	36.3	3.0	51	34.5	1.5
15	50	38.3	2.5	51	35.2	1.6
16	50	38.9	2.1	51	35.7	1.9
17	49	40.1	2.3	51	35.4	2.3
18	52	39.6	2.2	51	35.3	2.0
19–25	108	40.6	3.3	197	36.4	2.1

TABLE A-IV-27. *Nostril axis inclination, right and left (degrees)*

Age (years)	Male, right side			Female, right side		
	N	Mean	SD	*N*	Mean	SD
4	30	60.0	8.7	30	57.0	8.7
5	30	53.4	6.9	30	56.5	8.3
19–25	68	57.8	8.7	141	63.5	7.2
Age (years)	**Male, left side**			**Female, left side**		
	N	Mean	SD	*N*	Mean	SD
4	30	60.3	8.6	30	57.0	8.7
5	30	55.0	5.6	30	56.4	8.0
19–25	68	58.0	9.0	141	63.2	7.4

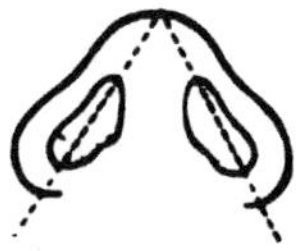

TABLE A-IV-28. *Alar-slope line inclination, right and left (degrees)*

Age (years)	Male, right side			Female, right side		
	N	Mean	SD	*N*	Mean	SD
19–25	28	76.8	7.2	45	63.1	4.5
Age (years)	**Male, left side**			**Female, left side**		
	N	Mean	SD	*N*	Mean	SD
19–25	28	76.8	7.2	45	62.4	4.1

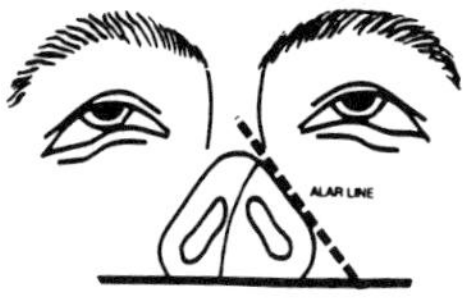

Lips and Mouth

LIPS AND MOUTH

TABLE A–V-1. *Width of the philtrum (cph-cph) (mm)*

Age (years)	Male *N*	Male Mean	Male SD	Female *N*	Female Mean	Female SD
1	18	6.7	1.0	20	6.5	1.1
2	31	6.7	1.3	31	6.8	1.0
3	30	7.0	1.0	30	6.8	1.0
4	30	8.1	1.0	30	7.8	1.3
5	30	8.0	1.1	30	7.9	1.0
6	50	8.2	1.2	50	8.4	1.3
7	50	8.7	1.2	50	8.4	1.4
8	51	8.8	1.1	51	8.7	1.1
9	51	8.9	1.0	50	8.7	1.1
10	50	8.9	1.3	49	9.0	1.3
11	50	9.0	1.3	51	8.9	1.1
12	52	9.3	1.2	53	9.3	1.2
13	50	10.3	1.3	49	9.5	1.2
14	49	10.2	1.3	51	9.4	1.0
15	50	10.5	1.5	51	9.4	1.2
16	50	10.9	1.6	50	9.3	1.3
17	49	11.0	1.8	51	9.7	1.5
18	52	10.9	1.7	51	9.9	1.4
19–25	108	10.4	1.4	200	9.7	1.5

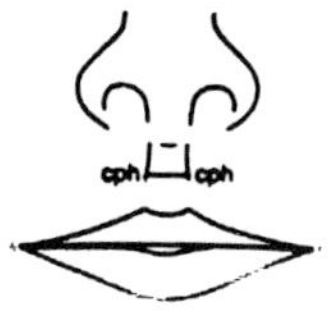

TABLE A–V-2. *Width of the mouth (ch-ch) (mm)*

Age (years)	Male *N*	Male Mean	Male SD	Female *N*	Female Mean	Female SD
0–5 months	8	30.0	1.6	5	30.6	2.4
6–12 months	20	33.1	2.2	8	33.0	1.6
1	18	34.8	2.6	28	33.3	2.5
2	31	35.2	2.6	31	35.0	1.8
3	30	36.7	2.4	30	36.3	2.5
4	30	38.9	2.5	30	37.9	2.2
5	30	40.7	2.4	30	39.5	2.7
6	50	41.7	2.8	50	41.2	2.9
7	50	42.7	2.7	50	42.4	2.2
8	51	44.6	2.2	51	43.1	2.8
9	51	45.5	3.1	50	44.6	2.9
10	50	45.9	2.9	49	44.9	3.1
11	50	46.4	3.5	50	45.9	3.3
12	52	48.2	3.4	53	46.5	3.3
13	50	49.1	3.0	49	48.1	2.8
14	49	50.1	3.5	51	47.5	2.7
15	50	51.8	3.3	51	49.1	3.9
16	50	52.1	3.2	51	48.9	3.8
17	49	53.5	3.6	51	49.4	3.1
18	52	53.3	3.3	51	49.8	3.2
19–25	109	54.5	3.0	200	50.2	3.5

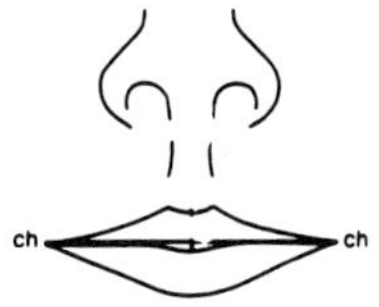

LIPS AND MOUTH

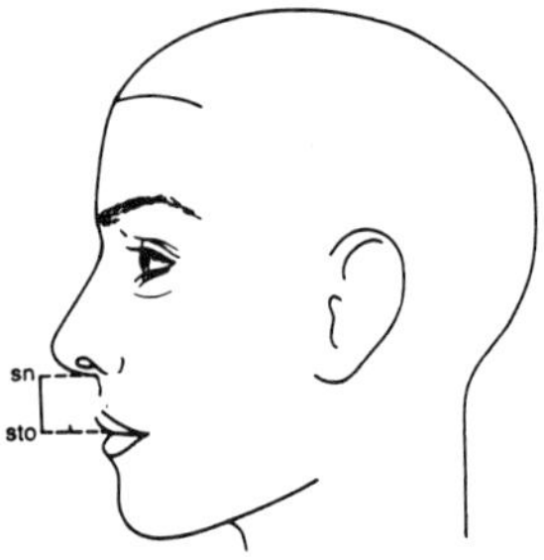

TABLE A–V-3. *Height of the upper lip (sn-sto) (mm)*

Age (years)	Male			Female		
	N	Mean	SD	*N*	Mean	SD
0–5 months	8	14.6	1.6	5	15.0	1.6
6–12 months	19	15.8	1.5	7	16.0	0.8
1	18	17.3	1.2	22	16.4	1.3
2	31	18.7	2.3	30	17.7	1.6
3	30	19.3	1.7	30	18.8	1.3
4	30	19.4	1.1	30	18.7	1.4
5	30	19.5	1.4	30	18.9	1.3
6	50	19.9	2.0	50	18.7	1.7
7	50	19.3	1.8	50	18.8	1.4
8	51	19.7	1.8	51	19.0	1.8
9	51	19.3	1.5	50	19.2	1.6
10	50	20.3	1.6	49	19.6	1.6
11	50	20.8	1.8	51	19.4	1.8
12	52	20.8	1.9	53	19.9	2.0
13	50	20.7	1.7	49	19.8	2.5
14	49	21.0	2.0	51	19.8	2.0
15	50	22.3	2.1	51	20.1	1.9
16	50	22.2	2.4	51	20.6	2.0
17	49	21.8	2.4	51	20.1	2.3
18	52	21.8	2.2	51	19.6	2.4
19–25	109	22.3	2.1	200	20.1	2.0

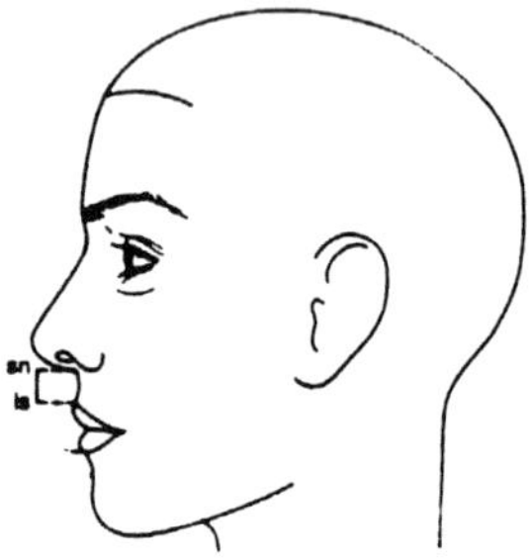

TABLE A–V-4. *Height of the cutaneous upper lip (sn-ls) (mm)*

Age (years)	Male			Female		
	N	Mean	SD	*N*	Mean	SD
0–5 months	8	10.8	1.7	5	10.2	0.8
6–12 months	19	11.4	1.3	7	10.7	1.1
1	18	11.8	0.9	21	10.9	1.1
2	31	12.6	1.7	20	12.0	1.7
3	30	13.2	1.6	30	12.7	1.8
4	30	13.8	1.3	30	12.6	1.5
5	30	13.6	1.1	30	13.0	1.7
6	21	14.4	2.1	20	12.6	1.3
7	19	13.7	1.8	29	12.9	1.4
8	29	14.0	2.2	21	13.2	2.3
9	34	13.3	1.8	24	13.4	1.9
10	21	14.4	1.9	18	12.7	1.6
11	15	14.5	1.6	9	13.0	2.4
12	10	14.4	2.0	5	11.6	1.5
13	25	14.2	1.8	12	12.3	2.0
14	34	15.0	2.4	26	12.8	1.9
15	23	15.1	2.2	9	13.3	0.9
16	24	14.8	1.8	5	14.4	2.1
17	16	14.4	1.3	12	13.5	2.2
18	13	14.8	2.4	8	12.5	2.8
19–25	109	15.9	1.9	199	13.8	4.6

TABLE A–V-5. *Vermilion height of the upper lip (ls-sto) (mm)*

Age (years)	Male			Female		
	N	Mean	SD	*N*	Mean	SD
0–5 months	8	3.9	1.1	5	4.8	1.3
6–12 months	19	4.4	1.0	7	5.3	1.4
1	18	5.4	0.9	21	5.6	1.0
2	31	5.9	1.2	30	5.4	1.3
3	30	5.7	1.0	30	5.7	1.0
4	30	7.5	1.0	30	7.6	1.0
5	30	7.6	0.9	30	7.5	1.1
6	50	7.9	1.1	50	8.0	1.1
7	50	8.0	1.1	50	7.9	1.2
8	51	8.0	1.2	51	8.0	0.9
9	51	8.4	1.2	50	8.3	1.1
10	50	8.7	1.3	49	8.3	1.2
11	50	8.3	1.1	51	8.5	1.4
12	52	8.6	1.3	53	8.9	1.2
13	50	9.1	1.5	49	8.5	1.4
14	49	8.9	1.3	51	9.0	1.2
15	50	9.3	1.5	51	8.8	1.3
16	50	9.3	1.1	51	8.8	1.3
17	49	9.5	1.5	51	8.6	1.6
18	52	8.9	1.5	51	8.4	1.3
19–25	109	8.0	1.4	200	8.7	1.3

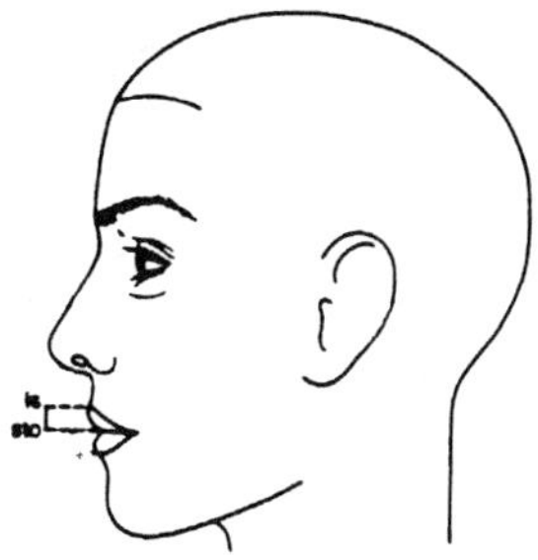

TABLE A–V-6. *Vermilion height of the lower lip (sto-li) (mm)*

Age (years)	Male			Female		
	N	Mean	SD	*N*	Mean	SD
1	18	6.6	1.2	20	6.2	1.2
2	31	6.6	1.4	30	5.7	1.4
3	30	6.3	1.3	30	5.9	1.3
4	30	7.0	1.3	30	6.9	1.3
5	30	7.4	1.1	30	7.1	1.3
6	50	8.2	1.4	50	7.7	1.4
7	50	8.6	1.5	50	8.8	1.5
8	51	9.2	1.3	51	9.1	1.3
9	51	9.8	1.6	50	9.3	1.8
10	49	9.9	1.5	49	9.5	1.3
11	50	9.8	1.6	51	9.5	1.5
12	52	9.9	1.5	53	9.9	1.3
13	50	10.0	1.5	49	10.1	1.8
14	49	10.2	1.5	51	10.4	1.5
15	50	10.7	1.5	51	10.2	1.7
16	50	10.8	1.5	51	10.0	1.9
17	49	11.0	2.2	51	10.0	1.5
18	52	10.4	1.9	51	9.7	1.6
19–25	109	9.3	1.6	200	9.4	1.5

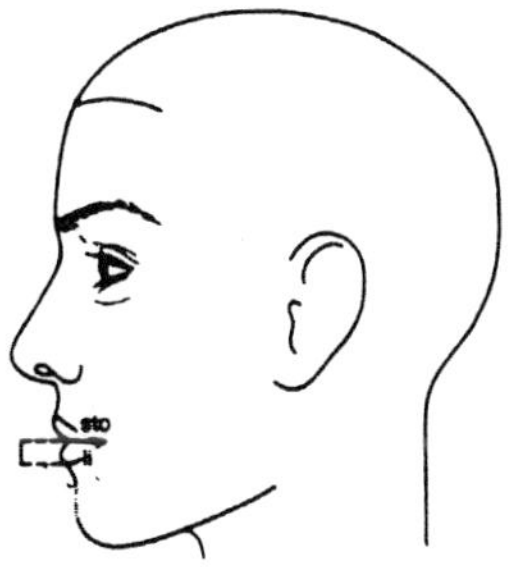

LIPS AND MOUTH

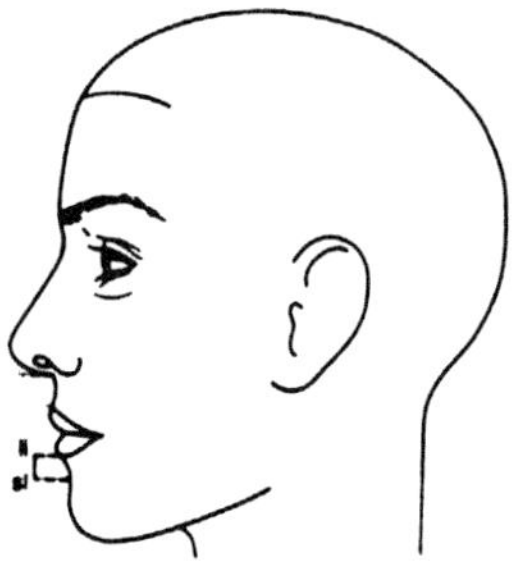

TABLE A–V-7. *Height of the cutaneous lower lip (li-sl) (mm)*

Age (years)	Male *N*	Male Mean	Male SD	Female *N*	Female Mean	Female SD
1	18	6.5	1.2	18	6.9	1.1
2	31	7.2	1.6	31	6.9	1.4
3	30	7.3	1.7	30	6.9	1.4
4	30	8.8	2.0	30	7.7	1.5
5	30	8.0	1.2	30	8.1	1.4
6	*	9.0	1.3	*	8.9	1.5
7	*	9.3	1.3	*	8.9	1.5
8	*	9.3	1.4	*	8.9	1.6
9	*	9.5	1.5	*	9.0	1.6
10	*	9.8	1.6	*	9.3	1.7
11	*	10.0	1.6	*	9.2	1.7
12	*	10.1	1.7	*	9.7	1.8
13	*	10.5	1.8	*	9.5	1.8
14	*	10.8	1.8	*	9.7	1.9
15	*	11.2	1.9	*	9.5	1.9
16	*	11.5	2.0	*	9.8	2.0
17	*	11.5	2.0	*	9.8	2.0
18	*	11.9	2.2	*	9.9	2.4
19–25	109	11.9	2.2	200	10.7	2.1

* Indirectly derived norms (for details see pages 141–142).

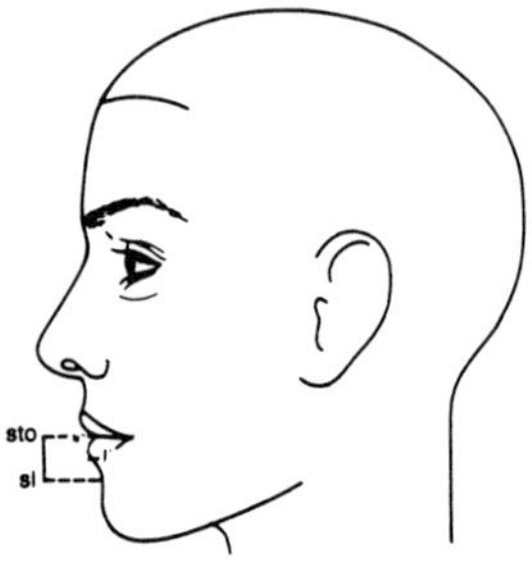

TABLE A–V-8. *Lower lip height (sto-sl) (mm)*

Age (years)	Male *N*	Male Mean	Male SD	Female *N*	Female Mean	Female SD
1	18	13.2	1.4	18	12.8	1.4
2	31	13.9	1.7	31	13.1	1.5
3	30	13.7	1.2	30	13.2	1.3
4	30	15.1	1.5	30	14.4	1.3
5	30	15.2	1.2	30	15.1	1.2
6	*	16.5	1.3	*	15.9	1.5
7	*	16.8	1.3	*	15.9	1.7
8	*	16.6	1.4	*	15.8	2.0
9	*	16.6	1.4	*	15.8	2.2
10	*	16.9	1.5	*	16.3	2.5
11	*	17.0	1.6	*	16.1	2.7
12	*	17.0	1.6	*	16.8	3.0
13	*	17.5	1.7	*	16.4	3.2
14	*	17.7	1.7	*	16.6	3.5
15	*	18.2	1.8	*	16.3	3.7
16	*	18.5	1.9	*	16.7	4.0
17	*	18.3	1.9	*	16.6	4.2
18	*	18.8	2.0	*	16.7	4.5
19–25	109	19.7	2.1	200	17.8	4.7

* Indirectly derived norms (for details see pages 141–142).

LIPS AND MOUTH

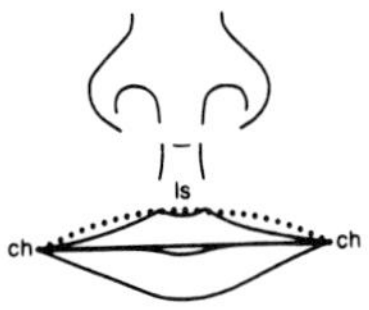

TABLE A-V-9. *Upper vermilion arc (ch-ls-ch) (mm)*

Age (years)	Male			Female		
	N	Mean	SD	*N*	Mean	SD
1	17	55.7	4.9	15	52.4	2.9
2	30	53.4	4.9	30	52.6	3.1
3	30	54.6	4.2	30	53.4	3.4
4	30	58.9	3.8	30	58.4	4.0
5	30	60.6	5.0	30	60.6	3.9
6	*	63.6	5.0	*	61.6	4.0
7	*	64.6	5.1	*	62.8	4.2
8	*	66.9	5.1	*	63.3	4.3
9	*	67.8	5.2	*	65.0	4.4
10	*	67.8	5.2	*	64.9	4.6
11	*	68.1	5.2	*	65.8	4.7
12	*	70.3	5.3	*	66.2	4.8
13	*	71.1	5.3	*	68.0	4.9
14	*	72.1	5.4	*	66.7	5.1
15	*	74.1	5.4	*	68.5	5.2
16	*	74.1	5.4	*	67.8	5.3
17	*	75.7	5.5	*	68.1	5.5
18	50	75.0	5.5	51	68.2	5.6
19–25	109	79.3	4.4	187	70.1	4.9

* Indirectly derived norms (for details see pages 141–142).

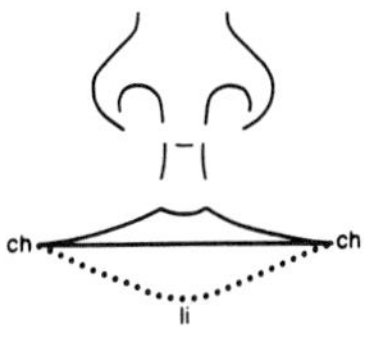

TABLE A-V-10. *Lower vermilion arc (ch-li-ch) (mm)*

Age (years)	Male			Female		
	N	Mean	SD	*N*	Mean	SD
1	17	45.6	4.5	15	43.5	2.7
2	30	45.2	4.8	30	44.4	3.0
3	30	45.9	3.9	30	44.1	3.3
4	30	47.7	4.1	30	47.0	3.5
5	30	50.3	4.0	30	49.1	3.3
6	*	54.3	4.0	*	53.1	3.4
7	*	55.5	4.1	*	54.6	3.5
8	*	57.9	4.1	*	55.3	3.6
9	*	59.0	4.2	*	57.2	3.7
10	*	59.5	4.2	*	57.4	3.9
11	*	60.1	4.2	*	58.6	4.0
12	*	62.3	4.3	*	59.3	4.1
13	*	63.4	4.3	*	61.2	4.2
14	*	64.7	4.4	*	60.3	4.3
15	*	66.8	4.4	*	62.3	4.4
16	*	67.2	4.4	*	61.9	4.5
17	*	68.9	4.5	*	62.5	4.6
18	50	68.6	4.0	50	62.9	6.3
19–25	109	71.6	4.5	187	64.0	4.8

* Indirectly derived norms (for details see pages 141–142).

LIPS AND MOUTH

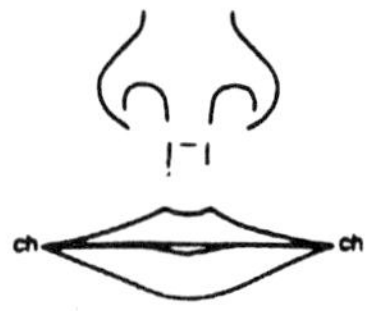

TABLE A-V-11. *Inclination of the labial fissure (ch-ch) (degrees)*

Age (years)	Male			Female		
	N	Mean	SD	*N*	Mean	SD
6	50	0.0	—	50	0.0	—
7	50	1.0	—	50	3.0	2.8
8	51	2.0	—	51	0.0	—
9	51	0.0	—	50	0.0	—
10	50	2.5	0.7	49	2.0	—
11	50	0.0	—	50	5.0	—
12	52	1.0	—	52	2.0	1.0
13	50	0.0	—	49	1.0	—
14	49	1.0	—	51	0.0	—
15	50	0.0	—	51	1.7	1.2
16	50	0.0	—	51	1.5	0.7
17	49	3.0	2.8	51	1.0	—
18	52	2.0	1.4	51	3.0	—

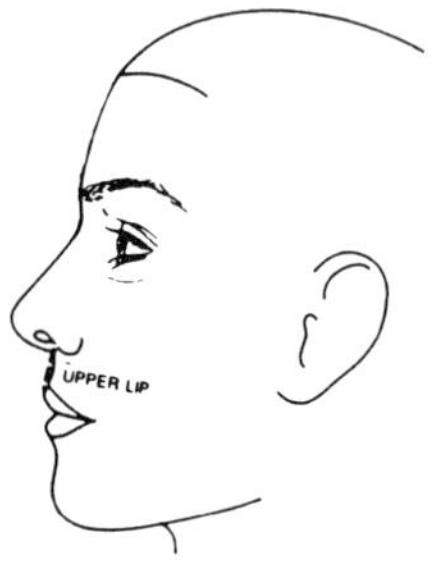

TABLE A-V-12. *Inclination of the upper lip (degrees)*

Age (years)	Male			Female		
	N	Mean	SD	*N*	Mean	SD
0–5 months	8	12.6	4.5	5	17.0	4.7
6–12 months	20	12.0	4.6	8	11.9	5.6
1	18	18.7	5.3	22	15.1	5.8
2	31	15.7	6.0	34	13.6	5.8
3	30	11.1	5.4	30	12.4	6.6
4	30	10.9	6.1	30	12.1	7.0
5	30	8.7	7.8	30	9.0	7.1
6	21	10.9	9.9	20	10.8	9.5
7	19	13.3	9.6	29	9.0	9.1
8	29	9.5	7.0	21	7.8	7.0
9	34	12.1	9.9	24	10.0	9.5
10	21	8.9	11.0	18	11.0	7.3
11	15	7.7	7.8	9	7.6	7.3
12	10	9.3	8.1	5	3.6	16.5
13	25	7.1	7.9	12	6.0	10.7
14	34	5.2	8.1	26	6.4	9.3
15	23	7.6	7.7	9	4.7	7.1
16	24	5.0	10.2	5	3.0	14.3
17	16	7.0	10.5	12	4.2	8.9
18	13	4.9	8.1	8	5.3	9.2
19–25	109	1.5	8.7	199	0.8	7.5

TABLE A-V-13. *Inclination of the lower lip (li-sl) (degrees)*

Age (years)	Male			Female		
	N	Mean	SD	*N*	Mean	SD
1	17	−74.4	8.1	18	−66.6	11.5
2	31	−61.1	23.2	34	−61.4	10.8
3	30	−62.1	12.4	30	−63.8	11.0
4	30	−58.6	12.2	30	−56.1	9.7
5	30	−61.7	11.2	30	−54.1	10.4
6	#	−60.6	11.7	#	−53.7	10.7
7	#	−59.5	12.1	#	−53.3	10.9
8	#	−58.5	12.6	#	−52.9	11.2
9	#	−57.4	13.1	#	−52.5	11.4
10	#	−56.3	13.6	#	−52.2	11.7
11	#	−55.2	14.0	#	−51.8	12.0
12	#	−54.1	14.5	#	−51.4	12.2
13	#	−53.1	15.0	#	−51.0	12.5
14	#	−52.0	15.4	#	−50.6	12.7
15	#	−50.9	15.9	#	−50.2	13.0
16	#	−49.8	16.4	#	−49.8	13.3
17	#	−48.7	16.8	#	−49.4	13.5
18	#	−47.7	17.3	#	−49.0	13.8
19–25	109	−46.6	17.8	199	−48.6	14.1

[#] Indirectly derived norms (for details see pages 141–142).

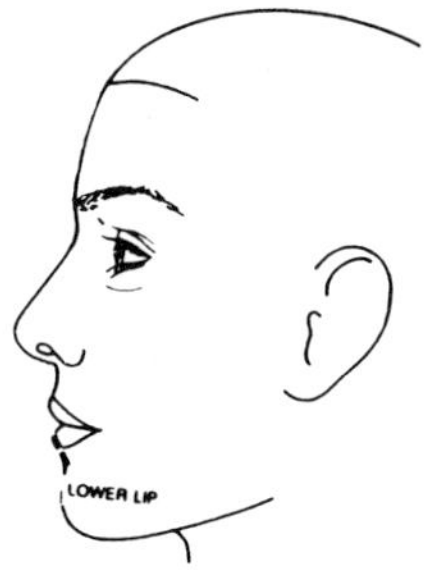

TABLE A-V-14. *Labiomental angle (degrees)*

Age (years)	Male			Female		
	N	Mean	SD	*N*	Mean	SD
1	18	110.7	15.0	21	124.0	15.3
2	31	123.4	12.5	34	124.1	13.0
3	30	124.8	13.3	30	124.9	19.0
4	30	123.8	18.1	30	126.6	10.6
5	30	121.7	13.0	30	134.1	13.3
6	#	121.1	13.6	#	133.2	13.4
7	#	124.4	14.2	#	132.3	13.5
8	#	123.7	14.8	#	131.4	13.5
9	#	123.1	15.4	#	130.5	13.6
10	#	124.9	16.0	#	129.6	13.7
11	#	127.8	16.5	#	128.6	13.8
12	#	130.6	17.1	#	127.7	13.9
13	#	133.3	17.7	#	126.8	13.9
14	#	136.1	18.3	#	125.9	14.0
15	#	138.9	18.9	#	125.0	14.1
16	#	141.7	19.5	#	124.1	14.2
17	#	144.5	20.1	#	123.2	14.3
18	#	147.2	20.7	#	122.3	14.3
19–25	44	113.5	20.7	45	121.4	14.4

[#] Indirectly derived norms (for details see pages 141–142).

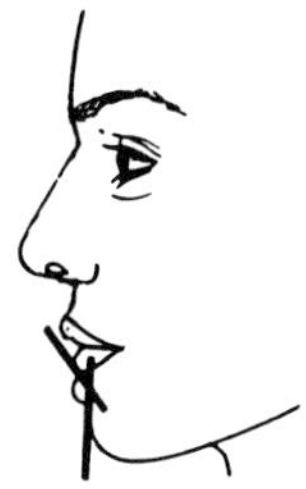

LIPS AND MOUTH

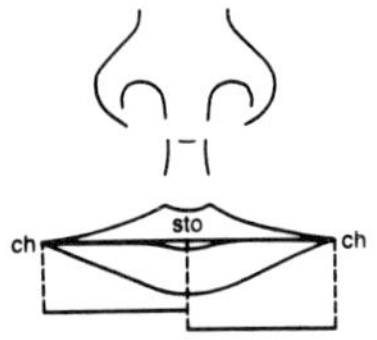

TABLE A–V-15. *Halves of the labial fissure length (ch-sto), right and left (mm)*

Age (years)	Male, right side			Female, right side		
	N	Mean	SD	*N*	Mean	SD
6	50	23.2	2.0	50	22.7	1.6
7	50	23.3	1.6	50	23.7	1.7
8	51	24.9	1.5	51	24.2	1.9
9	51	25.9	1.7	50	25.4	1.9
10	50	26.3	2.0	49	25.7	2.1
11	50	26.8	2.4	50	26.7	2.5
12	52	28.0	2.3	53	27.0	2.1
13	50	28.4	1.9	49	27.9	1.9
14	49	28.5	2.1	51	27.3	1.9
15	50	30.0	2.2	51	28.0	2.4
16	50	30.0	1.8	51	28.0	2.4
17	49	31.2	2.6	51	28.6	2.4
18	52	30.7	2.3	51	28.6	2.2
19–25	109	32.8	2.1	200	29.6	2.5

Age (years)	Male, left side			Female, left side		
	N	Mean	SD	*N*	Mean	SD
6	50	23.2	2.0	50	22.6	1.6
7	50	23.7	1.6	50	23.4	1.8
8	51	24.7	1.5	51	24.0	1.9
9	51	25.3	1.7	50	24.9	1.9
10	50	25.8	2.2	49	25.0	2.0
11	50	26.2	2.4	50	26.2	2.3
12	52	27.3	2.1	53	26.5	1.9
13	50	28.0	1.9	49	27.1	1.8
14	49	28.4	2.3	51	26.8	1.6
15	50	29.5	2.4	51	27.7	2.4
16	50	29.7	1.7	51	27.5	2.1
17	49	30.8	2.3	51	27.9	2.2
18	52	30.4	2.3	51	28.1	2.3
19–25	109	32.1	2.1	200	29.2	2.3

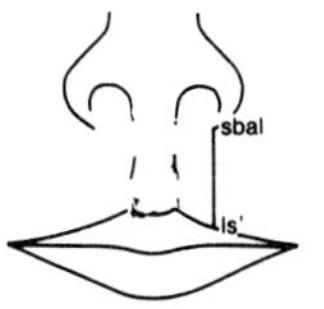

TABLE A-V-16. *Lateral upper-lip height (sbal-ls'), right and left (mm)*

Age (years)	Male, right side			Female, right side		
	N	Mean	SD	*N*	Mean	SD
1	18	10.8	0.8	18	10.3	1.3
2	31	12.2	1.7	30	11.5	1.3
3	30	12.5	1.1	30	12.0	1.3
4	30	13.9	1.2	30	13.1	1.2
5	30	13.9	1.4	30	13.3	1.9
6	50	14.3	1.6	50	13.5	1.6
7	50	14.0	1.6	50	14.0	1.5
8	51	14.5	1.5	51	13.8	1.8
9	51	13.8	1.8	50	14.0	1.4
10	50	14.9	1.3	49	14.2	1.5
11	50	15.2	1.9	51	14.0	1.7
12	52	15.3	1.7	53	14.2	2.0
13	50	15.0	1.5	49	14.2	1.9
14	49	15.6	2.0	51	14.2	1.7
15	50	15.9	1.7	51	14.6	1.6
16	50	16.2	1.8	51	15.2	1.9
17	49	15.6	2.1	51	15.1	2.0
18	52	16.1	2.0	51	14.6	2.1
19–25	109	17.0	2.1	200	14.7	1.7

Age (years)	Male, left side			Female, left side		
	N	Mean	SD	*N*	Mean	SD
1	18	10.8	0.8	18	10.3	1.3
2	31	12.2	1.8	30	11.5	1.3
3	30	12.5	1.2	30	12.0	1.3
4	30	14.0	1.2	30	13.1	1.2
5	30	13.9	1.3	30	13.3	1.8
6	50	14.3	1.6	50	13.5	1.6
7	50	14.0	1.6	50	14.0	1.4
8	51	14.5	1.5	51	13.8	1.8
9	51	13.8	1.7	50	14.0	1.4
10	50	14.9	1.3	49	14.2	1.4
11	50	15.1	1.9	51	13.9	1.8
12	52	15.3	1.6	53	14.2	2.0
13	50	15.0	1.5	49	14.3	1.9
14	49	15.4	2.0	51	14.2	1.8
15	50	15.9	1.8	51	14.5	1.5
16	50	16.1	1.9	51	15.2	1.8
17	49	15.4	2.1	51	15.1	2.1
18	52	16.1	2.0	51	14.7	2.2
19–25	109	17.0	2.1	200	14.7	1.8

LIPS AND MOUTH

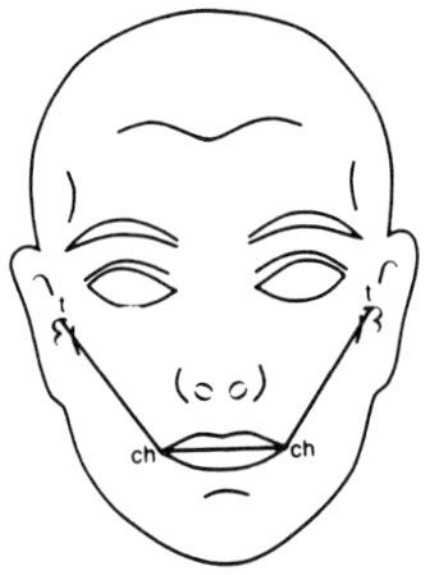

TABLE A-V-17. *Labio-aural distance (ch-t), right and left (mm)*

Age (years)	Male, right side			Female, right side		
	N	Mean	SD	*N*	Mean	SD
6	24	92.4	3.6	25	89.9	3.9
7	24	93.7	4.5	37	90.5	3.2
8	41	94.9	3.4	39	92.7	4.2
9	47	96.3	4.0	38	94.1	3.8
10	37	96.9	5.8	36	95.0	3.9
11	36	98.9	4.2	40	96.1	3.7
12	38	100.1	4.8	35	98.3	4.3
13	45	101.0	5.5	48	98.2	4.2
14	46	102.8	5.5	43	99.5	4.6
15	44	104.9	4.9	43	100.0	4.6
16	48	105.9	4.3	34	100.6	4.5
17	44	107.6	4.2	33	99.2	4.0
18	26	105.8	5.1	17	99.4	4.3
19–25	109	114.5	5.0	194	104.1	4.6

Age (years)	Male, left side			Female, left side		
	N	Mean	SD	*N*	Mean	SD
6	24	92.7	3.9	25	89.8	3.9
7	24	94.2	6.1	37	90.7	3.0
8	41	94.6	3.5	39	92.2	4.0
9	47	96.2	4.0	38	94.0	3.8
10	37	97.1	5.5	36	94.3	3.8
11	36	99.2	4.3	40	95.7	4.2
12	38	99.8	5.0	35	97.7	4.1
13	45	101.3	5.5	48	98.0	4.2
14	46	103.7	6.3	43	99.7	4.6
15	44	105.3	4.8	43	99.6	3.9
16	48	105.3	5.2	34	100.1	4.5
17	44	107.2	4.9	33	98.5	4.5
18	26	106.1	5.1	17	99.5	4.9
19–25	109	113.6	4.8	194	103.5	4.9

LIPS AND MOUTH

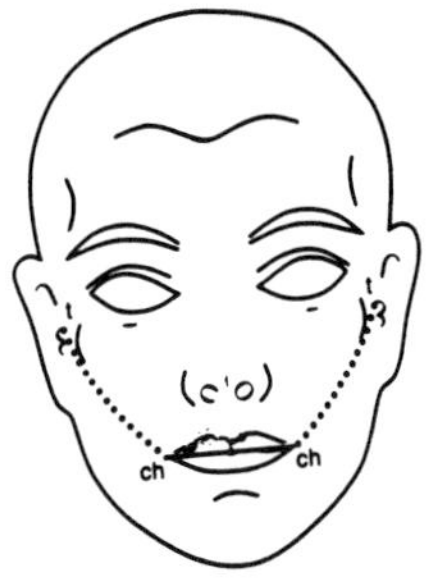

TABLE A-V-18. *Labio-aural surface distance (ch-t surf), right and left (mm)*

Age (years)	Male, right side			Female, right side		
	N	Mean	SD	*N*	Mean	SD
6	24	102.1	4.1	25	100.4	3.5
7	24	103.1	4.7	37	101.7	3.3
8	41	104.8	3.2	39	102.6	5.3
9	47	106.3	4.5	38	104.4	4.1
10	37	107.5	5.6	36	105.5	5.1
11	36	109.4	4.4	40	106.1	4.0
12	38	109.6	5.3	35	107.3	5.7
13	45	110.6	5.8	48	108.1	5.0
14	46	113.1	6.3	43	109.9	5.5
15	44	115.8	6.1	43	110.3	4.8
16	48	115.7	4.9	34	110.6	5.0
17	44	117.8	4.7	33	110.8	4.8
18	26	116.2	4.8	17	110.5	4.8
19–25	109	125.2	5.4	192	115.3	5.0

Age (years)	Male, left side			Female, left side		
	N	Mean	SD	*N*	Mean	SD
6	24	101.5	4.1	25	99.5	3.5
7	24	102.7	5.1	37	101.1	3.7
8	41	104.0	3.7	39	101.7	5.1
9	47	105.7	4.7	38	103.9	4.6
10	37	106.1	5.2	36	104.0	4.3
11	36	107.8	4.2	40	105.0	4.2
12	38	108.5	4.5	35	107.2	4.8
13	45	110.0	5.6	48	107.3	4.8
14	46	111.9	6.7	43	109.5	5.7
15	44	114.0	5.3	43	109.6	4.5
16	48	113.9	4.7	34	110.4	4.9
17	44	115.8	4.8	33	109.8	4.6
18	26	114.6	5.7	17	110.1	5.4
19–25	109	124.5	5.4	192	114.9	5.0

Ears

EARS

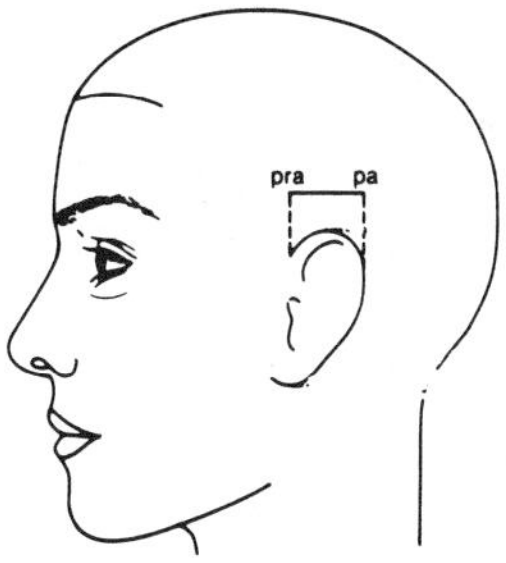

TABLE A–VI-1. *Width of the auricle (pra-pa), right and left (mm)*

Age (years)	Male, right side			Female, right side		
	N	Mean	SD	*N*	Mean	SD
1	18	32.7	1.6	18	31.7	2.4
2	31	33.4	1.9	31	31.8	1.7
3	30	34.3	1.7	30	32.7	1.5
4	30	34.8	1.8	30	31.6	2.1
5	30	34.1	2.2	30	32.5	2.7
6	50	34.3	2.3	50	33.4	1.9
7	50	34.6	2.0	50	33.5	1.9
8	51	34.8	2.1	51	33.6	1.7
9	51	35.1	2.2	50	33.4	2.1
10	50	35.5	2.7	49	34.0	1.8
11	50	35.3	2.3	51	33.2	2.4
12	52	35.3	2.3	53	34.1	2.0
13	50	35.9	2.3	49	33.5	2.2
14	49	35.6	2.5	51	33.1	2.1
15	50	36.3	2.1	51	33.5	2.1
16	50	35.4	2.7	51	33.6	2.2
17	49	36.0	2.6	51	33.5	2.6
18	52	36.2	2.1	50	34.1	2.6
19–25	109	36.9	2.5	200	33.5	2.3

Age (years)	Male, left side			Female, left side		
	N	Mean	SD	*N*	Mean	SD
1	18	32.8	1.6	18	31.7	2.4
2	31	33.6	1.9	31	31.8	1.4
3	30	34.2	1.7	30	32.8	1.5
4	30	34.8	2.0	30	31.8	1.8
5	30	33.9	2.1	30	32.7	2.0
6	50	33.9	2.2	50	33.0	1.7
7	50	34.3	2.3	50	33.2	1.7
8	51	34.7	2.2	51	33.3	2.1
9	51	34.9	2.1	50	33.1	1.9
10	50	35.4	2.4	49	33.6	2.0
11	50	34.6	2.5	51	32.8	2.4
12	52	35.2	2.3	53	33.7	2.3
13	50	35.7	2.0	49	33.3	2.2
14	49	35.3	2.5	51	32.4	1.7
15	50	35.8	2.1	51	33.3	1.9
16	50	35.3	2.8	51	33.1	2.2
17	49	35.9	2.2	51	33.2	2.7
18	52	35.4	2.2	51	33.5	2.1
19–25	109	36.4	2.4	200	33.7	2.2

EARS

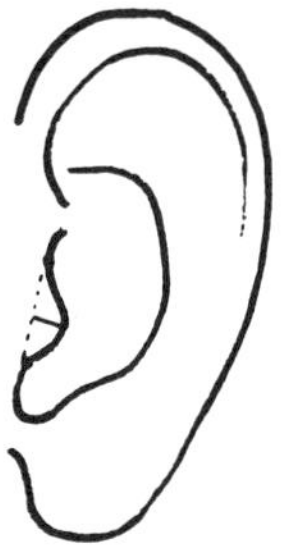

TABLE A–VI-2. *Height of the tragus, right and left (mm)*

Age (years)	Male, right side			Female, right side		
	N	Mean	SD	*N*	Mean	SD
6	50	2.3	0.5	50	2.4	0.6
7	50	2.4	0.6	50	2.4	0.6
8	51	2.5	0.5	51	2.6	0.8
9	51	2.3	0.7	50	2.5	0.6
10	50	2.7	0.8	49	2.5	0.6
11	50	2.9	0.8	51	2.5	0.6
12	52	3.0	0.7	53	2.8	0.6
13	50	2.9	0.8	49	2.9	0.7
14	49	2.7	0.8	51	2.7	0.6
15	50	3.2	0.9	51	3.0	0.6
16	50	3.1	0.6	51	2.9	0.8
17	49	3.1	0.7	51	2.9	0.7
18	52	2.9	0.6	51	2.9	0.7

Age (years)	Male, left side			Female, left side		
	N	Mean	SD	*N*	Mean	SD
6	50	2.4	0.5	50	2.4	0.6
7	50	2.5	0.6	50	2.5	0.7
8	51	2.6	0.6	51	2.5	0.8
9	51	2.5	0.7	50	2.6	0.7
10	50	2.7	0.8	49	2.6	0.6
11	50	2.8	0.8	51	2.6	0.6
12	52	3.0	0.8	53	2.8	0.7
13	50	3.0	0.8	49	3.1	0.7
14	49	2.8	0.9	51	2.8	0.7
15	50	3.3	0.8	51	3.2	0.8
16	50	3.1	0.7	51	2.8	0.8
17	49	3.2	0.8	51	3.0	0.7
18	52	3.0	0.7	51	2.8	0.7

EARS

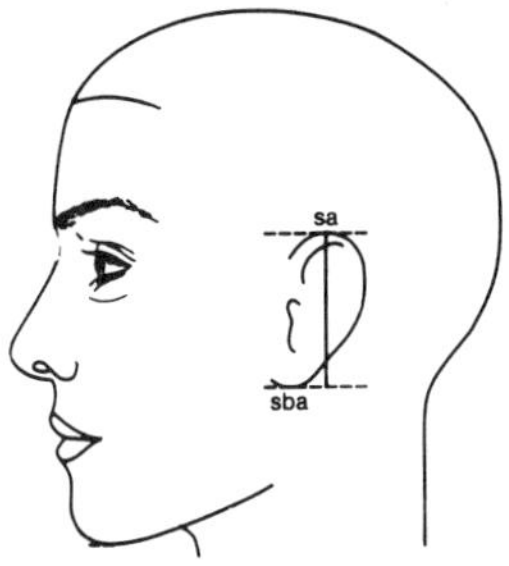

TABLE A–VI-3. *Length of the auricle (sa-sba), right and left (mm)*

Age (years)	Male, right side			Female, right side		
	N	Mean	SD	*N*	Mean	SD
1	18	46.9	2.3	18	45.4	2.0
2	31	48.2	2.1	31	47.3	2.3
3	30	50.5	2.8	30	48.6	2.1
4	30	53.5	2.9	30	51.0	2.9
5	30	54.2	2.6	30	51.8	2.4
6	50	55.0	3.0	50	53.8	2.9
7	50	55.4	3.1	50	54.1	2.4
8	51	56.5	3.0	51	54.8	2.8
9	51	57.4	3.1	50	55.7	3.6
10	50	58.5	3.0	49	56.0	3.2
11	50	58.7	3.1	51	55.6	3.3
12	52	59.6	3.6	53	57.7	3.5
13	50	60.8	4.3	49	58.4	3.6
14	49	61.4	3.6	51	57.6	3.3
15	50	63.2	3.2	51	58.0	3.8
16	50	61.2	3.6	51	57.8	3.1
17	49	62.0	3.7	51	58.5	3.3
18	52	62.2	3.5	50	58.3	3.2
19–25	109	62.7	3.6	200	59.6	3.4

Age (years)	Male, left side			Female, left side		
	N	Mean	SD	*N*	Mean	SD
1	18	46.9	2.3	18	45.4	2.0
2	31	48.1	2.1	31	47.5	2.2
3	30	50.7	2.7	30	48.9	2.4
4	30	52.9	2.7	30	50.1	2.7
5	30	53.6	2.8	30	51.0	2.6
6	50	55.0	2.9	50	54.2	2.9
7	50	55.0	3.7	50	54.0	2.4
8	51	56.7	3.2	51	55.0	3.0
9	51	57.3	3.5	50	56.0	3.3
10	50	58.6	3.1	49	56.2	2.9
11	50	58.7	3.7	51	55.8	3.6
12	52	59.1	3.6	53	57.4	3.6
13	50	61.0	4.0	49	58.5	3.9
14	49	61.0	3.7	51	57.4	3.3
15	50	62.4	2.9	51	57.9	3.5
16	50	61.4	4.0	51	57.9	2.8
17	49	62.0	4.1	51	59.0	3.6
18	52	62.4	3.7	51	58.5	3.4
19–25	109	62.9	3.5	200	59.9	3.5

EARS

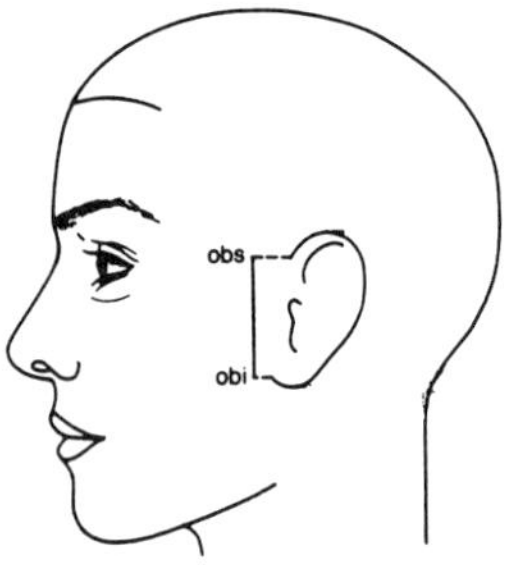

TABLE A–VI-4. *Morphological width of the ear (obs-obi), right and left (mm)*

Age (years)	Male, right side			Female, right side		
	N	Mean	SD	*N*	Mean	SD
1	18	35.5	1.1	18	34.3	1.5
2	31	36.3	1.5	31	35.6	1.5
3	30	37.5	1.9	30	36.4	1.4
4	30	41.8	2.5	30	39.5	1.8
5	30	41.1	2.7	30	41.1	2.8
6	27	41.5	2.7	29	41.2	3.4
7	24	42.7	2.8	38	41.9	2.4
8	43	42.8	3.2	39	43.0	3.1
9	47	44.3	3.0	38	43.9	3.8
10	37	45.0	2.9	36	43.7	3.5
11	37	46.1	2.6	41	44.7	3.0
12	45	45.6	3.1	43	46.3	2.9
13	46	47.8	4.1	49	47.5	3.4
14	46	48.1	3.5	43	46.0	3.1
15	44	49.3	4.0	43	46.9	3.8
16	49	48.9	3.0	34	48.0	2.9
17	46	50.9	4.0	33	49.0	3.1
18	34	50.2	3.9	17	48.5	4.6
Age (years)	Male, left side			Female, left side		
	N	Mean	SD	*N*	Mean	SD
1	18	35.3	1.0	18	34.3	1.5
2	31	36.2	1.6	31	35.8	1.4
3	30	37.5	1.7	30	36.4	1.6
4	30	42.3	2.2	30	39.9	1.8
5	30	41.8	2.6	30	40.9	2.6
6	27	42.4	2.8	29	41.6	3.0
7	24	43.0	2.8	38	42.4	2.4
8	43	43.8	3.5	39	43.3	3.1
9	47	44.8	2.6	38	44.3	3.1
10	37	46.0	3.3	36	44.4	3.1
11	37	46.6	2.7	41	45.7	3.1
12	45	45.5	3.4	43	46.7	3.1
13	46	48.5	3.4	49	48.0	3.2
14	46	49.4	3.7	43	47.2	3.0
15	44	49.7	3.2	43	46.8	3.5
16	49	49.6	3.3	34	48.6	2.8
17	46	51.3	3.7	33	49.6	3.0
18	34	50.2	4.2	17	48.9	4.5

TABLE A–VI-5. *Upper naso-aural distance (n-obs), right and left (mm)*

Age (years)	Male, right side			Female, right side		
	N	Mean	SD	*N*	Mean	SD
6	50	104.4	5.0	50	102.0	4.4
7	50	106.0	4.9	50	105.4	4.4
8	51	109.5	4.3	51	106.1	5.0
9	51	112.1	4.9	50	108.4	5.6
10	50	111.4	5.2	49	109.1	4.6
11	50	113.5	6.0	51	110.7	5.5
12	52	114.8	5.0	53	111.5	4.6
13	50	117.0	4.8	49	113.6	4.7
14	49	119.5	4.9	51	113.7	4.9
15	50	121.9	5.1	51	115.5	5.5
16	50	123.0	3.8	51	114.6	4.6
17	49	124.1	5.6	51	114.9	4.7
18	52	120.8	6.2	51	113.8	4.8

Age (years)	Male, left side			Female, left side		
	N	Mean	SD	*N*	Mean	SD
6	50	104.1	4.3	50	101.0	4.9
7	50	105.9	5.2	50	105.3	4.8
8	51	108.5	4.4	51	106.0	4.7
9	51	110.8	4.4	50	107.7	5.6
10	50	110.4	4.7	49	108.7	3.7
11	50	113.4	5.7	51	110.6	4.7
12	52	114.0	5.2	53	110.6	4.4
13	50	117.2	5.1	49	113.3	4.5
14	49	119.0	4.6	51	112.9	4.5
15	50	121.6	3.8	51	115.0	5.2
16	50	121.8	4.5	51	113.9	4.8
17	49	123.9	5.8	51	114.1	4.6
18	52	120.4	6.1	51	113.6	5.8

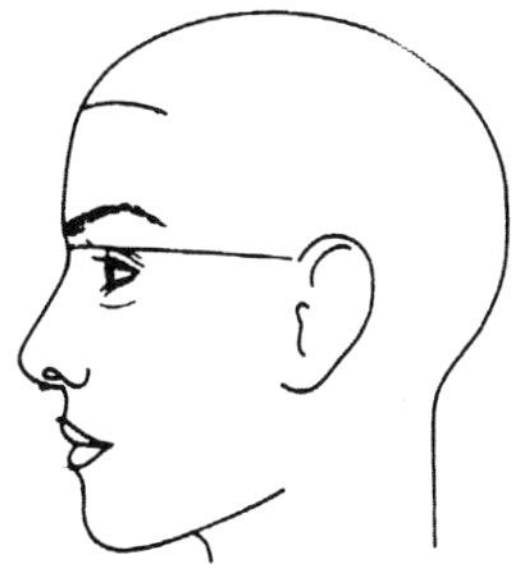

EARS

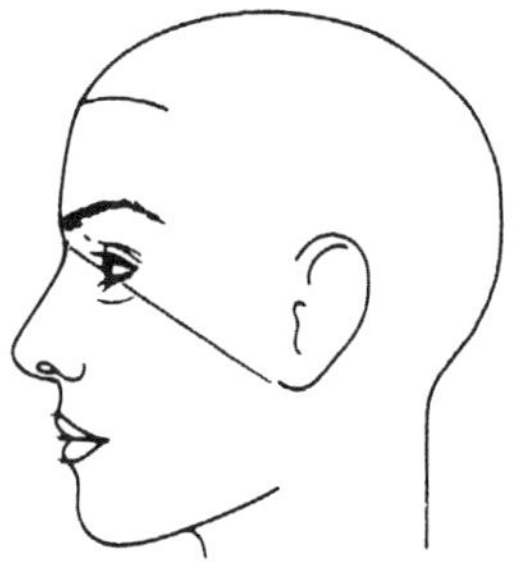

TABLE A–VI-6. *Lower naso-aural distance (n-obi), right and left (mm)*

Age (years)	Male, right side			Female, right side		
	N	Mean	SD	*N*	Mean	SD
6	50	101.4	3.8	50	98.4	3.0
7	50	101.6	4.5	50	100.4	3.5
8	51	104.6	4.1	51	101.3	3.7
9	51	105.3	4.2	50	103.3	4.8
10	50	106.9	4.5	49	105.1	4.1
11	50	109.4	4.7	51	107.3	4.4
12	52	109.8	4.1	53	109.0	4.4
13	50	111.6	5.2	49	110.5	4.5
14	49	114.4	5.2	51	110.1	3.6
15	50	117.1	5.0	51	111.6	5.0
16	50	118.7	4.2	51	111.7	3.8
17	49	120.4	4.1	51	111.8	3.7
18	52	118.1	4.4	51	113.1	4.5

Age (years)	Male, left side			Female, left side		
	N	Mean	SD	*N*	Mean	SD
6	50	101.8	4.2	50	99.3	2.9
7	50	102.8	4.2	50	102.0	3.7
8	51	105.5	4.0	51	102.8	3.5
9	51	106.6	4.1	50	104.9	4.5
10	50	108.3	3.9	49	106.5	3.5
11	50	110.8	4.9	51	107.9	4.1
12	52	110.9	4.5	53	110.6	4.4
13	50	113.6	5.2	49	111.7	3.9
14	49	115.6	5.7	51	111.7	3.8
15	50	118.9	4.3	51	112.7	4.7
16	50	119.4	4.5	51	113.1	4.1
17	49	122.0	4.4	51	112.9	3.4
18	52	119.1	4.0	51	114.2	4.9

EARS

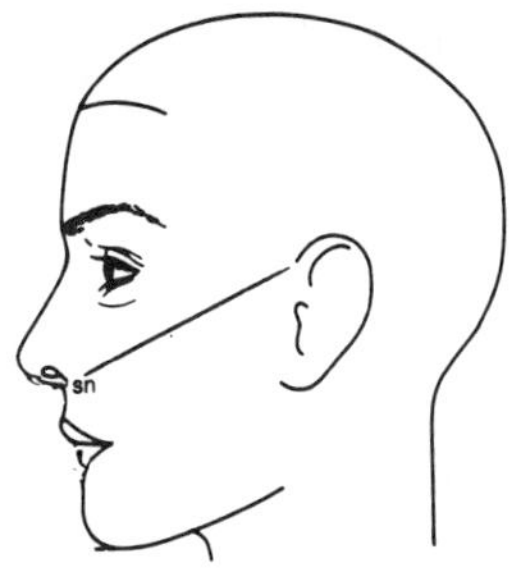

TABLE A–VI-7. *Upper subnasale–aural distance (sn-obs), right and left (mm)*

Age (years)	Male, right side			Female, right side		
	N	Mean	SD	*N*	Mean	SD
6	50	113.8	5.7	50	110.9	5.1
7	50	116.2	5.2	50	115.2	4.6
8	51	120.2	4.9	51	115.7	5.9
9	51	123.3	5.3	50	119.3	5.6
10	50	123.2	5.8	49	120.4	4.7
11	50	124.8	5.9	51	122.4	6.2
12	52	127.9	5.3	53	123.6	5.2
13	50	130.6	5.6	49	126.8	5.0
14	49	132.7	5.6	51	126.3	4.9
15	50	135.7	6.0	51	128.2	6.2
16	50	137.5	4.7	51	127.9	5.6
17	49	138.9	5.1	51	127.4	4.9
18	52	134.8	6.1	51	125.5	5.2

Age (years)	Male, left side			Female, left side		
	N	Mean	SD	*N*	Mean	SD
6	50	112.8	5.6	50	109.9	5.5
7	50	115.4	6.1	50	114.1	5.0
8	51	119.0	4.8	51	115.3	5.2
9	51	121.4	5.4	50	118.5	5.6
10	50	121.1	5.9	49	118.8	4.1
11	50	124.1	5.7	51	121.7	5.9
12	52	126.2	5.5	53	122.2	5.3
13	50	129.5	5.7	49	125.3	5.0
14	49	131.7	6.2	51	125.5	4.2
15	50	134.7	5.0	51	127.0	4.7
16	50	135.6	5.2	51	126.6	5.5
17	49	137.2	5.2	51	126.6	5.1
18	52	134.6	6.2	51	125.1	5.9

EARS

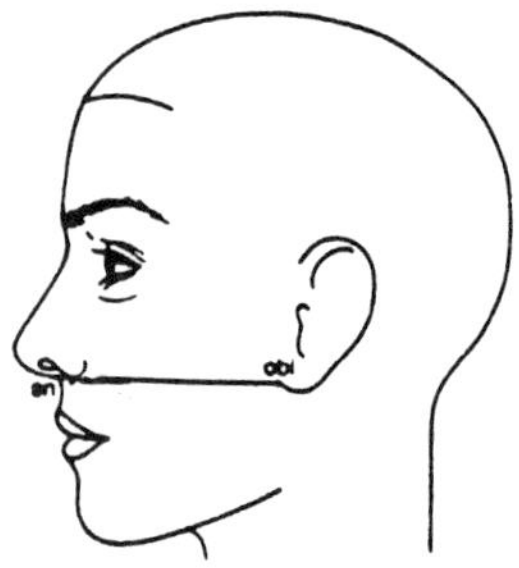

TABLE A–VI-8. *Lower subnasale–aural distance (sn-obi), right and left (mm)*

Age (years)	Male, right side			Female, right side		
	N	Mean	SD	*N*	Mean	SD
1	18	85.8	3.3	19	84.1	2.3
2	30	88.6	4.3	30	87.4	2.9
3	30	90.7	4.2	30	88.4	3.6
4	30	93.7	3.9	30	90.0	2.6
5	30	94.8	3.2	30	92.3	3.1
6	50	95.7	3.9	50	93.2	3.8
7	50	96.8	4.3	50	94.5	3.5
8	51	99.1	3.8	51	95.8	3.9
9	51	99.9	4.3	50	97.6	3.6
10	50	101.3	4.3	49	99.3	4.1
11	50	103.5	4.5	51	101.5	4.3
12	52	105.7	4.5	53	103.0	4.5
13	50	105.8	5.0	49	103.6	3.8
14	49	108.1	5.4	51	103.9	3.6
15	50	111.2	5.0	51	105.5	4.4
16	50	112.6	4.2	51	105.9	4.2
17	49	114.0	4.3	51	105.5	4.2
18	52	112.5	4.3	51	106.3	4.5
19–25	109	118.6	4.3	194	107.1	4.1

Age (years)	Male, left side			Female, left side		
	N	Mean	SD	*N*	Mean	SD
1	18	85.7	3.4	19	84.1	2.3
2	30	88.6	4.4	30	87.3	3.2
3	30	90.7	4.4	30	88.4	3.8
4	30	94.0	3.5	30	91.2	3.0
5	30	95.5	3.3	30	92.6	3.5
6	50	95.6	4.3	50	93.6	3.7
7	50	97.2	3.9	50	95.3	3.4
8	51	98.8	3.5	51	96.3	3.4
9	51	99.6	4.1	50	98.2	3.7
10	50	101.1	3.6	49	99.6	3.6
11	50	103.9	4.4	51	101.4	4.3
12	52	105.5	4.8	53	104.2	4.5
13	50	106.5	5.1	49	103.9	3.5
14	49	107.9	5.8	51	104.6	3.6
15	50	111.0	5.0	51	105.6	3.9
16	50	111.8	4.7	51	106.1	5.0
17	49	113.8	4.4	51	105.5	4.2
18	52	113.0	5.2	51	106.9	4.9
19–25	109	118.0	4.4	194	107.2	4.5

EARS

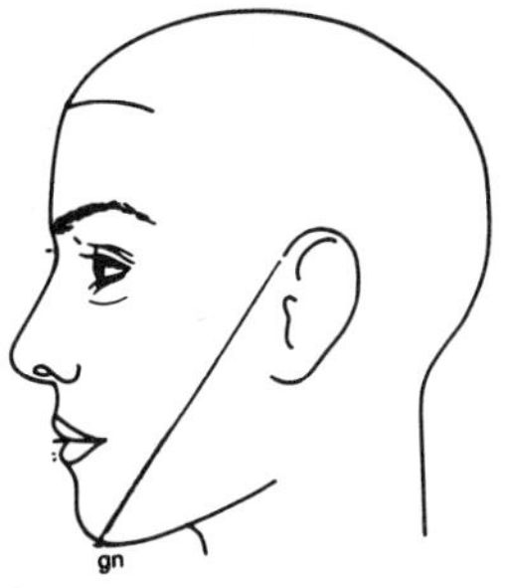

TABLE A–VI-9. *Upper gnathion–aural distance (gn-obs), right and left (mm)*

Age (years)	Male, right side			Female, right side		
	N	Mean	SD	*N*	Mean	SD
1	18	112.1	3.7	18	109.0	4.2
2	30	114.7	3.2	30	113.3	2.6
3	30	118.3	4.8	30	116.5	3.4
4	30	125.2	5.1	30	121.5	4.1
5	30	127.9	4.5	30	124.8	3.8
6	50	128.7	6.3	50	126.5	5.5
7	50	131.7	6.6	50	131.5	5.3
8	51	136.9	5.5	51	132.0	6.2
9	51	139.8	6.3	50	136.0	6.7
10	50	140.8	6.9	49	136.9	6.2
11	50	143.5	6.4	51	139.9	7.7
12	52	146.8	5.4	53	141.9	6.2
13	50	150.4	7.0	49	145.5	5.5
14	49	153.8	7.2	51	146.1	5.4
15	50	157.8	7.8	51	147.3	7.6
16	50	159.0	5.4	51	148.5	7.5
17	49	162.3	6.8	51	149.0	6.3
18	52	159.2	7.8	51	147.4	6.2

Age (years)	Male, left side			Female, left side		
	N	Mean	SD	*N*	Mean	SD
1	18	112.1	3.6	18	109.1	4.3
2	30	114.6	3.2	30	113.4	2.8
3	30	118.4	4.9	30	116.5	3.2
4	30	125.6	4.8	30	121.4	3.8
5	30	128.1	4.4	30	125.2	4.3
6	50	127.3	5.3	50	125.7	5.2
7	50	131.5	6.6	50	130.2	5.2
8	51	135.3	4.7	51	130.8	5.0
9	51	138.2	6.1	50	134.5	6.5
10	50	138.5	6.1	49	135.7	5.6
11	50	142.8	6.1	51	139.1	7.5
12	52	145.3	5.5	53	140.4	5.8
13	50	149.0	6.7	49	143.8	5.4
14	49	152.3	7.9	51	145.1	5.4
15	50	156.0	6.8	51	146.3	6.3
16	50	156.0	5.9	51	146.9	6.4
17	49	159.6	6.8	51	147.1	5.4
18	52	158.0	7.5	51	146.6	6.3

EARS

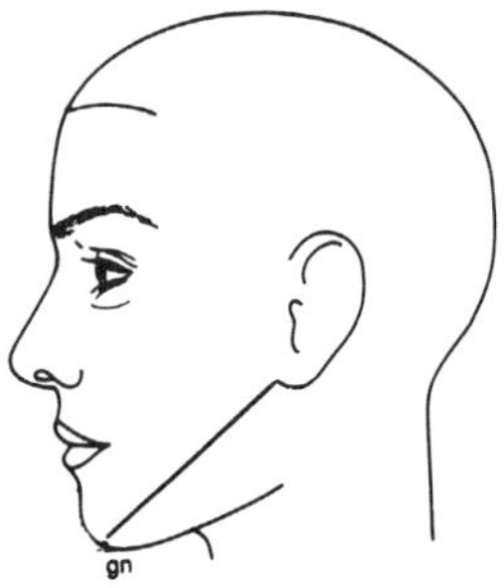

TABLE A-VI-10. *Lower gnathion–aural distance (gn-obi), right and left (mm)*

Age (years)	Male, right side			Female, right side		
	N	Mean	SD	*N*	Mean	SD
1	18	82.9	3.5	18	81.9	2.8
2	30	86.8	3.4	30	85.9	3.9
3	30	90.2	4.4	30	88.5	4.7
4	30	91.1	4.4	30	87.9	4.5
5	30	93.9	3.9	30	91.9	6.1
6	50	94.5	4.2	50	92.4	4.1
7	50	96.4	5.2	50	95.3	3.9
8	51	99.3	3.9	51	95.4	4.2
9	51	100.2	4.5	50	97.9	4.8
10	50	102.4	5.5	49	100.0	4.9
11	50	104.4	4.5	51	102.0	5.5
12	52	107.5	4.9	53	104.0	6.1
13	50	108.1	6.3	49	104.6	4.7
14	49	110.7	7.0	51	106.5	4.6
15	50	114.3	7.1	51	107.7	5.4
16	50	115.6	5.4	51	108.5	5.7
17	49	119.1	5.8	51	108.6	4.6
18	52	117.4	6.4	51	109.4	6.2
19–25	109	121.8	5.3	200	109.1	5.4

Age (years)	Male, left side			Female, left side		
	N	Mean	SD	*N*	Mean	SD
1	18	82.9	3.5	18	81.9	2.8
2	30	86.6	3.5	30	85.7	3.7
3	30	90.2	4.4	30	88.8	4.9
4	30	91.6	4.0	30	88.8	4.5
5	30	94.2	3.8	30	91.7	6.1
6	50	94.1	4.0	50	93.1	4.0
7	50	97.4	4.8	50	95.1	4.3
8	51	99.0	3.6	51	95.7	3.7
9	51	100.5	4.8	50	98.2	5.0
10	50	101.7	5.4	49	99.7	4.9
11	50	104.7	4.5	51	101.5	5.5
12	52	107.9	5.2	53	103.8	5.5
13	50	108.0	6.3	49	104.6	4.8
14	49	111.0	7.3	51	106.6	5.1
15	50	114.3	6.7	51	108.0	4.4
16	50	114.6	5.8	51	108.3	5.8
17	49	117.8	5.1	51	108.4	4.2
18	52	117.6	6.5	51	109.3	5.6
19–25	109	122.1	5.3	200	109.3	5.6

TABLE A–VI-11. *Occipito-aural distance (op-po), right and left (mm)*

Age (years)	Male, right side			Female, right side		
	N	Mean	SD	*N*	Mean	SD
6	31	88.8	6.6	36	87.6	5.7
7	33	89.2	6.3	42	87.7	5.0
8	45	90.3	5.5	40	89.1	4.2
9	47	89.1	4.3	38	89.5	4.6
10	37	91.0	3.3	37	89.2	4.9
11	39	91.6	5.3	41	90.5	4.4
12	50	93.2	5.3	50	91.9	4.1
13	47	93.6	4.5	49	91.0	4.5
14	46	91.6	4.6	43	90.9	4.1
15	44	93.6	4.7	43	89.4	5.1
16	49	93.9	4.3	35	91.6	4.2
17	48	94.9	5.8	36	91.8	5.2
18	50	95.4	5.0	38	93.4	5.6

Age (years)	Male, left side			Female, left side		
	N	Mean	SD	*N*	Mean	SD
6	31	89.3	6.8	36	87.6	5.7
7	33	89.2	6.2	42	88.1	5.2
8	45	90.2	5.4	40	89.4	4.5
9	47	89.4	4.2	38	89.7	4.4
10	37	91.8	4.2	37	89.7	4.8
11	39	92.0	5.2	41	91.0	4.5
12	50	93.3	5.7	50	92.0	4.0
13	47	93.7	4.4	49	90.8	4.5
14	46	92.0	4.4	43	90.8	4.0
15	44	93.7	5.0	43	89.4	5.1
16	49	93.9	4.5	35	91.7	4.2
17	48	95.6	5.4	36	91.8	5.2
18	50	95.3	5.1	38	93.7	5.5

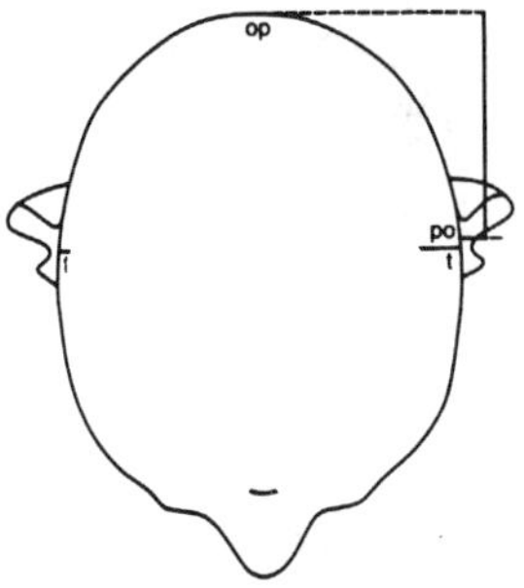

EARS

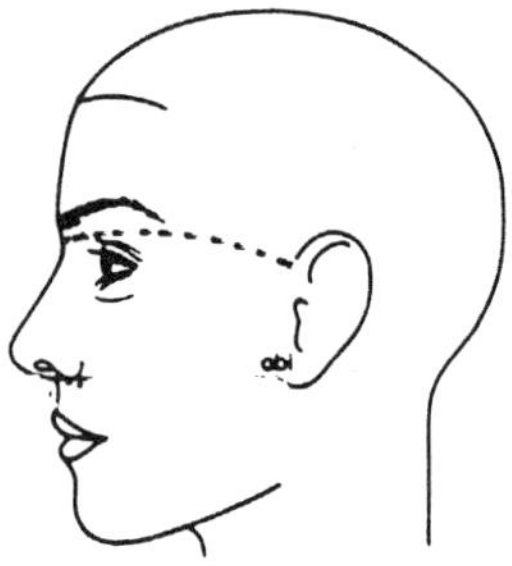

TABLE A-VI-12. *Upper naso-aural surface distance (n-obs surf), right and left (mm)*

Age (years)	Male, right side			Female, right side		
	N	Mean	SD	*N*	Mean	SD
6	50	124.9	5.9	50	121.5	4.9
7	50	126.5	5.3	50	124.7	5.2
8	51	129.9	4.5	51	125.1	5.2
9	51	131.9	5.8	50	128.7	7.1
10	50	132.3	5.8	49	129.7	4.7
11	50	133.4	6.3	51	130.4	5.4
12	52	135.7	5.5	53	132.5	5.8
13	50	137.3	5.6	49	134.4	5.5
14	49	139.8	6.6	51	134.6	5.3
15	50	142.5	6.0	51	136.1	5.4
16	50	143.6	4.7	51	136.0	5.8
17	49	145.6	5.7	51	135.2	4.7
18	52	142.4	5.2	51	135.7	4.8

Age (years)	Male, left side			Female, left side		
	N	Mean	SD	*N*	Mean	SD
6	50	123.7	5.9	50	120.3	5.2
7	50	124.6	6.0	50	124.6	5.4
8	51	128.7	5.3	51	124.9	5.7
9	51	131.1	5.7	50	128.5	7.2
10	50	131.3	5.5	49	128.8	4.3
11	50	133.0	5.5	51	129.8	6.2
12	52	134.8	5.3	53	131.5	4.8
13	50	138.6	5.6	49	134.1	5.6
14	49	140.2	6.5	51	133.9	5.2
15	50	142.8	5.7	51	136.4	5.5
16	50	142.3	5.7	51	135.4	5.9
17	49	145.3	6.0	51	135.1	5.3
18	52	141.5	5.5	51	134.6	6.0

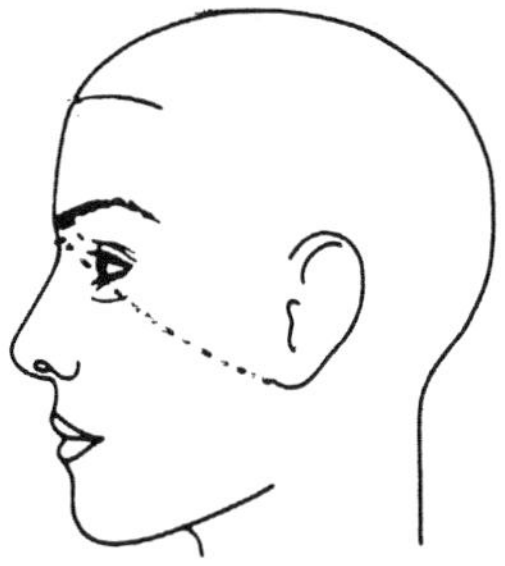

TABLE A–VI-13. *Lower naso-aural surface distance (n-obi surf), right and left (mm)*

Age (years)	Male, right side			Female, right side		
	N	Mean	SD	*N*	Mean	SD
6	50	122.8	5.5	50	120.4	3.8
7	50	123.2	5.4	50	122.6	4.1
8	51	127.1	4.8	51	124.7	4.8
9	51	128.2	5.6	50	126.9	5.7
10	50	130.1	5.2	49	129.1	5.0
11	50	133.3	5.3	51	130.9	5.2
12	52	132.8	5.7	53	133.9	5.4
13	50	135.5	5.5	49	134.8	4.8
14	49	137.5	5.8	51	135.1	5.3
15	50	140.8	5.7	51	136.8	5.0
16	50	141.7	5.0	51	137.5	4.6
17	49	144.9	5.5	51	137.1	4.7
18	52	142.5	5.0	51	138.6	5.2

Age (years)	Male, left side			Female, left side		
	N	Mean	SD	*N*	Mean	SD
6	50	121.3	5.4	50	119.5	3.8
7	50	122.7	5.2	50	122.3	4.5
8	51	125.8	4.0	51	123.9	5.1
9	51	127.2	5.2	50	126.3	5.8
10	50	128.9	5.4	49	127.9	4.7
11	50	131.6	5.2	51	129.4	5.4
12	52	132.0	5.0	53	132.9	5.1
13	50	134.6	5.5	49	134.2	4.7
14	49	136.0	6.3	51	134.2	4.6
15	50	139.8	5.5	51	135.3	5.2
16	50	139.8	5.5	51	136.1	4.8
17	49	143.9	6.1	51	136.6	4.4
18	52	141.0	5.3	51	137.9	5.7

EARS

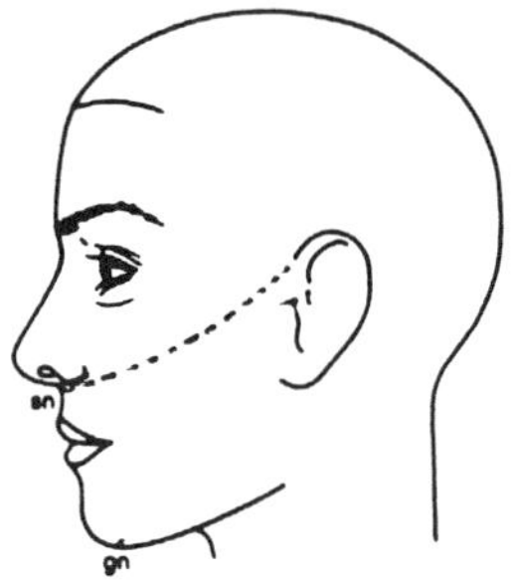

TABLE A–VI-14. *Upper subnasale–aural surface distance (sn-obs surf), right and left (mm)*

Age (years)	Male, right side			Female, right side		
	N	Mean	SD	*N*	Mean	SD
6	50	130.8	6.5	50	128.1	5.5
7	50	132.9	5.8	50	132.4	5.6
8	51	136.6	5.3	51	132.3	5.9
9	51	140.2	6.3	50	136.3	7.1
10	50	140.5	6.6	49	137.9	5.2
11	50	142.1	6.4	51	139.2	6.3
12	52	144.3	5.3	53	140.6	6.6
13	50	147.7	6.5	49	143.6	6.1
14	49	150.1	7.6	51	144.4	5.3
15	50	153.5	7.3	51	146.1	6.1
16	50	154.5	5.6	51	145.5	6.2
17	49	156.6	6.3	51	145.5	6.2
18	52	152.4	6.1	51	144.4	5.8

Age (years)	Male, left side			Female, left side		
	N	Mean	SD	*N*	Mean	SD
6	50	128.8	6.3	50	126.5	5.7
7	50	130.6	6.3	50	131.1	5.7
8	51	135.6	5.1	51	132.2	5.9
9	51	139.0	6.1	50	136.2	6.9
10	50	139.1	7.0	49	136.7	5.0
11	50	141.5	6.0	51	138.1	6.8
12	52	143.5	5.9	53	139.7	5.7
13	50	147.6	6.3	49	142.9	5.7
14	49	149.8	7.7	51	143.8	5.7
15	50	152.7	6.1	51	145.6	5.4
16	50	152.9	5.9	51	144.5	6.5
17	49	155.3	6.3	51	144.9	6.0
18	52	151.9	6.1	51	142.6	6.6

EARS

TABLE A–VI-15. *Lower subnasale–aural surface distance (sn-obi surf), right and left (mm)*

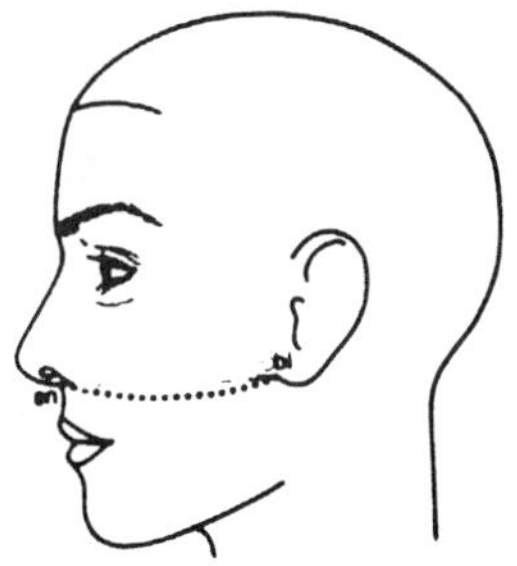

Age (years)	Male, right side			Female, right side		
	N	Mean	SD	*N*	Mean	SD
1	17	105.6	4.7	15	104.1	4.2
2	30	107.9	4.3	30	105.0	4.1
3	30	110.0	5.5	30	108.6	4.8
4	30	116.2	5.6	30	112.8	4.0
5	30	117.8	4.7	30	116.4	3.9
6	50	115.4	5.4	50	113.5	3.7
7	50	115.9	5.8	50	115.4	4.6
8	51	119.3	4.4	51	116.3	4.7
9	51	120.4	5.4	50	118.3	5.1
10	50	121.7	5.6	49	120.5	4.4
11	50	123.9	5.5	51	121.9	4.9
12	52	124.7	5.3	53	124.6	5.6
13	50	126.1	5.7	49	124.9	5.2
14	49	128.4	6.6	51	125.3	4.5
15	50	131.0	5.6	51	126.5	5.5
16	50	132.6	5.2	51	127.2	5.5
17	49	134.7	4.8	51	127.1	5.1
18	52	132.2	5.5	51	127.9	5.7
19–25	109	141.7	5.9	193	130.1	5.4

Age (years)	Male, left side			Female, left side		
	N	Mean	SD	*N*	Mean	SD
1	17	105.6	4.6	15	104.1	4.1
2	30	107.7	4.2	30	105.0	4.4
3	30	109.9	5.4	30	108.4	4.7
4	30	116.5	5.5	30	114.0	4.4
5	30	118.7	4.7	30	116.5	4.2
6	50	113.8	4.9	50	112.4	3.5
7	50	115.4	5.5	50	114.2	3.9
8	51	118.1	4.3	51	115.6	4.3
9	51	119.0	5.0	50	117.8	4.9
10	50	120.2	5.2	49	119.1	4.7
11	50	122.7	5.3	51	119.8	5.3
12	52	123.9	5.0	53	123.3	5.6
13	50	125.3	5.8	49	123.2	4.7
14	49	127.6	6.8	51	124.5	4.6
15	50	130.0	5.6	51	125.8	4.7
16	50	130.4	5.2	51	125.2	5.6
17	49	132.9	5.5	51	125.6	4.5
18	52	130.6	6.1	51	126.8	5.8
19–25	109	140.8	5.6	193	129.9	5.5

EARS

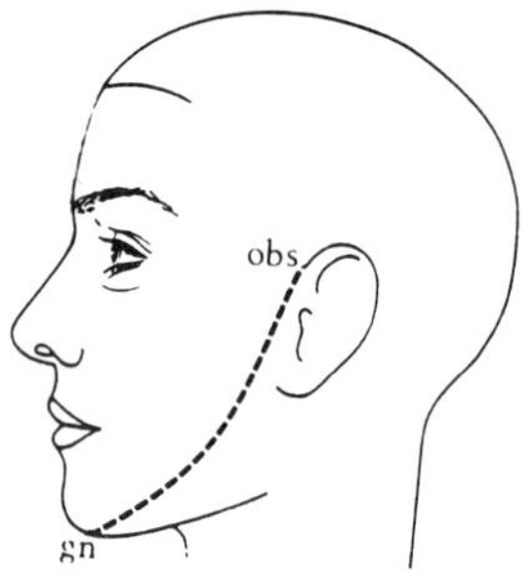

TABLE A–VI-16. *Upper gnathion–aural surface distance (gn-obs surf), right and left (mm)*

Age (years)	Male, right side			Female, right side		
	N	Mean	SD	*N*	Mean	SD
6	50	143.2	6.6	50	141.1	6.0
7	50	145.9	7.5	50	146.3	6.3
8	51	151.5	5.7	51	147.1	6.1
9	51	155.0	6.9	50	150.5	7.9
10	50	156.3	8.3	49	152.4	6.2
11	50	158.5	7.3	51	154.5	7.5
12	52	161.5	6.3	53	156.7	7.1
13	50	165.8	7.7	49	160.5	6.7
14	49	169.8	8.7	51	162.4	7.4
15	50	173.4	8.8	51	162.9	7.5
16	50	174.8	6.2	51	163.6	7.7
17	49	179.2	8.0	51	164.4	7.2
18	52	175.9	7.5	51	164.0	6.4
Age (years)	**Male, left side**			**Female, left side**		
	N	Mean	SD	*N*	Mean	SD
6	50	142.3	7.1	50	139.7	5.5
7	50	144.7	7.7	50	145.3	6.1
8	51	151.5	5.4	51	145.4	7.0
9	51	154.1	7.4	50	150.0	8.3
10	50	154.9	8.1	49	151.2	6.3
11	50	158.1	7.2	51	154.0	9.0
12	52	161.3	6.4	53	155.5	6.8
13	50	166.7	7.6	49	160.0	6.7
14	49	170.0	9.4	51	161.8	7.1
15	50	173.3	8.3	51	163.6	7.2
16	50	173.5	6.8	51	163.2	8.7
17	49	177.5	7.6	51	163.6	6.8
18	52	175.3	7.7	51	162.5	7.4

TABLE A-VI-17. *Lower gnathion–aural surface distance (gn-obi surf), right and left (mm)*

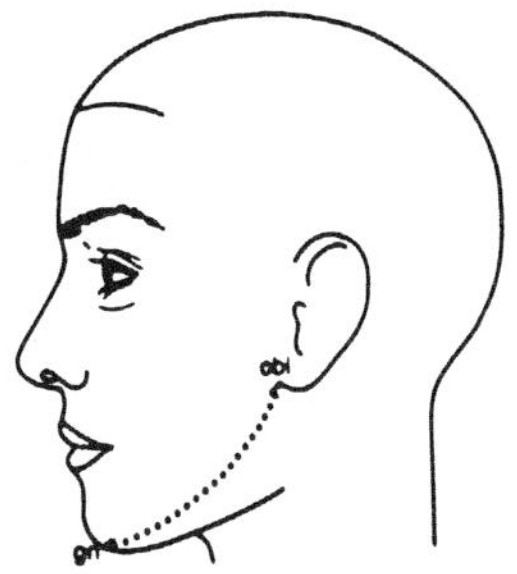

Age (years)	Male, right side			Female, right side		
	N	Mean	SD	*N*	Mean	SD
1	17	98.3	3.4	15	97.9	2.9
2	30	101.8	4.5	30	99.3	3.5
3	30	104.2	4.7	30	102.0	4.1
4	30	109.8	4.5	30	107.2	4.1
5	30	113.4	4.9	30	110.6	3.7
6	50	110.9	4.9	50	109.9	4.4
7	50	112.9	5.4	50	112.1	4.5
8	51	116.2	4.0	51	112.4	4.7
9	51	117.9	5.4	50	115.1	5.2
10	50	119.3	6.8	49	117.7	5.8
11	50	121.6	5.1	51	119.0	5.6
12	52	124.5	5.5	53	120.6	6.7
13	50	125.7	6.1	49	121.7	5.7
14	49	129.2	7.9	51	124.7	5.3
15	50	132.3	7.9	51	125.2	5.9
16	50	134.0	5.9	51	125.9	6.3
17	49	136.6	6.0	51	126.8	5.2
18	52	135.5	6.3	51	127.3	5.8
19–25	109	146.0	6.2	193	131.1	6.5

Age (years)	Male, left side			Female, left side		
	N	Mean	SD	*N*	Mean	SD
1	17	98.3	3.2	15	97.8	3.1
2	30	111.0	4.6	30	99.1	3.9
3	30	104.1	4.9	30	101.9	4.2
4	30	109.6	4.3	30	106.6	4.8
5	30	113.4	5.2	30	110.3	3.5
6	50	109.6	4.7	50	108.0	4.2
7	50	112.2	5.3	50	110.0	4.9
8	51	115.3	4.1	51	110.7	4.6
9	51	116.8	5.2	50	114.1	5.5
10	50	117.8	6.0	49	115.4	4.9
11	50	120.7	4.8	51	117.5	5.8
12	52	123.9	5.2	53	119.5	6.1
13	50	125.0	6.9	49	121.2	5.6
14	49	127.9	8.0	51	123.4	5.4
15	50	132.0	7.5	51	124.8	5.2
16	50	131.8	6.4	51	125.2	6.6
17	49	135.7	6.1	51	125.4	4.8
18	52	135.0	6.9	51	126.0	6.6
19–25	109	195.3	6.4	193	131.3	6.7

EARS

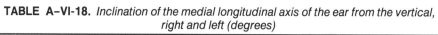

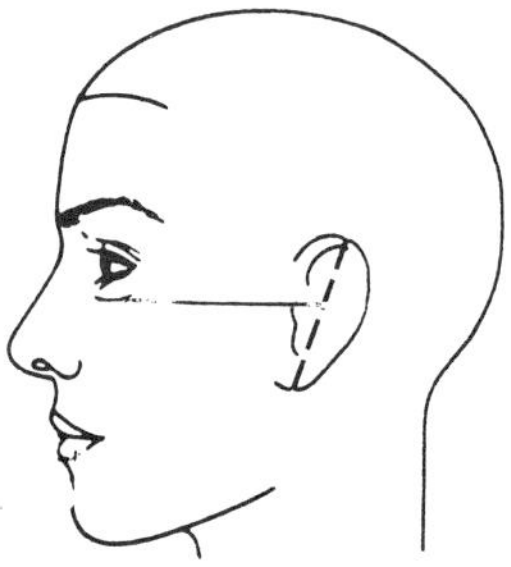

TABLE A-VI-18. *Inclination of the medial longitudinal axis of the ear from the vertical, right and left (degrees)*

Age (years)	Male, right side			Female, right side		
	N	Mean	SD	*N*	Mean	SD
1	17	19.2	5.3	18	17.9	4.4
2	31	21.1	5.7	34	18.7	4.1
3	30	18.4	5.5	30	20.0	7.4
4	—			—		
5	30	13.5	2.1	—		
6	50	18.9	4.7	49	19.2	5.3
7	50	21.3	4.6	48	20.0	6.0
8	51	20.4	5.7	51	19.3	5.2
9	51	21.6	5.3	48	18.9	4.8
10	49	19.2	4.7	49	18.4	5.5
11	47	17.5	5.0	51	17.6	5.6
12	52	21.8	5.2	53	18.3	5.1
13	50	19.3	5.4	49	18.3	5.3
14	49	19.5	5.1	50	18.0	5.6
15	50	20.9	5.3	50	19.6	4.7
16	50	19.5	4.7	51	18.1	5.4
17	49	18.9	4.3	51	16.9	5.2
18	52	20.1	4.6	50	18.7	5.3
19–25	45	19.4	3.8	200	18.0	5.0

Age (years)	Male, left side			Female, left side		
	N	Mean	SD	*N*	Mean	SD
1	17	19.1	5.4	18	18.0	4.5
2	31	20.9	5.4	34	18.7	4.1
3	30	18.4	5.6	30	20.4	7.4
4	30	20.3	5.5	30	21.0	7.3
5	30	19.6	5.0	30	20.3	5.0
6	50	19.5	5.0	49	20.1	6.5
7	50	21.5	4.8	48	21.2	5.5
8	51	21.2	5.4	51	19.9	5.0
9	51	22.3	5.4	49	20.0	3.7
10	49	19.6	4.7	49	19.1	5.7
11	48	18.9	5.6	51	18.4	5.6
12	52	21.8	4.9	53	18.2	5.4
13	50	20.6	4.8	49	17.8	4.6
14	49	20.0	4.9	50	18.3	5.5
15	50	21.1	5.4	50	19.9	4.3
16	50	19.2	4.6	50	17.6	5.1
17	49	20.2	4.8	51	17.4	5.5
18	52	20.0	5.0	51	18.4	4.8
19–25	108	19.7	4.8	200	17.5	4.6

EARS

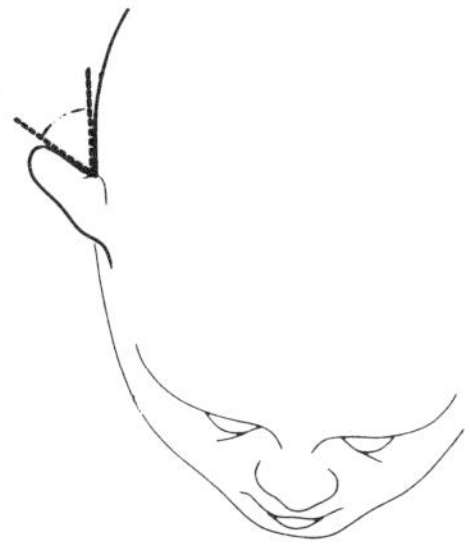

TABLE A–VI-19. *Protrusion of the ear from the head, right and left (degrees)*

Age (years)	Male, right side			Female, right side		
	N	Mean	SD	*N*	Mean	SD
1	18	16.3	7.2	20	15.6	6.5
2	31	14.5	6.6	34	16.2	7.3
3	30	18.7	7.6	30	16.0	7.6
4	30	19.3	5.6	30	20.3	5.9
5	30	19.8	6.9	30	20.4	7.3
6	50	21.6	6.5	50	19.9	6.1
7	50	21.6	6.2	50	21.5	7.6
8	51	21.3	6.9	51	19.3	6.1
9	51	21.3	6.8	50	21.1	6.7
10	50	21.4	8.1	49	22.2	7.7
11	50	22.0	6.3	51	20.7	6.2
12	52	21.7	6.1	53	22.0	6.9
13	49	22.9	6.6	49	22.4	8.1
14	49	23.4	6.7	50	23.9	6.5
15	50	22.0	8.4	51	23.2	5.4
16	50	25.0	7.1	51	22.7	5.2
17	49	24.5	6.0	51	23.6	6.8
18	52	22.9	5.4	51	23.2	7.7
19–25	95	19.6	4.6	199	20.8	5.8

Age (years)	Male, left side			Female, left side		
	N	Mean	SD	*N*	Mean	SD
1	18	16.3	7.2	20	15.4	6.6
2	31	14.4	6.3	34	16.1	7.1
3	30	18.3	7.5	30	16.0	7.5
4	30	19.2	5.7	30	20.5	6.0
5	30	19.7	6.9	30	19.9	6.7
6	50	22.8	7.3	50	19.9	5.8
7	50	21.3	6.1	50	21.3	8.1
8	51	21.7	6.3	51	19.9	5.7
9	51	20.6	6.9	50	20.7	6.4
10	50	21.5	7.1	49	22.2	7.2
11	50	21.3	6.2	51	19.6	5.4
12	52	21.9	6.4	53	21.9	6.3
13	49	22.1	6.1	49	22.1	8.1
14	49	22.1	6.3	50	22.7	6.1
15	50	22.2	8.4	51	23.8	5.5
16	50	24.8	6.9	51	23.5	5.6
17	49	23.4	6.7	51	22.5	6.5
18	52	23.0	5.5	51	22.3	6.6
19–25	95	19.6	4.5	199	20.7	5.8

Height and Weight

TABLE A–VII-1. *Body height (cm)*

Age (years)	Male			Female		
	N	Mean	SD	*N*	Mean	SD
0–5 months	8	65.9	4.5	5	61.6	2.2
6–12 months	18	70.0	4.0	8	71.4	3.4
1	18	81.6	5.2	20	81.9	5.0
2	31	93.6	4.2	30	93.0	4.2
3	30	99.6	4.6	30	99.5	5.1
4	28	109.7	4.9	30	108.2	4.3
5	30	115.6	4.2	28	114.4	4.5
6	49	118.3	6.2	50	119.3	9.9
7	50	124.6	7.1	50	123.9	6.8
8	51	132.8	6.1	51	128.9	6.5
9	51	136.2	8.0	50	134.0	6.5
10	50	139.7	7.0	49	141.4	7.1
11	50	144.6	7.9	51	146.3	7.6
12	52	150.9	7.1	53	154.1	5.8
13	50	157.5	7.7	48	157.3	8.3
14	48	160.9	8.2	51	159.5	6.4
15	50	170.6	9.8	51	161.6	6.4
16	50	173.5	6.9	51	159.4	8.4
17	49	176.8	6.2	50	161.9	6.0
18	51	176.6	8.1	51	162.7	6.9

TABLE A–VII-2. *Body weight (kg)*

Age (years)	Male			Female		
	N	Mean	SD	*N*	Mean	SD
0–5 months	8	7.5	1.5	5	5.8	0.4
6–12 months	20	9.3	1.9	8	8.6	2.0
1	18	12.4	1.4	21	11.8	1.7
2	31	14.4	2.2	31	13.9	1.6
3	30	16.1	1.9	29	15.7	1.9
4	28	18.7	4.9	28	17.2	2.3
5	29	20.1	2.4	25	18.9	2.4
6	49	22.3	3.0	50	22.1	3.0
7	50	24.2	3.4	50	24.5	3.2
8	51	28.4	5.0	51	27.1	4.5
9	51	31.0	5.4	50	30.0	5.0
10	50	35.1	6.8	49	34.3	6.0
11	50	39.1	8.4	51	39.6	8.8
12	52	42.6	6.9	53	45.1	9.1
13	50	47.1	8.5	48	48.1	9.7
14	48	52.8	9.5	51	50.6	10.2
15	50	62.7	11.2	51	54.7	8.7
16	50	63.5	11.5	51	57.6	8.4
17	49	72.6	10.5	51	56.5	7.7
18	51	68.9	9.0	51	55.0	9.6

Appendix B

Craniofacial Norms in 6-, 12-, and 18-year-old Chinese Subjects

Leslie G. Farkas, Rexon C. K. Ngim, and S. T. Lee

The Chinese population data about the head and face were obtained from randomly selected Chinese subjects in the Republic of Singapore during 1987. The measurements were taken by one of the authors (LGF). The subjects examined were healthy school boys and girls (6 and 12 years of age), army recruits, and university students (18 years of age). A total of 111 measurements (65 single and 46 paired) were taken from most of the subjects: 14 on the head, 31 on the face, 14 on the orbits, 23 on the nose, 17 on the lips and mouth, and 12 on the ears. In each of the age groups 30 males and 30 females were examined. This number applies for most of the measurements. The techniques of measurements are described in detail in Chapter 2.

Tables reporting the craniofacial norms in the Chinese population are marked with capital B. The anatomical region is identified by Roman numerals: I for the head, II for the face, III for the orbits, IV for the nose, V for the lips and mouth, and VI for the ears. In each table the number of the sample (*N*), the mean value of the measurement, and the standard deviation (SD) are given, separately for males and females.

TABLE B–I. *Measurements of the head in three age groups of the Chinese population*

Measurement	Age (years)	Male			Female		
		N	Mean	SD	*N*	Mean	SD
Width of the head (eu-eu)[a]	6	30	149.4	4.6	30	145.0	4.5
	12	30	153.2	5.8	30	148.7	5.7
	18	30	158.3	5.1	30	151.6	4.7
Width of the forehead (ft-ft)[a]	6	30	103.0	4.8	30	101.1	3.8
	12	30	108.9	4.4	30	108.5	5.6
	18	30	116.9	4.5	30	112.7	4.9
Skull-base width (t-t)[a]	6	30	135.0	4.5	30	130.0	4.9
	12	30	143.5	5.3	30	138.7	6.2
	18	30	151.8	5.3	30	141.9	5.0
Height of the calvarium (v-tr)[a]	6	30	49.3	7.6	30	49.0	8.0
	12	30	49.8	8.8	30	47.7	8.9
	18	30	55.5	9.3	30	48.7	8.7
Anterior height of the head (v-n)[a]	6	30	107.9	6.7	30	106.1	7.2
	12	30	110.7	8.5	30	108.9	6.8
	18	30	116.9	8.0	30	109.8	7.8
Special height of the head (v-en)[a]	6	30	118.0	6.4	30	115.4	7.2
	12	30	120.4	8.4	30	119.1	6.8
	18	30	126.9	7.5	30	119.4	7.4
Height of the head and nose (v-sn)[a]	6	30	153.1	9.7	30	149.5	7.4
	12	30	157.0	9.6	30	155.6	7.0
	18	30	165.4	19.4	30	159.5	8.7
Combined height of the head and face (v-gn)[a]	6	30	206.0	7.2	30	199.9	6.8
	12	30	219.9	8.8	30	215.8	7.4
	18	30	230.5	18.8	30	216.6	17.4
Height of the forehead I (tr-g)[a]	6	30	55.3	5.4	30	50.7	5.5
	12	30	54.7	6.9	30	54.8	7.8
	18	30	56.8	7.0	30	54.1	7.8
Height of the forehead II (tr-n)[a]	6	30	65.3	5.2	30	60.5	5.4
	12	30	64.8	7.0	30	64.8	8.1
	18	30	67.1	6.9	30	64.1	7.5
Length of the head (g-op)[a]	6	30	167.3	5.5	30	161.5	6.3
	12	30	172.0	4.3	30	169.5	5.1
	18	30	182.0	6.2	30	172.5	6.1
Circumference of the head (at g-op level)[a]	6	30	515.5	15.2	30	499.5	12.8
	12	30	532.2	13.6	30	525.8	14.8
	18	30	559.6	14.2	30	535.3	15.5
Inclination of the anterior surface of the forehead (general tr-g line)[b]	6	30	3.3	4.9	30	4.3	4.9
	12	30	−5.0	6.5	30	−2.7	4.9
	18	30	−13.7	5.5	30	−9.2	5.7
Auricular height of the head rt (v-po rt)[a]	6	2	126.5	2.1	10	124.8	6.4
	12	2	129.0	12.7	6	133.2	4.4
	18	25	137.4	6.0	15	129.1	6.0
Auricular height of the head lt (v-po lt)[a]	6	2	128.9	6.0	29	124.6	5.0
	12	30	131.5	6.1	29	130.2	4.7
	18	30	137.3	5.9	30	130.0	5.1

[a] In millimeters.
[b] In degrees.

TABLE B–II. *Measurements of the face in three age groups of the Chinese population*

Measurement	Age (years)	Male N	Male Mean	Male SD	Female N	Female Mean	Female SD
Width of the face (zy-zy)[a]	6	30	125.8	6.9	30	121.5	4.4
	12	30	132.2	5.8	30	130.8	5.1
	18	30	144.6	5.6	30	136.2	4.0
Width of the mandible (go-go)[a]	6	30	93.5	4.9	30	92.9	4.3
	12	30	101.9	4.4	30	98.7	5.1
	18	30	107.3	5.6	30	102.3	3.4
Physiognomical height of the face (tr-gn)[a]	6	30	164.0	5.5	30	157.3	5.6
	12	30	173.6	9.8	30	170.8	8.4
	18	30	187.3	7.2	30	176.2	8.3
Height of the face (n-gn)[a]	6	30	103.1	4.8	30	101.3	4.6
	12	30	112.8	5.3	30	110.5	4.7
	18	30	123.6	5.3	30	114.9	4.9
Height of the upper face (n-sto)[a]	6	30	66.0	3.9	30	63.7	3.4
	12	30	71.7	3.6	30	70.7	3.3
	18	30	78.2	4.0	30	71.8	5.5
Height of the lower face (sn-gn)[a]	6	30	61.9	4.5	30	59.5	3.5
	12	30	67.6	4.4	30	64.3	4.1
	18	30	72.7	5.2	30	66.4	5.6
Height of the mandible (sto-gn)[a]	6	30	43.3	3.7	30	42.5	3.3
	12	30	48.6	3.4	30	46.0	3.2
	18	30	53.4	4.2	30	47.2	3.4
Height of the chin (sl-gn)[a]	6	30	33.1	18.4	30	28.6	3.8
	12	30	32.4	3.1	30	31.1	2.9
	18	30	36.6	3.7	30	32.7	6.4
Height of the upper profile (tr-prn)[a]	6	30	102.6	5.1	30	96.2	6.5
	12	30	106.0	7.7	30	105.7	9.2
	18	30	114.3	8.3	30	108.5	8.3
Height of the lower profile (prn-gn)[a]	6	30	73.7	3.6	30	71.8	3.5
	12	30	81.8	5.4	30	77.7	4.0
	18	30	88.8	5.1	30	81.2	4.2
Lower half of the craniofacial height (en-gn)[a]	6	30	94.3	5.2	30	92.2	3.5
	12	30	104.7	5.1	30	100.5	3.9
	18	30	114.6	9.0	30	107.1	9.3
Distance between the glabella and the subnasale (g-sn)[a]	6	30	56.3	2.8	30	54.6	4.2
	12	30	60.8	3.6	30	60.4	3.6
	18	30	66.5	3.5	30	62.3	4.4
Supraorbital arc (t-g-t)[a]	6	30	274.4	9.2	30	264.5	8.8
	12	30	289.8	11.7	30	285.0	11.2
	18	30	308.2	10.4	30	293.0	10.1
Maxillary arc (t-sn-t)[a]	6	30	257.6	9.4	30	251.3	9.6
	12	30	277.9	10.7	30	275.8	11.1
	18	30	301.4	20.5	30	284.0	11.4
Mandibular arc (t-gn-t)[a]	6	30	267.2	13.5	30	261.4	10.1
	12	30	294.7	11.9	30	291.5	12.6
	18	30	320.5	9.0	30	304.5	11.0
Inclination of the upper face (g-sn line)[b]	6	30	0.3	3.8	30	0.0	3.2
	12	30	0.7	3.8	30	0.2	4.0
	18	30	1.9	3.6	30	0.4	3.2
Inclination of the Leiber line (g-ls line)[b]	6	30	3.3	3.1	30	3.4	2.8
	12	30	4.4	3.3	30	3.8	3.7
	18	30	5.2	3.7	30	3.6	3.0
Inclination of the lower face (sn-pg line)[b]	6	30	−17.4	5.4	30	−14.6	4.4
	12	30	−16.3	4.2	30	−14.7	6.2
	18	30	−15.3	5.5	30	−11.4	5.6
Inclination of the mandible (li-pg line)[b]	6	30	−31.0	8.1	30	−27.1	12.0
	12	30	−32.9	7.3	30	−30.6	8.3
	18	30	−30.8	8.0	30	−28.0	8.2
Inclination of the chin[b]	6	30	−10.9	11.9	30	−9.2	13.2
	12	30	−10.8	15.3	30	−11.4	16.9
	18	30	−8.6	11.3	30	−1.5	12.4

TABLE B–II. *Continued.*

Measurement	Age (years)	Male			Female		
		N	Mean	SD	*N*	Mean	SD
Inclination of the general profile line (g-pg line)[b]	6	30	−8.3	3.4	30	−7.7	2.5
	12	30	−8.4	3.1	30	−7.3	3.6
	18	30	−6.1	3.4	30	−2.2	17.8
Mentocervical angle[b]	6	21	98.0	10.6	21	95.0	8.1
	12	26	92.8	11.2	24	93.4	14.6
	18	—	—	—	14	96.2	14.4
Height of the mandibular ramus rt (go-cdl rt)[a]	6	30	54.7	5.6	30	53.0	4.4
	12	30	61.1	4.4	30	60.0	3.4
	18	30	69.3	4.8	30	62.5	4.8
Height of the mandibular ramus lt (go-cdl lt)[a]	6	30	54.7	5.7	30	52.8	4.6
	12	30	61.4	4.5	30	60.1	3.6
	18	30	69.5	4.8	30	62.3	4.7
Tragion–glabella depth rt (t-g rt)[a]	6	30	113.0	3.9	30	108.6	3.8
	12	30	119.2	4.9	30	115.6	4.7
	18	30	127.9	5.0	30	119.2	3.9
Tragion–glabella depth lt (t-g lt)[a]	6	30	112.3	3.7	30	108.4	3.3
	12	30	119.0	4.6	30	115.9	4.7
	18	30	127.2	4.5	30	118.4	4.0
Tragion–nasion depth rt (t-n rt)[a]	6	30	109.7	3.8	30	105.0	3.9
	12	30	115.7	4.6	30	112.3	4.4
	18	30	123.3	4.1	30	116.1	4.2
Tragion–nasion depth lt (t-n lt)[a]	6	30	108.5	3.1	30	104.1	3.8
	12	30	115.2	4.2	30	112.2	4.7
	18	30	122.1	4.3	30	114.6	4.3
Tragion–subnasale depth rt (t-sn rt)[a]	6	30	110.3	4.1	30	106.5	4.3
	12	30	119.9	4.2	30	116.1	4.1
	18	30	129.0	4.5	30	119.5	4.5
Tragion–subnasale depth lt (t-sn lt)[a]	6	30	109.9	3.6	30	106.0	3.6
	12	30	119.2	4.4	30	115.9	4.5
	18	30	128.4	4.4	30	118.7	4.7
Tragion–gnathion depth rt (t-gn rt)[a]	6	30	119.8	5.9	30	117.1	4.5
	12	30	132.2	5.1	30	129.5	5.4
	18	30	143.1	5.2	30	134.2	4.8
Tragion–gnathion depth lt (t-gn lt)[a]	6	30	120.0	5.7	30	117.3	4.5
	12	30	132.7	4.9	30	129.8	5.1
	18	30	143.4	4.4	30	134.8	4.9
Gonion–gnathion depth of mandible rt (go-gn rt)[a]	6	30	79.4	4.5	30	78.9	2.6
	12	30	90.7	6.0	30	88.2	5.1
	18	30	95.9	2.7	30	91.0	4.3
Gonion–gnathion depth of mandible lt (go-gn lt)[a]	6	30	79.5	4.6	30	78.9	2.6
	12	30	90.9	5.9	30	88.3	4.8
	18	30	96.0	2.8	30	91.0	4.2
Lateral surface half-arc in the upper third of the face rt (t-g surf rt)[a]	6	30	137.4	5.2	30	132.9	4.6
	12	30	146.1	6.3	30	142.7	5.8
	18	30	154.6	6.1	30	147.3	5.7
Lateral surface half-arc in the upper third of the face lt (t-g surf lt)[a]	6	30	137.0	4.4	30	131.6	4.4
	12	30	143.6	6.0	30	142.3	5.9
	18	30	153.6	4.8	30	145.6	4.9
Lateral surface half-arc in the middle third of the face rt (t-sn surf rt)[a]	6	30	128.0	5.1	30	125.1	5.1
	12	30	138.8	5.5	30	137.9	5.9
	18	30	148.3	4.9	30	142.5	5.3
Lateral surface half-arc in the middle third of the face lt (t-sn surf lt)[a]	6	30	129.6	4.7	30	126.2	4.9
	12	30	139.1	5.5	30	137.9	5.7
	18	30	149.7	5.7	30	141.7	5.9
Lateral surface half-arc in the lower third of the face rt (t-gn surf rt)[a]	6	30	133.6	7.2	30	131.2	5.4
	12	30	148.0	6.1	30	146.1	6.7
	18	30	160.0	5.2	30	152.9	5.7
Lateral surface half-arc in the lower third of the face lt (t-gn surf lt)[a]	6	30	134.0	6.8	30	130.3	5.2
	12	30	146.6	6.7	30	145.4	6.4
	18	30	160.5	4.2	30	150.9	5.5

[a] In millimeters.
[b] In degrees.

TABLE B–III. *Measurements of the orbits in three age groups of the Chinese population*

Measurement	Age (years)	Male			Female		
		N	Mean	SD	*N*	Mean	SD
Intercanthal width (en-en)[a]	6	30	34.4	2.7	30	34.0	2.3
	12	30	37.5	2.8	30	35.6	2.9
	18	30	37.6	3.3	30	36.5	3.2
Biocular width (ex-ex)[a]	6	30	81.4	3.4	30	79.6	4.5
	12	30	87.2	3.8	30	84.6	4.0
	18	30	91.7	4.0	30	87.3	5.2
Length of the eye fissure rt (ex-en rt)[a]	6	30	26.2	0.9	30	25.2	1.4
	12	30	27.4	1.4	30	26.9	1.4
	18	30	29.4	1.2	30	28.5	1.8
Length of the eye fissure lt (ex-en lt)[a]	6	30	26.2	0.9	30	25.2	1.4
	12	30	27.5	1.4	30	26.9	1.4
	18	30	29.4	1.3	30	28.4	1.7
Endocanthion–facial midline distance rt (en-se rt)[a]	6	30	19.3	1.8	30	19.0	1.6
	12	30	21.3	1.8	30	20.1	1.7
	18	30	22.4	1.7	30	21.1	1.9
Endocanthion–facial midline distance lt (en-se lt)[a]	6	30	19.0	1.7	30	18.6	1.3
	12	30	20.8	1.7	30	19.7	1.8
	18	30	21.9	1.8	30	20.3	1.8
Orbito-aural distance rt (ex-obs rt)[a]	6	30	77.3	3.7	30	75.9	4.1
	12	30	82.1	6.4	30	82.3	4.9
	18	30	87.4	5.1	30	84.6	4.9
Orbito-aural distance lt (ex-obs lt)[a]	6	30	77.1	4.0	30	74.8	3.5
	12	30	82.2	6.1	30	82.8	5.1
	18	30	87.1	4.6	30	83.5	5.1
Orbito-tragion distance rt (ex-t rt)[a]	6	30	79.1	6.4	30	73.9	2.9
	12	30	83.4	5.8	30	80.7	4.2
	18	30	87.3	3.8	30	82.5	3.1
Orbito-tragion distance lt (ex-t lt)[a]	6	30	79.1	6.5	30	73.7	2.9
	12	30	83.4	5.7	30	80.9	4.3
	18	30	86.9	4.0	30	82.0	3.5
Orbito-gonial distance rt (ex-go rt)[a]	6	30	87.2	4.6	30	83.1	3.9
	12	30	96.8	4.5	30	95.1	5.4
	18	30	108.2	5.3	30	98.6	4.7
Orbito-gonial distance lt (ex-go lt)[a]	6	30	86.9	4.7	30	83.1	3.9
	12	30	96.7	4.5	30	95.1	5.2
	18	30	107.6	5.5	30	98.6	4.8
Height of the orbit rt (or-os rt)[a]	6	30	25.3	3.1	30	24.2	2.2
	12	30	24.9	2.1	30	25.8	1.9
	18	30	26.5	2.5	30	26.4	3.4
Height of the orbit lt (or-os lt)[a]	6	30	25.3	3.1	30	24.2	2.2
	12	30	24.9	2.1	30	25.8	1.8
	18	30	26.5	2.5	30	26.6	3.1
Combined height of the orbit and the eyebrow rt (or-sci rt)[a]	6	30	35.3	2.4	30	33.8	2.8
	12	30	36.0	2.2	30	35.5	2.0
	18	30	39.0	3.0	30	37.2	3.2
Combined height of the orbit and the eyebrow lt (or-sci lt)[a]	6	30	35.3	2.4	30	33.9	2.8
	12	30	36.0	2.2	30	35.5	2.0
	18	30	38.8	3.3	30	37.2	3.2
Height of the eye fissure rt (ps-pi rt)[a]	6	30	8.6	0.9	30	8.8	0.8
	12	30	8.4	0.9	30	8.9	1.1
	18	30	9.4	0.7	30	9.5	1.2
Height of the eye fissure lt (ps-pi lt)[a]	6	30	8.7	0.8	30	8.8	0.8
	12	30	8.4	0.9	30	8.9	1.1
	18	30	9.4	0.7	30	9.5	1.2
Height of the upper lid rt (ps-os rt)[a]	6	30	10.7	1.3	30	9.8	1.4
	12	30	10.2	1.4	30	10.2	1.5
	18	30	10.0	1.0	30	10.7	1.8
Height of the upper lid lt (ps-os lt)[a]	6	30	10.7	1.3	30	9.8	1.4
	12	30	10.2	1.4	30	10.2	1.5
	18	30	10.0	1.0	30	10.7	1.8

TABLE B–III. *Continued.*

Measurement	Age (years)	Male			Female		
		N	Mean	SD	*N*	Mean	SD
Height of the lower lid rt (pi-or rt)[a]	6	30	6.9	1.0	30	6.3	1.0
	12	30	7.0	1.0	30	6.8	1.0
	18	30	7.0	1.2	30	7.1	1.7
Height of the lower lid lt (pi-or lt)[a]	6	30	6.9	1.0	30	6.3	1.0
	12	30	7.0	1.0	30	6.7	0.9
	18	30	7.0	1.2	30	7.1	1.7
Inclination of the eye fissure rt (ex-en line rt)[b]	6	30	9.0	2.2	30	11.5	2.6
	12	30	10.4	1.8	30	11.9	2.4
	18	30	12.1	2.1	30	12.4	2.0
Inclination of the eye fissure lt (ex-en line lt)[b]	6	30	9.1	2.1	30	11.4	2.6
	12	30	10.4	1.8	30	11.9	2.4
	18	30	12.1	2.1	29	12.5	2.0
Inclination of the orbital rims rt (os′/or′ line rt)[b]	6	29	10.2	2.7	30	8.9	2.6
	12	30	9.8	2.7	30	8.8	3.3
	18	30	10.6	2.9	28	8.6	3.2
Inclination of the orbital rims lt (os′/or′ line lt)[b]	6	30	10.6	3.3	30	8.9	2.6
	12	29	9.7	2.7	30	8.9	3.4
	18	30	10.6	2.9	30	8.6	3.2

[a] In millimeters.
[b] In degrees.

TABLE B–IV. *Measurements of the nose in three age groups of the Chinese population*

Measurement	Age (years)	Male			Female		
		N	Mean	SD	*N*	Mean	SD
Width of the nasal root (mf-mf)[a]	6	30	18.7	1.3	30	18.4	1.2
	12	30	20.1	1.6	30	19.2	1.4
	18	30	19.5	2.1	30	19.9	1.3
Width of the nose (al-al)[a]	6	30	33.0	2.0	30	31.8	2.4
	12	30	36.2	2.3	30	36.1	2.3
	18	30	39.2	2.9	30	37.2	2.1
Width between the facial insertion points of the alar base (ac-ac)[a]	6	30	33.3	2.1	30	32.4	1.8
	12	30	36.4	3.8	30	35.3	2.4
	18	30	38.6	2.3	30	36.5	2.3
Width of the columella (sn′sn′)[a]	6	30	7.1	0.7	30	6.9	0.9
	12	30	7.5	0.7	30	7.3	0.8
	18	30	7.8	0.5	30	7.3	0.8
Height of the nose (n-sn)[a]	6	30	44.8	3.0	30	44.7	3.1
	12	30	48.8	2.7	30	49.1	3.6
	18	30	53.5	2.8	30	51.7	3.3
Length of the nasal bridge (n-prn)[a]	6	30	36.9	3.1	30	36.8	3.1
	12	30	40.4	2.9	30	41.7	3.7
	18	30	46.2	2.8	30	44.3	3.7
Nasal tip protrusion (sn-prn)[a]	6	30	12.9	1.2	30	12.9	1.1
	12	30	14.6	1.3	30	14.6	1.4
	18	30	16.1	1.5	30	15.4	1.8
Inclination of the nasal bridge[b]	6	30	23.8	3.9	30	23.9	3.7
	12	30	22.9	9.0	30	22.9	9.9
	18	30	27.2	3.5	30	24.5	3.6
Inclination of the columella[b]	6	30	61.4	9.1	30	59.4	6.3
	12	30	61.6	6.2	30	63.3	8.4
	18	30	69.3	8.3	30	68.9	12.3
Glabellonasal angle[b]	6	21	159.2	4.5	21	160.7	7.7
	12	26	153.4	8.7	24	156.9	11.6
	18	—	—	—	15	144.5	39.8
Nasofrontal angle[b]	6	30	133.7	6.7	30	133.5	6.0
	12	30	134.3	7.5	30	137.3	6.7
	18	30	134.5	7.0	30	135.6	4.4

TABLE B–IV. *Continued.*

Measurement	Age (years)	Male *N*	Male Mean	Male SD	Female *N*	Female Mean	Female SD
Nasal tip angle[b]	6	30	90.2	6.7	30	88.1	4.2
	12	30	89.7	4.6	30	88.3	5.6
	18	30	83.9	6.1	30	82.7	6.6
Nasolabial angle[b]	6	30	91.1	18.5	30	92.6	9.2
	12	30	92.5	12.4	30	89.9	8.7
	18	30	86.9	12.2	30	88.5	11.2
Deviation of the nasal bridge[b]	6	28	0.8	4.0	29	0.0	0.0
	12	30	0.1	0.5	30	0.1	0.5
	18	29	0.2	0.8	30	0.3	1.0
Deviation of the columella[b]	6	28	2.3	12.3	29	0.0	0.0
	12	30	0.0	0.0	30	0.2	1.3
	18	29	0.2	0.9	30	0.0	0.0
Width of the nostril floor rt (sbal-sn rt)[a]	6	30	9.7	1.0	30	9.8	1.2
	12	30	10.8	1.8	30	11.1	1.5
	18	30	12.2	1.6	30	11.7	1.2
Width of the nostril floor lt (sbal-sn lt)[a]	6	30	9.7	1.0	30	9.8	1.2
	12	30	10.8	1.7	30	11.1	1.6
	18	30	12.2	1.6	30	11.7	1.2
Thickness of the ala rt (al′-al′ rt)[a]	6	30	6.0	0.6	30	6.3	0.7
	12	30	6.6	0.7	30	6.6	0.7
	18	30	7.2	0.8	30	6.8	0.8
Thickness of the ala lt (al′al′ lt)[a]	6	30	6.0	0.6	30	6.3	0.7
	12	30	6.8	0.7	30	6.6	0.7
	18	30	7.2	0.8	30	6.8	0.8
Distance between the facial insertion point and midpoint of columella rt (ac-sn rt)[a]	6	30	18.6	1.7	30	18.7	1.1
	12	30	21.0	1.1	30	20.9	1.4
	18	30	23.4	1.4	30	21.9	1.2
Distance between the facial insertion point and midpoint of columella lt (ac-sn lt)[a]	6	30	18.6	1.7	30	18.7	1.1
	12	30	21.0	1.1	30	20.8	1.4
	18	30	23.4	1.4	30	21.9	1.2
Length of the ala rt (ac-prn rt)[a]	6	30	24.1	1.5	30	23.1	1.2
	12	30	26.8	1.6	30	26.6	1.4
	18	30	30.8	1.5	30	28.0	1.4
Length of the ala lt (ac-prn lt)[a]	6	30	24.1	1.5	30	23.1	1.2
	12	30	26.8	1.6	30	26.6	1.5
	18	30	30.8	1.5	30	28.0	1.4
Length of the columella rt (sn-c′ rt)[a]	6	30	6.9	0.8	30	6.6	1.0
	12	30	7.2	1.3	30	7.2	1.2
	18	30	8.1	1.4	30	7.8	1.5
Length of the columella lt (sn-c′ lt)[a]	6	30	6.8	0.9	30	6.6	1.0
	12	30	7.2	1.3	30	7.1	1.2
	18	30	8.1	1.4	30	7.8	1.5
Depth of the nasal root rt (en-se sag rt)[a]	6	29	4.7	1.0	29	4.6	1.1
	12	30	6.1	1.9	30	5.9	1.4
	18	30	7.8	1.9	30	6.1	1.5
Depth of the nasal root lt (en-se sag lt)[a]	6	30	4.7	1.0	30	4.5	1.1
	12	30	6.1	1.8	30	6.0	1.3
	18	30	7.8	1.9	30	6.1	1.5
Surface length of the ala rt (ac-prn surf rt)[a]	6	30	29.1	2.2	29	27.7	1.4
	12	30	32.1	2.0	30	32.1	1.9
	18	30	36.2	2.0	30	33.2	1.9
Surface length of the ala lt (ac-prn surf lt)[a]	6	30	29.1	2.2	30	27.7	1.4
	12	30	32.1	2.0	30	32.1	1.9
	18	30	36.2	2.0	30	33.2	1.8
Inclination of the nostril axis rt[b]	6	30	29.3	7.6	30	27.5	9.7
	12	30	27.0	7.8	30	29.8	7.7
	18	30	33.9	10.5	30	31.9	11.3
Inclination of the nostril axis lt[b]	6	30	29.3	7.6	30	27.5	9.7
	12	30	27.0	7.8	30	29.2	8.0
	18	30	33.4	10.0	30	31.9	11.3

[a] In millimeters.
[b] In degrees.

TABLE B–V. *Measurements of the lips and mouth in three age groups of the Chinese population*

Measurement	Age (years)	Male *N*	Male Mean	Male SD	Female *N*	Female Mean	Female SD
Width of the philtrum (cph-cph)[a]	6	30	8.4	1.6	30	8.2	1.3
	12	30	9.5	1.4	30	9.0	1.2
	18	30	11.1	1.4	30	10.2	1.3
Width of the mouth (ch-ch)[a]	6	30	39.0	3.2	30	38.7	2.9
	12	30	44.5	2.8	30	43.2	3.0
	18	30	48.3	6.8	30	47.3	3.3
Height of the upper lip (sn-sto)[a]	6	30	20.1	1.5	30	19.6	1.7
	12	30	21.8	2.2	30	20.3	1.6
	18	30	23.5	2.2	30	21.6	2.1
Height of the cutaneous upper lip (sn-ls)[a]	6	30	13.3	1.6	30	12.8	2.0
	12	30	13.7	2.2	30	12.7	2.0
	18	30	14.2	2.0	30	13.4	2.0
Vermilion height of the upper lip (ls-sto)[a]	6	30	9.3	1.6	30	9.1	1.2
	12	30	10.2	1.6	30	10.0	1.4
	18	30	11.2	1.2	30	10.1	1.4
Vermilion height of the lower lip (sto-li)[a]	6	30	7.9	1.7	30	8.0	1.4
	12	30	10.3	1.5	30	10.0	1.8
	18	30	10.8	1.3	30	10.5	1.3
Height of the cutaneous lower lip (li-sl)[a]	6	30	7.4	1.3	30	7.9	1.5
	12	30	8.7	1.8	30	7.9	1.1
	18	30	9.2	1.4	30	8.7	2.0
Height of the lower lip (sto-sl)[a]	6	30	14.3	1.7	30	15.2	1.4
	12	30	17.8	2.1	30	16.9	1.5
	18	30	18.5	2.0	30	17.8	2.1
Vermilion surface arc of the upper lip (ch-ls-ch)[a]	6	30	59.7	4.0	30	56.4	4.7
	12	30	64.2	11.6	30	62.3	5.4
	18	30	73.8	4.5	30	66.8	4.8
Vermilion surface arc of the lower lip (ch-li-ch)[a]	6	30	48.6	4.3	30	47.0	3.4
	12	30	53.9	9.8	30	53.6	4.3
	18	30	63.4	4.0	30	59.7	4.8
Inclination of the upper lip[b]	6	30	17.6	7.2	30	19.7	5.0
	12	30	18.2	10.7	30	19.7	5.6
	18	30	16.8	10.1	30	18.5	6.9
Inclination of the lower lip[b]	6	30	−55.3	12.5	30	−54.0	11.9
	12	30	−65.0	11.0	30	−54.7	22.5
	18	30	−61.5	14.1	30	−58.5	8.6
Labiomental angle[b]	6	24	125.6	17.8	22	128.0	12.7
	12	26	110.2	15.9	25	125.1	18.1
	18	—	—	—	14	124.1	21.0
Half of the labial fissure rt (ch-sto rt)[a]	6	30	21.5	1.7	30	21.3	1.8
	12	30	25.2	2.1	30	24.3	2.7
	18	30	27.8	1.7	30	26.6	2.0
Half of the labial fissure lt (ch-sto lt)[a]	6	30	21.7	1.6	30	20.9	1.7
	12	30	24.9	2.0	30	23.8	2.7
	18	30	27.7	1.7	30	26.3	2.0
Lateral upper lip height rt (sbal-ls' rt)[a]	6	30	13.6	1.7	30	13.4	1.7
	12	30	14.8	2.2	30	13.9	1.5
	18	30	15.7	1.6	30	14.5	1.7
Lateral upper lip height lt (sbal-ls' lt)[a]	6	30	13.6	1.7	30	13.4	1.7
	12	30	14.8	2.2	30	13.9	1.5
	18	30	15.7	1.6	30	14.5	1.7
Labiotragial distance rt (ch-t rt)[a]	6	30	100.0	5.1	30	96.8	4.3
	12	30	110.2	4.4	30	107.0	4.4
	18	30	116.7	5.0	30	107.9	4.8
Labiotragial distance lt (ch-t lt)[a]	6	30	99.8	4.5	30	96.5	4.4
	12	30	109.9	4.6	30	106.9	3.7
	18	30	116.8	5.0	30	107.2	4.6
Lateral labiotragial arc rt (ch-t surf rt)[a]	6	30	109.8	5.8	30	108.3	5.8
	12	30	119.1	5.0	30	117.9	5.5
	18	30	125.7	5.6	30	119.5	5.8
Lateral labiotragial arc lt (ch-t surf lt)[a]	6	30	109.8	5.3	30	108.9	5.3
	12	30	119.0	5.0	30	118.3	5.8
	18	30	125.7	5.7	30	119.7	5.7

[a] In millimeters.
[b] In degrees.

TABLE B–VI. *Measurements of the ears in three age groups of the Chinese population*

Measurement	Age (years)	Male			Female		
		N	Mean	SD	*N*	Mean	SD
Width of the auricle rt (pra-pa rt)[a]	6	30	34.3	2.2	30	32.4	1.9
	12	30	35.8	2.5	30	34.0	2.7
	18	30	34.8	2.5	30	32.8	1.9
Width of the auricle lt (pra-pa lt)[a]	6	30	34.2	2.2	30	32.9	4.4
	12	30	35.8	1.9	30	33.5	2.7
	18	30	34.7	2.4	30	32.4	2.4
Length of the auricle rt (sa-sba rt)[a]	6	30	55.7	3.8	30	54.6	4.2
	12	30	59.2	2.6	30	58.6	3.7
	18	30	61.0	3.5	30	58.8	3.5
Length of the auricle lt (sa-sba lt)[a]	6	30	55.4	3.5	30	53.1	5.0
	12	30	58.2	2.8	30	57.5	3.5
	18	30	60.7	3.8	30	57.6	3.9
Morphological width of the ear rt (obs-obi rt)[a]	6	30	44.7	3.4	30	45.0	3.5
	12	30	48.6	2.7	30	49.1	4.4
	18	30	53.4	3.3	30	50.5	3.4
Morphological width of the ear lt (obs-obi lt)[a]	6	30	45.3	3.3	30	44.6	3.0
	12	30	49.0	3.1	30	48.7	4.5
	18	30	53.2	3.4	30	50.4	3.3
Upper naso-aural distance rt (n-obs rt)[a]	6	30	109.5	4.1	30	107.7	4.6
	12	30	114.2	9.2	30	115.1	5.2
	18	30	125.8	5.6	30	117.2	7.1
Upper naso-aural distance lt (n-obs lt)[a]	6	30	108.6	3.5	30	106.6	4.5
	12	30	114.0	8.8	30	115.3	4.8
	18	30	125.3	5.6	30	116.4	7.3
Lower naso-aural distance rt (n-obi rt)[a]	6	30	99.0	3.8	30	95.0	3.9
	12	30	106.7	4.3	30	102.7	4.4
	18	30	113.8	5.4	30	106.2	4.3
Lower naso-aural distance lt (n-obi lt)[a]	6	30	99.0	3.4	30	95.2	3.9
	12	30	106.2	6.7	30	103.2	4.5
	18	30	114.5	5.4	30	106.1	4.8
Lower subnasale–aural distance rt (sn-obi rt)[a]	6	30	99.0	3.8	30	95.0	3.9
	12	30	106.7	4.3	30	102.7	4.4
	18	30	113.8	5.4	30	106.2	4.3
Lower subnasale–aural distance lt (sn-obi lt)[a]	6	30	99.0	3.4	30	95.2	3.9
	12	30	106.2	6.7	30	103.2	4.5
	18	30	114.5	5.4	30	106.1	4.8
Upper gnathion–aural distance rt (gn-obs rt)[a]	6	30	132.7	6.5	30	131.2	5.3
	12	30	145.7	4.9	30	144.9	5.8
	18	30	159.5	5.1	30	149.6	5.1
Upper gnathion–aural distance lt (gn-obs lt)[a]	6	30	132.1	5.9	30	130.9	5.2
	12	30	146.2	4.6	30	144.9	5.7
	18	30	159.8	4.8	30	149.9	4.9
Lower gnathion–aural distance rt (gn-obi rt)[a]	6	30	98.2	7.5	30	95.5	3.7
	12	30	108.4	8.3	30	104.3	5.5
	18	30	115.8	5.8	30	109.7	4.3
Lower gnathion–aural distance lt (gn-obi lt)[a]	6	30	98.1	6.6	30	95.8	4.1
	12	30	109.0	8.1	30	104.8	5.5
	18	30	117.1	5.7	30	109.9	4.9
Lower subnasale–aural surface distance rt (sn-obi surf rt)[a]	6	30	119.9	6.7	30	117.8	4.4
	12	30	130.3	6.0	30	128.0	6.1
	18	30	137.7	6.1	30	132.3	5.1
Lower subnasale-aural surface distance lt (sn-obi surf lt)[a]	6	30	120.3	6.6	30	118.2	4.6
	12	30	130.5	6.0	30	128.6	6.4
	18	30	138.1	6.5	30	132.1	5.4
Lower gnathion–aural surface distance rt (gn-obi surf rt)[a]	6	30	113.9	6.7	30	111.9	5.4
	12	30	124.9	5.9	30	123.6	6.0
	18	30	136.5	6.8	30	128.8	5.3
Lower gnathion–aural surface distance lt (gn-obi surf lt)[a]	6	30	112.8	6.0	30	111.2	5.7
	12	30	125.2	6.1	30	122.8	7.2
	18	30	136.5	6.7	30	129.1	6.2

TABLE B–VI. *Continued.*

Measurement	Age (years)	Male			Female		
		N	Mean	SD	*N*	Mean	SD
Inclination of the medial axis of the ear rt[b]	6	28	21.8	5.8	28	21.6	5.0
	12	28	19.6	5.7	30	21.5	4.9
	18	28	19.5	5.0	29	19.3	3.2
Inclination of the medial axis of the ear lt[b]	6	30	21.7	5.7	29	21.5	4.8
	12	30	19.8	5.6	29	21.3	4.2
	18	30	19.8	5.3	29	19.0	3.0
Protrusion of the ear rt[b]	6	30	31.1	15.1	30	24.5	8.5
	12	30	26.9	6.7	30	22.1	7.7
	18	30	35.2	9.3	30	24.6	8.5
Protrusion of the ear lt[b]	6	30	31.0	15.1	30	24.7	8.5
	12	29	27.2	7.8	30	21.6	6.7
	18	30	35.2	9.3	30	26.4	10.3

[a] In millimeters.
[b] In degrees.

ACKNOWLEDGMENTS

The authors are grateful to all persons and institutes that have helped in the establishment of the facial anthropometric norms in the healthy Chinese population in Singapore.

The research was initially funded by the Ministry of Health of Singapore under the aegis of the Health and Manpower Development Plan (HMDP) visiting expert program.

The authors are indebted to Dr. Chew Chin Hin (Deputy Director of Medical Services, Republic of Singapore) and Dr. Kwa Soon Bee (Permanent Secretary, Health/ Director of Medical Services, Republic of Singapore) for their encouragement and support of this study.

The research project was supported also through the generous donations of the Lee Foundation, Shaw Foundation, and Singapore Totalisator Board.

The successful examination of children 6 and 12 years of age depended on the effective organization provided by the following individuals: Dr. Uma Rajan (Medical Director) and her staff of doctors and nurses in the School Health Services, Ministry of Health, Singapore; Mr. Chew Sang Chang and Mr. Patrick Sih Seah, the Principal and Vice Principal, respectively, of the former Silat Primary School; Mrs. Lilian Quek, the Principal of the Zhangde Primary School; and the teachers at both schools. The collection of the facial data in young adults was made possible by the help of the following medical representatives of the Singapore Armed Forces in the Headquarters Medical Services (HQMS): LTC Dr. Lim Meng Kin, the Chief Medical Officer, HQMS; LTC Dr. Tan Kang Ping, Commanding Officer, Medical Classification Centre; and Major Dr. Low Wye Mun, Secretary, Medical Research Committee, HQMS.

The authors are grateful to the following individuals: Dr. S. C. Emmanuel, Director of the Research and Evaluation Department, Ministry of Health, Singapore, for providing professional help; Mrs. Aylianna Phe, statistician, for statistical assistance; the doctors and staff of the R & E Department, for prompt data entry, processing, and generation of the population norms; Miss Priscilla Loh, Secretary, Department of Plastic Surgery, Singapore General Hospital, Ministry of Health, for secretarial assistance; and Mr. Ung Eu Chong, for his volunteer services in the project.

Appendix C

Craniofacial Norms in Young Adult African-Americans

Leslie G. Farkas, Govindasarma Venkatadri, and Ananda V. Gubbi

The study group consisted of 50 males and 50 females between 18 and 25 years of age. A total number of 108 measurements (79 linear and 29 angular; 71 single and 37 paired) were taken by one of the authors (LGF): 14 on the head, 29 on the face, 11 on the orbits, 28 on the nose, 18 on the lips and mouth, and 8 on the ears.

The interview of the probands revealed no other race in the families in 28 of the 100 subjects (in 20 males and 8 females). In 72% (72 of 100; 29 males and 43 females), of the families the presence of other races was acknowledged. The most frequent race was that of the American Indians (28%), followed by a combination of American Indians and Caucasians (20%), Caucasians only (13%), and other races (11%).

In the tables the capital C indicates the African-Americans. The regions of the craniofacial complex are identified by Roman numerals: I for the head, II for the face, III for the orbits, IV for the nose, V for the lips and mouth, and VI for the ears. The tables report the mean values and the standard deviations (SD) of the measurements.

TABLE C–I. *Measurements of the head in young adult African-Americans*

Measurement	Male		Female	
	Mean	SD	Mean	SD
Width of the head (eu-eu)[a]	148.8	6.7	141.4	6.0
Width of the forehead (ft-ft)[a]	116.3	5.6	111.4	5.2
Skull-base width (t-t)[a]	143.4	6.1	136.1	4.7
Height of the calvarium (v-tr)[a]	43.3	8.5	40.5	8.8
Anterior height of head (v-n)[a]	117.1	9.5	107.8	10.1
Special height of head (v-en)[a]	126.7	9.3	117.7	10.2
Height of head and nose (v-sn)[a]	167.5	9.9	155.4	9.5
Combined height of the head and face (v-gn)[a]	237.1	10.6	218.7	9.9
Height of forehead I (tr-g)[a]	61.8	6.8	55.7	6.4
Height of forehead II (tr-n)[a]	72.0	7.7	67.1	5.9
Length of head (g-op)[a]	199.2	6.0	186.6	6.5
Circumference of the head[a]	573.6	15.8	547.0	16.2
Inclination of the anterior surface of the forehead (tr-g)[b]	−9.9	7.1	−9.5	7.0
Auricular height of the head (v-po lt)[a]	132.9	6.8	124.8	6.9

[a] In millimeters.
[b] In degrees.

TABLE C–II. *Measurements of the face in young adult African-Americans*

Measurement	Male		Female	
	Mean	SD	Mean	SD
Width of the face (zy-zy)[a]	138.7	5.6	130.5	4.8
Width of the mandible (go-go)[a]	104.2	6.0	96.7	5.0
Physiognomical height of the face (tr-gn)[a]	194.6	10.3	180.1	7.5
Height of the face (n-gn)[a]	125.9	8.2	116.5	6.1
Height of the upper face (n-sto)[a]	78.0	4.8	72.7	4.5
Height of the lower face (sn-gn)[a]	78.9	6.7	71.5	5.2
Height of the mandible (sto-gn)[a]	57.5	5.2	52.1	5.5
Height of the chin (sl-gn)[a]	36.0	5.6	35.2	5.7
Height of the upper profile (tr-prn)[a]	116.1	15.8	107.5	7.6
Height of the lower profile (prn-gn)[a]	92.9	11.3	87.2	5.7
Lower half of the craniofacial height (en-gn)[a]	118.9	7.9	108.3	6.2
Distance between the glabella and the subnasale (g-sn)[a]	68.8	8.6	64.6	5.2
Supraorbital arc (t-g-t)[a]	308.6	12.0	292.3	18.2
Maxillary arc (t-sn-t)[a]	301.2	14.3	287.0	19.8
Mandibular arc (t-gn-t)[a]	332.8	18.0	311.3	12.7
Inclination of the upper face profile from the vertical (g-sn)[b]	2.5	4.7	4.7	4.1
Inclination of the Leiber line from the vertical (g-ls)[b]	7.6	3.6	8.9	3.9
Inclination of the lower face profile from the vertical (sn-pg)[b]	−12.6	9.8	−13.0	6.8
Inclination of the mandible from the vertical (li-pg)[b]	−27.3	17.0	−29.8	18.6
Inclination of the chin from the vertical[b]	−6.4	13.8	−6.3	12.7
Inclination of the general profile line from the vertical (g-pg)[b]	−7.6	12.3	−4.2	5.5
Mentocervical angle[b]	89.3	12.9	92.5	12.9

TABLE C–II. *Continued.*

Measurement	Male Mean	Male SD	Female Mean	Female SD
Tragion–glabella depth (t-g)[a]				
Right	128.9	5.0	121.9	4.6
Left	128.3	4.5	120.8	4.8
Tragion–nasion depth (t-n)[a]				
Right	122.9	5.0	116.7	4.2
Left	123.2	4.9	116.5	4.2
Tragion–subnasale depth (t-sn)[a]				
Right	132.6	5.8	125.4	4.9
Left	132.6	5.6	124.8	4.5
Tragion–gnathion depth (t-gn)[a]				
Right	149.5	7.3	138.1	6.0
Left	150.2	6.9	138.8	5.6
Lateral surface half-arc in the upper third of the face (t-g surf)[a]				
Right	153.6	6.3	147.0	6.0
Left	155.2	6.2	147.3	5.9
Lateral surface half-arc in the middle third of the face (t-sn surf)[a]				
Right	149.6	7.7	143.8	6.1
Left	151.5	7.1	145.2	5.8
Lateral surface half-arc in the lower third of the face (t-gn surf)[a]				
Right	167.3	9.6	156.1	7.0
Left	165.5	9.3	155.4	6.5

[a] In millimeters.
[b] In degrees.

TABLE C–III. *Measurements of the orbits in young adult African-Americans*

Measurement	Male Mean	Male SD	Female Mean	Female SD
Intercanthal width (en-en)[a]	35.8	2.8	34.4	3.4
Biocular width (ex-ex)[a]	96.8	4.6	92.9	5.3
Length of the eye fissure (ex-en)[a]				
Right	32.9	1.7	32.4	2.4
Left	32.9	1.6	32.2	2.0
En–facial midline distance (en-se)[a]				
Right	23.3	2.1	22.2	2.1
Left	23.3	2.0	22.0	2.0
Orbit–tragion distance (ex-t)[a]				
Right	83.8	4.3	79.9	4.0
Left	83.6	4.2	79.5	3.7
Orbito–gonial distance (ex-go)[a]				
Right	103.4	6.8	95.8	5.3
Left	103.4	6.7	95.7	5.0
Orbito–eyebrow height (or-sci)[a]				
Right	39.8	3.3	39.1	3.4
Left	39.5	3.2	38.9	3.3
Height of the eye fissure (ps-pi)[a]				
Right	10.0	1.1	10.4	1.2
Left	10.0	1.1	10.4	1.2
En/ex sagittal level difference				
Left	7.8	2.8	7.2	2.4
Eye fissure inclination (en-ex line)[b]				
Right	7.0	3.4	9.1	2.5
Left	7.0	3.3	9.0	2.6
Inclination of the line placed over the upper and lower orbital rims (os'/or')[b]				
Right	16.4	3.9	12.5	3.8
Left	16.4	3.9	12.5	3.8

[a] In millimeters.
[b] In degrees.

TABLE C–IV. *Measurements of the nose in young adult African-Americans*

Measurement	Male		Female	
	Mean	SD	Mean	SD
Width of the nasal root (mf-mf)[a]	22.8	2.8	21.3	2.3
Width of the nose (al-al)[a]	44.1	3.4	40.1	3.2
Anatomical width of the nose (ac-ac)[a]	40.0	4.5	37.7	3.5
Width between the labial insertions of the alar base (sbal-sbal)[a]	23.1	3.4	22.7	3.5
Width of the columella (sn′-sn′)[a]	8.0	1.0	7.5	0.8
Height of the nose (n-sn)[a]	51.9	3.0	48.8	3.7
Length of the nasal bridge (n-prn)[a]	45.6	3.5	42.6	3.7
Nasal tip protrusion (sn-prn)[a]	17.6	2.0	16.1	2.1
Inclination of the nasal bridge[b]	32.2	5.0	33.4	5.7
Inclination of the columella[b]	72.0	17.6	71.2	13.1
Nasal tip inclination[b]	55.1	8.7	54.1	8.0
Glabellonasal angle[b]	149.6	11.6	159.8	8.9
Nasofrontal angle[b]	126.5	12.0	127.6	8.1
Nasal root–slope angle[b]	84.2	11.5	84.9	12.7
Nasal tip angle[b]	69.2	10.3	70.1	11.7
Alar–slope angle[b]	96.7	18.0	90.2	13.8
Nasolabial angle[b]	71.4	14.5	73.9	14.5
Bridge deviation[b]				
Right	0.3	1.2	0.0	0.0
Left	0.2	0.9	0.0	0.0
Columella deviation[b]				
Right	0.0	0.0	0.0	0.0
Left	0.0	0.0	0.0	0.0
Width of the nostril floor (sbal-sn)[a]				
Right	12.4	1.7	12.5	2.8
Left	12.5	1.8	12.6	2.7
Thickness of the ala (al′-al′)[a]				
Right	6.9	1.0	6.5	1.1
Left	6.9	1.0	6.5	1.0
Facial alar base–subnasale distance (ac-sn)[a]				
Right	24.1	4.0	22.9	2.6
Left	24.1	4.1	22.9	2.6
Length of the ala (ac-prn)[a]				
Right	34.0	2.2	31.5	1.8
Left	34.0	2.2	31.6	1.8
Length of the columella (sn-c′)[a]				
Right	9.8	1.6	9.2	2.0
Left	9.8	1.6	9.1	1.9
Depth of the nasal root (en-se sag)[a]				
Right	12.6	1.8	10.5	2.4
Left	12.7	1.7	10.5	2.4
Surface length of the ala (ac-prn surf)[a]				
Right	40.8	3.2	37.5	2.4
Left	40.8	3.2	37.6	2.4
Inclination of the longitudinal nostril axis[b]				
Right	23.5	8.0	25.2	12.4
Left	23.7	8.0	25.1	12.4
Inclination of the alar–slope line (al-prn)[b]	42.0	7.1	43.0	6.8

[a] In millimeters.
[b] In degrees.

TABLE C–V. *Measurements of the lips and mouth in young adult African-Americans*

Measurement	Male		Female	
	Mean	SD	Mean	SD
Width of the philtrum (cph-cph)[a]	13.0	2.1	12.0	1.7
Width of the mouth (ch-ch)[a]	54.6	4.1	53.6	4.0
Height of the upper lip (sn-sto)[a]	26.1	2.5	24.5	3.0
Height of the cutaneous upper lip (sn-ls)[a]	16.4	2.0	14.0	2.0
Vermilion height of the upper lip (ls-sto)[a]	13.6	2.1	13.3	1.9
Vermilion height of the lower lip (sto-li)[a]	13.8	2.1	13.2	1.9
Height of the cutaneous lower lip (li-sl)[a]	11.8	2.5	10.7	2.4
Height of the lower lip (sto-sl)[a]	22.1	2.4	20.2	2.4
Vermilion surface arc of the upper lip (ch-ls-ch)[a]	79.8	8.2	76.5	6.0
Vermilion surface arc of the lower lip (ch-li-ch)[a]	72.6	7.3	69.9	6.6
Inclination of the labial fissure[b]	0.0	0.0	0.0	0.0
Inclination of the upper lip[b]	27.5	12.3	29.4	13.2
Inclination of the lower lip[b]	−66.9	15.8	−65.6	21.3
Labiomental angle[b]	101.5	17.7	101.6	18.0
Halves of the labial fissure length (ch-sto)[a]				
Right	33.1	2.8	32.4	3.2
Left	33.0	2.7	32.3	3.0
Lateral upper lip height (sbal-ls')[a]				
Right	16.9	2.4	15.6	1.9
Left	16.8	2.4	15.5	1.8
Labiotragial distance (ch-t)[a]				
Right	121.3	5.7	113.4	5.2
Left	121.3	5.4	112.8	4.9
Lateral labiotragial arc (ch-t surf)[a]				
Right	130.2	6.4	121.9	6.1
Left	130.8	5.6	122.4	5.5

[a] In millimeters.
[b] In degrees.

TABLE C–VI. *Measurements of the ears in young adult African-Americans*

Measurement	Male		Female	
	Mean	SD	Mean	SD
Width of the auricle (pra-pa)[a]				
Right	36.7	3.5	34.6	2.9
Left	36.2	3.2	34.2	2.9
Length of the auricle (sa-sba)[a]				
Right	60.5	4.1	57.2	3.5
Left	59.8	4.0	57.0	3.3
Lower subnasale–aural distance (sn-obi)[a]				
Right	118.9	6.4	112.2	4.4
Left	119.3	6.3	112.2	4.2
Lower gnathion–aural distance (gn-obi)[a]				
Right	124.4	7.5	115.8	7.0
Left	125.1	7.7	115.7	5.6
Lower subnasale–aural surface distance (sn-obi surf)[a]				
Right	140.2	8.8	135.0	5.7
Left	141.3	8.8	135.6	4.8
Lower gnathion–aural surface distance (gn-obi surf)[a]				
Right	146.4	8.5	135.3	6.4
Left	147.0	8.4	135.5	6.2
Inclination of the ear[b]				
Right	17.4	5.2	20.9	5.7
Left	17.4	5.2	20.9	5.7
Protrusion of the ear[b]				
Right	16.2	6.8	12.6	4.9
Left	16.2	6.8	12.6	4.9

[a] In millimeters.
[b] In degrees.

ACKNOWLEDGMENTS

The collection of the craniofacial data in young adult African-Americans was made possible by a private donation made by the following individuals: Joseph Bernat, D.D.S., and Paul R. Creighton, D.D.S., Children's Hospital, Buffalo, New York; and Manuel H. Castillo, M.D., Veterans Administration Hospital, Buffalo, New York. The authors are grateful to Professors Joyce E. Sirianni, Ph.D., and A. T. Steegmann, Jr., Ph.D., Department of Anthropology, University of Buffalo, New York, for technical assistance provided during the examinations. Statistical analysis of the materials was carried out by Ananda V. Gubbi, Ph.D., biostatistician, University of Pittsburgh.

Appendix D

Statistical Appendix

Curtis K. Deutsch and Ralph B. D'Agostino, Jr.

Leslie Farkas, M.D., has outlined elegant methods for producing reliable, objective craniofacial measurements. Reviewing the normative data, one is impressed by individual differences in these measurements. Though biological constraints limit the range of possible values, it is clear that there is considerable variation in the species. So how can we judge the magnitude of these measurements, equilibrating for subject differences in age, sex, and ethnicity?

What is needed is a common metric that allows these comparisons. In this appendix we describe a commonly used means of standardization, one based on the "normal distribution." In addition, we (a) describe normal distributions and how they relate to standardized scores, (b) define standardized scores (here, *z*-scores) and describe their application to linear measurements, angles of inclination, and proportions, and (c) provide worked examples.

PROPERTIES OF NORMAL DISTRIBUTIONS

When data are normally distributed, a plot of the values of the data (on the x axis) versus the relative frequency of the values observed (on the y axis) will be bell-shaped and symmetric. All standard normal distributions have a mean of 0 and a standard deviation of 1, regardless of the mean and standard deviation of the data from which they were originally calculated. One can determine the precise percentage of observations expected to fall within any range of particular values. For instance, it is known that in any normal distribution, 68% of observations will fall within one standard deviation of the mean, 95% will fall within two standard deviations of the mean, and 99.7% will fall within three standard deviations of the mean (1).

Using the normal curve, one can ask questions about the relative frequencies in a distribution by calculating

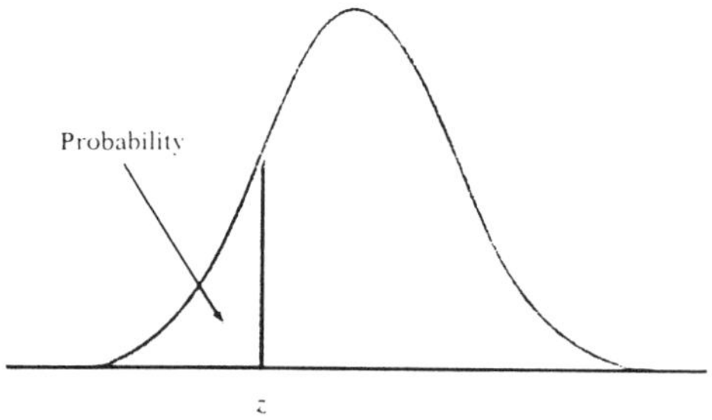

Table entry is probability at or below z.

TABLE D-1. *Cumulative standard normal probabilities*[a]

z	.00	.01	.02	.03	.04	.05	.06	.07	.08	.09
−3.4	.0003	.0003	.0003	.0003	.0003	.0003	.0003	.0003	.0003	.0002
−3.3	.0005	.0005	.0005	.0004	.0004	.0004	.0004	.0004	.0004	.0003
−3.2	.0007	.0007	.0006	.0006	.0006	.0006	.0006	.0005	.0005	.0005
−3.1	.0010	.0009	.0009	.0009	.0008	.0008	.0008	.0008	.0007	.0007
−3.0	.0013	.0013	.0013	.0012	.0012	.0011	.0011	.0011	.0010	.0010
−2.9	.0019	.0018	.0018	.0017	.0016	.0016	.0015	.0015	.0014	.0014
−2.8	.0026	.0025	.0024	.0023	.0023	.0022	.0021	.0021	.0020	.0019
−2.7	.0035	.0034	.0033	.0032	.0031	.0030	.0029	.0028	.0027	.0026
−2.6	.0047	.0045	.0044	.0043	.0041	.0040	.0039	.0038	.0037	.0036
−2.5	.0062	.0060	.0059	.0057	.0055	.0054	.0052	.0051	.0049	.0048
−2.4	.0082	.0080	.0078	.0075	.0073	.0071	.0069	.0068	.0066	.0064
−2.3	.0107	.0104	.0102	.0099	.0096	.0094	.0091	.0089	.0087	.0084
−2.2	.0139	.0136	.0132	.0129	.0125	.0122	.0119	.0116	.0113	.0110
−2.1	.0179	.0174	.0170	.0166	.0162	.0158	.0154	.0150	.0146	.0143
−2.0	.0228	.0222	.0217	.0212	.0207	.0202	.0197	.0192	.0188	.0183
−1.9	.0287	.0281	.0274	.0268	.0262	.0256	.0250	.0244	.0239	.0233
−1.8	.0359	.0351	.0344	.0336	.0329	.0322	.0314	.0307	.0301	.0294
−1.7	.0446	.0436	.0427	.0418	.0409	.0401	.0392	.0384	.0375	.0367
−1.6	.0548	.0537	.0526	.0516	.0505	.0495	.0485	.0475	.0465	.0455
−1.5	.0668	.0655	.0643	.0630	.0618	.0606	.0594	.0582	.0571	.0559
−1.4	.0808	.0793	.0778	.0764	.0749	.0735	.0721	.0708	.0694	.0681
−1.3	.0968	.0951	.0934	.0918	.0901	.0885	.0869	.0853	.0838	.0823
−1.2	.1151	.1131	.1112	.1093	.1075	.1056	.1038	.1020	.1003	.0985
−1.1	.1357	.1335	.1314	.1292	.1271	.1251	.1230	.1210	.1190	.1170
−1.0	.1587	.1562	.1539	.1515	.1492	.1469	.1446	.1423	.1401	.1379
−0.9	.1841	.1814	.1788	.1762	.1736	.1711	.1685	.1660	.1635	.1611
−0.8	.2119	.2090	.2061	.2033	.2005	.1977	.1949	.1922	.1894	.1867
−0.7	.2420	.2389	.2358	.2327	.2296	.2266	.2236	.2206	.2177	.2148
−0.6	.2743	.2709	.2676	.2643	.2611	.2578	.2546	.2514	.2483	.2451
−0.5	.3085	.3050	.3015	.2981	.2946	.2912	.2877	.2843	.2810	.2776
−0.4	.3446	.3409	.3372	.3336	.3300	.3264	.3228	.3192	.3156	.3121
−0.3	.3821	.3783	.3745	.3707	.3669	.3632	.3594	.3557	.3520	.3483
−0.2	.4207	.4168	.4129	.4090	.4052	.4013	.3974	.3936	.3897	.3859
−0.1	.4602	.4562	.4522	.4483	.4443	.4404	.4364	.4325	.4286	.4247
−0.0	.5000	.4960	.4920	.4880	.4840	.4801	.4761	.4721	.4681	.4641

the areas under the curve (2). Here, we present these areas as a table of standard normal probabilities (Table D–1). This table provides values for the cumulative area under the normal curve up to and including the *z*-score. These areas define proportions of patients who have measurements at that point on the distribution or less.

Table D–1 relates the *z*-scores (on a continuum) to the area delineated under the normal curve. A *z*-score of 0 occurs when a patient's measurements are *exactly* the same as the average value found in the normative data for patients of a given age, gender, and ethnicity.

One can plot the distribution of a given measurement (e.g., ch-ch) and can describe it by its mean and standard deviation. Means and standard deviations are statistics calculated from the *samples* of individuals included in the study. Each sex-, age-, and ethnicity-specific sample is taken from the *population* of available patients with the same sex, age, and ethnic status. The *sample mean,* $\bar{X}$, is an estimate of the *population mean,* μ, and the *sample standard deviation,* SD, is an estimate of the *population standard deviation,* σ.

Population parameters, μ and σ, are based on calculations made using *every patient* in a population. Because it is impossible to measure every patient in the popula-

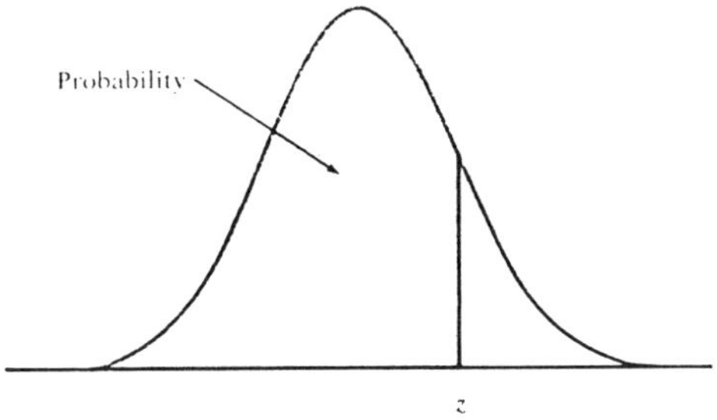

Table entry is probability at or below z.

TABLE D–1. *Continued.*

z	.00	.01	.02	.03	.04	.05	.06	.07	.08	.09
0.0	.5000	.5040	.5080	.5120	.5160	.5199	.5239	.5279	.5319	.5359
0.1	.5398	.5438	.5478	.5517	.5557	.5596	.5636	.5675	.5714	.5753
0.2	.5793	.5832	.5871	.5910	.5948	.5987	.6026	.6064	.6103	.6141
0.3	.6179	.6217	.6255	.6293	.6331	.6368	.6406	.6443	.6480	.6517
0.4	.6554	.6591	.6628	.6664	.6700	.6736	.6772	.6808	.6844	.6879
0.5	.6915	.6950	.6985	.7019	.7054	.7088	.7123	.7157	.7190	.7224
0.6	.7257	.7291	.7324	.7357	.7389	.7422	.7454	.7486	.7517	.7549
0.7	.7580	.7611	.7642	.7673	.7704	.7734	.7764	.7794	.7823	.7852
0.8	.7881	.7910	.7939	.7967	.7995	.8023	.8051	.8078	.8106	.8133
0.9	.8159	.8186	.8212	.8238	.8264	.8289	.8315	.8340	.8365	.8389
1.0	.8413	.8438	.8461	.8485	.8508	.8531	.8554	.8577	.8599	.8621
1.1	.8643	.8665	.8686	.8708	.8729	.8749	.8770	.8790	.8810	.8830
1.2	.8849	.8869	.8888	.8907	.8925	.8944	.8962	.8980	.8997	.9015
1.3	.9032	.9049	.9066	.9082	.9099	.9115	.9131	.9147	.9162	.9177
1.4	.9192	.9207	.9222	.9236	.9251	.9265	.9279	.9292	.9306	.9319
1.5	.9332	.9345	.9357	.9370	.9382	.9394	.9406	.9418	.9429	.9441
1.6	.9452	.9463	.9474	.9484	.9495	.9505	.9515	.9525	.9535	.9545
1.7	.9554	.9564	.9573	.9582	.9591	.9599	.9608	.9616	.9625	.9633
1.8	.9641	.9649	.9656	.9664	.9671	.9678	.9686	.9693	.9699	.9706
1.9	.9713	.9719	.9726	.9732	.9738	.9744	.9750	.9756	.9761	.9767
2.0	.9772	.9778	.9783	.9788	.9793	.9798	.9803	.9808	.9812	.9817
2.1	.9821	.9826	.9830	.9834	.9838	.9842	.9846	.9850	.9854	.9857
2.2	.9861	.9864	.9868	.9871	.9875	.9878	.9881	.9884	.9887	.9890
2.3	.9893	.9896	.9898	.9901	.9904	.9906	.9909	.9911	.9913	.9916
2.4	.9918	.9920	.9922	.9925	.9927	.9929	.9931	.9932	.9934	.9936
2.5	.9938	.9940	.9941	.9943	.9945	.9946	.9948	.9949	.9951	.9952
2.6	.9953	.9955	.9956	.9957	.9959	.9960	.9961	.9962	.9963	.9964
2.7	.9965	.9966	.9967	.9968	.9969	.9970	.9971	.9972	.9973	.9974
2.8	.9974	.9975	.9976	.9977	.9977	.9978	.9979	.9979	.9980	.9981
2.9	.9981	.9982	.9982	.9983	.9984	.9984	.9985	.9985	.9986	.9986
3.0	.9987	.9987	.9987	.9988	.9988	.9989	.9989	.9989	.9990	.9990
3.1	.9990	.9991	.9991	.9991	.9992	.9992	.9992	.9992	.9993	.9993
3.2	.9993	.9993	.9994	.9994	.9994	.9994	.9994	.9995	.9995	.9995
3.3	.9995	.9995	.9995	.9996	.9996	.9996	.9996	.9996	.9996	.9997
3.4	.9997	.9997	.9997	.9997	.9997	.9997	.9997	.9997	.9997	.9998

From Moore and McCabe, 1993.
[a] This table is condensed from Table 1 of ref. 7, which was adapted from ref. 2.
[b] The first two digits of the z-score are in the first column. The third digit is in the top row.

tion, we calculate the statistics, $\overline{X}$ and SD, using the data from the sample. In this book, we assume that the underlying distribution of the sample mean, $\overline{X}$, is normal. This assumption is reasonable because each age-, sex-, and ethnicity-specific sample consisted of about 50 patients, which, for most applications, is considered large. Even if the individual measurements were not "normal", their average should approximate a normal distribution.

Next, we describe how to use these data to make inferences about an *individual* patient's measurements.

STANDARDIZED SCORES

How can the clinician judge which measurements are extreme? For instance, if an 18-year-old male's head circumference and mouth width were both 8 mm larger than average, the clinician might believe that the 8-mm increase in mouth width was more interesting than a 8-mm increase in head circumference. The difficulty the clinician faces here is that the value "8 mm" does mean the same thing for each measurement (i.e., 8 mm is a larger proportion of the total mouth width than of the total head circumference). This problem in scaling measurements is expressed against a background of age, sex, and ethnicity differences (see above). A technique commonly used to handle these problems is to calculate *standardized scores* for individuals and then use these for comparisons. A standardized score, referred to as a *z*-score when the underlying distribution of the measurement is normal, is the number of standard deviations a measurement is from the mean value for that particular measurement. To calculate this score for a given measurement one uses the following formula:

$$z = \frac{\text{patient score minus normative group mean}}{\text{normative group standard deviation}}$$

Here, normative group refers to either (a) the sex-, age-, and ethnicity-specific norms found for measurements in the preceding appendixes or (b) functions of these measurements (e.g., proportions) (3). A positive *z*-score indicates that a patient's score was larger than the normative group mean, whereas a negative *z*-score indicates that a patient's score was smaller than the normative group mean.

WHY USE STANDARDIZED SCORES?

The answer to this question depends on the application:

Medical Genetics and Teratology

A useful property of *z*-scores is that they provide a metric for describing individual differences in craniofacial morphology. For instance, a medical geneticist may wish to describe dysmorphic features in patients with a specific but rare cytogenetic abnormality. Alternatively, a dysmorphologist may attempt to assess malformations associated with a teratogen. In both cases, it may be difficult to assemble more than a modest sample of patients, and the patients may be variable with respect to age, sex, and ethnicity. By using standardized scores, one can express individual differences using a common metric. Cutoff points can be applied to determine the prevalence of dysmorphic signs in each sample.

Because *z*-scores provide standard means of operationally defining dysmorphic signs, they may find application in systematics. One can describe the frequency of specific dysmorphic features in identifiable craniofacial syndromes or delineate previously unidentified syndromes.

Plastic Surgery

Like medical geneticists and teratologists, plastic surgeons can use these metrics to quantify the extent of dysmorphology. Furthermore, they use them to plan surgical interventions and perform postoperative assessments.

In planning a procedure, the surgeon is essentially conducting a single-subject experiment. He or she utilizes not only indices but also individual linear measurements and angles of inclination. The computation of *z*-scores is helpful, because it provides a statistical description of craniofacial features for patients of like age, sex, and ethnicity that can be used as a quantitative guide. Leslie Farkas, M.D., has suggested a qualitative grading system for these purposes (ref. 3, pages 5–8).

Physical Anthropology

Anthropologists use *z*-scores to obtain common dependent measures (e.g., a *z*-score for ch-ch) which they can compare across experimental conditions. Whereas plastic surgeons and medical geneticists often express their findings as dichotomous variables (i.e., extreme scores, present or absent), the physical anthropologist expresses the variable on a continuous scale (i.e., including all possible values of the measure, from minimum to maximum). In many cases the clinician sacrifices valuable information when outcome measures are dichotomized.

PROPORTIONS

In medical anthropology, dysmorphic signs conventionally have been diagnosed by proportions—indices that provide operational definitions of the anomalies. These ratio-based definitions gauge the extreme placement of a single measurement on a continuum by contrasting it against another measurement. For instance, the intercanthal index is commonly used to diagnose hypertelorism (3) (see example below).

This approach finds broad application in anthropometrics. A medical geneticist or plastic surgeon might operationally define craniofacial anomalies by using anthropometrics to reproduce and quantify his/her pattern recognition methods. The surgeon is presented with a further challenge of restoring harmony to the dysmorphic face; in some regions of the craniofacial complex, anatomical–physical constraints may limit surgical adjustment of individual dimensions, the components of the proportion index.

To illustrate how to use Table D–1, we offer the following two examples:

Example 1. A 10-year-old Caucasian male patient has an intercanthal index [defined as intercanthal width (en-

en) × 100, divided by biocular width (ex-ex)] equal to 42. We would like to determine if his orbits are hyperteloric (wide-set), which occurs when this ratio is very large. To determine this we refer to the population values for the mean and standard deviation for this measurement in 10-year-old Caucasian boys. We find that the mean is 37.7 with a standard deviation of 1.9; thus this patient's *z*-score is

$$z = \frac{42.0 - 37.7}{1.9} = 2.3$$

This score is more than 2 standard deviations above the population mean, and thus there is evidence that he has hypertelorism. We see from Table D–1 that the value of 2.3 corresponds to the probability .9893. This means that over 98% of Caucasian boys of his age would have a smaller intercanthal index.

Example 2. A 16-year-old Caucasian female has a left palpebral fissure inclination equal to −1 degree. Because the inclination is negative, the clinician would term this fissure "antimongoloid." We would like to determine how abnormal this degree of rotation is. First, we would look up the mean and standard deviation for 16-year-old females and find that the mean is 4.0 degrees and the standard deviation is 2.5. Our patient's *z*-score is −2, which indicates that her angle is much more negative than the majority of women of her age and race. From Table D-1, we see what percentage of 16-year-old women had measurements above or below −2:.0228. Thus, our patient is in the bottom 2.3% of Caucasian women of the same age. In other words, we expect 97.7% of Caucasian women of the same age to have a higher, more positive inclination.

CUTOFF POINTS FOR *z*-SCORES

Though cutoff points used to indicate abnormality are arbitrary, there is a convention in the physical and medical anthropology literatures to use a *z*-score cutoff of 2. That is, a high value would be defined as a *z*-score greater than or equal to +2, and a low value would be defined as a *z*-score less than or equal to −2. We see from the Table D-1 that 2.28% of values in the normal distribution lie above the *z*-score of +2. Similarly, 2.28% of values lie below the *z*-score of −2.

Why has the convention of a *z*-score cutoff of 2 been so widely adopted in the literature? Though there is no formal answer, it seems likely that it was chosen because 2.00 lies so close to the 1.96 cutoff point used in scientific hypothesis testing. The *z*-score of 1.96 delineates the "$p < .05$" cutoff point—the rule of the game for determining whether or not a finding is *statistically* significant. But the clinician should not feel bound to this figure, because this cutoff point does not confer *clinical* significance.

Ultimately, it is up to the clinician/scientist to decide how extreme a *z*-score should be before it is considered abnormal. For instance, if a clinician believes that a mouth should be considered "wide" if 90% of individuals of the same age, gender, and ethnicity have smaller mouths, then a *z*-score of greater than 1.29 would indicate a "wide" mouth.

OPERATIONAL DEFINITIONS BASED ON PROPORTIONS

To assess a wide mouth, for instance, one may contrast the linear measurement ch-ch against the neighboring horizontal measurement, maxillary width (zy-zy). This contrast serves as an internal control for size (see below). If one operationally defined a wide mouth merely by an extreme value of *z* (ch-ch), one would fail to detect positive diagnoses in faces with generally small craniofacial proportions—for example, in a microcephalic individual with a *z* (ch-ch) of 1.5.

Using the proportion approach, one computes the *z*-score for this index, *z* [(ch-ch × 100)/(zy-zy)], and determines whether or not an individual has an extreme placement for the index (e.g., $z > 2.0$). But using this proportion introduces an obvious conundrum: Does this extreme score operationally define a wide mouth or narrow maxilla? A partial solution can be found by adding a secondary criterion to the operational definition of a wide mouth—a stipulation that the maxilla is not narrow. But this approach is successful in only a subset of cases, and provides a fallible indicator in other cases.

Are there alternative solutions to this conundrum? While a *z*-score less than −2 for ch-ch has been described in the literature as "small" in absolute terms for an age-, sex-, and ethnicity-specific group, a more interesting question is whether the linear measurement ch-ch is small in relative terms. That is, can ch-ch appear small for some individuals even when their *z*-score is greater than −2? The answer is *yes,* if the patient's general craniofacial dimensions were very large. The choice of an operational definition is conditioned on one's theoretical model (Deutsch et al., *in preparation*).

For instance, one way to frame the question of "What is a small mouth?" is by adopting an embryological approach. One might ask how a feature is configured in relation to the dimensions of its embryonic primordia (*Anlagen*)—for example, the frontonasal, maxillary, and mandibular processes (4,5). One would then contrast the measurement of interest against an intra-*Anlage* measurement. Because the ch-ch lies at the border of the maxilla (4), the zy-zy would meet this criterion (ch-ch lies at the interface of the maxilla and mandible). The embryologist might be more interested in the configuration of the maxillary process than in the mouth width per se; here, an index alone might prove informative.

Alternatively, one might assess the extreme values of anthropometric measurements after adjusting for *general* craniofacial size. This is in contrast to assessing the magnitude of a linear distance (e.g., ch-ch) relative to a single measurement (e.g., zy-zy).

The surface scanning technology allows straightforward identification of landmarks in a three-dimensional coordinate system (see Chapters 17 and 18).* In our example, we would adjust the magnitude of ch-ch for *general* facial dimensions, computed from surface measurements. We are currently developing these general size adjustments using structured light scanning (Deutsch et al., *in preparation*).

* Digitized coordinate data can also be used to describe shape and size (*morphometrics*) in a way that simple linear measurements and angles cannot. For an excellent technical discussion of morphometry and estimation of generalized size factors based on coordinate data, the reader is referred to Bookstein (6).

REFERENCES

1. Winkler RL, Hays WL. *Statistics: probability, inference, and decision,* 2nd ed. New York: Holt, Rinehart, & Winston, 1975.
2. Moore DS, McCabe GP. *Introduction to the practice of statistics,* 2nd ed. New York: WH Freeman, 1993.
3. Farkas LG, Munro IR. *Anthropometric facial proportions in medicine.* Springfield, IL: Charles C Thomas, 1987.
4. Sperber GH. *Craniofacial Embryology,* 4th ed. London: Wright, 1989.
5. Johnston MC. Embryology of the head and neck. In: McCarthy J, ed. *Plastic surgery.* Philadelphia: WB Saunders, 1990;4:2451–2495.
6. Bookstein FL. *Morphometric tools for landmark data: geometry and biology.* New York: Cambridge University Press, 1991.
7. Pearson ES, Hartley HO. *Biometrika tables for statisticians,* vol. 1. Cambridge, UK: Cambridge University, Press, 1966.
8. Rosner B. *Fundamentals of biostatistics,* 3rd ed. Boston: PWS–Kent, 1990;535–539.

Appendix A–1

Norms of the Craniofacial Asymmetries in North American Caucasians

Tania A. Hreczko and Leslie G. Farkas

Asymmetries were calculated in 56 linear and angular surface paired measurements of the head and face. In linear measurements (both projective and tangential) and angular measurements (inclinations and angles), 1-mm difference and 1-degree difference, respectively, found between the right and left sides, were regarded as the smallest asymmetries. Asymmetries were recorded in 2 cranial, 8 facial, 13 orbital, 10 nasal, 5 labio-oral, and 18 auricular paired measurements. In 48.2% of the subjects (27 of 56) the asymmetries were reported in 18 age groups (1–18 years), in 33.9% (19 of 56) they were reported in 13 age groups (6–18 years), and in 17.9% (10 of 56) they were reported in 4 and 5 age groups (1–4 and 1–5 years, respectively).

In the given age group the differences between the individual paired measurements were expressed as means with their standard deviations (SD), separately in both sexes. The frequency of the asymmetrical findings was calculated for each age category.

In the tables, A–1 indicates the norms of asymmetries in the various age groups of North American Caucasians. The regions of the craniofacial complex are identified by Roman numerals: I for the head, II for the face, III for the orbits, IV for the nose, V for the lips and mouth, and VI for the ears. *N* represents the total number of measurements in a given age, *n* represents the number of asymmetrical findings, mean value is expressed in millimeters or degrees of asymmetry, SD stands for standard deviation, and the "percent" column lists percentage of asymmetry.

TABLE A–1-I/1. *Auricular height of the head (v-po rt,lt)*

Age (years)	N	Asymmetry (male) n	Percent	Mean	SD	N	Asymmetry (female) n	Percent	Mean	SD
6	50	6	12.0	2.5	0.8	50	2	4.0	4.0	0.0
7	50	4	8.0	2.0	0.0	50	5	10.0	3.8	1.3
8	51	8	15.7	2.9	1.5	51	10	19.6	3.4	2.0
9	51	9	17.6	3.2	1.5	50	6	12.0	3.8	2.2
10	50	15	30.0	3.7	1.6	49	7	14.3	3.3	2.6
11	48	13	27.1	3.2	1.3	51	17	33.3	3.7	2.1
12	52	17	32.7	3.4	1.5	53	15	28.3	4.1	2.6
13	50	8	16.0	3.1	1.6	49	22	44.9	3.5	1.9
14	49	10	20.4	3.1	1.4	51	7	13.7	2.6	0.5
15	50	6	12.0	3.2	1.9	51	6	11.8	5.3	1.5
16	50	10	20.0	3.7	2.1	50	5	10.0	3.8	1.5
17	49	16	32.7	2.9	1.1	51	6	11.8	3.0	1.5
18	52	7	13.5	4.0	1.4	51	5	9.8	3.8	1.6

TABLE A–1-I/2. *Distance between the vertex and each tragion (v-t rt,lt)*

Age (years)	N	Asymmetry (male) n	Percent	Mean	SD	N	Asymmetry (female) n	Percent	Mean	SD
6	26	3	11.5	3.0	1.7	22	5	22.7	2.0	0.0
7	21	3	14.3	2.7	1.2	36	11	30.6	3.1	1.4
8	41	8	19.5	2.9	1.1	38	14	36.8	3.2	1.6
9	47	10	21.3	2.9	1.0	38	11	28.9	3.1	1.8
10	37	17	45.9	3.9	2.3	35	11	31.4	3.3	2.4
11	36	14	38.9	3.1	1.4	40	16	40.0	4.2	2.3
12	34	15	44.1	3.4	1.5	28	11	39.3	2.8	1.0
13	44	10	22.7	2.7	0.9	46	22	47.8	3.3	1.5
14	46	9	19.6	3.1	1.8	43	9	20.9	2.8	0.8
15	44	7	15.9	2.1	0.4	43	9	20.9	4.3	2.1
16	48	18	37.5	3.4	1.6	34	6	17.6	3.7	2.1
17	44	19	43.2	3.2	1.4	33	7	21.2	2.9	1.2
18	26	9	34.6	3.8	1.6	17	5	29.4	4.0	2.0

TABLE A–1-II/1. *Tragion–glabellar depth of the upper third of the face (t-g rt,lt)*

Age (years)	N	Asymmetry (male) n	Percent	Mean	SD	N	Asymmetry (female) n	Percent	Mean	SD
1	18	5	27.8	1.2	0.4	20	1	5.0	1.0	0.0
2	30	7	23.3	1.0	0.0	32	12	37.5	1.5	0.5
3	30	17	66.7	1.2	0.4	30	11	36.7	1.2	0.4
4	30	4	13.3	3.8	1.0	30	2	6.7	3.0	0.0
5	30	5	16.7	3.0	0.0	30	1	3.3	3.0	0.0

TABLE A-1-II/2. *Tragion-nasion depth of the upper third of the face (t-n rt,lt)*

Age		Asymmetry (male)					Asymmetry (female)			
(years)	*N*	*n*	Percent	Mean	SD	*N*	*n*	Percent	Mean	SD
1	18	2	11.1	1.0	0.0	20	5	25.0	1.2	0.4
2	30	9	30.0	1.2	0.4	32	6	18.7	1.2	0.4
3	30	9	30.0	1.2	0.7	30	14	46.7	1.5	0.7
4	30	2	6.7	3.0	0.0	30	7	23.3	3.1	0.4
5	30	4	13.3	3.5	1.0	30	1	3.3	3.0	0.0
6	50	27	54.0	3.0	1.5	50	19	38.0	2.5	0.5
7	50	27	54.0	3.0	1.1	50	24	48.0	3.3	1.2
8	51	21	41.2	3.0	1.1	51	24	47.1	2.8	0.9
9	51	25	49.0	2.9	1.0	50	25	50.0	2.6	1.0
10	50	26	52.0	2.7	1.0	49	20	40.8	2.6	0.8
11	50	22	44.0	3.0	1.1	51	19	37.3	2.7	1.1
12	52	31	59.6	3.0	1.0	53	23	43.4	2.7	0.8
13	50	22	44.0	3.1	1.2	49	19	38.8	3.0	1.2
14	49	21	42.9	3.2	1.4	51	27	52.9	3.3	1.5
15	51	33	64.7	2.8	1.1	50	23	46.0	3.1	2.0
16	50	22	44.0	2.8	0.8	51	21	41.2	2.8	1.3
17	49	24	49.0	3.0	1.1	51	32	62.7	3.3	1.6
18	52	36	69.1	2.8	1.1	51	24	47.1	3.5	1.7

TABLE A-1-II/3. *Tragion-subnasale depth of the middle third of the face (t-sn rt,lt)*

Age		Asymmetry (male)					Asymmetry (female)			
(years)	*N*	*n*	Percent	Mean	SD	*N*	*n*	Percent	Mean	SD
1	18	4	22.2	1.3	0.5	20	4	20.0	1.3	0.5
2	30	12	40.0	1.3	0.5	32	12	37.5	1.3	0.5
3	30	16	53.3	1.1	0.2	30	11	36.7	1.4	0.5
4	30	3	10.0	3.0	0.0	30	4	13.3	3.5	1.0
5	30	4	13.3	3.3	0.5	30	5	16.7	3.0	0.0
6	50	27	54.0	3.2	1.6	50	22	44.0	2.9	1.0
7	50	26	52.0	2.8	1.0	50	20	40.0	3.1	1.2
8	51	23	45.1	2.9	1.7	51	23	45.1	2.7	1.1
9	51	22	43.1	2.7	1.2	50	26	52.0	2.9	1.1
10	50	32	64.0	2.9	1.2	49	26	53.1	2.8	1.1
11	50	35	70.0	2.9	1.2	51	24	47.1	2.8	1.2
12	52	34	65.4	3.1	1.2	53	28	52.8	2.9	1.1
13	50	20	40.0	2.9	1.1	49	27	55.1	2.9	1.0
14	49	25	51.0	3.0	1.1	51	31	60.8	3.1	1.8
15	51	25	49.0	3.2	1.2	50	25	50.0	3.4	2.0
16	50	24	48.0	3.2	1.2	51	29	56.9	3.1	1.5
17	49	31	63.3	3.1	1.2	51	28	54.9	3.0	1.6
18	52	28	53.8	2.9	1.0	51	26	51.0	3.7	1.6

TABLE A-1-II/4. *Tragion–gnathion depth of the lower third of the face (t-gn rt,lt)*

Age (years)	*N*	Asymmetry (male)				*N*	Asymmetry (female)			
		n	Percent	Mean	SD		*n*	Percent	Mean	SD
1	18	1	5.6	1.0	0.0	19	3	15.8	1.0	0.0
2	30	14	46.7	1.1	0.4	31	11	35.5	1.2	0.4
3	30	17	66.7	1.1	0.2	30	10	33.3	1.3	0.5
4	30	3	10.0	3.0	0.0	30	5	16.7	3.2	0.4
5	30	5	16.7	4.0	0.7	30	4	13.3	3.3	0.5
6	50	31	62.0	3.2	1.3	50	32	64.0	3.1	1.2
7	50	30	60.0	3.2	1.3	50	28	56.0	3.1	1.3
8	51	29	56.9	3.0	1.1	51	34	66.7	3.1	1.1
9	51	34	66.7	3.1	1.1	50	25	50.0	3.0	0.9
10	50	31	62.0	3.4	1.1	49	31	60.8	3.5	1.1
11	50	27	54.0	3.3	1.4	51	28	54.9	3.3	1.3
12	52	30	57.7	3.4	1.3	53	34	64.2	3.1	1.2
13	50	25	50.0	2.9	1.0	49	32	65.3	3.2	1.5
14	49	28	57.1	3.0	1.2	51	28	54.9	3.4	1.7
15	51	31	60.8	3.2	1.1	50	27	54.0	3.3	1.4
16	50	35	70.0	3.7	1.4	51	30	58.8	3.3	1.3
17	49	31	63.3	3.8	1.8	51	28	54.9	3.2	1.5
18	52	33	63.5	3.2	1.2	51	26	51.0	3.7	1.6

TABLE A-1-II/5. *Gonion–gnathion depth of the mandible (go-gn rt,lt)*

Age (years)	*N*	Asymmetry (male)				*N*	Asymmetry (female)			
		n	Percent	Mean	SD		*n*	Percent	Mean	SD
1	18	0	0	0	0	18	2	11.1	1.0	0.0
2	30	9	30.0	1.1	0.3	30	6	20.0	1.0	0.0
3	30	9	30.0	1.2	0.4	30	5	16.6	1.2	0.4
4	30	0	0.0	0.0	0.0	30	0	0.0	0.0	0.0
5	30	1	3.3	3.0	0.0	30	0	0.0	0.0	0.0

TABLE A-1-II/6. *Lateral surface half-arc in the upper third of the face (t-g surf rt,lt)*

Age (years)	*N*	Asymmetry (male)				*N*	Asymmetry (female)			
		n	Percent	Mean	SD		*n*	Percent	Mean	SD
1	17	5	29.4	1.6	0.5	15	6	40.0	1.3	0.5
2	30	18	60.0	1.3	0.5	30	12	40.0	1.8	0.5
3	30	19	63.3	1.4	0.8	30	15	50.0	1.2	0.4
4	30	8	26.7	4.4	1.5	30	5	16.7	4.6	1.1
5	30	7	23.3	6.1	2.8	30	8	26.7	4.4	1.4

TABLE A-1-II/7. *Lateral surface half-arc in the middle third of the face (t-sn surf rt,lt)*

Age (years)		Asymmetry (male)					Asymmetry (female)			
	N	*n*	Percent	Mean	SD	*N*	*n*	Percent	Mean	SD
1	17	7	41.2	1.6	0.5	15	5	33.3	1.2	0.4
2	30	17	56.7	1.5	0.6	30	13	43.3	1.2	0.4
3	30	16	53.3	1.4	0.8	30	13	43.3	1.2	0.4
4	30	17	56.7	3.3	2.0	30	15	50.0	3.6	1.6
5	30	14	46.7	4.2	2.4	30	10	33.3	2.8	0.9
6	50	24	48.0	3.4	1.7	50	24	48.0	2.9	0.9
7	50	27	54.0	3.1	1.1	50	23	46.0	2.8	1.2
8	51	31	60.8	3.5	1.5	51	22	43.1	3.2	1.4
9	51	29	56.9	3.1	1.3	50	25	50.0	3.0	0.8
10	50	34	68.0	3.6	1.8	49	28	57.1	3.3	1.3
11	50	31	62.0	3.6	2.0	51	23	45.1	3.7	1.5
12	52	34	65.4	2.9	1.1	53	29	54.7	3.0	1.1
13	50	33	66.0	3.2	0.9	49	31	63.3	3.5	1.1
14	49	27	55.1	3.6	1.9	51	21	41.2	3.6	2.5
15	51	33	64.7	3.8	1.3	50	33	66.0	3.8	1.9
16	50	35	70.0	3.7	1.7	51	36	70.6	3.4	1.4
17	49	36	73.5	4.1	2.4	51	28	54.9	3.5	1.4
18	52	28	53.8	3.7	1.8	51	31	60.8	3.5	1.9

TABLE A-1-II/8. *Lateral surface half-arc in the lower third of the face (t-gn surf rt,lt)*

Age (years)		Asymmetry (male)					Asymmetry (female)			
	N	*n*	Percent	Mean	SD	*N*	*n*	Percent	Mean	SD
1	17	3	17.6	1.0	0.0	15	3	20.0	1.7	0.6
2	30	6	20.0	1.5	0.5	30	12	40.0	1.1	0.3
3	30	19	63.3	1.3	0.6	30	20	66.7	1.5	0.5
4	30	12	40.0	2.9	2.3	30	15	50.0	2.4	0.8
5	30	13	43.3	3.2	1.2	30	13	43.3	2.8	0.9
6	50	32	64.0	4.3	1.7	50	30	60.0	3.9	1.7
7	50	32	64.0	3.5	1.7	50	30	60.0	4.1	1.6
8	51	33	64.7	3.9	1.5	51	34	66.7	3.6	1.7
9	51	35	68.6	3.7	1.6	50	29	58.0	3.5	1.5
10	50	38	76.0	3.9	2.0	49	31	63.3	4.1	2.2
11	50	39	78.0	3.5	1.7	51	37	72.5	4.1	1.6
12	52	41	78.8	3.5	1.6	53	34	64.2	4.1	1.8
13	50	33	66.0	4.2	2.2	49	35	71.4	3.9	1.4
14	49	34	69.4	3.6	2.4	51	37	72.5	4.1	1.8
15	51	39	76.5	3.4	1.5	50	36	72.0	3.9	1.9
16	50	40	80.0	4.2	1.9	51	40	78.4	3.9	1.6
17	49	40	81.6	5.0	2.8	51	44	86.3	3.8	2.4
18	52	38	73.1	4.2	2.0	51	36	70.6	3.8	1.5

TABLE A–1-III/1. *Length of the eye fissure (ex-en rt,lt)*

Age (years)	*N*	Asymmetry (male)				*N*	Asymmetry (female)			
		n	Percent	Mean	SD		*n*	Percent	Mean	SD
1	18	0	0.0	0.0	0.0	22	0	0.0	0.0	0.0
2	31	0	0.0	0.0	0.0	31	0	0.0	0.0	0.0
3	30	1	3.3	1.0	0.0	30	0	0.0	0.0	0.0
4	30	1	3.3	1.0	0.0	30	1	3.3	3.0	0.0
5	30	1	3.3	1.0	0.0	30	1	3.3	1.0	0.0
6	50	1	2.0	1.0	0.0	50	2	4.0	1.0	0.0
7	50	1	2.0	1.0	0.0	50	1	2.0	1.0	0.0
8	51	0	0.0	0.0	0.0	51	1	2.0	1.0	0.0
9	51	1	2.0	1.0	0.0	50	0	0.0	0.0	0.0
10	50	1	2.0	1.0	0.0	49	0	0.0	0.0	0.0
11	50	0	0.0	0.0	0.0	51	1	2.0	1.0	0.0
12	52	2	3.8	1.5	0.7	53	1	1.9	1.0	0.0
13	50	0	0.0	0.0	0.0	49	1	2.0	2.0	0.0
14	49	1	2.0	1.0	0.0	51	0	0.0	0.0	0.0
15	50	2	4.0	1.0	0.0	51	1	2.0	2.0	0.0
16	50	0	0.0	0.0	0.0	51	0	0.0	0.0	0.0
17	49	1	2.0	1.0	0.0	51	0	0.0	0.0	0.0
18	52	1	1.9	1.0	0.0	51	0	0.0	0.0	0.0

TABLE A–1-III/2. *Endocanthion–facial midline distance (en-se rt,lt)*

Age (years)	*N*	Asymmetry (male)				*N*	Asymmetry (female)			
		n	Percent	Mean	SD		*n*	Percent	Mean	SD
1	17	0	0.0	0.0	0.0	15	0	0.0	0.0	0.0
2	30	0	0.0	0.0	0.0	30	1	3.3	1.0	0.0
3	30	0	0.0	0.0	0.0	30	0	0.0	0.0	0.0
4	30	3	10.0	1.2	0.6	30	3	10.0	1.0	0.0
5	30	1	3.3	2.0	0.0	30	4	13.3	1.0	0.0
6	50	8	16.0	1.1	0.4	50	4	8.0	1.3	0.5
7	50	9	18.0	1.7	0.5	50	8	16.0	1.4	0.5
8	51	6	11.8	1.7	0.8	51	6	11.8	1.5	0.5
9	51	8	15.7	1.5	0.8	50	8	16.0	1.8	0.5
10	50	6	12.0	1.3	0.5	49	5	10.2	1.4	0.5
11	50	6	12.0	1.5	0.5	51	13	25.5	1.4	0.7
12	52	10	19.2	1.4	0.5	53	18	34.0	1.4	0.5
13	50	14	28.0	1.4	0.5	49	7	14.3	1.3	0.5
14	49	13	26.5	2.3	1.2	51	15	29.4	1.2	0.4
15	50	13	26.0	1.7	0.9	51	12	23.5	1.4	0.5
16	50	10	20.0	1.4	0.5	51	7	13.7	2.1	1.1
17	49	17	34.7	1.6	0.7	51	12	23.5	1.4	0.5
18	52	18	34.6	1.7	0.8	51	7	13.7	1.7	0.8

TABLE A-1-III/3. *Orbito-aural distance (ex-obs rt,lt)*

Age (years)	N	Asymmetry (male)				N	Asymmetry (female)			
		n	Percent	Mean	SD		*n*	Percent	Mean	SD
6	35	18	51.4	3.7	1.3	39	18	46.2	3.4	1.5
7	35	15	45.5	3.5	2.1	42	20	47.6	3.6	1.3
8	45	17	37.8	2.7	1.0	41	13	31.7	3.4	1.4
9	48	17	35.4	3.9	1.3	37	14	37.8	3.4	1.9
10	37	21	56.8	4.0	1.7	37	13	35.1	2.8	1.1
11	39	13	33.3	3.8	1.0	41	21	51.2	3.0	0.9
12	50	22	44.0	3.0	1.1	50	18	36.0	3.0	1.2
13	47	30	63.8	3.9	1.7	49	29	59.2	3.0	0.8
14	46	25	54.3	3.6	1.3	43	17	39.5	3.2	1.2
15	44	18	40.9	3.7	1.8	43	17	39.5	3.2	1.2
16	49	16	32.7	3.6	1.6	35	9	25.7	3.7	1.7
17	48	17	35.4	3.6	1.9	36	14	38.9	3.1	1.1
18	50	13	26.0	3.5	1.3	38	7	18.4	3.4	1.4

TABLE A-1-III/4. *Orbito-tragion distance (ex-t rt,lt)*

Age (years)	N	Asymmetry (male)				N	Asymmetry (female)			
		n	Percent	Mean	SD		*n*	Percent	Mean	SD
1	18	1	5.6	1.0	0.0	18	0	0.0	0.0	0.0
2	30	8	26.7	1.1	0.4	30	7	23.3	1.1	0.4
3	30	10	33.3	1.0	0.3	30	7	23.3	1.3	0.8
4	30	7	23.3	2.3	0.8	30	11	36.7	2.5	0.9
5	30	7	23.3	2.3	0.8	30	6	20.0	3.0	1.1
6	50	9	18.0	3.6	1.8	50	12	24.0	3.0	1.0
7	50	10	20.0	3.1	0.9	50	13	26.0	3.5	1.5
8	51	16	31.4	2.8	1.2	51	13	25.5	3.0	1.1
9	51	14	27.5	2.8	0.9	50	8	16.0	2.8	0.9
10	50	15	30.0	3.1	1.2	49	15	30.6	2.8	0.8
11	50	9	18.0	3.3	0.9	51	10	19.6	2.6	1.0
12	52	15	28.8	2.5	0.6	53	19	35.8	2.9	0.9
13	50	25	50.0	3.0	1.1	49	16	32.7	2.8	0.8
14	49	15	30.6	3.3	1.6	51	13	25.5	3.5	1.3
15	50	13	26.0	3.5	1.3	51	13	25.5	3.5	2.0
16	50	14	28.0	2.7	0.7	51	12	23.5	3.2	1.1
17	49	21	42.9	3.0	1.0	51	9	17.6	3.3	1.2
18	52	11	21.2	2.9	1.3	51	13	25.5	3.2	1.1

TABLE A-1-III/5. *Orbito-gonial distance (ex-go rt,lt)*

Age (years)	N	Asymmetry (male)				N	Asymmetry (female)			
		n	Percent	Mean	SD		*n*	Percent	Mean	SD
6	50	10	20.0	2.8	1.1	50	13	26.0	2.5	0.8
7	50	10	20.0	3.0	0.9	50	22	44.0	2.8	1.0
8	51	17	33.3	3.0	1.1	51	15	29.4	2.7	1.0
9	51	19	37.3	2.8	1.1	50	11	22.0	2.7	1.3
10	50	13	26.0	3.2	1.6	49	18	36.7	2.9	0.7
11	50	15	30.0	2.8	1.1	51	22	43.1	2.7	1.0
12	52	19	36.5	3.1	1.3	53	15	28.3	2.9	0.8
13	50	22	44.0	2.6	0.6	49	17	34.7	3.1	1.4
14	49	11	22.4	2.9	1.4	51	23	45.1	3.1	1.2
15	50	13	26.0	3.1	1.3	51	19	37.3	2.6	1.1
16	50	19	38.0	3.5	1.5	51	13	25.5	3.7	1.0
17	49	20	40.8	3.1	0.9	51	17	33.3	3.5	1.5
18	52	17	32.7	3.6	1.6	51	13	25.5	3.8	1.6

TABLE A–1-III/6. *Height of the orbit (or-os rt,lt)*

Age (years)	Asymmetry (male)					Asymmetry (female)				
	N	*n*	Percent	Mean	SD	*N*	*n*	Percent	Mean	SD
1	18	0	0.0	0.0	0.0	19	0	0.0	0.0	0.0
2	31	0	0.0	0.0	0.0	31	2	6.5	1.0	0.0
3	30	0	0.0	0.0	0.0	31	1	3.3	1.0	0.0
4	30	1	3.3	3.0	0.0	30	2	6.7	1.5	0.7
5	30	0	0.0	0.0	0.0	30	1	3.3	1.0	0.0

TABLE A–1-III/7. *Combined height of the orbit and the eyebrow (or-sci rt,lt)*

Age (years)	Asymmetry (male)					Asymmetry (female)				
	N	*n*	Percent	Mean	SD	*N*	*n*	Percent	Mean	SD
1	18	0	0.0	0.0	0.0	19	0	0.0	0.0	0.0
2	31	0	0.0	0.0	0.0	31	1	3.2	1.0	0.0
3	30	1	3.3	1.0	0.0	30	1	3.3	1.0	0.0
4	30	1	3.3	3.0	0.0	30	1	3.3	2.0	0.0

TABLE A–1-III/8. *Height of the eye fissure (ps-pi rt,lt)*

Age (years)	Asymmetry (male)					Asymmetry (female)				
	N	*n*	Percent	Mean	SD	*N*	*n*	Percent	Mean	SD
1	18	0	0.0	0.0	0.0	22	0	0.0	0.0	0.0
2	31	2	6.5	1.0	0.0	31	2	6.5	1.0	0.0
3	30	0	0.0	0.0	0.0	30	1	3.3	1.0	0.0
4	30	2	6.7	1.0	0.0	30	0	0.0	0.0	0.0
5	30	0	0.0	0.0	0.0	30	2	6.7	1.0	0.0
6	50	1	2.0	2.0	0.0	50	0	0.0	0.0	0.0
7	50	0	0.0	0.0	0.0	50	0	0.0	0.0	0.0
8	51	0	0.0	0.0	0.0	51	0	0.0	0.0	0.0
9	51	0	0.0	0.0	0.0	50	0	0.0	0.0	0.0
10	50	0	0.0	0.0	0.0	49	0	0.0	0.0	0.0
11	50	0	0.0	0.0	0.0	51	0	0.0	0.0	0.0
12	52	0	0.0	0.0	0.0	53	0	0.0	0.0	0.0
13	50	1	2.0	1.0	0.0	49	0	0.0	0.0	0.0
14	49	0	0.0	0.0	0.0	51	0	0.0	0.0	0.0
15	50	0	0.0	0.0	0.0	51	0	0.0	0.0	0.0
16	50	0	0.0	0.0	0.0	51	0	0.0	0.0	0.0
17	49	2	4.3	1.0	0.0	51	0	0.0	0.0	0.0
18	52	1	1.9	1.0	0.0	51	0	0.0	0.0	0.0

TABLE A–1-III/9. *Height of the upper lid (os-ps rt,lt)*

Age (years)	Asymmetry (male)					Asymmetry (female)				
	N	*n*	Percent	Mean	SD	*N*	*n*	Percent	Mean	SD
1	18	0	0.0	0.0	0.0	19	0	0.0	0.0	0.0
2	31	0	0.0	0.0	0.0	31	1	3.2	1.0	0.0
3	30	0	0.0	0.0	0.0	30	1	3.3	1.0	0.0
4	30	1	3.3	1.0	0.0	30	0	0.0	0.0	0.0

TABLE A–1-III/10. *Height of the lower lid (pi-or rt,lt)*

Age (years)	N	Asymmetry (male)				N	Asymmetry (female)			
		n	Percent	Mean	SD		n	Percent	Mean	SD
1	18	0	0.0	0.0	0.0	19	0	0.0	0.0	0.0
2	31	0	0.0	0.0	0.0	31	0	0.0	0.0	0.0
3	30	0	0.0	0.0	0.0	30	1	3.3	1.0	0.0
4	30	1	3.3	1.0	0.0	30	0	0.0	0.0	0.0
5	30	0	0.0	0.0	0.0	30	1	3.3	1.0	0.0

TABLE A–1-III/11. *Endocanthion–facial midline surface distance (en-se surf rt,lt)*

Age (years)	N	Asymmetry (male)				N	Asymmetry (female)			
		n	Percent	Mean	SD		n	Percent	Mean	SD
6	50	1	2.0	2.0	0.0	50	2	4.0	2.0	0.0
7	50	5	10.0	2.4	0.9	50	3	6.0	2.3	0.6
8	51	4	7.8	2.3	0.5	51	1	2.0	3.0	0.0
9	51	3	5.9	2.0	0.0	50	6	12.0	2.0	0.0
10	50	3	6.0	2.3	0.6	49	3	6.1	3.0	1.7
11	50	4	8.0	2.3	0.5	51	6	11.8	2.2	0.4
12	52	5	9.6	2.0	0.0	52	9	17.3	2.1	0.3
13	50	6	12.0	2.0	0.0	49	2	4.1	2.5	0.7
14	49	10	20.4	2.9	1.0	51	4	7.8	2.3	0.5
15	50	7	14.0	2.6	0.8	51	2	3.9	2.5	0.7
16	50	6	12.0	2.3	0.5	51	4	7.8	2.5	1.0
17	49	6	12.2	2.2	0.4	51	5	9.8	2.0	0.0
18	52	11	21.2	2.3	0.5	51	2	3.9	3.5	0.7

TABLE A–1-III/12. *Inclination of the line connecting the endocanthion and exocanthion of the eye fissure (en-ex line rt,lt)*

Age (years)	N	Asymmetry (male)				N	Asymmetry (female)			
		n	Percent	Mean	SD		n	Percent	Mean	SD
1	18	0	0.0	0.0	0.0	22	2	9.1	1.0	0.0
2	31	3	9.7	1.0	0.0	34	2	5.9	1.0	0.0
3	30	4	13.3	1.0	0.0	30	6	20.0	1.3	0.5
4	30	1	3.3	1.0	0.0	30	1	3.3	2.0	0.0
5	30	0	0.0	0.0	0.0	30	1	3.3	2.0	0.0
6	50	0	0.0	0.0	0.0	49	1	2.0	4.0	0.0
7	50	0	0.0	0.0	0.0	50	0	0.0	0.0	0.0
8	51	0	0.0	0.0	0.0	51	2	3.9	2.0	0.0
9	51	1	2.0	1.0	0.0	48	1	2.1	1.0	0.0
10	50	2	4.0	1.5	0.7	49	1	2.0	2.0	0.0
11	50	0	0.0	0.0	0.0	50	3	6.0	3.0	1.4
12	52	2	3.8	2.5	0.7	53	0	0.0	0.0	0.0
13	50	0	0.0	0.0	0.0	49	0	0.0	0.0	0.0
14	48	0	0.0	0.0	0.0	51	0	0.0	0.0	0.0
15	50	0	0.0	0.0	0.0	51	0	0.0	0.0	0.0
16	48	3	6.3	1.0	0.0	51	1	2.0	1.0	0.0
17	49	1	2.0	1.0	0.0	51	0	0.0	0.0	0.0
18	50	0	0.0	0.0	0.0	51	0	0.0	0.0	0.0

TABLE A–1-III/13. *Inclination of the line placed over the upper and lower orbital rims (os′/or′ rt,lt)*

Age (years)	N	Asymmetry (male) n	Percent	Mean	SD	N	Asymmetry (female) n	Percent	Mean	SD
1	17	0	0.0	0.0	0.0	15	0	0.0	0.0	0.0
2	31	1	3.2	1.0	0.0	33	1	3.0	1.0	0.0
3	30	3	10.0	1.0	0.0	30	2	6.7	1.0	0.0
4	30	1	3.3	4.0	0.0	30	3	10.0	2.7	0.6
5	30	1	3.3	2.0	0.0	30	1	3.3	4.0	0.0

TABLE A–1-IV/1. *Width of the nostril floor (sbal-sn rt,lt)*

Age (years)	N	Asymmetry (male) n	Percent	Mean	SD	N	Asymmetry (female) n	Percent	Mean	SD
1	18	0	0.0	0.0	0.0	18	0	0.0	0.0	0.0
2	31	1	3.2	1.0	0.0	30	1	3.3	1.0	0.0
3	30	2	6.7	1.0	0.0	30	2	6.7	1.0	0.0
4	30	1	3.3	2.0	0.0	30	1	3.3	1.0	0.0
5	30	1	3.3	2.0	0.0	30	0	0.0	0.0	0.0
6	50	1	2.0	1.0	0.0	50	6	12.0	1.3	0.5
7	50	7	14.0	1.1	0.4	50	3	6.0	1.3	0.6
8	51	8	15.7	1.8	0.5	51	4	7.8	1.3	0.5
9	51	2	3.9	2.0	0.0	49	4	8.2	1.0	0.0
10	50	9	18.0	1.2	0.7	49	7	14.3	1.4	0.8
11	50	9	18.0	1.2	0.4	51	7	13.7	1.3	0.5
12	52	15	28.8	1.3	0.5	53	9	17.0	1.4	0.7
13	50	9	18.0	1.1	0.3	49	9	18.4	1.1	0.3
14	49	9	18.4	1.6	0.9	51	7	13.7	1.3	0.8
15	50	13	26.0	1.2	0.4	51	10	19.6	1.9	1.1
16	50	7	14.0	1.4	0.8	51	6	11.8	1.0	0.0
17	49	10	20.4	1.6	0.7	51	7	13.7	1.4	1.1
18	52	8	15.4	1.8	1.4	51	8	15.7	1.6	0.7

TABLE A–1-IV/2. *Thickness of the ala (al′-al′ rt,lt)*

Age (years)	N	Asymmetry (male) n	Percent	Mean	SD	N	Asymmetry (female) n	Percent	Mean	SD
1	18	0	0.0	0.0	0.0	18	0	0.0	0.0	0.0
2	31	0	0.0	0.0	0.0	31	1	3.2	1.0	0.0
3	30	0	0.0	0.0	0.0	30	0	0.0	0.0	0.0
4	30	1	3.3	1.0	0.0	30	1	3.3	1.0	0.0
5	30	1	3.3	1.0	0.0	30	1	3.3	1.0	0.0
6	50	0	0.0	0.0	0.0	50	0	0.0	0.0	0.0
7	50	1	2.0	1.0	0.0	50	0	0.0	0.0	0.0
8	51	0	0.0	0.0	0.0	51	0	0.0	0.0	0.0
9	51	0	0.0	0.0	0.0	50	1	2.0	1.0	0.0
10	50	1	2.0	1.0	0.0	49	2	4.1	1.5	0.7
11	50	0	0.0	0.0	0.0	51	1	2.0	2.0	0.0
12	52	2	3.8	1.0	0.0	53	1	1.9	1.0	0.0
13	50	0	0.0	0.0	0.0	49	2	4.1	1.0	0.0
14	49	0	0.0	0.0	0.0	51	1	2.0	1.0	0.0
15	50	0	0.0	0.0	0.0	51	1	2.0	1.0	0.0
16	50	1	2.0	1.0	0.0	51	1	2.0	1.0	0.0
17	49	2	4.1	1.5	0.7	51	0	0.0	0.0	0.0
18	52	0	0.0	0.0	0.0	51	0	0.0	0.0	0.0

TABLE A–1-IV/3. *Alar base–subnasale distance (ac-sn rt,lt)*

Age (years)	N	Asymmetry (male)				N	Asymmetry (female)			
		n	Percent	Mean	SD		n	Percent	Mean	SD
1	18	0	0.0	0.0	0.0	18	0	0.0	0.0	0.0
2	31	2	6.5	1.0	0.0	31	1	3.2	1.0	0.0
3	30	2	6.7	1.0	0.0	30	2	6.7	1.0	0.0
4	30	1	3.3	1.0	0.0	30	1	3.3	1.0	0.0
5	30	1	3.3	1.0	0.0	30	—	—	—	—

TABLE A–1-IV/4. *Length of the ala (ac-prn rt,lt)*

Age (years)	N	Asymmetry (male)				N	Asymmetry (female)			
		n	Percent	Mean	SD		n	Percent	Mean	SD
1	18	0	0.0	0.0	0.0	18	0	0.0	0.0	0.0
2	31	0	0.0	0.0	0.0	31	4	12.9	1.3	0.5
3	30	3	10.0	1.0	0.0	30	0	0.0	0.0	0.0
4	30	0	0.0	0.0	0.0	30	0	0.0	0.0	0.0
5	30	0	0.0	0.0	0.0	30	0	0.0	0.0	0.0
6	50	1	2.0	1.0	0.0	50	0	0.0	0.0	0.0
7	50	2	4.0	1.0	0.0	50	0	0.0	0.0	0.0
8	51	2	3.9	1.5	0.7	51	1	2.0	1.0	0.0
9	51	1	2.0	1.0	0.0	50	2	4.0	1.5	0.7
10	50	2	4.0	2.5	2.1	49	4	8.2	1.8	1.0
11	50	1	2.0	4.0	0.0	51	2	3.9	1.0	0.0
12	52	3	5.8	2.0	0.0	53	0	0.0	0.0	0.0
13	50	0	0.0	0.0	0.0	49	4	8.2	1.0	0.0
14	49	0	0.0	0.0	0.0	51	1	2.0	3.0	0.0
15	50	1	2.0	2.0	0.0	51	4	7.8	1.0	0.0
16	50	1	2.0	2.0	0.0	51	3	5.9	1.3	0.6
17	49	6	12.2	1.2	0.4	51	0	0.0	0.0	0.0
18	52	1	1.9	1.0	0.0	51	2	3.9	1.0	0.0

TABLE A–1-IV/5. *Length of the columella (sn-c' rt,lt)*

Age (years)	N	Asymmetry (male)				N	Asymmetry (female)			
		n	Percent	Mean	SD		n	Percent	Mean	SD
1	18	0	0.0	0.0	0.0	21	0	0.0	0.0	0.0
2	31	6	19.4	1.2	0.4	30	1	3.3	1.0	0.0
3	30	4	13.3	1.3	0.5	30	4	13.3	1.3	0.5
4	30	0	0.0	0.0	0.0	30	1	3.3	1.0	0.0
5	30	4	13.3	1.0	0.0	30	0	0.0	0.0	0.0
6	50	2	4.0	1.0	0.0	50	0	0.0	0.0	0.0
7	50	4	8.0	1.0	0.0	50	5	10.0	1.0	0.0
8	51	4	7.8	1.0	0.0	51	1	2.0	1.0	0.0
9	51	2	3.9	1.0	0.0	49	5	10.2	1.4	0.5
10	50	4	8.0	1.0	0.0	49	7	14.3	1.0	0.0
11	50	4	8.0	1.5	1.0	51	5	9.8	1.0	0.0
12	52	5	9.6	1.2	0.4	53	4	7.5	1.0	0.0
13	50	8	16.0	1.3	0.5	49	3	6.1	1.7	0.6
14	49	3	6.1	1.0	0.0	51	5	9.8	1.0	0.0
15	50	10	20.0	1.1	0.3	51	7	13.7	1.1	0.4
16	50	8	16.0	1.0	0.0	51	4	7.8	1.3	0.5
17	49	7	14.3	1.3	0.5	51	5	9.8	1.3	0.5
18	52	5	9.6	1.2	0.4	51	4	7.8	1.0	0.0

TABLE A–1-IV/6. *Depth of the nasal root (en-se sag rt,lt)*

Age (years)	Asymmetry (male)					Asymmetry (female)				
	N	*n*	Percent	Mean	SD	*N*	*n*	Percent	Mean	SD
1	17	1	5.9	1.0	0.0	15	0	0.0	0.0	0.0
2	31	0	0.0	0.0	0.0	32	4	12.5	1.0	0.0
3	30	7	23.3	1.1	0.4	30	5	16.7	1.0	0.0
4	30	1	3.3	2.0	0.0	30	2	6.7	1.0	0.0
5	30	0	0.0	0.0	0.0	30	1	3.3	1.0	0.0
6	34	0	0.0	0.0	0.0	39	0	0.0	0.0	0.0
7	33	0	0.0	0.0	0.0	42	0	0.0	0.0	0.0
8	45	0	0.0	0.0	0.0	41	0	0.0	0.0	0.0
9	47	1	2.2	1.0	0.0	38	0	0.0	0.0	0.0
10	37	1	2.7	1.0	0.0	37	1	2.7	3.0	0.0
11	39	0	0.0	0.0	0.0	41	1	2.4	1.0	0.0
12	50	3	6.0	1.7	0.6	50	0	0.0	0.0	0.0
13	47	0	0.0	0.0	0.0	49	0	0.0	0.0	0.0
14	46	1	2.2	1.0	0.0	43	0	0.0	0.0	0.0
15	41	2	4.9	2.5	0.7	43	1	2.3	1.0	0.0
16	48	1	4.2	3.0	0.0	35	0	0.0	0.0	0.0
17	48	1	2.1	2.0	0.0	36	2	5.6	2.0	0.0
18	48	0	0.0	0.0	0.0	37	1	2.7	1.0	0.0

TABLE A–1-IV/7. *Surface length of the ala (ac-prn surf rt,lt)*

Age (years)	Asymmetry (male)					Asymmetry (female)				
	N	*n*	Percent	Mean	SD	*N*	*n*	Percent	Mean	SD
1	17	0	0.0	0.0	0.0	15	0	0.0	0.0	0.0
2	30	1	3.3	3.0	0.0	30	0	0.0	0.0	0.0
3	30	0	0.0	0.0	0.0	30	1	3.3	1.0	0.0
4	30	0	0.0	0.0	0.0	30	0	0.0	0.0	0.0
5	30	0	0.0	0.0	0.0	30	0	0.0	0.0	0.0
6	50	1	2.0	1.0	0.0	50	0	0.0	0.0	0.0
7	50	5	10.0	2.4	0.9	50	3	6.0	2.3	0.6
8	51	2	3.9	2.0	0.0	51	2	3.9	1.0	0.0
9	51	0	0.0	0.0	0.0	50	3	6.0	1.7	0.6
10	50	4	8.0	1.3	0.5	49	5	10.2	1.8	0.8
11	50	3	6.0	1.7	0.6	51	6	11.8	1.5	0.5
12	52	6	11.5	1.7	0.8	53	1	1.9	2.0	0.0
13	50	3	6.0	1.3	0.6	49	5	10.2	1.6	0.9
14	49	1	2.0	3.0	0.0	51	0	0.0	0.0	0.0
15	50	4	8.0	1.3	0.5	51	5	9.8	1.4	0.5
16	50	2	4.0	1.0	0.0	51	6	11.8	1.5	0.5
17	49	5	10.2	2.6	1.3	51	1	2.0	1.0	0.0
18	52	2	3.8	2.5	0.7	51	3	5.9	1.7	1.1

TABLE A–1-IV/8. *Nostril axis asymmetry*

Age (years)	Asymmetry (male)					Asymmetry (female)				
	N	*n*	Percent	Mean	SD	*N*	*n*	Percent	Mean	SD
1	17	5	29.4	5.8	4.0	18	4	22.2	3.7	1.9
2	28	15	53.6	2.4	1.4	32	14	43.8	1.8	0.7
3	28	11	39.3	3.1	4.0	30	15	50.0	2.5	1.1
4	30	1	3.3	8.0	0.0	30	0	0.0	0.0	0.0
5	30	3	10.0	16.3	4.7	30	3	10.0	7.3	4.6

TABLE A-1-IV/9. *Deviation of the nasal bridge*

Age (years)		Asymmetry (male)					Asymmetry (female)			
	N	*n*	Percent	Mean	SD	*N*	*n*	Percent	Mean	SD
6	50	3	6.0	2.3	1.5	50	2	4.0	2.5	0.7
7	50	6	12.0	2.3	0.8	49	3	6.1	2.7	1.1
8	51	2	3.9	2.0	0.0	51	2	3.9	3.0	1.4
9	51	6	11.8	3.7	2.7	49	6	12.2	3.0	0.6
10	50	8	16.0	3.5	1.1	49	7	14.3	3.3	1.6
11	50	3	6.0	2.3	0.6	50	6	12.0	3.2	1.2
12	52	7	13.5	5.0	2.4	53	12	22.6	3.4	1.2
13	49	13	26.5	4.0	1.1	49	9	18.4	3.4	1.1
14	49	17	34.7	4.3	2.3	51	15	29.4	3.7	1.5
15	50	16	32.0	3.7	1.1	51	14	27.5	4.6	2.4
16	50	11	22.0	3.9	1.3	51	11	21.6	2.8	1.3
17	49	17	34.7	3.8	1.4	51	7	13.7	3.1	1.1
18	52	15	28.8	4.6	2.5	51	13	25.5	3.4	1.3

TABLE A-1-IV/10. *Deviation of the columella*

Age (years)		Asymmetry (male)					Asymmetry (female)			
	N	*n*	Percent	Mean	SD	*N*	*n*	Percent	Mean	SD
6	50	0	0.0	0.0	0.0	50	0	0.0	0.0	0.0
7	50	8	16.0	3.3	1.4	50	10	20.0	3.7	2.4
8	51	5	9.8	4.2	2.3	51	5	9.8	5.4	2.9
9	51	5	9.8	3.4	1.1	49	2	4.1	2.0	0.0
10	50	7	14.0	3.4	1.3	49	3	6.1	3.7	1.5
11	50	6	12.0	3.8	1.7	50	7	14.0	5.0	2.4
12	52	6	11.5	6.0	2.0	53	4	7.5	3.5	1.3
13	49	11	22.4	3.6	1.1	49	6	12.2	4.8	2.6
14	49	11	22.4	4.5	1.9	51	9	17.6	3.4	1.5
15	50	18	36.0	4.5	2.1	51	12	23.5	4.5	1.7
16	50	13	26.0	3.7	1.5	51	10	19.6	4.2	1.2
17	49	14	28.6	4.5	2.1	51	5	9.8	2.6	0.9
18	52	10	19.2	4.7	3.1	51	7	13.7	4.4	1.5

TABLE A-1-V/1. *Halves of the labial fissure length (ch-sto rt,lt)*

Age (years)		Asymmetry (male)					Asymmetry (female)			
	N	*n*	Percent	Mean	SD	*N*	*n*	Percent	Mean	SD
6	50	2	4.0	2.0	0.0	50	2	4.0	2.5	0.7
7	50	2	4.0	2.0	0.0	50	7	14.0	2.4	0.8
8	51	5	9.8	2.2	0.4	51	4	7.8	2.5	1.0
9	51	11	21.6	2.5	1.2	50	8	16.0	2.8	1.8
10	50	14	28.0	2.4	0.6	49	12	24.5	3.1	1.1
11	50	12	24.0	2.7	1.0	50	7	14.0	3.4	1.3
12	52	14	26.9	2.6	0.9	53	12	22.6	2.5	0.7
13	50	5	10.0	3.0	1.2	49	12	24.5	2.8	1.0
14	49	7	14.3	2.0	0.0	51	7	13.7	2.7	1.1
15	50	9	18.0	2.6	0.9	51	8	15.7	2.5	0.5
16	50	6	12.0	2.5	0.8	51	11	21.6	3.0	1.2
17	49	8	16.3	3.4	1.5	51	11	21.6	2.8	1.2
18	52	4	7.7	3.0	0.8	51	8	15.7	2.9	1.0

TABLE A–1-V/2. *Lateral upper lip height (sbal-ls' rt,lt)*

Age (years)	N	Asymmetry (male)				N	Asymmetry (female)			
		n	Percent	Mean	SD		n	Percent	Mean	SD
1	18	0	0.0	0.0	0.0	18	0	0.0	0.0	0.0
2	31	2	6.5	1.0	0.0	30	0	0.0	0.0	0.0
3	30	4	13.3	1.0	0.0	30	0	0.0	0.0	0.0
4	30	3	10.0	1.0	0.0	30	1	3.3	1.0	0.0
5	30	1	3.3	1.0	0.0	30	2	6.7	1.0	0.0
6	50	0	0.0	0.0	0.0	50	0	0.0	0.0	0.0
7	50	5	10.0	1.0	0.0	50	4	8.0	1.3	0.5
8	51	2	3.9	2.0	1.4	51	3	5.9	1.3	0.6
9	51	2	3.9	1.0	0.0	50	1	2.0	1.0	0.0
10	50	4	8.0	1.0	0.0	49	4	8.2	1.8	1.0
11	50	4	8.0	1.5	0.6	51	11	21.6	1.4	0.7
12	52	9	17.3	1.4	0.7	53	4	7.5	1.8	0.5
13	50	2	4.0	1.5	0.7	49	7	14.3	1.3	0.5
14	49	4	8.2	1.3	0.5	51	4	7.8	1.8	1.0
15	50	9	18.0	1.2	0.4	51	5	9.8	1.6	0.5
16	50	1	2.0	2.0	0.0	51	5	9.8	1.8	0.8
17	49	4	8.2	1.3	0.5	51	7	13.7	1.6	0.5
18	52	4	7.7	1.8	1.0	51	4	7.8	2.0	1.2

TABLE A–1-V/3. *Labio-tragial distance (ch-t rt,lt)*

Age (years)	N	Asymmetry (male)				N	Asymmetry (female)			
		n	Percent	Mean	SD		n	Percent	Mean	SD
6	24	9	37.5	2.6	1.1	25	8	32.0	2.6	0.8
7	24	15	62.5	4.6	3.5	37	21	56.8	3.0	0.8
8	41	22	53.7	2.5	0.7	39	18	46.2	2.6	0.9
9	47	21	44.7	3.2	1.6	38	18	47.4	2.9	1.1
10	37	20	54.1	2.5	0.6	36	17	47.2	2.8	0.9
11	36	20	55.6	3.3	1.2	40	20	50.0	2.6	0.9
12	38	18	47.4	2.8	0.8	35	16	45.7	2.6	0.7
13	45	24	53.3	2.7	0.9	48	22	45.8	2.7	1.0
14	46	27	58.7	2.9	1.1	48	24	50.0	2.7	1.1
15	44	25	56.8	3.3	1.4	43	26	60.5	3.3	1.2
16	48	26	54.2	3.4	1.2	34	19	55.9	3.1	1.1
17	44	22	50.0	3.5	1.4	33	19	57.6	3.5	1.5
18	26	14	53.8	2.6	0.6	17	9	52.9	2.4	0.7

TABLE A–1-V/4. *Lateral labio-tragial arc (ch-t surf rt,lt)*

Age (years)	N	Asymmetry (male)				N	Asymmetry (female)			
		n	Percent	Mean	SD		n	Percent	Mean	SD
6	24	13	54.2	3.1	1.3	25	14	56.0	3.2	1.1
7	24	16	66.7	4.4	2.9	37	24	64.9	3.7	1.3
8	41	20	48.8	3.3	1.7	39	20	51.3	3.4	1.4
9	47	31	66.0	3.4	1.2	38	22	57.9	3.0	1.2
10	37	20	54.1	4.0	1.9	36	21	58.3	3.6	1.5
11	36	20	55.6	3.8	1.8	40	18	45.0	3.2	1.1
12	38	19	50.0	3.3	1.5	35	19	54.3	3.3	1.6
13	45	29	64.4	2.9	0.9	48	25	52.1	3.1	1.1
14	46	30	65.2	3.4	1.8	48	24	50.0	3.3	1.6
15	44	28	63.6	3.9	2.7	43	27	62.8	3.1	1.2
16	48	34	70.8	3.3	1.4	34	16	47.1	2.8	1.0
17	44	27	61.4	4.1	2.1	33	20	60.6	3.3	1.2
18	26	18	69.2	3.6	1.8	17	10	58.8	2.9	0.7

TABLE A-1-V/5. *Inclination of the labial fissure*

Age (years)	N	Asymmetry (male) n	Percent	Mean	SD	N	Asymmetry (female) n	Percent	Mean	SD
6	50	0	0.0	0.0	0.0	50	0	0.0	0.0	0.0
7	50	1	2.0	1.0	0.0	50	2	4.0	3.0	2.8
8	51	2	3.9	2.0	0.0	51	0	0.0	0.0	0.0
9	51	0	0.0	0.0	0.0	50	0	0.0	0.0	0.0
10	50	2	4.0	2.5	0.7	49	4	8.2	2.0	0.0
11	50	0	0.0	0.0	0.0	50	1	2.0	5.0	0.0
12	52	1	1.9	1.0	0.0	52	3	5.8	2.0	1.0
13	50	0	0.0	0.0	0.0	49	1	2.0	1.0	0.0
14	49	1	2.0	1.0	0.0	51	0	0.0	0.0	0.0
15	50	0	0.0	0.0	0.0	51	3	5.9	1.7	1.2
16	50	0	0.0	0.0	0.0	51	2	3.9	1.5	0.7
17	49	2	4.1	3.0	2.8	51	1	2.0	1.0	0.0
18	52	2	3.8	2.0	1.4	51	1	2.0	3.0	0.0

TABLE A-1-VI/1. *Width of the auricle (pra-pa rt,lt)*

Age (years)	N	Asymmetry (male) n	Percent	Mean	SD	N	Asymmetry (female) n	Percent	Mean	SD
1	18	1	5.6	2.0	0.0	18	0	0.0	0.0	0.0
2	31	3	9.7	1.7	0.6	31	4	12.9	1.3	0.5
3	30	11	36.7	1.2	0.4	30	11	36.7	1.0	0.0
4	30	7	23.3	2.1	0.4	30	6	20.0	2.3	0.5
5	30	2	6.7	2.0	0.0	30	7	23.3	3.3	2.6
6	50	9	18.0	1.2	0.4	50	14	28.0	1.5	0.7
7	50	9	18.0	1.4	0.7	50	11	22.0	1.5	0.7
8	51	9	17.6	1.3	0.7	51	14	27.5	1.3	0.5
9	51	9	17.6	1.7	0.7	50	9	18.0	1.0	0.0
10	50	5	10.0	1.2	0.4	49	12	24.5	2.3	1.1
11	50	8	16.0	1.9	0.8	51	8	15.7	1.3	0.7
12	52	11	21.2	1.7	0.9	53	11	20.8	1.3	0.5
13	50	11	22.0	1.4	0.5	49	7	14.3	1.7	0.8
14	49	5	10.2	1.6	0.9	51	9	17.6	1.7	1.0
15	50	7	14.0	1.4	0.8	51	12	23.5	1.6	1.0
16	50	9	18.0	1.7	1.0	51	7	13.7	1.3	0.5
17	49	5	10.2	1.8	1.1	51	7	13.7	1.3	0.5
18	52	12	23.1	1.9	0.7	50	6	12.0	1.7	0.8

TABLE A–1-VI/2. *Length of the auricle (sa-sba rt,lt)*

Age		Asymmetry (male)					Asymmetry (female)			
(years)	*N*	*n*	Percent	Mean	SD	*N*	*n*	Percent	Mean	SD
1	18	0	0.0	0.0	0.0	18	0	0.0	0.0	0.0
2	31	6	19.4	1.7	1.2	31	8	25.8	1.5	0.9
3	30	8	26.7	1.3	0.5	30	9	30.0	1.2	0.4
4	30	9	30.0	2.3	0.5	30	10	33.3	2.5	1.3
5	30	8	26.7	2.1	0.4	30	11	36.7	2.5	0.7
6	50	13	26.0	1.4	0.7	50	15	30.0	1.5	1.0
7	50	10	20.0	1.5	1.3	50	10	20.0	1.3	0.7
8	51	17	33.3	1.9	0.7	51	12	23.5	1.6	0.7
9	51	13	25.5	1.5	0.8	50	8	16.0	1.6	0.9
10	50	9	18.0	1.4	0.7	49	9	18.4	1.8	0.7
11	50	8	16.0	2.0	1.4	51	6	11.8	1.8	1.0
12	52	12	23.1	1.4	0.7	53	7	13.2	1.1	0.4
13	50	11	22.0	1.5	0.7	49	15	30.6	2.0	1.1
14	49	9	18.4	1.6	0.7	51	7	13.7	2.3	0.8
15	50	15	30.0	2.4	1.3	51	12	23.5	2.6	1.7
16	50	11	22.0	1.5	1.0	51	12	23.5	1.3	0.5
17	49	13	26.5	1.8	0.8	51	9	17.6	2.1	1.3
18	52	11	21.2	1.5	0.5	50	13	26.0	1.9	1.0

TABLE A–1-VI/3. *Morphological width of the ear (obs-obi rt,lt)*

Age		Asymmetry (male)					Asymmetry (female)			
(years)	*N*	*n*	Percent	Mean	SD	*N*	*n*	Percent	Mean	SD
1	18	3	16.7	1.0	0.0	18	0	0.0	0.0	0.0
2	31	6	19.4	1.2	0.4	31	7	22.6	1.3	0.5
3	30	10	33.3	1.2	0.4	30	7	23.3	1.1	0.4
4	30	8	26.7	2.3	0.5	30	7	23.3	2.3	0.5
5	30	11	36.7	2.5	1.2	30	9	30.0	2.4	0.5
6	28	12	42.9	3.3	0.9	29	8	27.6	2.5	0.5
7	24	15	62.5	2.5	0.8	38	21	55.3	2.4	0.6
8	43	22	51.2	2.9	0.8	39	16	41.0	2.5	0.8
9	47	21	44.7	2.5	0.5	38	16	42.1	2.3	0.5
10	37	20	54.1	2.8	1.0	36	20	55.6	2.5	0.7
11	37	12	32.4	2.4	0.5	41	18	43.9	3.2	1.2
12	45	16	35.6	2.8	0.9	43	21	48.8	2.9	1.2
13	46	26	56.5	2.8	0.9	49	22	44.9	2.6	1.0
14	46	25	54.3	2.6	0.8	43	29	67.4	3.3	1.3
15	44	22	50.0	3.0	1.4	43	22	51.2	2.6	0.9
16	49	16	32.7	2.8	1.2	34	15	44.1	2.7	1.1
17	46	22	47.8	3.0	1.0	33	20	60.6	2.8	1.2
18	34	16	47.1	2.3	0.4	17	11	64.7	2.5	0.7

TABLE A-1-VI/4. *Upper naso-aural distance (n-obs rt,lt)*

Age (years)	N	Asymmetry (male)				N	Asymmetry (female)			
		n	Percent	Mean	SD		*n*	Percent	Mean	SD
6	50	29	58.0	3.5	2.2	50	34	68.0	3.5	1.1
7	50	28	56.0	3.6	1.5	50	30	60.0	3.5	1.4
8	51	39	76.5	3.6	1.6	51	26	51.0	3.1	1.1
9	51	33	64.7	3.8	2.1	50	30	60.0	3.3	1.2
10	50	32	64.0	3.4	2.0	49	33	67.3	3.1	1.4
11	50	35	70.0	3.3	1.6	51	26	51.0	3.7	1.8
12	52	33	63.5	3.2	1.3	53	42	79.2	3.2	1.4
13	50	33	66.0	3.7	2.1	49	33	67.3	3.3	1.5
14	49	32	65.3	3.3	1.7	51	32	62.7	3.3	1.3
15	50	30	60.0	3.6	1.5	51	30	58.8	3.3	1.3
16	50	30	60.0	3.5	1.4	51	29	56.9	3.1	1.3
17	49	33	67.3	4.0	1.8	51	36	70.6	3.8	1.4
18	52	30	57.7	3.1	1.2	51	36	70.6	3.5	1.4

TABLE A-1-VI/5. *Lower naso-aural distance (n-obi rt,lt)*

Age (years)	N	Asymmetry (male)				N	Asymmetry (female)			
		n	Percent	Mean	SD		*n*	Percent	Mean	SD
6	50	32	64.0	3.2	1.4	50	27	54.0	3.2	1.3
7	50	30	60.0	3.2	1.3	50	26	52.0	3.2	1.7
8	51	33	64.7	2.7	0.8	51	30	58.8	2.8	1.0
9	51	32	60.8	3.2	1.0	50	37	74.0	3.0	1.1
10	50	33	66.0	3.3	1.2	49	34	69.4	3.2	1.3
11	50	27	54.0	3.4	1.7	51	21	41.2	2.8	1.1
12	52	28	53.8	3.1	1.0	53	31	58.5	3.3	1.6
13	50	38	76.0	3.3	1.3	49	27	55.1	3.0	1.2
14	49	30	61.2	3.4	1.2	51	38	74.5	3.0	1.2
15	50	31	62.0	3.7	2.7	51	23	45.1	3.4	1.1
16	50	29	58.0	2.8	0.8	51	26	51.0	3.0	1.3
17	49	31	63.3	3.7	1.6	51	31	60.8	3.3	1.5
18	52	24	47.1	3.5	1.6	51	31	60.8	3.0	1.0

TABLE A-1-VI/6. *Upper subnasale-aural distance (sn-obs rt,lt)*

Age (years)	N	Asymmetry (male)				N	Asymmetry (female)			
		n	Percent	Mean	SD		*n*	Percent	Mean	SD
6	50	34	68.0	3.9	1.4	50	29	58.0	3.2	1.4
7	50	34	68.0	3.7	1.7	50	31	62.0	3.7	2.0
8	51	28	54.9	3.9	1.7	51	30	58.8	2.9	1.0
9	51	35	68.6	3.7	1.5	50	35	70.0	3.0	1.1
10	50	34	68.0	3.8	1.7	49	31	63.3	3.3	1.3
11	50	32	64.0	3.6	1.6	51	30	58.8	3.1	1.2
12	52	37	71.2	3.7	1.4	53	36	67.9	3.3	1.5
13	50	33	66.0	4.1	2.2	49	35	71.4	3.3	1.7
14	49	25	51.0	3.6	1.9	51	33	64.7	3.9	2.1
15	50	33	66.0	3.4	1.8	51	33	64.7	3.8	1.7
16	50	37	74.0	3.5	1.2	51	37	72.5	2.9	1.2
17	49	33	67.3	4.1	2.0	51	35	68.6	3.5	1.2
18	52	29	55.8	3.6	1.8	51	30	58.8	3.9	1.7

TABLE A-1-VI/7. *Lower subnasale–aural distance (sn-obi rt,lt)*

Age (years)	N	Asymmetry (male)				N	Asymmetry (female)			
		n	Percent	Mean	SD		n	Percent	Mean	SD
1	18	2	11.1	1.5	0.7	19	2	10.5	1.0	0.0
2	30	9	30.0	1.2	0.4	30	15	50.0	1.1	0.5
3	30	11	36.7	1.4	0.7	30	9	30.0	1.2	0.4
4	30	4	13.3	3.8	1.0	30	5	16.7	4.2	0.8
5	30	2	6.7	5.0	0.0	30	2	6.7	5.0	0.0
6	50	23	46.0	3.1	1.2	50	30	60.0	2.9	0.9
7	50	27	54.0	3.3	1.4	50	19	38.0	3.2	1.8
8	51	22	43.1	2.8	1.2	51	30	58.8	3.0	0.9
9	51	32	62.7	2.8	1.0	50	25	50.0	3.0	1.2
10	50	32	64.0	3.0	1.5	49	29	59.2	2.9	1.0
11	50	29	58.0	3.4	1.7	51	18	35.3	2.9	1.1
12	52	34	65.4	3.1	1.1	53	32	60.4	3.3	1.3
13	50	30	60.0	3.5	1.9	49	30	61.2	2.9	1.1
14	49	29	59.2	3.3	1.6	51	22	43.1	3.3	2.0
15	50	23	46.0	3.4	1.5	51	23	45.1	3.7	1.1
16	50	37	74.0	3.0	1.1	51	28	54.9	3.1	1.1
17	49	27	55.1	3.7	1.6	51	32	62.7	3.2	1.2
18	52	29	55.8	2.9	1.1	51	30	58.8	3.4	1.4

TABLE A-1-VI/8. *Upper gnathion–aural distance (gn-obs rt,lt)*

Age (years)	N	Asymmetry (male)				N	Asymmetry (female)			
		n	Percent	Mean	SD		n	Percent	Mean	SD
1	18	1	5.6	1.0	0.0	18	1	5.6	2.0	0.0
2	30	10	33.3	1.3	0.7	30	8	26.7	1.1	0.4
3	30	8	26.7	1.5	0.5	30	12	40.0	1.0	0.0
4	30	9	30.0	2.0	0.0	30	9	30.0	2.8	1.6
5	30	7	23.3	2.6	0.8	30	9	30.0	2.8	0.7
6	50	35	70.0	3.9	1.8	50	34	68.0	3.1	1.1
7	50	35	70.0	3.6	2.0	50	34	68.0	3.6	1.9
8	51	36	70.6	3.5	1.5	51	32	62.7	3.3	1.3
9	51	34	66.7	4.0	1.9	50	30	60.0	3.9	1.5
10	50	35	70.0	4.1	1.9	49	34	69.4	3.3	1.1
11	50	33	66.0	3.5	1.3	51	30	58.8	3.5	1.2
12	52	34	65.4	3.4	1.2	53	37	69.8	3.4	1.3
13	50	36	72.0	4.2	1.5	49	34	69.4	3.2	1.4
14	49	39	79.6	3.2	1.4	51	35	68.6	3.6	1.7
15	50	34	68.0	3.5	2.0	51	36	70.6	3.7	1.7
16	50	42	84.0	3.8	1.6	51	35	66.7	3.8	1.8
17	49	38	77.6	4.4	2.1	51	35	68.6	3.8	1.4
18	52	28	53.9	3.9	1.9	51	34	66.7	3.3	1.3

TABLE A-1-VI/9. *Lower gnathion–aural distance (gn-obi rt,lt)*

Age (years)	N	Asymmetry (male)				N	Asymmetry (female)			
		n	Percent	Mean	SD		n	Percent	Mean	SD
1	18	3	16.7	1.0	0.0	18	2	11.1	1.0	0.0
2	30	12	40.0	1.3	0.5	30	7	23.3	1.4	0.5
3	30	14	46.7	1.1	0.4	30	11	36.7	1.1	0.3
4	30	7	23.3	3.1	0.4	30	5	16.7	3.4	0.5
5	30	3	10.0	5.0	2.0	30	5	16.7	4.8	1.5
6	50	26	52.0	2.6	0.7	50	28	56.0	3.2	1.4
7	50	27	54.0	3.5	1.6	50	23	46.0	3.0	1.1
8	51	24	47.1	2.7	0.9	51	24	47.1	3.0	1.0
9	51	24	47.1	3.5	1.5	50	30	60.0	3.2	1.0
10	50	34	68.0	3.0	1.2	49	20	40.8	2.9	1.0
11	50	33	66.0	3.2	1.2	51	28	54.9	3.3	1.2
12	52	26	50.0	2.9	1.1	53	27	50.9	3.5	1.5
13	50	27	54.0	3.4	1.3	49	26	53.1	2.8	1.0
14	49	25	51.0	3.1	1.0	51	33	64.7	3.2	1.5
15	50	31	62.0	3.0	1.2	51	33	64.7	3.3	1.5
16	50	33	66.0	3.3	1.3	51	23	45.1	3.1	1.4
17	49	31	63.3	3.8	1.5	51	27	52.9	3.7	1.8
18	52	30	57.7	3.2	1.3	51	35	68.6	2.9	0.9

TABLE A-1-VI/10. *Occipito-aural distance (op-po rt,lt)*

Age (years)	N	Asymmetry (male)				N	Asymmetry (female)			
		n	Percent	Mean	SD		n	Percent	Mean	SD
6	31	4	12.9	4.0	1.6	36	0	0.0	0.0	0.0
7	33	2	6.1	2.0	0.0	42	4	9.5	4.0	2.7
8	45	3	6.7	3.0	1.0	39	4	10.3	4.8	2.4
9	47	3	6.4	3.3	1.2	38	3	7.9	2.3	0.6
10	37	7	18.9	3.7	1.8	37	3	8.1	6.0	1.7
11	39	3	7.7	5.0	1.0	41	2	4.9	6.5	2.1
12	50	3	6.0	7.0	5.0	50	1	2.0	1.0	0.0
13	47	1	2.1	2.0	0.0	49	3	6.1	2.7	1.2
14	46	7	15.2	3.1	0.9	43	2	4.7	2.5	0.7
15	44	3	6.8	3.0	1.7	43	4	9.3	3.3	1.0
16	49	6	12.3	3.3	1.4	35	2	5.7	2.5	0.7
17	48	5	10.4	6.4	1.1	36	0	0.0	0.0	0.0
18	50	1	2.0	3.0	0.0	38	1	2.6	9.0	0.0

TABLE A-1-VI/11. *Upper naso-aural surface distance (n-obs surf rt,lt)*

Age (years)	N	Asymmetry (male)				N	Asymmetry (female)			
		n	Percent	Mean	SD		n	Percent	Mean	SD
6	50	34	68.0	3.2	1.5	50	31	62.0	3.6	1.7
7	50	43	86.0	4.1	2.1	50	25	50.0	3.4	1.5
8	51	35	68.6	4.2	1.9	51	37	72.5	3.4	1.2
9	51	32	62.7	4.4	2.1	50	38	76.0	3.6	1.6
10	50	33	66.0	3.5	1.5	49	28	57.1	3.6	1.5
11	50	36	72.0	4.1	1.7	51	30	58.8	3.7	1.8
12	52	32	61.5	3.4	1.7	53	35	66.0	3.7	1.7
13	50	41	82.0	3.6	1.5	49	31	63.3	4.0	1.5
14	49	40	81.6	3.9	1.8	51	37	72.5	3.9	2.3
15	50	31	62.0	4.4	2.4	51	31	60.8	4.0	2.1
16	50	31	62.0	4.3	2.3	51	38	74.5	4.0	1.5
17	49	37	75.5	3.8	2.1	51	38	74.5	4.7	2.2
18	52	29	55.8	3.8	1.7	51	35	68.6	4.3	2.3

TABLE A–1-VI/12. *Lower naso-aural surface distance (n-obi surf rt,lt)*

Age (years)	Asymmetry (male)					Asymmetry (female)				
	N	*n*	Percent	Mean	SD	*N*	*n*	Percent	Mean	SD
6	50	33	66.0	3.5	1.4	50	26	52.0	3.1	1.1
7	50	32	64.0	3.2	1.2	50	25	50.0	3.4	1.7
8	51	31	60.8	3.6	1.5	51	24	47.1	3.8	1.3
9	51	36	70.6	3.7	1.9	50	30	60.0	3.6	1.5
10	50	36	72.0	4.2	1.7	49	32	65.3	3.5	1.6
11	50	27	54.0	3.4	1.3	51	33	64.7	3.4	1.5
12	52	30	57.7	3.0	1.2	53	33	62.3	2.9	1.3
13	50	25	50.0	3.5	1.3	49	29	59.2	3.4	1.6
14	49	28	57.1	3.4	1.5	51	31	60.8	3.5	1.9
15	50	35	70.0	4.4	2.0	51	34	66.7	3.5	1.6
16	50	32	64.0	3.4	1.5	51	31	60.8	3.2	1.1
17	49	34	69.4	4.2	1.9	51	35	68.6	3.2	1.2
18	52	34	65.4	3.7	1.8	51	32	62.7	3.3	1.4

TABLE A–1-VI/13. *Upper subnasale–aural surface distance (sn-obs surf rt,lt)*

Age (years)	Asymmetry (male)					Asymmetry (female)				
	N	*n*	Percent	Mean	SD	*N*	*n*	Percent	Mean	SD
6	50	35	70.0	3.5	1.7	50	34	68.0	3.9	1.6
7	50	35	70.0	4.1	2.2	50	35	70.0	3.4	1.5
8	51	35	68.6	3.9	1.8	51	36	70.6	3.4	0.9
9	51	30	58.8	4.1	2.0	50	35	70.0	3.7	1.8
10	50	38	76.0	4.0	2.0	49	35	71.4	3.6	1.7
11	50	33	66.0	3.6	1.7	51	36	70.0	3.9	1.5
12	52	27	52.0	4.1	2.0	53	39	73.6	3.8	1.5
13	50	35	70.0	3.6	1.7	49	30	61.2	3.4	1.5
14	49	34	69.4	4.0	1.9	51	33	64.7	3.8	1.7
15	50	34	68.0	3.6	1.8	51	33	64.7	4.4	2.3
16	50	37	74.0	4.0	1.9	51	34	66.7	4.1	1.8
17	49	34	69.4	4.5	2.2	51	41	80.4	4.2	1.9
18	52	32	61.5	3.8	1.5	51	39	76.5	4.4	1.9

TABLE A–1-VI/14. *Lower subnasale–aural surface distance (sn-obi surf rt,lt)*

Age (years)	Asymmetry (male)					Asymmetry (female)				
	N	*n*	Percent	Mean	SD	*N*	*n*	Percent	Mean	SD
1	17	2	11.8	1.5	0.7	15	3	20.0	1.7	0.6
2	30	13	43.3	1.5	0.7	30	11	36.7	1.5	0.9
3	30	15	50.0	1.4	0.6	30	13	43.3	1.2	0.4
4	30	10	33.3	2.5	0.5	30	15	50.0	3.0	2.6
5	30	10	33.3	3.2	1.2	30	9	30.0	3.0	1.0
6	50	31	62.0	3.7	1.7	50	26	52.0	3.2	0.9
7	50	39	78.0	3.3	1.2	50	37	74.0	3.7	1.6
8	51	30	58.8	3.6	2.0	51	25	49.0	3.4	1.7
9	51	35	68.6	3.4	1.6	50	31	62.0	2.8	1.0
10	50	26	52.0	4.3	1.9	49	35	71.4	3.1	1.1
11	50	32	64.0	4.0	1.6	51	43	84.3	3.6	1.5
12	52	36	69.2	3.2	1.3	53	29	54.7	3.7	2.1
13	50	33	66.0	3.5	1.5	49	37	75.5	3.8	1.7
14	49	33	67.4	3.5	1.7	51	30	58.8	3.3	1.9
15	50	31	62.0	3.8	1.5	51	35	68.6	4.0	2.1
16	50	35	70.0	4.2	1.8	51	37	72.5	3.3	1.5
17	49	38	77.6	4.1	2.2	51	33	64.7	3.7	1.7
18	52	39	75.0	3.7	1.7	51	35	68.6	3.7	1.7

TABLE A-1-VI/15. *Upper gnathion–aural surface distance (gn-obs surf rt,lt)*

Age		Asymmetry (male)					Asymmetry (female)			
(years)	*N*	*n*	Percent	Mean	SD	*N*	*n*	Percent	Mean	SD
6	50	31	62.0	3.5	1.5	50	36	72.0	3.6	1.2
7	50	37	74.0	3.5	1.5	50	37	74.0	3.8	2.0
8	51	44	86.3	3.5	2.0	51	38	74.5	3.9	2.2
9	51	36	70.6	3.6	2.1	50	33	66.0	3.7	1.3
10	50	31	62.0	4.3	2.2	49	30	61.2	3.5	1.8
11	50	33	66.0	3.6	1.7	51	36	70.6	3.8	2.1
12	52	33	63.5	3.4	1.2	53	35	66.0	4.2	2.3
13	50	32	64.0	3.9	2.1	49	36	73.5	3.7	1.4
14	49	33	67.4	3.6	1.7	51	34	66.7	3.9	1.6
15	50	39	78.0	4.2	2.0	51	34	66.7	3.7	2.1
16	50	37	74.0	4.4	2.3	51	35	68.6	4.3	2.7
17	49	32	65.3	5.1	2.0	51	35	68.6	4.0	2.1
18	52	33	63.5	3.9	1.7	51	40	78.4	3.8	1.9

TABLE A-1-VI/16. *Lower gnathion–aural surface distance (gn-obi surf rt,lt)*

Age		Asymmetry (male)					Asymmetry (female)			
(years)	*N*	*n*	Percent	Mean	SD	*N*	*n*	Percent	Mean	SD
1	17	2	11.8	2.0	0.0	15	4	26.7	1.3	0.5
2	30	11	36.7	1.3	0.5	30	10	33.3	2.0	0.8
3	30	18	60.0	1.4	0.8	30	11	36.7	1.4	0.5
4	30	10	33.3	2.9	1.4	30	10	33.3	3.5	1.8
5	30	12	40.0	3.2	1.2	30	11	36.7	3.3	2.1
6	50	34	68.0	3.7	1.8	50	29	58.0	4.2	1.5
7	50	35	70.0	3.6	1.9	50	34	68.0	4.0	2.0
8	51	29	56.9	3.3	1.4	51	34	66.7	3.9	1.9
9	51	33	64.7	3.4	1.7	50	26	52.0	4.1	1.5
10	50	39	78.0	4.0	2.1	49	29	59.2	4.6	2.6
11	50	34	68.0	3.9	1.8	51	32	62.7	4.1	2.1
12	52	35	67.3	3.3	1.4	53	38	71.7	3.7	1.5
13	50	35	70.0	3.9	1.7	49	33	65.3	3.2	1.2
14	49	30	61.2	3.7	2.0	51	27	52.9	3.6	1.7
15	50	31	62.0	3.7	1.8	51	34	66.7	3.9	1.7
16	50	34	68.0	4.3	2.1	51	32	62.7	3.6	1.8
17	49	35	71.4	4.1	2.2	51	32	62.7	3.5	1.6
18	52	31	59.6	4.0	1.8	51	38	74.5	3.9	1.7

TABLE A-1-VI/17. *Inclination of the medial longitudinal axis of the ear*

Age (years)	*N*	Asymmetry (male)				*N*	Asymmetry (female)			
		n	Percent	Mean	SD		*n*	Percent	Mean	SD
1	17	1	5.9	1.0	0.0	18	1	5.6	2.0	0.0
2	31	4	12.9	2.7	1.7	34	7	20.6	2.0	0.8
3	30	9	30.0	1.0	0.0	30	8	26.7	1.5	1.4
4	—	—	—	—	—	—	—	—	—	—
5	—	—	—	—	—	—	—	—	—	—
6	50	22	44.0	4.8	2.5	49	22	44.9	5.3	2.4
7	50	25	50.0	5.4	2.7	48	27	56.3	4.6	1.2
8	51	24	47.1	4.5	1.6	51	27	52.9	5.3	2.6
9	51	25	49.0	5.1	2.2	48	29	60.4	4.1	1.7
10	50	27	54.0	3.8	1.7	49	29	59.2	5.0	2.6
11	48	23	47.9	4.9	2.3	51	28	54.9	3.9	1.7
12	52	23	44.2	4.8	2.1	53	19	35.8	3.8	1.8
13	50	30	60.0	4.6	2.5	49	33	67.3	4.6	2.2
14	49	21	42.9	4.9	2.3	50	28	56.0	4.0	1.9
15	50	31	62.0	4.4	2.3	50	27	54.0	4.0	1.9
16	50	26	52.0	4.0	2.3	50	22	44.0	4.7	2.5
17	49	30	61.2	4.6	1.8	51	24	47.1	4.4	2.2
18	52	20	38.5	4.4	3.0	50	22	44.0	4.1	2.7

TABLE A-1-VI/18. *Protrusion of the ear*

Age (years)	*N*	Asymmetry (male)				*N*	Asymmetry (female)			
		n	Percent	Mean	SD		*n*	Percent	Mean	SD
1	18	0	0.0	0.0	0.0	20	3	15.0	3.0	0.0
2	31	12	38.7	2.5	0.9	34	11	32.4	1.9	0.7
3	30	12	40.0	2.0	0.7	30	12	40.0	2.4	1.0
4	—	—	—	—	—	—	—	—	—	—
5	—	—	—	—	—	—	—	—	—	—
6	50	15	30.0	6.7	3.5	50	12	24.0	4.6	2.9
7	50	11	22.0	5.0	1.3	50	17	34.0	4.5	1.9
8	51	19	37.3	4.3	2.1	51	20	39.2	4.8	2.1
9	51	20	39.2	5.5	2.4	50	13	26.0	7.7	3.7
10	50	16	32.0	7.2	5.1	49	15	30.6	6.4	2.7
11	50	23	46.0	6.0	2.4	51	17	33.3	6.0	2.4
12	52	18	34.6	5.3	1.9	53	12	22.6	6.1	2.4
13	49	13	26.5	6.3	3.2	49	16	32.7	6.2	2.5
14	49	22	44.9	6.4	3.3	50	23	46.0	5.0	2.0
15	50	12	24.0	5.9	3.6	51	10	19.6	7.0	4.6
16	50	28	56.0	5.1	2.3	51	8	15.7	7.1	4.2
17	49	17	34.7	6.0	2.4	51	14	27.5	7.7	3.2
18	52	13	25.0	4.7	2.1	51	13	25.5	6.0	2.3

Appendix B–1

Asymmetry in the Face of Young Adult Chinese Males and Females

Rexon C. K. Ngim, Leslie G. Farkas, and S. T. Lee

Quantitative evaluation of the asymmetries in each region of the face is essential for harmonious surgical restoration between the sides of the face. In a healthy individual there are likely to be normal differences between both sides of the face. Studies devoted to anthropometric examination of paired measurements of the face are few and mainly limited to Caucasians (1–4). Quantitative anthropometric differences between paired measurements of the Chinese face are not available in the literature.

MATERIALS

The study group consisted of 60 healthy young adult Chinese Singaporeans, aged 18–19 years, equally divided between sexes.

METHOD

Using standard anthropometric methods (see Chapter 2), 111 measurements were taken from each subject by one of the authors (LGF). Of the 48 paired measurements, 35 were chosen for the study: 23 linear projective and 7 tangential (arc) measurements and 5 angular measurements (2 inclinations and 3 angles). They were as follows:

Face (7). Projective linear measurements: g-t, n-t, sn-t, and gn-t. Tangential measurements: g-t surf, sn-t surf, gn-t surf.

Orbits (7). Projective linear measurements: en-ex, en-se, ex-t, ex-go, or-sci, ps-pi. Inclination: eye fissure axis.

Nose (10). Projective linear measurements:

sbal-sn, ac-prn, ac-sn, sn-c′, al′-al′. Tangential measurements: ac-prn surf. Sagittal measurements: en-se sag. Inclination: nostril axis. Deviation (angle): nasal bridge, columella

Lips and mouth (4). Projective linear measurements: sbal-ls′, ch-sto, ch-t. Tangential measurements: ch-t surf.

Ears (7). Projective linear measurements: pra-pa, sa-sba, sn-obi, gn-obi. Tangential measurements: sn-obi surf, gn-obi surf. Angle: ear protrusion.

QUANTITATIVE ASSESSMENT OF ASYMMETRY

Paired linear projective and tangential measurements differing by 1 mm or more and angular measurements differing by at least 1 degree were regarded as asymmetrical (3).

An anthropometric system with specification was designed by a statistician to capture the data which were recorded in a standardized research protocol. A statistical analysis software package (SAS) was used to store and analyze the data. A separate program was prepared to sort and tabulate the data for analysis of the frequency and degree of asymmetry for each paired measurement taken. As to the former, the mean value of the difference between the paired measurements and the standard deviation were recorded separately in both sexes. The differences in frequencies and the extent of asymmetries in the sexes and the differences between the asymmetries in the individual facial regions were analyzed using the statistical methods of Student's *t* test and the standard error of differences (SED) (5).

In the tables, B–1 indicates the asymmetries in the Singapore Chinese. The regions of the craniofacial complex are identified by Roman numerals: I for the head, II for the face, III for the orbits, IV for the nose, V for the lips and mouth, and VI for the ears. The sample size is 30 males and 30 females (changes in the size of the sample are given in the footnotes), *n* represents the number of asymmetrical findings, mean value is expressed in millimeters or degrees of asymmetry, SD stands for standard deviation, and the "percent" column lists percentage of asymmetry.

RESULTS

Males

Face

Frequency of asymmetries. In the projective paired measurements the frequency of asymmetry was between 56.7% and 76.7%. The smallest was in the upper third of the face (t-g), and the largest was in the middle third or maxillary region of the face (t-sn). In this region, the tangential half-arc measurement (t-sn surf) also showed the highest percentage (73.3%) of asymmetries.

Extent of asymmetries. The projective paired measurements differed by 2–2.5 mm, and the tangential paired measurements differed by 3–3.9 mm. The greatest differences were observed in the upper third of the face in both the projective and tangential t-g measurements (Table B–1-II).

Orbits

Frequency of asymmetries. There was one subject (3.3%) with an asymmetrical eye fissure length (ex-en). There was also one subject (3.3%) with asymmetry in the combined orbit and eyebrow height (or-sci). The percentage of asymmetries increased in inter-areal projective measurements determining the position of orbits in the face. Medially to the nasal root midpoint (en-se), asymmetry was 30%; distally to the angle of the mandible (ex-go), it was 30%; and laterally to the ear canal (ex-t), it was 60%. Here frequency was significantly higher (SED 12.2, difference 30) than in the earlier two asymmetries. The eye fissure heights and inclinations were symmetrical.

Extent of asymmetries. The mean differences between the asymmetrical measurements ranged from 1 to 6 mm. The smallest asymmetry was between the eye fis-

TABLE B–1-II. *Asymmetries in the face of young adult Chinese*

Paired measurement (right and left)[a]	Sex[b]	Asymmetry			
		n	Percent	Mean	SD
Tragion–glabella depth (t-g)	M	17	56.7	2.5	1.7
	F	23	76.7	2.3	1.0
Tragion–nasion depth (t-n)	M	20	66.7	2.1	1.4
	F	21	70.0	2.5	1.5
Tragion–subnasale depth (t-sn)	M	23	76.7	2.0	1.1
	F	16	53.3	2.8	1.5
Tragion–gnathion depth (t-gn)	M	21	70.0	2.1	1.2
	F	22	73.3	2.4	1.3
Lateral surface half-arc in the upper third of the face (t-g surf)	M	18	60.0	3.9	1.7
	F	18	62.1	3.9	2.3
Lateral surface half-arc in the middle third of the face (t-sn surf)	M	22	73.3	3.7	2.1
	F	25	86.2	3.1	2.3
Lateral surface half-arc in the lower third of the face (t-gn surf)	M	19	63.3	3.0	2.2
	F	23	79.3	3.7	3.2

[a] In t-g surf, t-sn surf, and t-gn surf the *N* in females is 29.
[b] M, males; F, females.

sure lengths (ex-en: 1 mm), and the largest was between the combined heights of the orbit and eyebrow (or-sci: 6 mm). The mean difference (3.1 mm) found between the ex-go measurements was significantly greater ($p = 0.01$) than that between the paired ex-t measurements (1.8 mm) (Table B–1-III).

Nose

Frequency of asymmetries. Six of the 10 measurements were symmetrical in males. Asymmetrical nasal root depth, nostril inclination, and columella deviation were found, in one subject each. In two males the nasal bridge was deviated to the side (6.9%).

Extent of asymmetries. The difference between the asymmetrical root depths was minimal (1 mm). The range of asymmetries between the angular measurements was between 3 and 12 degrees: The largest was between the nostril inclinations (12 degrees), and the smallest was in the nasal bridge deviation (3 degrees) (Table B–1-IV).

Lips and Mouth

Frequency of asymmetries. The range of asymmetries was between 3.3% and 65.5%. The smallest was in one subject with asymmetrical halves of mouth (ch-sto), and the largest was in the labio-aural surface arc (ch-t surf). Both measurements locating the mouth from the ear canal showed asymmetry in almost two-thirds of the subjects, with greater frequency of asymmetry between the tangential (ch-t surf: 65.5%) than between the projective (ch-t: 56.7%) measurements.

Extent of asymmetries. The differences between the asymmetrical measurements ranged from 1.7 mm (ch-t) to 2.4 mm (ch-t surf). The two halves of the mouth (ch-sto) differed by 2 mm (Table B–1-V).

Ears

Frequency of asymmetries. The asymmetries ranged from 53.3% to 80%. Size asymmetries showed the highest percentages: 70% in length (sa-sba) and 80% in width (pra-pa). Moderately smaller were the frequencies of the projective and tangential measurements defining the position of the auricles in relation to the midaxis (sn and gn midpoints) of the face. Surprisingly, the asymmetries in the tangential measurements (sn-obi surf, gn-obi surf) were less frequent (mean 55%) than the projective ones (sn-obi and gn-obi: 63.3%).

TABLE B–1-III. *Asymmetries in the orbits of young adult Chinese*

Paired measurement (right and left)	Sex[a]	Asymmetry			
		n	Percent	Mean	SD
Length of the eye fissure (ex-en)	M	1	3.3	1.0	—
	F	1	3.3	2.0	—
Endocanthion–facial midline distance (en-se)	M	9	30.0	1.8	0.8
	F	12	40.0	2.3	0.9
Orbitotragial distance (ex-t)	M	18	60.0	1.8	1.1
	F	18	60.0	1.9	0.9
Orbitogonial distance (ex-go)	M	9	30.0	3.1	1.3
	F	11	36.7	2.3	1.2
Combined height of the orbit and the eyebrow (or-sci)	M	1	3.3	6.0	—
	F	0	0.0	—	—
Height of the eye fissure (ps-pi)	M	0	0.0	—	—
	F	0	0.0	—	—
Inclination of the longitudinal axis of the eye fissure (ex-en line) (in degrees)	M	0	0.0	—	—
	F	0	0.0	—	—

[a] M, males; F, females.

TABLE B–1-IV. *Asymmetries in the nose of young adult Chinese*

Paired measurement (right and left)[a]	Sex[b]	Asymmetry			
		n	Percent	Mean	SD
Width of the nasal floor (sbal-sn)	M	0	0.0	—	—
	F	0	0.0	—	—
Thickness of the ala (al'al')	M	0	0.0	—	—
	F	0	0.0	—	—
Position of the facial insertion points of the alae in relation to the midpoint in columella base (ac-sn)	M	0	0.0	—	—
	F	0	0.0	—	—
Length of the ala (ac-prn)	M	0	0.0	—	—
	F	0	0.0	—	—
Length of the columella (sn-c')	M	0	0.0	—	—
	F	0	0.0	—	—
Depth of the nasal root (en-se sag)	M	1	3.3	1.0	—
	F	0	0.0	—	—
Surface length of the ala (ac-prn surf)	M	0	0.0	—	—
	F	1	3.3	2.0	—
Inclination of the longitudinal nostril axis (in degrees)	M	1	3.4	12.0	—
	F	0	0.0	—	—
Deviation of the nasal bridge (in degrees)	M	2	6.9	3.0	0.0
	F	2	6.7	4.0	1.4
Deviation of the columella (in degrees)	M	1	3.5	5.0	0.0
	F	0	0.0	—	—

[a] In the nostril inclination, nasal bridge deviation, and columella deviation the *N* in males is 29.
[b] M, males; F, females.

TABLE B–1-V. *Asymmetries in the lip and mouth of young adult Chinese*

Paired measurement (right and left)[a]	Sex[b]	Asymmetry			
		n	Percent	Mean	SD
Halves of the labial fissure length	M	1	3.3	2.0	—
(ch-sto)	F	7	23.3	1.3	0.5
Lateral upper lip	M	0	0.0	—	—
height (sbal-ls')	F	0	0.0	—	—
Labiotragial distance	M	17	56.7	1.7	0.9
(ch-t)	F	18	60.0	2.7	1.6
Lateral labiotragial	M	19	65.5	2.4	1.8
arc (ch-t surf)	F	18	62.1	2.1	1.4

[a] In ch-t surf measurment the *N* in both sexes is 29.
[b] M, males; F, females.

Extent of asymmetries. The degrees of asymmetries were smaller in the width and length of the ears (mean 1.9 mm) than in measurements showing the position of the ears to the face (mean 2.5 mm in projective and 2.6 mm in tangential measurements). The differences were moderately greater in measurements related to the gnathion (mean 2.9 mm) than in those related to subnasale (mean 2.2 mm) (Table B–1-VI).

Females

Face

Frequency of asymmetries. The percentage of asymmetrical depth measurements of the face was between 53.3% and 86.2%. The pattern of asymmetries differed at the three levels of the face. The *projective* paired measurements in the middle third or maxillary region (t-sn) was the least asymmetrical (53.3%). Next was in the lower third or mandibular region (t-gn) at 73.3%. The upper third of the face (t-g) showed the highest frequency of asymmetry (76.7%). This significantly differed from the finding in the middle third (SED 11.9, difference 23.4). The *tangential* paired measurements (t-sn surf) in the middle third of the face displayed the highest frequency of asymmetry (86.2%). This is followed by the measurements (t-gn surf) in the lower third (79.3%), and the smallest frequency of asymmetry was in the upper third (t-g surf: 62.1%). Again, it significantly differed from the frequency of the measurement in the middle third of the face (SED 11.1, difference 24.1).

Extent of asymmetries. The mean difference between the tangential measurements was moderately larger (3.6 mm) than in the projective measurements (2.5 mm). In the projective distances the maxillary depth (t-sn) showed the highest differences (2.8 mm), and in the tangential ones the upper-third depth showed the highest differences (t-g surf: 3.9 mm) (Table B–1-II).

Orbits

Frequency of asymmetries. Among the paired measurements of the soft tissues of the orbits, only the length of the eye fissure was asymmetrical. And this was in one subject (3.3%). The frequency of asymmetry was high in the measurements determining the position of the orbits in the face. The greatest frequency of asymmetries was in the measurement directed to the ear canal (ex-t: 60%) and was followed by the measurement oriented towards the nasal root (en-se: 40%). The smallest percentage of asymmetry in measurement was related to the angle of the mandible (ex-go: 36.7%).

Extent of asymmetries. The mean difference between the asymmetries was in the range of 1.9 mm (ex-t) and 2.3 mm (en-se and ex-go) (Table B–1-III).

Nose

Frequency of asymmetries. Of the seven linear measurements, only the surface length of the nasal ala (ac-prn surf) present in one subject was asymmetrical (3.3%). The bridge deviation observed in two subjects (6.7%) was the only asymmetry of the three angular measurements.

Extent of asymmetries. The difference between the asymmetrical tangential ala lengths was 2 mm. The mean deviation of the nasal bridge was 4 degrees (Table B–1-IV).

TABLE B–1-VI. *Asymmetries in the ears of young adult Chinese*

Paired measurement (right and left)	Sex[a]	Asymmetry			
		n	Percent	Mean	SD
Width of the auricle	M	24	80.0	1.7	0.8
(pra-pa)	F	19	63.3	1.8	1.7
Length of the auricle	M	21	70.0	2.0	1.2
(sa-sba)	F	27	90.0	1.6	1.1
Lower subnasale–aural distance	M	19	63.3	2.2	1.2
(sn-obi)	F	15	50.0	2.5	1.6
Lower gnathion–aural distance	M	19	63.3	2.8	2.0
(gn-obi)	F	16	53.3	2.2	1.3
Lower subnasale–aural surface distance	M	17	56.7	2.2	1.1
(sn-obi surf)	F	14	46.7	2.9	2.2
Lower gnathion–aural surface distance	M	16	53.3	3.0	1.9
(gn-obi surf)	F	17	65.7	3.5	2.3
Ear protrusion	M	0	0.0	—	—
	F	3	10.0	17.7	11.7

[a] M, males; F, females.

Lips and Mouth

Frequency of asymmetries. Asymmetrical halves (ch-sto) of the mouth were recorded in 23.3% of the subjects. Significantly higher was the percentage of asymmetries in measurements establishing the position of the mouth in relation to the ear canal, both projectively (60%; SED 11.8, difference 36.7) and tangentially (62.1%; SED 11.9, difference 38.8).

Extent of asymmetries. The asymmetry was minimal between the halves of the mouth (mean 1.3 mm). The differences between the measurements establishing the position of the mouth in the face were moderately larger, more in projective (mean 2.7 mm) than in tangential (mean 2.1 mm) measurements (Table B–1-V).

Ears

Frequency of asymmetries. The percentage of asymmetries in the length of the ears was significantly higher (90%) than in width (63.3%) (SED 10.4, difference 26.7). Among the measurements determining the position of the ears (obi) in relation to the facial midaxis points (sn and gn), the mean frequencies were identical (51.7%) in the projective and tangential measurements. Individually, the mean percentage of tangential asymmetries was moderately larger (3.2%) than that of the projective ones (2.4%). Asymmetry in the angle of the protrusion was recorded in 10%.

Extent of asymmetries. Mean differences between the asymmetrical width and length measurements of the ears were minimal (1.8 mm and 1.6 mm). The mean differences between the tangential measurements (sn-obi surf and gn-obi surf) were moderately larger (2.9 mm and 3.5 mm) than those between the projective measurements (sn-obi and gn-obi: 2.5 mm and 2.2 mm). The extent of asymmetry was significantly larger tangentially (gn-obi surf: 3.5 mm) than projectively (gn-obi: 2.2 mm) in measurements directed to the lower third of the face (p = 0.01) (Table B–1-VI).

Analysis of the Gender-Related Asymmetries

Face

Differences in frequencies of asymmetries. The percentages of asymmetry of the two projective depth measurements (t-g and t-n) in the upper third of the face were more frequent in females than in males. In the maxillary region the projective depth asymmetries (t-sn) were significantly fewer in females (53.3%) than in males (76.7%) (SED 9.1, difference 23.4). In the mandibular region the frequency of the asymmetrical t-gn measurements was slightly higher in females than in males. The asymmetries between the tangential (half-arc) measurements were more frequent in females than in males at all three levels of the face.

Differences in the extent of asymmetries. The mean differences between the projective depth measurements in both sexes were below 3 mm, and they were slightly larger in females (range 2.3–2.8 mm) than in males (range 2.0–2.5 mm). The extent of asymmetry was greater between the tangential (half-arc) measurements, reaching almost 4 mm: In males it ranged from 3 mm to 3.9 mm, and in females it ranged from 3.1 mm to 3.9 mm. The mean difference from all projective depth asymmetries (in both sexes) was smaller (2.3 mm) than the mean difference of all tangential asymmetries (3.6 mm) (Table B–1-II).

Orbits

Differences in frequencies of asymmetries. Asymmetrical eye fissure lengths (ex-en) were observed in both sexes, one each. Asymmetrical vertical measurement of the combined orbit and eyebrow height (or-sci) was found only in one man. The asymmetries of the projective measurements establishing the position of the orbits in the face in relation to the nasal root (en-se) were more frequent in females (40%) than in males (30%), and the frequency of asymmetry in relation to the mandible angle (ex-go) is 36.7% in females versus 30% in males. The frequencies of asymmetries between the measurements defining the position of the orbits in relation to the ear canal (ex-t) were identical in both sexes (60%).

Differences in the extent of asymmetries. The mean differences between the asymmetrical eye fissure lengths and between the distances between the orbit and the nasal root and the ear canal, respectively, were slightly higher in females than in males. The extent of asymmetry between the measurements establishing the location of the orbit (ex) in relation to the mandible angle (go) was moderately larger in males than in females (Table B–1-III).

Nose

Differences in frequencies of asymmetries. Asymmetrical nasal root depth (en-se sag) and nostril inclination were seen only in males (both in one subject). The surface length of the nasal ala was asymmetrical in only one female. Columella deviation was found in one male, whereas nasal bridge deviation was observed in two subjects of each sex. The nostril floor widths, ala thickness, ala lengths, columella lengths, and the ac-sn distances were symmetrical in both sexes.

Differences in the extent of asymmetries. The mean deviation of the nasal bridge was slightly greater in females (4 degrees) than in males (3 degrees) (Table B–1-IV).

Lips and Mouth

Differences in frequencies of asymmetries. Asymmetrical mouth halves were significantly more frequent in females (23.3%) than in males (3.3%) (SED 8.4, difference 20.0). The percentage of the asymmetrical projective distances between the mouth and the ear canal (ch-t) was slightly higher in females. The arc measurements between the same landmarks (ch-t surf) revealed slightly higher percentage of asymmetries in males (65.5%) than in females (62.1%). Asymmetries were not found in lateral heights of the upper lip.

Differences in the extent of asymmetries. The differences between the paired asymmetrical measurements were minimal in both males and females. Males had slightly higher means between the halves of the mouth (ch-sto) and between the arc measurements taken from the mouth to the ear canal (ch-t surf) than did females. The projective measurements taken between the same features (ch-t) showed slightly higher mean value in females than in males (Table B–1-V).

Ears

Difference in frequencies of asymmetries. The ear-width asymmetry was more frequent in males (females 63.3%, males 80%), whereas the ear-length asymmetry was significantly higher in females (females 90%, males 70%) (SED 10.0, difference 20.0). Asymmetries in projective measurements (sn-obi and gn-obi) defining the position of the ear in the face were more frequent in males (63.3%) than in females (50.0% and 53.3%). The frequencies of asymmetrical tangential measurements directed to the subnasale point were higher in males (56.7%) than in females (46.7%), whereas the measurements related to the chin point showed more asymmetries in females (56.7%) than in males (53.3%). Asymmetrical ear protrusion angles were recorded only in females.

Differences in the extent of asymmetries. Differences between the extents of asymmetries in the size of the ears were minimal between the sexes. The extent of asymmetries in the sexes between the projective and tangential measurements establishing the position of the ears in the face was minimal (less than 1 mm). Both asymmetrical tangential measurements to the subnasale (sn-obi surf) and the chin landmark (gn-obi surf) were marginally larger in females than in males (Table B–1-VI).

DISCUSSION AND CONCLUSIONS

The frequency and/or extent of asymmetries was greatly influenced by the anatomical position of the facial structures to the midline of the face. The relatively wide central zone on the anterior surface of the face, which was described as anthroposcopically "sensitive" (6), revealed less and milder asymmetries compared to the paired measurements positioned mainly on the lateral aspects of the face.

Located in the "sensitive" zone, the orbits, the nose, and the upper lip and mouth showed the mildest and the least frequent asymmetries rarely present in both sexes. Thus, a mild asymmetry between the combined heights of the orbit and eyebrow was recorded in one man only (3.3%). A very mild eye fissure length asymmetry was found in one man and one woman. Of the 10 linear and angular measurements of the nose a mild deviation of the nasal bridge was the only disfigurement seen in both sexes (2 and 2). Mild uneven nasal root depth, mild nostril axis asymmetry, and columella deviation were recorded in one male. In one female, a mildly asymmetrical surface length of the alae was noted. In the orolabial region, an uneven length of the halves of the mouth was seen in one male (3.3%) but was found in significantly higher percentage (23.3%) in females (SED 8.4, difference 20.0). This asymmetry was the most frequent in the "sensitive" zone.

Compared with the features on the central and anterior surface of the face, the paired measurements of anatomic structures on the sides of the face exhibited much higher frequencies and greater extents of asymmetries. The linear and tangential distances defined the projection (depth) of the face at three levels and established the location of the orbits, mouth, and ears in the facial framework. The frequencies of asymmetries in projective depth measurements of the face were in range of 56.7–76.7% in males and 53.3–86.2% in females. The asymmetries were more frequent in tangential than in projective linear measurements, in both sexes. Of the three levels of the face, the middle (maxillary region) displayed the most frequent depth asymmetry (t-sn) in males (76.7%), and this is significantly higher than in females (53.3%) (SED 9.1, difference 23.4). At this level, females also had the highest percentage of tangential asymmetries (t-sn surf) (86.2%), higher than in males (73.3%).

Analysis of the extent of asymmetries in the three levels of the face revealed that in *projective* paired measurements the largest mean difference of 2.8 mm was found in the middle third (maxillary) region of the face (t-sn). In the *tangential* (arc) measurements the largest mean difference was recorded in the upper third of the face (3.9 mm) between the surface t-g half-arcs. It can be inferred that the frequency of facial asymmetries might thus be influenced by the position of the ears in measurements running between the ear canal and the axial or para-axial landmarks of the face. Uneven levels of the ear canals are not rare findings in the healthy population. Other factors causing asymmetries such as the dislocation of midline landmarks from the vertical was not observed in this population sample.

Of the three paired measurements establishing the position of the orbits in the face, the laterally (ex-t) locating the orbits in relation to the ear canal exhibited in both sexes the highest percentage of asymmetries (60%). This is more frequent than in measurements related medially (en-se: males 30%, females 40%) or distally (ex-go: males 30%, females 36.7%). The frequencies of asymmetries in the latter two measurements (medial and distal) are significantly different from the frequency of asymmetries in the lateral measurements (ex-t) only in males (SED 12.2, different 30). The frequencies of projective and tangential measurement asymmetries establishing the position of the mouth in relation to the ear canal (ch-t and ch-t surf) showed similar ranges in males (56.7–65.5%) and females (60–62.1%).

In the ears, areal measurements showed higher percentages of asymmetry compared to the inter-areal measurements. The percentages of asymmetries in width (80% in males and 63.3% in females) and length (70% in males and 90% in females) were higher than those in the measurements indicating the position of the lower insertion point (obi) of the auricles in relation to the subnasale and gnathion points of the facial midline (Table B–1-VI). Generally, the mean differences between the asymmetrical projective measurements were smaller (by 1–3.1 mm) than those in tangential measurements (2.1–3.9 mm), in both sexes.

The mean extent of asymmetries was smaller in the "sensitive" area of the face (1 mm up to 1.7 mm) than in the paired measurements locating the orbits and mouth in the facial framework, or in depth measurements of the face (mean 2.2 mm up to 3 mm). This was seen in both sexes.

In both sexes, the largest number of symmetrical measurements were found in the soft nose: the widths of the nostril floor, length of the alae and columella, thickness of the alae, and level of the facial insertions of the alae. In the orbits the heights and the inclinations of the eye fissures were symmetrical. In the upper lip the lateral heights (sbal-ls′) exhibited symmetry.

The knowledge of normal occurrence of frequencies and extents of asymmetrical measurements in a given population helps in the evaluation of abnormal asymmetries in patients with facial syndromes or post-trauma facial deformities. Knowing the degree and extent of asymmetries assists in surgical planning and restoration of the balance between the halves of the face. Symmetry is one of the requirements on visual assessment of the soft tissue around the orbits and in soft nose, lips, and mouth.

In conclusion, our study documented the frequency and extent of normal range of facial asymmetry. Visual judgment may miss minor differences which will otherwise be objectively documented. This is particularly evident in asymmetries of measurements determining the position of the ears, orbits, mouth, and nose in the healthy face.

REFERENCES

1. Hajniš K, Hajnišová M. Timing of corrective operations of the face according to the dynamics of growth [in Czechoslovakian]. *Rozhl Chir* 1966;8.
2. Figalová P. Asymmetry of the face. *Anthropologie* 1969;7:31–34.
3. Farkas LG. *Anthropometry of the head and face in medicine.* New York: Elsevier, 1981.
4. Farkas LG, Cheung G. Facial asymmetry in healthy North-American Caucasians. *Angle Orthod* 1981;51:70–77.
5. Hughes DR. Finger dermatoglyphics from Nuristan, Afganistan. *Man* 1967;2:119–125.
6. Farkas LG, Kolar JC. Anthropometrics and art in the aesthetics of women's faces. *Clin Plast Surg* 1987;14:599–616.

Appendix C–1

Asymmetry of the Face in Young Adult African-Americans

Govindasarma Venkatadri, Leslie G. Farkas, and Ananda V. Gubbi

Asymmetry can directly influence the impression obtained from visual inspection of the face. The degree of asymmetry in subjects having aberrations such as birth defects or post-traumatic disfigurements can be assessed by measurement and comparison against the symmetrical norms found in healthy individuals. However, facial asymmetry has not been found to be decisive in distinguishing the "attractive" from the "unattractive" face in healthy North American Caucasians (see Chapter 13).

In the last 60 years, papers devoted to discussing facial asymmetry in healthy subjects were limited to Caucasian populations, based on examination of photographs (1–3) or on direct anthropometric examination of individuals (4–7). Quantitative determination of the differences between paired measurements of the face in black population has not, so far, been reported in anthropologic or medical literature.

MATERIALS

The study group consisted of 50 males and 50 females, all healthy young adult African-Americans aged between 18 and 25 years.

METHOD

In each subject, 108 measurements were taken from the head and face by one of the authors (LGF) using standard anthropometric methods (see Chapter 2). For quantitative determination of asymmetry, the differences between the right and left were calculated in a selected group of 35 paired measurements (30 linear and 5 angular measurements): 7 on the face, 7 on the orbits, 10 on the nose, 4 on the lips and mouth, and 7 on the ears.

QUANTITATIVE ASSESSMENT OF ASYMMETRY

Paired linear measurements differing by 1 mm or more, or by 1 degree or more in angular measurements, were regarded as asymmetrical (6). In tables, the frequency (related to 50 males and 50 females) and the extent of asymmetries are reported in both sexes separately. The differences in the frequency and extent of asymmetries in paired measurements are established in each region of the face. Gender-related differences are also reported for the individual regions of the face. The statistical analysis of the data was carried out with the help of Student's *t* test and the method of the standard error of difference (SED) (8).

In the tables, C–1 indicates the asymmetries in African-Americans. The regions of the craniofacial complex are identified by Roman numerals: II for the face, III for the orbits, IV for the nose, V for the lips and mouth, and VI for the ears. Mean values are expressed in millimeters or in degrees of asymmetry, *n* represents the number of asymmetrical findings (the total number of measurements in a given age in 50 males and 50 females), SD stands for standard deviation, and the "percent" column lists percentage of asymmetry.

RESULTS

Males

Face

Frequency of asymmetries. The frequency of asymmetries was in the range of 62% and 78% in the projective measurements: It was smallest in the tragion–glabella depths (t-g rt and lt) in the upper third of the face, and it was largest in the tragion–subnasale depths (t-sn rt and lt) in the middle third (maxillary region) of the face. The frequency of asymmetries in the tangential half-arcs was higher, with increasing percentage from the upper third of the face (t-g surf rt and lt: 68%) through the middle third (t-sn rt and lt: 80%) to the lower third (t-gn surf rt and lt: 86%) (Table C–1-II).

Extent of asymmetries. The mean difference between the asymmetrical projective depth measurements was in the range of 1.8 mm (t-n and t-sn) and 2.5 mm (t-gn). The mean differences were larger between the asymmetrical half-arc measurements, ranging from 3.6 mm (t-sn surf) to 4.7 mm (t-gn surf). In the mandibular region (lower third of the face) the mean difference (4.7 mm) between the half-arc measurements (t-gn surf rt and lt) was significantly larger ($p = 0.001$) than the mean difference between the depth measurements (t-gn rt and lt).

TABLE C–1-II. *Asymmetries in the face of young adult African-Americans*

Paired measurement (rt and lt)	Sex[a]	Asymmetry			
		n	Percent	Mean	SD
Tragion–glabella depth (t-g)	M	31	62.0	2.2	1.3
	F	41	82.0	2.0	1.4
Tragion–nasion depth (t-n)	M	37	74.0	1.8	1.0
	F	34	68.0	1.7	0.9
Tragion–subnasale depth (t-sn)	M	39	78.0	1.8	1.0
	F	31	62.0	2.3	1.1
Tragion–gnathion depth (t-gn)	M	35	70.0	2.5	1.5
	F	40	80.0	2.5	1.6
Lateral surface half-arc in the upper third of the face (t-g surf)	M	34	68.0	4.3	2.2
	F	44	88.0	3.0	1.6
Lateral surface half-arc in the middle third of the face (t-sn surf)	M	40	80.0	3.6	2.5
	F	39	78.0	3.6	2.3
Lateral surface half-arc in the lower third of the face (t-gn surf)	M	43	86.0	4.7	4.4
	F	41	82.0	3.3	2.0

[a] M, male; F, female.

Orbits

Frequency of asymmetries. The range of asymmetries was between 2% and 56%: It was smallest between the lengths (ex-en) and the heights (ps-pi) of the eye fissures, and it was largest between the projective orbitotragial distances (ex-t). The total average frequency of all asymmetrical measurements in the soft-tissue orbits (ex-en, en-se, sci-or and the inclination of the eye fissures) was significantly smaller (7%) than the total average frequency of the projective orbitotragial (ex-t) and orbitogonial (ex-go) asymmetries (52%) (SED 7.9, difference 45.0) (Table C–1-III).

Extent of asymmetries. The smallest mean differences (1 mm) was observed between the asymmetrical lengths (ex-en) of the eye fissures. The largest mean difference (2.8 mm) was noted in males between the projective orbitogonial distances (ex-go). The mean differences between the asymmetrical eye fissure inclinations was 2 degrees (Table C–1-III).

Nose

Frequency of asymmetries. The percentage of asymmetries was low, between 2% and 8%. Asymmetries in the nostril floor width, ala length, and columella length, as well as in the direction of the nasal bridge, showed the smallest percentages (2%). The largest frequency of asymmetries (8%) was seen in the depth measurements

TABLE C–1-III. *Asymmetries in the orbits of young adult African-Americans*

Paired measurement (rt and lt)	Sex[a]	n	Percent	Mean	SD
Length of the eye fissure (ex-en)	M	1	2.0	1.0	0.0
	F	0	0.0	0.0	0.0
Endocanthion–facial midline distance (en-se)	M	2	4.0	2.0	1.0
	F	4	8.0	2.0	0.5
Orbitotragial distance (ex-t)	M	28	56.0	1.8	0.9
	F	31	62.0	2.0	1.0
Orbitogonial distance (ex-go)	M	24	48.0	2.8	2.2
	F	25	50.0	1.7	1.0
Combined height of the orbit and the eybrow (or-sci)	M	6	12.0	1.7	0.7
	F	6	12.0	2.5	0.8
Height of the eye fissure (ps-pi)	M	0	0.0	0.0	0.0
	F	1	2.0	1.0	0.0
Inclination of the longitudinal axis of the eye fissure (ex-en line) (in degrees)	M	5	10.0	2.0	0.6
	F	0	0.0	0.0	0.0

[a] M, male; F, female.

of the nasal root and in tangential lengths of the nasal alae (ac-prn surf). The symmetry was not affected in the ala thickness, in the positions of the facial alar bases (ac-sn), and in the direction of the columella (Table C–1-IV).

Extent of asymmetries. The smallest mean differences between the linear measurements (1 mm) were found in the depth of the nasal root, in the projective length of the columella, and in both the projective and tangential lengths of the nasal alae. The largest difference in linear measurements was 3 mm, found between the asymmetrical nostril floor widths. In angular measurements, the deviation of the nasal bridge was minimal (mean 2 degrees). The mean difference between the asymmetrical nostril axis was 8 degrees (Table C–1-IV).

Lips and Mouth

Frequency of asymmetries. Among the projective asymmetrical measurements, the lateral upper lip heights (sbal-ls′ rt and lt) revealed the smallest frequency (6%), followed by the halves of labial fissure lengths (ch-sto rt and lt) (10%), with a significantly higher percentage of asymmetries (60%) in the labiotragial distances (ch-t rt and lt), one of the measurements displaying the position of the mouth in relation to the ear canal (SED 8.1, difference 50.0). Frequency of asymmetries in paired tangential measurements (ch-t surf rt and lt), also seeking to determine the position of the mouth in the face, was moderately higher (72%) (Table C–1-V).

TABLE C–1-IV. *Asymmetries in the nose of young adult African-Americans*

Paired measurement (rt and lt)	Sex[a]	n	Percent	Mean	SD
Width of the nasal floor (sbal-sn)	M	1	2.0	3.0	0.0
	F	4	8.0	1.0	0.0
Thickness of the ala (al′-al′)	M	0	0.0	0.0	0.0
	F	1	2.0	1.0	0.0
Position of the facial insertion points of the alae in relation to the midpoint in columella base (ac-sn)	M	0	0.0	0.0	0.0
	F	2	4.0	2.0	0.0
Length of the ala (ac-prn)	M	1	2.0	1.0	0.0
	F	0	0.0	0.0	0.0
Length of the columella (sn-c′)	M	1	2.0	1.0	0.0
	F	4	8.0	1.8	0.4
Depth of the nasal root (en-se sag)	M	4	8.0	1.0	0.0
	F	0	0.0	0.0	0.0
Surface length of the ala (ac-prn surf)	M	4	8.0	1.0	0.0
	F	4	8.0	1.0	0.0
Inclination of the longitudinal nostril axis (in degrees)	M	2	4.0	8.0	2.0
	F	0	0.0	0.0	0.0
Deviation of the nasal bridge (in degrees)	M	1	2.0	2.0	0.0
	F	0	0.0	0.0	0.0
Deviation of the columella (in degrees)	M	0	0.0	0.0	0.0
	F	1	2.0	2.0	0.0

[a] M, male; F, female.

Extent of asymmetries. In projective paired measurements the mean values of differences in asymmetrical measurements were in the range of 1.7–2.0 mm. The extent of asymmetry was significantly greater (mean 3.2 mm) in the pair of lateral labiotragial (tangential) arcs (ch-t surf rt and lt) than in the projective labiotragial distances (ch-t rt and lt) (mean 1.8 mm) ($p = 0.01$) (Table C–1-V).

TABLE C-1-V. *Asymmetries in the lips and mouth of young adult African-Americans*

Paired measurement (rt and lt)	Sex[a]	n	Percent	Mean	SD
Halves of the labial fissure length (ch-sto)	M	5	10.0	1.7	0.5
	F	9	18.0	2.3	0.9
Lateral upper lip height (sbal-ls′)	M	3	6.0	2.0	0.8
	F	6	12.0	1.1	0.6
Labio-tragial distance (ch-t)	M	30	60.0	1.8	1.1
	F	36	72.0	2.3	1.7
Lateral labio-tragial arc (ch-t surf)	M	36	72.0	3.2	2.7
	F	39	78.0	2.4	1.4

[a] M, male; F, female.

Ears

Frequency of asymmetries. The frequency of asymmetries was significantly greater in length (66%) than in width (44%) of the ears (SED 9.7, difference 22.0). Percentages of asymmetries in distances determining the position of the ears (actually their lower position: obi) in relation to the subnasale landmark (sn) of the nose, as well as to the chin point (gnathion, gn), were moderately higher, ranging from 66% to 70%. The mean frequency of both tangential measurements (sn-obi surf and gn-obi surf rt and lt) was slightly higher (72%) than in both projective measurements (sn-obi and gn-obi rt and lt) (68%). Protrusions of the ear did not reveal asymmetries (Table C–1-VI).

Extent of asymmetries. The mean difference between the asymmetrical width and length measurements of the ears was minimal (0.6 mm). The mean differences between the tangential paired measurements were moderately larger than those between projective paired measurements showing the position of the ears in the face (Table C–1-VI).

Females

Face

Frequency of asymmetries. The percentage of asymmetry in the projective depth measurements was the lowest (62%) in the middle third of the face (t-sn rt and lt), followed by the two depth measurements in the upper third (68% and 82% in t-n and t-g rt and lt, respectively), and the highest in the lower third (88% in t-gn rt and lt). The percentages of asymmetries in tangential half-arcs were higher in the upper and middle third (88% in t-g surf and 78% in t-sn surf rt and lt, respectively) of the face than were the appropriate projective depth measurements (Table C–1-II).

Extent of asymmetries. The total mean value of differences in the asymmetrical depth measurements in all thirds of the face was 2.1 mm; it was moderately more, 3.3 mm, in the three tangential half-arcs.

TABLE C–1-VI. *Asymmetries in the ears of young adult African-Americans*

Paired measurement (rt and lt)	Sex[a]	Asymmetry			
		n	Percent	Mean	SD
Width of the auricle (pra-pa)	M	22	44.0	1.6	0.8
	F	22	44.0	1.5	0.6
Length of the auricle (sa-sba)	M	33	66.0	2.0	1.2
	F	27	54.0	2.1	1.5
Lower subnasale–aural distance (sn-obi)	M	33	66.0	2.1	1.2
	F	32	64.0	2.2	1.2
Lower gnathion–aural distance (gn-obi)	M	35	70.0	2.5	2.5
	F	33	66.0	2.4	2.3
Lower subnasale–aural surface distance (sn-obi surf)	M	37	74.0	2.7	2.5
	F	34	68.0	2.4	2.5
Lower gnathion–aural surface distance (gn-obi surf)	M	35	70.0	3.1	2.6
	F	36	72.0	3.1	3.2
Ear protrusion (in degrees)	M	0	0.0	0.0	0.0
	F	0	0.0	0.0	0.0

[a] M, male; F, female.

Orbits

Frequency of asymmetries. The soft-tissue orbits disclosed the smallest percentages of asymmetries: 2% of the eye fissure heights (ps-pi rt and lt), 8% in the endocanthion–facial midline distance (en-se rt and lt), and 12% in the combined height of the orbit and the eyebrow (or-sci rt and lt). No asymmetry was observed in the width (ex-en rt and lt) and inclination of the eye fissures. Asymmetries defining the position of the orbits in the face, laterally in relation to the ear canal (ex-t rt and lt) and distally to the angle of the mandible (ex-go rt and lt), revealed higher frequencies than did the orbital soft-tissue asymmetries (56% and 62%). The total mean frequency of orbital soft-tissue asymmetries (7.7%) was significantly smaller than the mean frequency of asymmetries (56%) reported in the two measurements determining the position of the orbits in the face (SED 8.0, difference 48.3) (Table C–1-III).

Extent of asymmetries. The smallest difference (mean 1 mm) was found in the asymmetrical heights of the eye fissures (ps-pi rt and lt), and the highest difference (mean 2.5 mm) was found between the combined heights of the orbits and eyebrows (or-sci rt and lt). The asymmetries found in the paired measurements defining the position of the orbits (en-se, ex-t, ex-go) were also small in extent (1.7–2 mm) (Table C–1-III).

Nose

Frequency of asymmetries. In females, the ala thickness asymmetry and the columella deviation revealed the smallest frequency (2%), and the asymmetrical nostril floor widths, columella lengths, and tangential nasal ala lengths revealed the largest ones (8%). Asymmetries between the lengths of the nasal alae, depths of the nasal root, inclinations of the nostril axis, and changes in vertical direction of the nasal bridge were not recorded (Table C–1-IV).

Extent of asymmetries. The mean between the asymmetrical measurements was minimal: 1 mm between the nostril floor width, thickness of alae, and tangential ala lengths; 1.8 mm between the length of the columella; and 2 mm between the levels of the facial alar bases (ac-sn). A 2-degree columella deviation was found in one subject (Table C–1-IV).

Lips and Mouth

Frequency of asymmetries. Asymmetrical lateral upper lip heights (sbal-ls′ rt and lt) were observed in 12% of the subjects. Asymmetry between the halves of the labial fissure lengths (ch-sto rt and lt) was recorded at 18%. The percentage of asymmetries in paired measurements determining the position of the mouth relative to the ear canal was quite high: projectively (ch-t rt and lt), 72%; tangentially (ch-t surf rt and lt), 78%. The mean frequency of asymmetry in these two measurements (75%) substantially exceeded that found in the upper lip and mouth (15%) (SED 7.9, difference 60.0) (Table C–1-V).

Extent of asymmetries. The mean differences between the asymmetrical measurements were generally mild, in the range of 1.1 mm and 2.4 mm; they were smallest in the upper lip (sbal-ls′ rt and lt) and largest in the tangential lateral labiotragial arcs (ch-t surf rt and lt) (Table C–1-V).

Ears

Frequency of asymmetries. Asymmetries in width of the ears (pra-pa rt and lt) were moderately less frequent (44%) than those in length of the ears (sa-sba rt and lt: 54%). Asymmetries in the protrusions of the ears were not observed. Paired projective measurements showing the position of the ears (the ears' lower insertion points: obi rt and lt) in relation to the major landmarks located in the mid-axis of the lower face, the subnasale (sn) and the gnathion (gn), revealed higher percentages of asymmetries ranging from 64% to 66%. The tangential paired measurements (sn-obi surf and gn-obi surf rt and lt) showed slightly higher percentages of asymmetries (68–72%) than did the projective distances between the same landmarks. The mean frequency of the asymmetries in sizes of ears (49%) was only moderately smaller than the mean percentage of the asymmetrical projective and tangential measurements defining the position of the ears in the face (68%) (Table C–1-VI).

Extent of asymmetries. The mean difference in ear widths was minimal (1.5 mm) and only slightly greater for ear length measurements (2.1 mm). Mean differences in the asymmetrical paired projective and tangential measurements were mild (range 2.2–3.1 mm), with the latter slightly exceeding the former (Table C–1-VI).

Analysis of Gender-Related Asymmetries

Face

Differences in frequencies of asymmetries. In the upper third of the face, asymmetry in tragion–glabella depth was significantly more frequent in females (82%) than in males (62%) (SED 8.8, difference 20.0). The same relationship existed between the sexes in the tangential lateral surface half-arcs (t-g surf; in males, 68%; in females, 88%; SED 8.0, difference 20.0). A more frequent incidence of asymmetry in the upper third of the face in males (74%) than in females (68%) was noted in the measurements of tragion–nasion depth. In the middle third of the face, asymmetries of both projective depth (t-sn) and the lateral surface half-arc (t-sn surf) were found to occur somewhat more frequently in males (78% and 80%, respectively) than in females (62% and 78%, respectively). In the lower third of the face, the frequency of asymmetrical depth measurements (t-gn) was significantly higher in females (88%) than in males (68%) (SED 8.0, difference 20.0). Compared with those in the middle third of the face, the frequency of asymmetries in depth measurements of the mandibular region moderately decreased in males (from 78% to 68%) but significantly increased in females (from 62% to 88%) (SED 8.3, difference 26) (Table C–1-II).

Differences in the extent of asymmetries. Generally, the mean differences between asymmetrical projective depth measurements were smaller (range 1.7–2.5 mm) than those of the tangential half-arcs (range 3.0–4.7 mm). In the three thirds of the face, in both sexes, the mean differences between the asymmetrical projective depth measurements were in the range of 1.7 mm (in the upper third) and 2.5 mm (in the lower third). The mean differences in the upper third and lower third of the face were slightly greater in males than in females. In the middle third of the face, the degree of asymmetries in females (2.3 mm) significantly exceeded the findings in males (1.8 mm) ($p = 0.05$). The extent of asymmetries between the asymmetrical half-arcs exceeded those in the depth measurements in all three thirds of the face, in both sexes, ranging from 3.0 mm to 4.7 mm. In the upper and lower third of the face, the asymmetrical half-arcs revealed significantly greater mean differences (4.3 mm and 4.7 mm, respectively) in males than in females (3.0 mm and 3.3 mm, respectively) ($p = 0.001$ and $p = 0.05$, respectively). The comparison of the tangential and projective asymmetries in the lower third of the face of both sexes disclosed significantly larger differences between the tangential half-arcs (t-gn surf rt and lt) (4.0 mm) than between the projective depth measurements (t-gn rt and lt) (2.5 mm) ($p = 0.01$) (Table C–1-II).

Orbits

Differences in the frequency of asymmetries. Asymmetrical eye fissure inclinations (10%) and asymmetry in length of the eye fissures (2%) were recorded only in males. Asymmetry in the height of the eye fissures was observed in one female (2%). The paired projective measurements outlining the position of the orbits in the face, medially to the mid-axis of the face (sellion point), laterally to the ear canal (tragion landmark), and distally to the angle of the mandible (gonion), revealed slightly higher percentages of asymmetries in females than in males (Table C–1-III).

Differences in the extent of asymmetries. Generally, the mean differences were mild, slightly differing between the sexes. Only the orbitogonial distances, determining the position of the orbits in relation to the gonion of the mandible, revealed greater differences between the sexes, significantly greater in males (mean 2.8 mm) than in females (mean 1.7 mm) ($p = 0.02$) (Table C–1-III).

Nose

Differences in the frequency of asymmetries. Asymmetrical and symmetrical measurements were almost equally divided between the sexes. The frequency of asymmetrical measurements ranged from 2% to 8%. Asymmetrical nostril floor widths and columella lengths were more frequent in females (8%) than in males (2%). Asymmetrical tangential (surface) ala lengths were found in equal percentage in both sexes (8%). Asymmetries in projective ala lengths, nasal root depths, inclinations of the nostril axis, and bridge deviation were recorded only in males. Asymmetries seen in females only included the ala thickness, the facial insertions of alar bases, and the deviation of the columella (Table C–1-IV).

Differences in the extent of asymmetries. The mean difference between the asymmetrical nostril floor widths was moderately larger in males (3 mm) than in females (1 mm). In measurements where findings were symmetrical in one sex and asymmetrical in the other, the mean differences were minimal (1–2 mm). Only the inclination of the longitudinal nostril axis revealed a noteworthy difference: a mean 8-degree asymmetry in males compared to the symmetrical inclinations in females (Table C–1-IV).

Lips and Mouth

Differences in the frequency of asymmetries. The percentages of asymmetries between the lateral upper lip heights, the halves of the labial fissure, and the projective (ch-t) and tangential (ch-t surf) measurements defining the position of the mouth in relation to the ear canal (actually the tragion landmark) were moderately higher in females than in males (Table C–1-V).

Differences in the extent of asymmetries. The mean differences in the asymmetrical lateral upper lip heights and in tangential measurements between the mouth (cheilion) and the ear canals (tragion) were moderately larger in males than in females. In comparison with males, the females showed moderately larger differences between the asymmetrical halves of the mouth and in projective distances between the mouth (cheilion) and the ear canal (Table C–1-V).

Ears

Differences in the frequency of asymmetries. Asymmetrical lengths of the ears were more often observed in males than in females. Likewise, measurements defining the position of the ears in the face, both projectively (sn-obi rt and lt and gn-obi rt and lt) and tangentially (obi-sn surf rt and lt and obi-gn surf rt and lt), revealed moderately higher percentages of asymmetry in males than in females (Table C–1-VI).

Differences in the extent of asymmetries. Asymmetry in the width and length of the ears, as well as asymmetries of ear position, differed only slightly between the sexes (Table C–1-VI).

DISCUSSION AND CONCLUSIONS

The frequency and extent of asymmetries in paired measurements was greatly influenced by their anatomical location. It was found that paired distances (e.g., the eye fissure lengths) within the central zone outlined by lines connecting the frontotemporale point of the forehead with the gonion landmark of the mandible on each side (9) showed fewer asymmetries and of milder degree when compared to the paired measurements taken between the landmarks located inside the "sensitive" central zone and outside it (e.g., lower gnathion–aural distance rt and lt). Thus, the paired measurements taken from the soft-tissue orbits, the soft nose, the upper lip and mouth, and the halves of the mouth revealed the smallest percentages (range 2–18%), with the smallest differences (range 1–3 mm). In contrast, the depth measurements of the face and the measurements determining the position of the orbits, mouth, and ears disclosed high percentages of asymmetry (range 44–88%) with greater differences (range 1.7–4.7 mm). Generally, the tangential (surface) half-arcs, compared to the projective linear distances, were asymmetrical in much higher percentage and exhibited greater differences (e.g., tragion–

glabella depth and the lateral surface half-arc in the upper third of the face).

While the differences between the sexes were not significant in those asymmetries found within the central zone of the face, statistical differences in both percentage and extent of asymmetrical measurement located outside the central zone were occasionally significant. The depth measurements of the face were significantly more often asymmetrical in females than in males: The percentages of asymmetries in the tragion–glabella depth (82%), in the lateral surface half-arcs in the upper third of the face (88%), and in the tragion–gnathion depth of the lower face (88%) were significantly higher in females (SED 8.8, difference 20.0; SED 8.0, difference 20.0; SED 8.0, difference 20.0) than in males (62%, 68%, and 68%, respectively). The mean difference (2.3 mm) found between the asymmetrical tragion–subnasale depth measurements of the face was significantly greater in females ($p = 0.05$) than in males (mean 1.8 mm). In the facial upper third, males were found to have a significantly greater degree of asymmetry in the tangential (surface) half-arcs (compare: male mean 4.7 mm, female mean 3.3 mm; $p = 0.001$ and 0.05, respectively). It was further noted that the extent of orbitogonial asymmetry in males (mean 2.8 mm) significantly exceeded that found in females (mean 1.7 mm) ($p = 0.02$).

The following symmetrical features were recorded amongst the males: the heights of the eye fissures, the thickness of alae, and the position of the facial insertions of alae and undeviated columella. In females, symmetries were noted in the length and inclination of the eye fissures, alae length, nasal root depth, inclination of the nostrils, and undeviated nasal bridge.

Neither the mild asymmetries of the central facial zone nor the greater asymmetries found more frequently outside the "sensitive" zone caused any visible change in healthy facial harmony.

REFERENCES

1. Busse H. Über normale Asymmetrien des Gesichts und im Körperbau des Menschen. *Z Morphol Anthropol* 1936;35:412–445.
2. Peck H, Peck S. A concept of facial esthetics. *Angle Orthod* 1970;40: 284–318.
3. Roggendorf von E. Die Symmetrieanalyze des Gesichts. *Anat Anz* 1972;132:178–188.
4. Hajniš K, Hajnišová M. Timing of corrective operations of the face according to the dynamics of growth [in Czech]. *Rozhl Chir* 1966;45:533–544
5. Figalova P. Asymmetry of the face. *Anthropologie* 1969;7:31–34.
6. Farkas LG. *Anthropometry of the head and face in medicine.* New York: Elsevier, 1981.
7. Farkas LG, Cheung G. Facial asymmetry in healthy North-American Caucasians. *Angle Orthod* 1981;51:70–77.
8. Hughes DR. Finger dermatoglyphics from Nuristan, Afganistan. *Man* 1967;2:119–125.
9. Farkas LG, Kolar JC. Anthropometrics and art in the aesthetics of woman's faces. *Clin Plast Surg* 1987;14:599–616.

Subject Index

Subject Index